Frequency Handbook
for
Radio Monitoring
HF

Edition 2025

EACH SIGNAL IS INFORMATION

Dipl.- Ing. Roland Proesch

Frequency Handbook
for
Radio Monitoring
HF

Edition 2025

Bibliografische Information der Deutschen Nationalbibliothek
Die Deutsche Nationalbibliothek verzeichnet diese Publikation in der Deutschen Nationalbibliografie; detaillierte
bibliografische Daten sind im Internet über http://dnb.d-nb.de abrufbar.

Email: roland@proesch.net
Publisher: BoD · Books on Demand GmbH, Überseering 33,
22297 Hamburg, bod@bod.de
Print: Libri Plureos GmbH, Friedensallee 273, 22763 Hamburg
Cover design: Anne Proesch

Web page: www.frequencymanager.de

ISBN: 978-3-7693-7684-5

Disclaimer:

The information in this book have been collected over years. The main problem is that there are not many open sources to get information about this sensitive field. Although I tried to verify the information from different sources it may be that there are mistakes. Please do not hesitate to contact me if you discover any wrong description.

Content

General

Our software Frequency Manager is meanwhile available since 1994. Very often we heard the question from our customer if it may be possible to have it in printed form.

Here it is!

The amount of information in our electronic database is very huge. We tried to combine them in a way that the book is not exploding in number of pages.

Disclaimer

Please keep in mind that listening to radio transmitter outside the broadcast or amateur radio bands may be forbidden by law.

Table description

Frequency	**The frequency in kHz of the transmission**
Mode	This field contains the used mode during tansmission. For an explanation to the different modes please refer to our "Technical Handbook for Radio Monitoring HF"
Mode Parameter	This field contains a combination of signal parameter: B: the symbolrate of the signal S: the shift in Hz Ch: the number of channels ChS : the frequency distance between channels
Callsign	The callsign of the user
User	The user of this frequency
Country	The country of the user

There are several rows in the database which are written in italics and where the background has a grey shade. The locations of these stations have been confirmed by a professional direction finder system.

Frequency	Mode	Mode Parameter	Callsign	User	Country
2207,50	*STANAG 4285*	*B=2400Bd*		*NOR N Stavanger*	*NOR*
2210,00	SSB		3VB	Bizerte R	TUN
2210,00	SSB, RTTY			D army	D
2212,00	PACTOR III	B=100Bd S=200Hz	9Z4DH	Sailmail Chaguaramas	TRD
2215,00	ALE USB	B=125Bd Ch=8 ChS=250Hz		E MOI net	E
2216,00	*ALE USB*	*B=125Bd Ch=8 ChS=250Hz*		*G MIL net*	*G*
2219,00	ALE USB	B=125Bd Ch=8 ChS=250Hz		D CG net	D

Frequency List

0 – 1000 kHz

Frequency	Mode	Mode Parameter	Callsign	User	Country
,08	CW , MSK	S=2,3Hz		RUS ELF transmitter ZEVS Murmansk	RUS
11,90	CW			Alpha Novosibirsk	RUS
11,90	CW			Alpha Komsomolsk na Amure	RUS
12,64	CW			Alpha Krasnodar	RUS
12,64	CW			Alpha Novosibirsk	RUS
12,64	CW			Alpha Komsomolsk na Amure	RUS
14,88	CW			Alpha Krasnodar	RUS
14,88	CW			Alpha Novosibirsk	RUS
14,88	CW			Alpha Komsomolsk na Amure	RUS
16,40	MSK	B=200Bd S=100Hz	JXN	NOR N Novik	NOR
17,20	CW		SAQ	Grimeton R	S
17,80	MSK	B=100Bd S=200Hz	NAA	USA N Cuttler , ME	USA
18,10	T600	B=50Bd S=75Hz	RDL	RUS N Moscow	RUS
18,20	MSK	B=50Bd S=100Hz	VTX3	IND N Vijayanarayanam	IND
18,30	MSK	B=200Bd S=100Hz	HWU	F N Le Blanc	F
18,50	MSK	B=200Bd S=100Hz	DHO38	D N Saterland-Ramsloh	D
19,58	MSK	B=200Bd S=100Hz Ch=1	GBZ	G N Anthorn	G
20,20	T600	B=50Bd S=75Hz	RDL	RUS N Moscow	RUS
20,27	MSK	B=200Bd S=100Hz	ICV	I N Tavolara	I
20,50	CW		RAB99	Khabarovsk TS	RUS
20,50	CW		RJH69	Molodeco TS	BLR
20,50	CW		RJH63	Krasnodar TS	RUS
20,50	CW		RJH99	Nizhnij Novgorod TS	RUS
20,50	CW		RJH77	Archangelsk TS	RUS
20,50	CW		RJH66	Bishkek TS	KGZ
20,76	MSK	B=200Bd S=100Hz	ICV	I N Tavolara	I
20,90	MSK	B=200Bd S=100Hz	HWU	F N Le Blanc	F
21,05	MSK	B=200Bd S=100Hz	HWU	F N Le Blanc	F
21,10	CW		RDL	RUS N Moscow , TX Tashkent	UZB
21,10	MSK	B=50Bd S=50Hz	RDL	RUS N Moscow , TX Tashkent	UZB
21,10	CW		RDL	RUS N Moscow , TX Krasnodar	RUS
21,10	MSK	B=200Bd S=100Hz	RDL	RUS N Moscow , TX Krasnodar	RUS
21,40	MSK	B=200Bd S=100Hz	NPM	USA N Hawaii	HWA
21,70	MSK	B=200Bd S=100Hz	HWU	F N Le Blanc	F
22,10	MSK	B=50Bd S=50Hz	GBZ	G N Skelton	G
22,10	MSK	B=50Bd S=50Hz	JJI	J N Ebino	J
23,00	CW		RAB99	Khabarovsk TS	RUS
23,00	CW		RJH63	Krasnodar TS	RUS
23,00	CW		RJH69	Molodeco TS	BLR
23,00	CW		RJH99	Nizhnij Novgorod TS	RUS
23,00	CW		RJH77	Archangelsk TS	RUS
23,00	CW		RJH66	Bishkek TS	KGZ
23,40	2FSK	B=50Bd S=50Hz	DHO38	D N Ramsloh	D
23,40	*MSK*	*B=200Bd S=100Hz*	*DHO38*	*D N Ramsloh*	*D*
24,00	MSK	B=200Bd S=100Hz	NAA	USA N Cuttler , ME	USA
24,60	CW		NLK	USA N Seattle , WA	USA
25,00	CW		RJH63	Krasnodar TS	RUS
25,00	CW		RAB99	Khabarovsk TS	RUS
25,00	CW		RJH66	Bishkek TS	KGZ
25,00	CW		RJH69	Molodeco TS	BLR
25,00	CW		RJH90	Nizhnij Novgorod TS	RUS
25,00	CW		RJH77	Archangelsk TS	RUS
25,10	CW		RAB99	Khabarovsk TS	RUS
25,10	CW		RJH63	Krasnodar TS	RUS
25,10	CW		RJH69	Molodeco TS	BEL
25,10	CW		RJH90	Nizhnij Novgorod TS	RUS

Frequency	Mode	Mode Parameter	Callsign	User	Country
25,10	CW		RJH77	Archangelsk TS	RUS
25,10	CW		RJH66	Bishkek TS	KGZ
25,20	*MSK*	*B=200Bd S=100Hz*	*NML4*	*USA N Lamoure , ND*	*USA*
25,50	CW		RAB99	Khabarovsk TS	RUS
25,50	CW		RJH63	Krasnodar TS	RUS
25,50	CW		RJH69	Molodeco TS	BLR
25,50	CW		RJH90	Nizhnij Novgorod TS	RUS
25,50	CW		RJH77	Archangelsk TS	RUS
25,50	CW		RJH66	Bishkek TS	KGZ
26,00	MSK	B=50Bd S=100Hz		ISR N TX Dimona	ISR
26,70	M/S 1,0	B=50Bd S=85Hz	TBB	TUR N Ankara	TUR
27,30	M/S 1,0	B=50Bd S=75Hz	RDL	RUS N Moscow	RUS
28,50	CW		NCA	Aguada TS	PTR
29,70	MSK	B=100Bd S=50Hz		ISR N Tx Dimona	ISR
36,00	MSK	B=200Bd S=100Hz	DHO37	D N Ramsloh	D
37,50	MSK	B=200Bd S=100Hz	TFK	ISL N Grindavik	ISL
38,00	M/S 1,0	B=100Bd S=100Hz	SHR	S N TX Grimeton	S
40,00	CW		JJY	Ootakadoya-yama TS	J
40,40	CW		SRC	S N TX Grimeton	S
40,40	M/S 1,0	B=200Bd S=100Hz	SRC	S N TX Grimeton	S
40,75	MSK	B=200Bd S=100Hz	NAU	USA N Aguada	PTR
42,00	QPSK	B=104Bd		Cavelink System	SUI
42,50	M/S 1,0	B=100Bd S=100Hz	SAS2	S N TX Grimeton	S
44,10	M/S 1,0	B=200Bd S=100Hz		S N TX Grimeton	S
44,20	MSK	B=200Bd S=100Hz	SAS3	S N TX Grimeton	S
45,90	MSK	B=200Bd S=100Hz	NSY	USA N Niscemi , Sicily	I
49,00	*MSK*	*B=200Bd S=100Hz*	*SXA*	*GRC N Marathon*	*GRC*
50,00	CW		RTZ	Irkutsk TS	RUS
51,95	M/S 1,0	B=100Bd S=50Hz	GYW1	G N Crimond	G
54,00	MSK	B=200Bd S=100Hz	NDI	USA N Awase , Okinawa City	J
57,40	MSK	B=100Bd S=75Hz	TFK	ISL N Grindavik	ISL
57,70	MSK	B=50Bd S=100Hz	LBH	NOR N TX Gossa	NOR
60,00	ASK		MSF	Anthorn TS	G
60,00	ASK		WWVB	Fort Collins TS , CO	USA
60,00	CW		JJY	Mt. Hagane, Kyushu Island	J
61,83	MSK	B=200Bd S=100Hz		G N Inskip	G
62,60	MSK	B=200Bd S=100Hz	FUG	F N Saissac	F
63,85	MSK	B=200Bd S=100Hz	FTA63	F N St. Assise	F
65,80	MSK	B=200Bd S=100Hz	FUE	F N Brest	F
66,70	CW		RBU	Moskow TS	RUS
68,00	MSK	B=200Bd S=100Hz	GYN1	G N Northwood	G
68,50	CW		BPC	CHN time service	CHN
77,50	ASK		DCF77	Mainflingen TS	D
77,50	CW		BSF	Taiwan TS	TWN
78,00	CW FH			MARS-75 system	RUS
81,00	MSK	B=100Bd S=85Hz		G N Inskip	G
82,80	STANAG 5065	B=75Bd S=85Hz	MKL	G F AMCC Northwood	G
100,00	CW			CHAYKA chains	RUS
100,00	CW			Bryansk, master	RUS
100,00	CW			Petrozavodsk, minor W	RUS
100,00	CW			Slonim, minor X	BLR
100,00	CW			Simferopol, minor Y	RUS
100,00	CW			Syzran , minor Z	RUS
100,00	CW			Okhotsk , minor Z	RUS
100,00	CW			Tokachibuto , minor Y	J
100,00	CW			Ussuriysk , minor X	RUS
100,00	CW			Petropavlovsk-Kamchatsky , minor W	RUS
100,00	CW			Alexandrovsk-Sakhalinsky , master	RUS
100,00	CW			Petropavlovsk-Kamchatsky , master	RUS
100,00	CW			Attu Island , minor W	USA
100,00	CW			Alexandrovsk-Sakhalinsky, minor X	RUS
100,00	CW			Pankratiev Island , minor X	RUS

Frequency	Mode	Mode Parameter	Callsign	User	Country
100,00	CW			Tumanny , minor W	RUS
100,00	CW			Inta , master	RUS
100,00	CW			Yuzhno-Uralsk , master	RUS
100,00	CW			Yekaterinburg , minor W	RUS
100,00	CW			Kurgan , minor X	RUS
100,00	CW			LORAN-C	
100,00	CW			Helong , minor Y	CHN
100,00	CW			Rongcheng , master	CHN
100,00	CW			Xuancheng , minor X	CHN
100,00	CW			Rongcheng , minor Y	CHN
100,00	CW			Raoping , minor X	CHN
100,00	CW			Xuancheng , master	CHN
100,00	CW			Hexian , master	CHN
100,00	CW			Raoping , minor X	CHN
100,00	CW			Chongzuo , minor Y	CHN
100,00	CW			Pohang, master	KOR
100,00	CW			Kwang Ju, minor W	KOR
100,00	CW			Ussuriisk, minor Z	RUS
100,00	CW			Anthorn, master	
100,00	CW			Anthorn, minor Y	
100,00	CW		BPL	Lingtong TS	CHN
100,00	CW			Kotelnikovo, master	RUS
100,00	CW			Poltavskaya, minor O	RUS
100,00	CW			Astrakhan minor P	RUS
100,00	CW			Balashov, minor Q	RUS
100,00	CW			Afif, master	ARS
100,00	CW			Salwa, minor W	ARS
100,00	CW			Ash Shaykh Humayd, minor Y	ARS
100,00	CW			Pucheng, master	CHN
100,00	CW			Havre, minor Y	CAN
100,00	CW			George, master	CAN
100,00	CW			Fallon, minor X	USA
122,50	BAUDOT	B=50Bd S=85Hz	CFH	CAN N Halifax	CAN
124,50	MSK	B=100Bd S=85Hz	SXV	GRC N Kriti	GRC
125,00		B=2000Bd		RFID	
129,10	ASCII	B=200Bd S=340Hz	DCF49	EFR Mainflingen	D
134,20		B=2000Bd		RFID	
135,60	ASCII	B=200Bd S=340Hz	HGA22	EFR Lakihegy	HNG
135,80	MSK	B=100Bd S=85Hz	SXJ	GRC N Marathon	GRC
136,00	WSPR	B=1,46Bd Ch=4 ChS=1,46Hz		WSPR net	
136,13	FT8	B=5,86Bd Ch=8 ChS=5,86Hz		FT8 channel	
139,00	ASCII	B=200Bd S=340Hz	DCF39	EFR Burg	D
147,30	*BAUDOT*	*B=50Bd S=85Hz*	*DDH47*	*DWD TX Pinneberg*	*D*
162,00	PSK	B=250Bd	TDF	Allouis TS	F
293,00	AM		KSR	Erra Chidia P	MRC
305,00	CW		UMA4	RUS Leningradskaya Base	ANT
306,00	CW		RCV	RUS N HQ Sevastopol	UKR
315,00	AM		FEZ	Fes P	MRC
331,00	AM		HRM	Hassi R. Mel P	ALG
346,00	AM		CH	Chambery P	F
380,00	CW		UMA4	RUS Leningradskaya Base	ANT
424,00	SITOR B	B=100Bd S=170Hz	JNX	Kushiro R	J
424,00	SITOR B	B=100Bd S=170Hz	JGC	Yokohama R	J
424,00	SITOR B	B=100Bd S=170Hz	JNR	Moji R	J
424,00	SITOR B	B=100Bd S=170Hz	JNL	Otura R	J
424,00	SITOR B	B=100Bd S=170Hz	JNB	Naha R	J
426,00	CW		KPH	San Francisco R , CA (Bolinas/Port Reyes)	USA
430,00	CW		PKK	Kupang R , Timor	INS
430,00	CW		PKD	Surabaya R , Djawa	INS
430,00	CW		PKP2	Telukbayur R , Sumatera	INS

Frequency	Mode	Mode Parameter	Callsign	User	Country
430,00	CW		PKC4	Panjang R , Sumatera	INS
436,50	CW		FFC	Bordeaux Arcachon	F
438,00	CW		PKD3	Lembar R , Lombok	INS
438,00	CW		PKM	Bitung R , Sulawesi	INS
438,00	CW		PKA	Sabang R , We	INS
440,00	CW		DI2BO	Peine	D
440,00	CW		NMA	Miami M FL	USA
440,04	CW		DI2AG	Dormitz	D
440,20	*STANAG 4285*	*B=2400Bd*	*JXU*	*NOR N TX Bodo*	*NOR*
448,00	CW		PKM44	Pantolan R , Sulawesi	INS
448,00	CW		PKP	Dumai R , Sumatera	INS
448,00	CW		PKC	Palembang R , Sumatera	INS
452,00	CW		PKI2	Jakarta R	INS
452,00	CW		PKC5	Pangkal Balam R	INS
452,00	CW		PKC2	Plaju R , Sumatera	INS
456,00	CW		PKR	Semarang R , Djawa	INS
456,00	CW		PKG	Bandjarmasin R , Kalimantan	INS
458,00	CW		PKY5	Merauke R , Irian Jaya	INS
458,00	CW		PKY4	Sorong R , Irian Jaya	INS
462,00	CW		XSZ	Dalian R	CHN
465,00	CW		PKS	Pontianak R , Kalimantan	INS
465,00	CW		PNK	Jayapura R , Irian Jaya	INS
465,00	CW		PKF	Makassar R , Sulawesi	INS
470,00	CW		PKE	Amboina R , Ceram	INS
473,00	CW		RDL	RUS N Moscow	RUS
474,00	CW		PKE5	Ternate R , Maluku	INS
474,00	CW		PKR3	Cilacap R , Jawa	INS
474,00	CW		PKZ2	Cirebon R , Jawa	INS
474,00	CW		PKB	Belawan R , Sumatera	INS
474,00	CW		PKB3	Sibolga R , Sumatera	INS
474,20	WSPR	B=1,46Bd Ch=4 ChS=1,46Hz		WSPR net	
474,20	FT8	B=5,86Bd Ch=8 ChS=5,86Hz		FT8 channel	
479,00	CW		SV1AYC	nr Athens	GRC
487,50	CW		PKD5	Benoa R , Bali	INS
490,00	SITOR B	B=100Bd S=170Hz	GNI	Niton R	G
490,00	SITOR B	B=100Bd S=170Hz	TFA	Reykjavik R	ISL
490,00	SITOR B	B=100Bd S=170Hz	GCC	Cullercoats R	G
490,00	SITOR B	B=100Bd S=170Hz		Selia Marina R	I
490,00	SITOR B	B=100Bd S=170Hz	CTV	Monsanto R	POR
490,00	SITOR B	B=100Bd S=170Hz	CTH	POR N Horta	AZR
490,00	SITOR B	B=100Bd S=170Hz	CTQ	Porto Santo R	MDR
490,00	SITOR B	B=100Bd S=170Hz	FRL	La Garde R Toulon	F
490,00	SITOR B	B=100Bd S=170Hz	VFF	CAN CG Iqaluit	CAN
490,00	SITOR B	B=100Bd S=170Hz	CWM27	La Paloma R	URG
490,00	SITOR B	B=100Bd S=170Hz		Chukpyong R	KOR
490,00	SITOR B	B=100Bd S=170Hz		Pyongsan R	KOR
490,00	SITOR B	B=100Bd S=170Hz	XVG	Hai Phong R	VTN
490,00	SITOR B	B=100Bd S=170Hz	FRC	Corsen R	F
490,00	SITOR B	B=100Bd S=170Hz			WW
490,00	SITOR B	B=100Bd S=170Hz		La Maddalena R	I
490,00	SITOR B	B=100Bd S=170Hz		Mondolfo R	I
490,00	*SITOR B*	*B=100Bd S=170Hz*	*DDK*	*DWD TX Pinneberg*	*D*
490,00	SITOR B	B=100Bd S=170Hz		Mondolfo R	I
490,00	SITOR B	B=100Bd S=170Hz		Istanbul R	TUR
490,00	SITOR B	B=100Bd S=170Hz		Izmir R	TUR
490,00	SITOR B	B=100Bd S=170Hz		Antalya R	TUR
490,00	SITOR B	B=100Bd S=170Hz	EAV	Valencia R	E
490,00	SITOR B	B=100Bd S=170Hz		Constanta R	ROU
490,00	SITOR B	B=100Bd S=170Hz	FRC	Corsen R	F
490,00	SITOR B	B=100Bd S=170Hz	TAN	Izmir R	TUR

Frequency	Mode	Mode Parameter	Callsign	User	Country
490,00	SITOR B	B=100Bd S=170Hz	TAF	Samsun R	TUR
500,00	64QAM	B=7900Bd		High speed international NAVTEX planned	WW
500,00	CW		UUO	Kerch R	RUS
500,00	CW		LOK	ARG N Laurie Island , Orcadas	ANT
500,00	CW		PKD5	Benoa R , Bali	INS
500,00	CW		PKE5	Ternate R , Maluku	INS
500,00	CW		XSG	Shanghai R	CHN
500,00	CW		XSZ	Dalian R	CHN
500,00	CW		XSV	Tianjin R	CHN
500,00	CW		PKR3	Cilacap R , Jawa	INS
500,00	CW		PKZ2	Cirebon R , Jawa	INS
500,00	CW		PKI2	Jakarta R	INS
500,00	CW		PKR	Semarang R , Djawa	INS
500,00	CW		PKD	Surabaya R , Djawa	INS
500,00	CW		PKK	Kupang R , Timor	INS
500,00	CW		PKG	Bandjarmasin R , Kalimantan	INS
500,00	CW		PKS	Pontianak R , Kalimantan	INS
500,00	CW		PKD3	Lembar R , Lombok	INS
500,00	CW		PNK	Jayapura R , Irian Jaya	INS
500,00	CW		PKY5	Merauke R , Irian Jaya	INS
500,00	CW		PKY4	Sorong R , Irian Jaya	INS
500,00	CW		PKE	Amboina R , Ceram	INS
500,00	CW		PKM	Bitung R , Sulawesi	INS
500,00	CW		PKF	Makassar R , Sulawesi	INS
500,00	CW		PKM44	Pantolan R , Sulawesi	INS
500,00	CW		PKB	Belawan R , Sumatera	INS
500,00	CW		PKP	Dumai R , Sumatera	INS
500,00	CW		PKC	Palembang R , Sumatera	INS
500,00	CW		PKC5	Pangkal Balam R	INS
500,00	CW		PKC4	Panjang R , Sumatera	INS
500,00	CW		PKC2	Plaju R , Sumatera	INS
500,00	CW		PKA	Sabang R , We	INS
500,00	CW		PKB3	Sibolga R , Sumatera	INS
500,00	CW		PKP2	Telukbayur R , Sumatera	INS
500,00	CW		KPH	San Francisco R , CA (Bolinas/Port Reyes)	USA
501,00	CW		GI4DPE	Londonderry	UK
505,00	CW		OK0EMW	Breclav	CZE
505,00	CW		DI2BO	Peine	D
505,00	CW		DI2BE	Muensterdorf/Itzehoe	D
505,00	CW		DI2AG	Dormitz	D
505,00	CW		DI2KA	Cuxhaven	D
505,10	CW		SM6BHZ	Gothenburg	S
505,18	CW		DI2AM	Rostock	D
512,00	CW		XSZ	Dalian R	CHN
518,00	SITOR B	B=100Bd S=170Hz	PBK	NCG Ijmuiden	HOL
518,00	SITOR B	B=100Bd S=170Hz	SDJ	Stockholm R , Gislövshammar	S
518,00	SITOR B	B=100Bd S=170Hz	SAH	Bjuröklubb R	S
518,00	SITOR B	B=100Bd S=170Hz	SAA	Gislövshammar R	S
518,00	SITOR B	B=100Bd S=170Hz	SAG	Grimeton R	S
518,00	SITOR B	B=100Bd S=170Hz	LGD	Orlandet R	NOR
518,00	SITOR B	B=100Bd S=170Hz	TFA	Reykjavik R	ISL
518,00	SITOR B	B=100Bd S=170Hz	EAL	Las Palmas R	CNR
518,00	SITOR B	B=100Bd S=170Hz	CTQ	Porto Santo R	MDR
518,00	SITOR B	B=100Bd S=170Hz	CNP	Casablanca R	MRC
518,00	SITOR B	B=100Bd S=170Hz	UJB	Astrakhan R	RUS
518,00	SITOR B	B=100Bd S=170Hz	EAV	Cabo de La Nao R	E
518,00	SITOR B	B=100Bd S=170Hz	VCK	CAN CG Sept Iles R	CAN
518,00	SITOR B	B=100Bd S=170Hz	NMB	USCG Savannah , GA	USA
518,00	SITOR B	B=100Bd S=170Hz	XMJ329	Prescott R , ONT	CAN
518,00	SITOR B	B=100Bd S=170Hz	VCO	CAN CG Sydney	CAN
518,00	SITOR B	B=100Bd S=170Hz	VAR3	Yarmouth R	CAN
518,00	SITOR B	B=100Bd S=170Hz	VON	S. John`s R	CAN

Frequency	Mode	Mode Parameter	Callsign	User	Country
518,00	SITOR B	B=100Bd S=170Hz	XLJ895	CAN CG Thunder Bay , ONT	CAN
518,00	SITOR B	B=100Bd S=170Hz	VFF	CAN CG Iqaluit	CAN
518,00	SITOR B	B=100Bd S=170Hz	VOK	CAN CG Labrador , NFL	CAN
518,00	SITOR B	B=100Bd S=170Hz	OXI	Nuuk R	GRL
518,00	SITOR B	B=100Bd S=170Hz	PJC	Curacao R	ATN
518,00	SITOR B	B=100Bd S=170Hz	L3K	ARG N Ushuaia	ARG
518,00	SITOR B	B=100Bd S=170Hz	L3D	ARG N Rio Gallegos	ARG
518,00	SITOR B	B=100Bd S=170Hz	L3D	ARG N Rio Gallegos	ARG
518,00	SITOR B	B=100Bd S=170Hz	L2W	ARG N Comodoro Rivadavia	ARG
518,00	SITOR B	B=100Bd S=170Hz	L2I	ARG N Bahia Blanca	ARG
518,00	SITOR B	B=100Bd S=170Hz	L2I	ARG N Bahia Blanca	ARG
518,00	SITOR B	B=100Bd S=170Hz	L2P	ARG N Mar del Plata	ARG
518,00	SITOR B	B=100Bd S=170Hz	L2B	ARG N Buenos Aires	ARG
518,00	SITOR B	B=100Bd S=170Hz	V5W	Walvis Bay R	NMB
518,00	SITOR B	B=100Bd S=170Hz	ZSQ	Port Elizabeth R	AFS
518,00	SITOR B	B=100Bd S=170Hz	3BM	Port Louis R	MAU
518,00	SITOR B	B=100Bd S=170Hz	EQM	Bushehr R	IRN
518,00	SITOR B	B=100Bd S=170Hz	EQI	Abbas R	IRN
518,00	SITOR B	B=100Bd S=170Hz	SUZ	Serapeum R	EGY
518,00	SITOR B	B=100Bd S=170Hz	PNK	Jayapura R , Irian Jaya	INS
518,00	SITOR B	B=100Bd S=170Hz	PKE	Amboina R , Ceram	INS
518,00	SITOR B	B=100Bd S=170Hz	PKF	Makassar R , Sulawesi	INS
518,00	SITOR B	B=100Bd S=170Hz	HSA	Bangkok R	THA
518,00	SITOR B	B=100Bd S=170Hz		Puerto Princesa R	PHL
518,00	SITOR B	B=100Bd S=170Hz		Manila R	PHL
518,00	SITOR B	B=100Bd S=170Hz		Davao R	PHL
518,00	SITOR B	B=100Bd S=170Hz	XSI	Sanya R	CHN
518,00	SITOR B	B=100Bd S=170Hz	XSL	Fuzhou R	CHN
518,00	SITOR B	B=100Bd S=170Hz	XVT	Da Nang R	VTN
518,00	SITOR B	B=100Bd S=170Hz	XSV	Tianjin R	CHN
518,00	SITOR B	B=100Bd S=170Hz	9WH21	Sandakan R	MLA
518,00	SITOR B	B=100Bd S=170Hz	9WW21	Miri R	MLA
518,00	SITOR B	B=100Bd S=170Hz	9MG	Penang R	MLA
518,00	SITOR B	B=100Bd S=170Hz	XVG	Hai Phong R	VTN
518,00	SITOR B	B=100Bd S=170Hz		ICRC Kigali	RRW
518,00	SITOR B	B=100Bd S=170Hz		Chukpyong R	KOR
518,00	SITOR B	B=100Bd S=170Hz	XVS	Ho Chi Minh Ville R	VTN
518,00	SITOR B	B=100Bd S=170Hz		Pyongsan R	KOR
518,00	SITOR B	B=100Bd S=170Hz	NOJ	USCG Kodiak	ALS
518,00	SITOR B	B=100Bd S=170Hz	VAJ	Prince Rupert R	CAN
518,00	SITOR B	B=100Bd S=170Hz	NMQ	USCG Long Beach R , CA	USA
518,00	SITOR B	B=100Bd S=170Hz	XLK835	CAN CG Tofino	CAN
518,00	SITOR B	B=100Bd S=170Hz	HCG	Guayaquil R	EQA
518,00	SITOR B	B=100Bd S=170Hz	OBY2	Paita R	PRU
518,00	SITOR B	B=100Bd S=170Hz	OBF4	Mollendo R	PRU
518,00	SITOR B	B=100Bd S=170Hz	CBA	Antofagasta R	CHL
518,00	SITOR B	B=100Bd S=170Hz	CBV	Playa Ancha Zonal R	CHL
518,00	SITOR B	B=100Bd S=170Hz	CBT	CHL N Talcahuano	CHL
518,00	SITOR B	B=100Bd S=170Hz	CBP	Puerto Montt Zonal R	CHL
518,00	SITOR B	B=100Bd S=170Hz	CBP	Puerto Montt Zonal R	CHL
518,00	SITOR B	B=100Bd S=170Hz	CBM	Magallanes Zonal R	CHL
518,00	SITOR B	B=100Bd S=170Hz	CBY	I. de Pascua R (Easter Iland)	CHL
518,00	SITOR B	B=100Bd S=170Hz	CBY	I. de Pascua R (Easter Iland)	CHL
518,00	SITOR B	B=100Bd S=170Hz	LGS	Svalbard R	NOR
518,00	SITOR B	B=100Bd S=170Hz	UHY	Murmansk R	RUS
518,00	SITOR B	B=100Bd S=170Hz			WW
518,00	SITOR B	B=100Bd S=170Hz		Sous CROSS / Soulac	F
518,00	SITOR B	B=100Bd S=170Hz		S. Georges R	BER
518,00	SITOR B	B=100Bd S=170Hz	FRC	Sous CROSS / Corse	F
518,00	SITOR B	B=100Bd S=170Hz	FRL	Sous CROSS / La Garde	F
518,00	SITOR B	B=100Bd S=170Hz		Rosario R	ARG
518,00	SITOR B	B=100Bd S=170Hz	5BA	Cyprus R	CYP

Frequency	Mode	Mode Parameter	Callsign	User	Country
518,00	SITOR B	B=100Bd S=170Hz	9AS	Split R	HRV
518,00	SITOR B	B=100Bd S=170Hz	9HD	Malta R	MLT
518,00	SITOR B	B=100Bd S=170Hz	9VG49	Singapore R	SNG
518,00	SITOR B	B=100Bd S=170Hz	A4M	Muscat R	OMA
518,00	SITOR B	B=100Bd S=170Hz	A9M	Bahrain R	BHR
518,00	SITOR B	B=100Bd S=170Hz	CBA	Antofagasta R	CHL
518,00	SITOR B	B=100Bd S=170Hz	CBM	Magallanes Zonal R	CHL
518,00	SITOR B	B=100Bd S=170Hz	CBP	Puerto Montt R	CHL
518,00	SITOR B	B=100Bd S=170Hz	CBT	CHL N Talcahuano	CHL
518,00	SITOR B	B=100Bd S=170Hz	CBV	Playa Ancha Zonal R	CHL
518,00	SITOR B	B=100Bd S=170Hz	CTH	POR N Horta R	AZR
518,00	SITOR B	B=100Bd S=170Hz	CTV	Monsanto R	POR
518,00	SITOR B	B=100Bd S=170Hz	EAC	Tarifa R	E
518,00	SITOR B	B=100Bd S=170Hz	EAF	Cabo Finisterre R	E
518,00	SITOR B	B=100Bd S=170Hz	EAR	La Coruna R	E
518,00	SITOR B	B=100Bd S=170Hz	ESA	Tallin R	EST
518,00	SITOR B	B=100Bd S=170Hz	FFU	Brest-Le Conquet R	F
518,00	SITOR B	B=100Bd S=170Hz	GCC	Cullercoats R	G
518,00	SITOR B	B=100Bd S=170Hz	GNI	Niton R	G
518,00	SITOR B	B=100Bd S=170Hz	GPK	Portpatrick R	G
518,00	SITOR B	B=100Bd S=170Hz	HZG	Dammam R	ARS
518,00	SITOR B	B=100Bd S=170Hz	HZH	Jeddah R	ARS
518,00	SITOR B	B=100Bd S=170Hz	JGC	Yokohama R	J
518,00	SITOR B	B=100Bd S=170Hz	JNB	Naha R	J
518,00	SITOR B	B=100Bd S=170Hz	JNL	Otura R	J
518,00	SITOR B	B=100Bd S=170Hz	JNR	Moji R	J
518,00	SITOR B	B=100Bd S=170Hz	JNX	Kushiro R	J
518,00	SITOR B	B=100Bd S=170Hz	L2B	ARG N Buenos Aires	ARG
518,00	SITOR B	B=100Bd S=170Hz	L2N	ARG N Bahia Blanca	ARG
518,00	SITOR B	B=100Bd S=170Hz	L2P	ARG N Mar del Plata	ARG
518,00	SITOR B	B=100Bd S=170Hz	L2T	ARG N Mar del Plata	ARG
518,00	SITOR B	B=100Bd S=170Hz	L2W	ARG N Comodoro Rivadavia	ARG
518,00	SITOR B	B=100Bd S=170Hz	L3I	Rio Gallegos R	ARG
518,00	SITOR B	B=100Bd S=170Hz	L3P	ARG N Ushuaia Prefectura	ARG
518,00	SITOR B	B=100Bd S=170Hz	LGP	Bodoe R	NOR
518,00	SITOR B	B=100Bd S=170Hz	LGQ	Rogaland R	NOR
518,00	SITOR B	B=100Bd S=170Hz	LGV	Vardo R	NOR
518,00	SITOR B	B=100Bd S=170Hz	LZW	Varna R	BUL
518,00	SITOR B	B=100Bd S=170Hz	NMA	USCG Miami FL	USA
518,00	SITOR B	B=100Bd S=170Hz	NMC	USA CG Pt. Reyes , CA	USA
518,00	SITOR B	B=100Bd S=170Hz	NMF	USCG Boston MA	USA
518,00	SITOR B	B=100Bd S=170Hz	NMG	USCG New Orleans LA	USA
518,00	SITOR B	B=100Bd S=170Hz	NMN	USCG Portsmouth VA	USA
518,00	SITOR B	B=100Bd S=170Hz	NMO	USA CG Honolulu	HWA
518,00	SITOR B	B=100Bd S=170Hz	NMQ9	USCG Long Beach	USA
518,00	SITOR B	B=100Bd S=170Hz	NMR	USA CG San Juan	PTR
518,00	SITOR B	B=100Bd S=170Hz	NMW	USCG Astoria OR	USA
518,00	SITOR B	B=100Bd S=170Hz	NOJ	USCG Kodiak ALS	USA
518,00	SITOR B	B=100Bd S=170Hz	OST	Oostende R	BEL
518,00	SITOR B	B=100Bd S=170Hz	PKX	Jakarta R , Jawa	INS
518,00	SITOR B	B=100Bd S=170Hz	SDJ	Stockholm R , Bjuröklubb	S
518,00	SITOR B	B=100Bd S=170Hz	SUH	Alexandria R	EGY
518,00	SITOR B	B=100Bd S=170Hz	SVH	Iraklion Kritis R	GRC
518,00	SITOR B	B=100Bd S=170Hz	SVK	Kerkyra R Korfu	GRC
518,00	SITOR B	B=100Bd S=170Hz	SVL	Limnos R	GRC
518,00	SITOR B	B=100Bd S=170Hz	TAF	Samsun R	TUR
518,00	SITOR B	B=100Bd S=170Hz	TAH	Istanbul R	TUR
518,00	SITOR B	B=100Bd S=170Hz	TAL	Antalya R	TUR
518,00	SITOR B	B=100Bd S=170Hz	ASK	Karachi R	PAK
518,00	SITOR B	B=100Bd S=170Hz	SUK	Kosseir R	EGY
518,00	SITOR B	B=100Bd S=170Hz	TAN	Izmir R	TUR
518,00	SITOR B	B=100Bd S=170Hz	TFA	Reykjavik R	ISL

Frequency	Mode	Mode Parameter	Callsign	User	Country
518,00	SITOR B	B=100Bd S=170Hz	UBE	Petropavlovsk Kamchatskii R	RUS
518,00	SITOR B	B=100Bd S=170Hz	UCT2	Beringovskiy R	RUS
518,00	SITOR B	B=100Bd S=170Hz	UDN	Novorossiysk R	RUS
518,00	SITOR B	B=100Bd S=170Hz	UFO	Kholmsk R	RUS
518,00	SITOR B	B=100Bd S=170Hz	UGE	Arkhangelsk R	RUS
518,00	SITOR B	B=100Bd S=170Hz	UMN	Murmansk R	RUS
518,00	SITOR B	B=100Bd S=170Hz	UPB	Providenya Bukhta R	RUS
518,00	SITOR B	B=100Bd S=170Hz	URL8	Sevastopol R	RUS
518,00	SITOR B	B=100Bd S=170Hz	UTT	Odessa R	UKR
518,00	SITOR B	B=100Bd S=170Hz	UTW	Mariupol R	UKR
518,00	SITOR B	B=100Bd S=170Hz	UUI	Odessa R	UKR
518,00	SITOR B	B=100Bd S=170Hz	UVD	Magadan R	RUS
518,00	SITOR B	B=100Bd S=170Hz	VAR3	Yarmouth R	CAN
518,00	SITOR B	B=100Bd S=170Hz	VCK	Sept Iles R	CAN
518,00	SITOR B	B=100Bd S=170Hz	VCO	CAN CG Sydney	CAN
518,00	SITOR B	B=100Bd S=170Hz	VRX	Hong Kong R	CHN
518,00	SITOR B	B=100Bd S=170Hz	VWB	Mumbai R	IND
518,00	SITOR B	B=100Bd S=170Hz	VWM	Madras R	IND
518,00	SITOR B	B=100Bd S=170Hz	XSG	Shanghai R	CHN
518,00	SITOR B	B=100Bd S=170Hz	XSQ	Guangzhou R	CHN
518,00	SITOR B	B=100Bd S=170Hz	XSZ	Dalian R	CHN
518,00	SITOR B	B=100Bd S=170Hz	ZBM	Bermuda Harbour R	BER
518,00	SITOR B	B=100Bd S=170Hz	ZSC	Capetown R	AFS
518,00	SITOR B	B=100Bd S=170Hz	ZSD	Durban R	AFS
518,00	SITOR B	B=100Bd S=170Hz	CWM27	La Paloma R	URG
518,00	SITOR B	B=100Bd S=170Hz	L3K	ARG N Ushuaia	ARG
518,00	SITOR B	B=100Bd S=170Hz		Cabo de la Nao R	E
518,00	*SITOR B*	*B=100Bd S=170Hz*	*DDK*	*DWD TX Pinneberg*	*D*
518,00	SITOR B	B=100Bd S=170Hz	OST	Oostende R	BEL
518,00	SITOR B	B=100Bd S=170Hz	TFA	Reykjavik R	ISL
518,00	SITOR B	B=100Bd S=170Hz	EJK	Valentia R	IRL
518,00	SITOR B	B=100Bd S=170Hz	4XO	Haifa R	ISR
518,00	SITOR B	B=100Bd S=170Hz		Ismaila R	EGY
518,00	SITOR B	B=100Bd S=170Hz		Heraklion R	GRC
518,00	SITOR B	B=100Bd S=170Hz		Niton R	G
518,00	SITOR B	B=100Bd S=170Hz	SDJ	Stockholm R , Grimeton	S
521,50	CW		UDK	Murmansk R	RUS
522,00	CW		XSG	Shanghai R	CHN
523,00	T600	B=50Bd S=200Hz	RMP	RUS N HQ Kaliningrad	RUS
530,00	AM			USA Traveler Information Service	USA
570,00	T600	B=50Bd S=200Hz	RMP	RUS N HQ Kaliningrad	RUS

1000 – 5000 kHz

Frequency	Mode	Mode Parameter	Callsign	User	Country
1359,00	USB		XVG	Hai Phong R	VTN
1607,00	CW			CS Ch 201	R I
1607,50	CW			CS Ch 202	R I
1608,00	CW			CS Ch 203	R I
1608,50	CW			CS Ch 204	R I
1609,00	CW			CS Ch 205	R I
1609,50	CW			CS Ch 206	R I
1610,00	CW			CS Ch 207	R I
1610,00	AM			USA Traveler Information Service	USA
1610,50	CW			CS Ch 208	R I
1611,00	CW			CS Ch 209	R I
1611,50	CW			CS Ch 210	R I
1612,00	CW			CS Ch 211	R I
1612,50	CW			CS Ch 212	R I

Frequency	Mode	Mode Parameter	Callsign	User	Country
1613,00	CW			CS Ch 213	R I
1613,50	CW			CS Ch 214	R I
1614,00	CW			CS Ch 215	R I
1614,50	CW			CS Ch 216	R I
1615,00	CW			CS Ch 217	R I
1615,50	USB		SVH	Iraklion Kritis R	GRC
1615,50	CW			CS Ch 218	R I
1616,00	CW			CS Ch 219	R I
1616,50	CW			CS Ch 220	R I
1617,00	CW			CS Ch 221	R I
1617,50	CW			CS Ch 222	R I
1618,00	CW			CS Ch 223	R I
1618,50	SITOR , CW	B=100Bd S=170Hz	SUQ	Ismaila R	EGY
1618,50	CW			CS Ch 224	R I
1619,00	CW			CS Ch 225	R I
1619,50	CW			CS Ch 226	R I
1620,00	CW			CS Ch 227	R I
1620,50	CW			CS Ch 228	R I
1621,00	CW			CS Ch 229	R I
1621,50	USB			Floro R	NOR
1621,50	USB			Rogaland R	NOR
1621,50	USB			Tjome R	NOR
1621,50	USB		LGV	Vardo R	NOR
1621,50	CW			CS Ch 230	R I
1621,50	USB			Bodo R	NOR
1622,00	CW			CS Ch 231	R I
1622,20	*STANAG 4285*	*B=2400Bd*	*DHJ58*	*D N Glücksburg TX Neuharlingersiel*	*D*
1622,50	CW			CS Ch 232	R I
1623,00	CW			CS Ch 233	R I
1623,50	CW			CS Ch 234	R I
1624,00	CW			CS Ch 235	R I
1624,50	USB		OXZ	Lyngby R	DNK
1624,50	CW			CS Ch 236	R I
1635,00	USB			Vardo R via Hammerfest	NOR
1635,00	SSB			CS Ch 241	R I
1636,40	USB		HZH	Jeddah R	ARS
1638,00	USB			Vaasa MRSC	FNL
1638,00	SSB			CS Ch 242	R I
1641,00	SSB			CS Ch 243	R I
1641,00	SSB		OXJ	Torshaven R	FRI
1642,50	SSB			Den Helder CG	HOL
1644,00	USB		EJM	Malin Head R	IRL
1644,00	SSB			CS Ch 244	R I
1644,00	USB		EAL	Las Palmas R	CNR
1647,00	SSB			CS Ch 245	R I
1650,00	USB		TYA	Cotonou R	BEN
1650,00	SSB			CROSS Grez-Niz	F
1650,00	SSB			CROSS Jobourg	F
1650,00	SSB			CROSS Corsen	F
1650,00	SSB			CS Ch 246	R I
1650,00	USB		ESA	Tallin R	EST
1653,00	SSB			CS Ch 247	R I
1656,00	SSB			CS Ch 248	R I
1656,00	USB			Malaga R via Tarfifa	E
1659,00	USB			Bodo R via Andenes	NOR
1659,00	SSB			CS Ch 249	R I
1660,00	USB		DCA	MRCC Bremen Rescue	D
1660,40	USB		3VZ	Zarzis R	TUN
1662,00	SSB			CS Ch 250	R I
1665,00	USB		LZW	Varna R	BUL
1665,00	USB			Tjome R	NOR
1665,00	SSB			CS Ch 251	R I

Frequency	Mode	Mode Parameter	Callsign	User	Country
1668,00	SSB			CS Ch 252	R I
1671,00	SSB			CS Ch 253	R I
1671,00	SSB			Rogaland R via Farsund	NOR
1672,40	USB		3VS	Sfax R	TUN
1674,00	SSB			CS Ch 254	R I
1674,00	SSB		SDJ	Stockholm R , Tingstädte	S
1677,00	SSB			CS Ch 255	R I
1677,00	SSB		EJM	Malin Head R	IRL
1677,00	SSB		OFH	Mariehamn R	FNL
1677,00	USB			Bilbao R via Machichaco	E
1678,20	STANAG 4285	B=2400Bd	OSN	BEL N Oostende	BEL
1680,00	SSB			CS Ch 256	R I
1680,00	USB			Floro R	NOR
1680,80	CW			RUS BRAS system	RUS
1683,00	SSB			CS Ch 257	R I
1686,00	SSB			CS Ch 258	R I
1687,40	SSB		3VB	Bizerte R	TUN
1689,00	SSB			CS Ch 259	R I
1689,00	SSB		EAL	Las Palmas R	CNR
1692,00	SSB			CS Ch 260	R I
1692,00	USB			Rogaland R	NOR
1695,00	USB			Vardo R via Berlevag	NOR
1695,00	SSB			CS Ch 261	R I
1696,00	USB			CROSS La Garde	F
1696,00	SSB		SVK	Kerkyra R	GRC
1696,40	SSB		3VM	Mahdia R	TUN
1698,00	USB		SXE	Aspropyrgos Attikis R	GRC
1698,00	USB			Peiraias CG R	GRC
1698,00	SSB			CS Ch 262	R I
1698,00	USB			Coruna R via Finisterre	E
1699,40	USB		3VS	Sfax R	TUN
1701,00	SSB			CS Ch 263	R I
1704,00	SSB			CS Ch 264	R I
1704,00	USB			Malaga R via Tarfifa	E
1704,00	USB			Skamlebak R	DNK
1707,00	SSB			CS Ch 265	R I
1707,00	USB			Coruna R	E
1707,70	CW			RUS BRAS system	RUS
1710,00	USB			Bodo R via Sandnessjoen	NOR
1710,00	SSB			CS Ch 266	R I
1713,00	USB		LGV	Vardo R	NOR
1713,00	SSB			CS Ch 267	R I
1716,00	SSB			CS Ch 268	R I
1719,00	USB			Hailuoto R	FNL
1719,00	SSB			CS Ch 269	R I
1720,40	SSB		4OB	Bar R	MNE
1722,00	SSB			CS Ch 270	R I
1722,00	SSB		LJB	Bodo R via Björnöya	NOR
1725,00	SSB			CS Ch 271	R I
1725,00	SSB		LGN	Rogaland R via Vigre	NOR
1726,00	SSB		HZH	Jeddah R	ARS
1726,40	SSB		SVH	Iraklion Kritis R	GRC
1728,00	SSB			CS Ch 272	R I
1728,00	USB			Rogaland R via Bergen	NOR
1731,00	SSB			CS Ch 273	R I
1731,00	USB			Bodo R via Svalbard	NOR
1734,00	USB		SXE	Aspropyrgos Attikis R	GRC
1734,00	SSB			CS Ch 274	R I
1734,00	USB		OXZ	Lyngby R TX Blavand R	DNK
1737,00	SSB			CS Ch 275	R I
1740,00	SSB			CS Ch 276	R I
1741,40	SSB		SVH	Iraklion Kritis R	GRC

Frequency	Mode	Mode Parameter	Callsign	User	Country
1741,40	USB		3VL	Kelibia R	TUN
1743,00	SSB			CS Ch 277	R I
1743,00	USB			Bodo R via Jan Mayen	NOR
1743,00	USB			Stornoway CG MRCC	G
1746,00	USB		EJK	Valentia R	IRL
1746,00	SSB			CS Ch 278	R I
1749,00	SSB			CS Ch 279	R I
1750,00	USB			Hopen M	NOR
1752,00	SSB			CS Ch 280	R I
1752,00	SSB		EJK	Valentia R	IRL
1755,00	USB			Valencia R via Cabo de la Nao	E
1755,00	SSB			CS Ch 281	R I
1755,00	USB		EAO	Palma R	E
1757,00	USB		LJB	Bjornoya M	NOR
1758,00	SSB			CS Ch 282	R I
1758,00	SSB		OXJ	Torshaven R	FRI
1758,00	USB		OXZ	Lyngby R TX Skagen R	DNK
1761,00	SSB			CS Ch 283	R I
1764,00	SSB			CS Ch 284	R I
1767,00	USB			Valencia R via Cabo de la Nao	E
1767,00	SSB			CS Ch 285	R I
1767,00	SSB		SVN	Olympia R	GRC
1767,00	USB			Milford Haven CG MRCC	G
1767,00	USB			Bovbjerg R	DNK
1768,20	*STANAG 4285*	*B=2400Bd*	*DHJ58*	*D N Glücksburg TX Marlow*	*D*
1768,40	SSB		3VT	Tunis R	TUN
1770,00	SSB			CS Ch 286	R I
1770,00	SSB			Bodo R	NOR
1770,00	USB			Shetland CG MRCC	G
1770,70	CW			RUS BRAS system	RUS
1771,00	SSB		3VM	Mahdia R	TUN
1773,00	SSB		IPD	Civiatavecchia R	I
1776,00	SSB		IQA	Augusta R	I
1776,00	SSB			CS Ch 288	R I
1779,00	SSB			CS Ch 289	R I
1779,00	SSB		SDJ	Stockholm R , Bjuroklubb	S
1780,40	USB		3VK	Tabarka R	TUN
1782,00	SSB			CS Ch 290	R I
1782,00	SSB			Floro R via Orlandet	NOR
1785,00	SSB			CS Ch 291	R I
1785,00	USB			Rogaland R via Farsund	NOR
1785,85	CW			RUS BRAS system	RUS
1788,00	SSB			CS Ch 292	R I
1791,00	SSB			CS Ch 293	R I
1792,00	SSB		7TA	Alger R	ALG
1794,00	SSB			CS Ch 294	R I
1797,00	SSB		SDJ	Stockholm R , Gislövshammar	S
1797,07	CW			RUS BRAS system	RUS
1802,50	CW		W1AW	ARRL morse training	USA
1803,00	USB			Bodo R via Andenes	NOR
1812,10	CW			RUS BRAS system	RUS
1813,00	USB		J2A	Djibouti R	DJI
1824,00	SSB		SVR	Rhodos R	GRC
1831,00	CW		OSN41	BEL N Oostende	BEL
1836,60	WSPR	B=1,46Bd Ch=4 ChS=1,46Hz		WSPR net	
1838,00	PSK31	B=31Bd		PSK31 channel	
1838,00	JT65	B=2,69Bd Ch=65 ChS=2,69Hz		JT65 channel	
1838,50	MFSK			MFSK modes user	
1839,00	JT9	B=1,736Bd Ch=9 ChS=1,736Hz		JT9 channel	

Frequency	Mode	Mode Parameter	Callsign	User	Country
1840,00	ROS USB	B=1Bd Ch=144 ChS=15,625Hz		ROS frequency	
1840,00	FT8	B=5,86Bd Ch=8 ChS=5,86Hz		FT8 channel	
1842,00	JS8	B=6,25Bd Ch=8 ChS=6,25Hz		JS8 user	
1843,00	ALE USB	B=125Bd Ch=8 ChS=250Hz		HFLINK network	
1849,00	SSB		SUK	Kosseir R	EGY
1850,00	SSB		TAT	Canakkala R	TUR
1852,00	USB			Palermo R	I
1854,00	CW		OKM1	Hradec Kralove - Stezery P	CZE
1856,00	SSB		HZH	Jeddah R	ARS
1856,00	USB			Lampedusa R	I
1858,20	STANAG 4285	B=2400Bd	DHJ58	D N Glücksburg	D
1869,00	USB			Yarmouth CG MRCC	G
1873,00	LSB			Ham radio emergency frequency	R1
1875,00	*STANAG 4285*	*B=2400Bd*		*G MIL Crimond*	*G*
1876,00	USB		TFA	Reykjavik R	ISL
1880,00	USB			Falmouth CG MRCC	G
1880,00	USB			Holyhead CG MRCC	G
1883,00	USB			Clyde CG MRCC	G
1883,00	USB			Belfast CG MRCC	G
1888,00	USB		IPD	Civiatavecchia R	I
1890,00	USB		PBB	NLD CG Den Helder	HOL
1897,20	*STANAG 4285*	*B=2400Bd*	*DHJ58*	*D N Glücksburg*	*D*
1904,00	USB			Peiraias CG R	GRC
1909,00	ALE USB	B=125Bd Ch=8 ChS=250Hz		HFLINK network	
1911,00	SSB		7TB	Annaba R	ALG
1911,00	SSB		CND	Agadir R	MRC
1919,00	PACTOR III	B=100Bd S=200Hz	FOHXM	Sailmail Manihi Atoll	OCE
1925,00	SSB		IPL	Livorno R	I
1925,00	USB			Hunber CG MRCC	G
1930,00	ALE USB	B=125Bd Ch=8 ChS=250Hz		POL MIL net	POL
1940,00	RADAR			MF/HF SA Radar Kolhapur	IND
1940,00	RADAR			MF/HF SA Radar Juliusruh	D
1955,00	RADAR			MF/HF SA Radar Wakkanai	J
1957,50	SSB			Den Helder Rescue	HOL
1980,00	RADAR			MF/HF SA Radar Adelaide	AUS
1980,00	RADAR			MF/HF SA Radar Andenes	NOR
1980,00	RADAR			MF/HF SA Radar Davis Station	ANT
1980,00	RADAR			MF/HF SA Radar Pontianak	INS
1980,00	RADAR			MF/HF SA Radar Qujing	CHN
1980,00	RADAR			MF/HF SA Radar Tirunelveli	IND
1990,00	RADAR			MF/HF SA Radar Langfang	CHN
1996,00	ALE USB	B=125Bd Ch=8 ChS=250Hz		HFLINK network	
2000,00	USB			VOLMET New York R	USA
2000,20	*STANAG 4285*	*B=2400Bd*		*NOR N TX Stavanger*	*NOR*
2008,00	OOK			MF/HF SA Radar Pameungpeuk	INS
2009,00	USB		HLU	Ulreung R	KOR
2009,00	ALE USB	B=125Bd Ch=8 ChS=250Hz		S MIL net	
2020,00	USB		VJJ	RFDS Charleville	AUS
2020,00	USB		VJD	RFDS Alice Springs	AUS
2020,00	USB		VNZ	RFDS Pt. Augusta	AUS
2020,00	USB		VJN	RFDS Cairns	AUS
2020,00	USB		VJI	RFDS Mount Isa	AUS
2025,00	ALE USB	B=125Bd Ch=8 ChS=250Hz		S MIL net	S

Frequency	Mode	Mode Parameter	Callsign	User	Country
2037,00	ALE USB	B=125Bd Ch=8 ChS=250Hz		S MIL net	
2037,50	SSB		VWP	Port Blair R	ADM
2038,00	ALE USB	B=125Bd Ch=8 ChS=250Hz		POL MIL net	POL
2044,00	USB		HLY	Yeosu R	KOR
2045,00	SSB			ship/shore working frequency	R I
2045,00	SSB		OXJ	Torshaven R	FRI
2048,00	SSB			ship/ship working frequency	R I
2051,00	SSB			ship/shore working frequency	R I
2053,00	USB				RUS
2054,00	SSB			ship/shore working frequency	R I
2054,00	USB		VAE	Tofino R	CAN
2054,00	USB		VAJ	Prince Rupert R	CAN
2054,00	FAX 120/576		NOJ	USA CG Kodiak	USA
2055,00	STANAG 4285	B=2400Bd	IDN	I N Naples	I
2056,00	USB		VMW	Wiluna M	AUS
2057,00	SSB			ship/shore working frequency	R I
2060,00	SSB			SS Ch 241	R I
2060,00	SSB			SS Ch 268	R I
2060,00	SSB			SS Ch 295	R I
2061,40	BAUDOT	B=75Bd S=850Hz	ARK		
2063,00	SSB			SS Ch 242	R I
2063,00	SSB			SS Ch 269	R I
2063,70	STANAG 4285	B=2400Bd	FUG	F N Saissac	F
2065,00	USB		L2A	ARG N Buenos Aires	ARG
2065,00	USB		L2V	ARG N Commodoro Rivadavia	ARG
2065,00	USB		L3J	ARG N Ushuaia Prefectura	ARG
2065,00	USB		L2O	ARG N Mar del Plata	ARG
2066,00	SSB			SS Ch 243	R I
2066,00	SSB			SS Ch 270	R I
2066,80	F N FSK	B=50Bd S=850Hz Ch=1	FUG	F N Saissac	F
2069,00	SSB			SS Ch 244	R I
2069,00	SSB			SS Ch 271	R I
2072,00	CW		VTG3	IND N Mumbai	IND
2072,00	SSB			SS Ch 245	R I
2072,00	SSB			SS Ch 272	R I
2075,00	SSB			SS Ch 246	R I
2075,00	SSB			SS Ch 273	R I
2078,00	SSB			SS Ch 247	R I
2078,00	SSB			SS Ch 274	R I
2080,00	USB		A3A	Nukualofa R	TON
2081,00	SSB			SS Ch 248	R I
2081,00	SSB			SS Ch 275	R I
2082,20	STANAG 4539	B=2400Bd		Croughton	G
2084,00	SSB			SS Ch 249	R I
2084,00	SSB			SS Ch 276	R I
2087,00	SSB			SS Ch 250	R I
2087,00	SSB			SS Ch 277	R I
2088,20	STANAG 4539	B=2400Bd		Croughton	G
2090,00	SSB			SS Ch 251	R I
2090,00	SSB			SS Ch 278	R I
2091,00	CW		HLC	Incheon R	KOR
2091,00	CW		HLK	Kangnung R	KOR
2091,00	CW		HLN	Gunsan R	KOR
2091,00	CW		HLU	Ulreung R	KOR
2093,00	SSB			SS Ch 252	R I
2093,00	SSB			SS Ch 279	R I
2096,00	SSB			SS Ch 253	R I
2096,00	SSB			SS Ch 280	R I
2097,20	STANAG 4539	B=2400Bd		Croughton	G

Frequency	Mode	Mode Parameter	Callsign	User	Country
2097,30	CW		A		USA
2099,00	SSB			SS Ch 254	R I
2099,00	SSB			SS Ch 281	R I
2101,00	CW			RUS BRAS system	RUS
2102,00	SSB			SS Ch 255	R I
2102,00	SSB			SS Ch 282	R I
2105,00	SSB			SS Ch 256	R I
2105,00	SSB			SS Ch 283	R I
2108,00	SSB			SS Ch 257	R I
2108,00	SSB			SS Ch 284	R I
2110,00	USB			Cheju MRCC	KOR
2110,00	USB		HLC	Incheon MRCC	KOR
2110,00	USB			Mokpo MRCC	KOR
2110,00	USB			Busan MRCC	KOR
2110,00	USB		HLF	Tonghae MRCC	KOR
2111,00	SSB			SS Ch 258	R I
2111,00	SSB			SS Ch 285	R I
2111,00	USB		3DP	Suva R	FJI
2114,00	SSB			SS Ch 259	R I
2114,00	SSB			SS Ch 286	R I
2114,00	USB		EAL	Las Palmas R	CNR
2115,50	STANAG 4285	B=2400Bd		G MIL St. Eval	G
2116,00	USB		OYR	Aasiaat R	GRL
2117,00	SSB			SS Ch 260	R I
2117,00	SSB			SS Ch 287	R I
2120,00	SSB			SS Ch 261	R I
2120,00	SSB			SS Ch 288	R I
2121,40	USB		PBC	NLD N Goeree Island	HOL
2122,50	USB			G N Sea Cadets	G
2122,50	STANAG 4285	B=2400Bd		G MIL St. Eval	G
2123,00	SSB			SS Ch 262	R I
2123,00	SSB			SS Ch 289	R I
2126,00	SSB			SS Ch 263	R I
2126,00	SSB			SS Ch 290	R I
2129,00	USB		OYR	Aasiaat R	GRL
2129,00	SSB			SS Ch 264	R I
2129,00	SSB			SS Ch 291	R I
2130,00	NFM			RUS railroad net	RUS
2132,00	SSB			SS Ch 265	R I
2132,00	SSB			SS Ch 292	R I
2135,00	SSB			SS Ch 266	R I
2135,00	SSB			SS Ch 293	R I
2138,00	SSB			SS Ch 267	R I
2138,00	SSB			SS Ch 294	R I
2138,00	RADAR			MF/HF SA Radar Quijing	CHN
2141,00	SSB			USCG rescue operation channel	USA
2142,00	SITOR	B=100Bd S=170Hz		SS Ch 201	R I
2142,50	SITOR	B=100Bd S=170Hz		SS Ch 202	R I
2142,50	R&S MODEM	B=2400Bd		D CG	D
2143,00	SITOR	B=100Bd S=170Hz		SS Ch 203	R I
2143,50	SITOR	B=100Bd S=170Hz		SS Ch 204	R I
2144,00	SITOR	B=100Bd S=170Hz		SS Ch 205	R I
2144,50	SITOR	B=100Bd S=170Hz		SS Ch 206	R I
2145,00	SITOR	B=100Bd S=170Hz		SS Ch 207	R I
2145,50	SITOR	B=100Bd S=170Hz		SS Ch 208	R I
2146,00	SITOR	B=100Bd S=170Hz		SS Ch 209	R I
2146,50	SITOR	B=100Bd S=170Hz		SS Ch 210	R I
2147,00	SITOR	B=100Bd S=170Hz		SS Ch 211	R I
2147,50	SITOR	B=100Bd S=170Hz		SS Ch 212	R I
2148,00	SITOR	B=100Bd S=170Hz		SS Ch 213	R I
2148,50	SITOR	B=100Bd S=170Hz		SS Ch 214	R I
2149,00	SITOR	B=100Bd S=170Hz		SS Ch 215	R I

Frequency	Mode	Mode Parameter	Callsign	User	Country
2149,50	SITOR	B=100Bd S=170Hz		SS Ch 216	R I
2150,00	SSB			SS Ch 217	R I
2150,00	NFM			RUS railroad net	RUS
2150,50	SITOR	B=100Bd S=170Hz		SS Ch 218	R I
2151,00	SITOR	B=100Bd S=170Hz		SS Ch 219	R I
2151,50	SITOR	B=100Bd S=170Hz		SS Ch 220	R I
2152,00	SITOR	B=100Bd S=170Hz		SS Ch 221	R I
2152,50	SITOR	B=100Bd S=170Hz		SS Ch 222	R I
2153,00	SITOR	B=100Bd S=170Hz		SS Ch 223	R I
2153,50	SITOR	B=100Bd S=170Hz		SS Ch 224	R I
2154,00	SITOR	B=100Bd S=170Hz		SS Ch 225	R I
2154,50	SITOR	B=100Bd S=170Hz		SS Ch 226	R I
2155,00	SITOR	B=100Bd S=170Hz		SS Ch 227	R I
2155,50	SITOR	B=100Bd S=170Hz		SS Ch 228	R I
2156,00	SITOR	B=100Bd S=170Hz		SS Ch 229	R I
2156,50	SITOR	B=100Bd S=170Hz		SS Ch 230	R I
2157,00	SITOR	B=100Bd S=170Hz		SS Ch 231	R I
2157,50	SITOR	B=100Bd S=170Hz		SS Ch 232	R I
2158,00	SITOR	B=100Bd S=170Hz		SS Ch 233	R I
2158,50	SITOR	B=100Bd S=170Hz		SS Ch 234	R I
2159,00	SITOR	B=100Bd S=170Hz		SS Ch 235	R I
2159,50	SITOR	B=100Bd S=170Hz		SS Ch 236	R I
2162,00	SSB		ZBP	Pitcairn Island R	PTC
2163,70	4-DPSK	B=250Bd		DGPS Portsmouth	G
2166,00	CW		RMP	RUS N HQ Kaliningrad	RUS
2167,00	USB		H4H	Honiara R	SLM
2168,00	USB		YJM	Port Vila R	VUT
2169,00	*STANAG 4285*	*B=2400Bd*		*G MIL Inskip*	*G*
2174,50	SITOR , CW	B=100Bd S=170Hz	JNN	Shiogama R	J
2174,50	.			Distress/safety frequency	WW
2174,50	SITOR , CW	B=100Bd S=170Hz	VRC	Hong Kong R	CHN
2174,50	SITOR , CW	B=100Bd S=170Hz	JNL	Otura R	J
2174,50	SITOR , CW	B=100Bd S=170Hz	JGD	Kobe R	J
2174,50	SITOR , CW	B=100Bd S=170Hz	JNE	Hiroshima R	J
2174,50	SITOR , CW	B=100Bd S=170Hz	JNJ	Kagoshima R	J
2174,50	SITOR , CW	B=100Bd S=170Hz	JNC	Maizuru R	J
2174,50	SITOR , CW	B=100Bd S=170Hz	JNR	Moji R	J
2174,50	SITOR , CW	B=100Bd S=170Hz	JNT	Nagoya R	J
2174,50	SITOR , CW	B=100Bd S=170Hz	JNV	Niigata R	J
2174,50	SITOR , CW	B=100Bd S=170Hz	JNB	Naha R	J
2174,50	SITOR B	B=100Bd S=170Hz	ZLM	Taupo Maritime R	NZL
2174,50	SITOR , CW	B=100Bd S=170Hz	JGC	Yokohama R	J
2174,50	SITOR , CW	B=100Bd S=170Hz	PKD5	Benoa R , Bali	INS
2174,50	SITOR , CW	B=100Bd S=170Hz	PKZ34	Cigading R , Jawa	INS
2174,50	SITOR , CW	B=100Bd S=170Hz	PKR3	Cilacap R , Jawa	INS
2174,50	SITOR , CW	B=100Bd S=170Hz	PKZ2	Cirebon R , Jawa	INS
2174,50	SITOR , CW	B=100Bd S=170Hz	PKR	Semarang R , Djawa	INS
2174,50	SITOR , CW	B=100Bd S=170Hz	PKD	Surabaya R , Djawa	INS
2174,50	SITOR , CW	B=100Bd S=170Hz	PKN	Balikpapan R , Kalimantan	INS
2174,50	SITOR , CW	B=100Bd S=170Hz	PKS	Pontianak R , Kalimantan	INS
2174,50	SITOR , CW	B=100Bd S=170Hz	PKO	Tarakan R , Kalimantan	INS
2174,50	SITOR , CW	B=100Bd S=170Hz	PKE37	Sanana R , Kep Sula	INS
2174,50	SITOR , CW	B=100Bd S=170Hz	PKD3	Lembar R , Lombok	INS
2174,50	SITOR , CW	B=100Bd S=170Hz	PKY2	Biak R , Irian Jaya	INS
2174,50	SITOR , CW	B=100Bd S=170Hz	PKY23	Fak Fak R	INS
2174,50	SITOR , CW	B=100Bd S=170Hz	PNK	Jayapura R , Irian Jaya	INS
2174,50	SITOR , CW	B=100Bd S=170Hz	PKY3	Manokwari R , Irian Jaya	INS
2174,50	SITOR , CW	B=100Bd S=170Hz	PKY5	Merauke R , Irian Jaya	INS
2174,50	SITOR , CW	B=100Bd S=170Hz	PKY4	Sorong R , Irian Jaya	INS
2174,50	SITOR , CW	B=100Bd S=170Hz	PKM	Bitung R , Sulawesi	INS
2174,50	SITOR , CW	B=100Bd S=170Hz	PKF3	Kendari R	INS
2174,50	SITOR , CW	B=100Bd S=170Hz	PKM44	Pantolan R , Sulawesi	INS

Frequency	Mode	Mode Parameter	Callsign	User	Country
2174,50	SITOR , CW	B=100Bd S=170Hz	PKM25	Tahuna R	INS
2174,50	SITOR , CW	B=100Bd S=170Hz	PKP38	Batu Ampar R	INS
2174,50	SITOR , CW	B=100Bd S=170Hz	PKP	Dumai R , Sumatera	INS
2174,50	SITOR , CW	B=100Bd S=170Hz	PKC4	Panjang R , Sumatera	INS
2174,50	SITOR , CW	B=100Bd S=170Hz	PKJ28	Sei Kolak Kijang R	INS
2174,50	SITOR , CW	B=100Bd S=170Hz	PKB3	Sibolga R , Sumatera	INS
2174,50	SITOR , CW	B=100Bd S=170Hz		Benete R , Sumbawa	INS
2174,50	SITOR , CW	B=100Bd S=170Hz	PKK	Kupang R , Timor	INS
2175,40	USB		SPS	Witowo R	POL
2177,00	USB		TFT	Hornafjördur R	ISL
2177,00	USB			Floro R	NOR
2177,00	USB			Rogaland R	NOR
2177,00	USB			Tjome R	NOR
2177,00	USB		LGV	Vardo R	NOR
2177,00	USB			Bodo R	NOR
2177,00	USB		TFX	Siglufjördur R	ISL
2177,00	USB		TFZ	Isafjördur R	ISL
2177,00	DSC	B=100Bd S=170Hz	JNL	Otura R	J
2177,00	DSC	B=100Bd S=170Hz	JGD	Kobe R	J
2177,00	DSC	B=100Bd S=170Hz	JNE	Hiroshima R	J
2177,00	DSC	B=100Bd S=170Hz	JNT	Nagoya R	J
2177,00	DSC	B=100Bd S=170Hz	JNB	Naha R	J
2177,00	USB		TFA	Reykjavik R	ISL
2177,00	USB		TYA	Cotonou R	BEN
2177,00	SSB		TFV	Vestmannaeyar R	ISL
2177,00	DSC	B=100Bd S=170Hz	JNJ	Kagoshima R	J
2177,00	DSC	B=100Bd S=170Hz	JNC	Maizuru R	J
2177,00	DSC	B=100Bd S=170Hz	JNR	Moji R	J
2177,00	DSC	B=100Bd S=170Hz	JNV	Niigata R	J
2177,00	DSC	B=100Bd S=170Hz	JNN	Shiogama R	J
2177,00	SSB		OST	Oostende R	BEL
2177,00	DSC	B=100Bd S=170Hz	PKD5	Benoa R , Bali	INS
2177,00	DSC	B=100Bd S=170Hz	JGC	Yokohama R	J
2177,00	DSC	B=100Bd S=170Hz	PKE5	Ternate R , Maluku	INS
2177,00	DSC	B=100Bd S=170Hz	PKZ34	Cigading R , Jawa	INS
2177,00	DSC	B=100Bd S=170Hz	PKR3	Cilacap R , Jawa	INS
2177,00	DSC	B=100Bd S=170Hz	PKZ2	Cirebon R , Jawa	INS
2177,00	DSC	B=100Bd S=170Hz	PKX	Jakarta R , Jawa	INS
2177,00	DSC	B=100Bd S=170Hz	PKR	Semarang R , Djawa	INS
2177,00	DSC	B=100Bd S=170Hz	PKS	Pontianak R , Kalimantan	INS
2177,00	DSC	B=100Bd S=170Hz	PKO	Tarakan R , Kalimantan	INS
2177,00	DSC	B=100Bd S=170Hz	PKE37	Sanana R , Kep Sula	INS
2177,00	DSC	B=100Bd S=170Hz	PKD3	Lembar R , Lombok	INS
2177,00	DSC	B=100Bd S=170Hz	PKY2	Biak R , Irian Jaya	INS
2177,00	DSC	B=100Bd S=170Hz	PKY23	Fak Fak R	INS
2177,00	DSC	B=100Bd S=170Hz	PNK	Jayapura R , Irian Jaya	INS
2177,00	DSC	B=100Bd S=170Hz	PKY3	Manokwari R , Irian Jaya	INS
2177,00	DSC	B=100Bd S=170Hz	PKY5	Merauke R , Irian Jaya	INS
2177,00	DSC	B=100Bd S=170Hz	PKY4	Sorong R , Irian Jaya	INS
2177,00	DSC	B=100Bd S=170Hz	PKE	Amboina R , Ceram	INS
2177,00	DSC	B=100Bd S=170Hz	PKM	Bitung R , Sulawesi	INS
2177,00	DSC	B=100Bd S=170Hz	PKF3	Kendari R	INS
2177,00	DSC	B=100Bd S=170Hz	PKF	Makassar R , Sulawesi	INS
2177,00	DSC	B=100Bd S=170Hz	PKM44	Pantolan R , Sulawesi	INS
2177,00	DSC	B=100Bd S=170Hz	PKM25	Tahuna R	INS
2177,00	DSC	B=100Bd S=170Hz	PKP38	Batu Ampar R	INS
2177,00	DSC	B=100Bd S=170Hz	PKB	Belawan R , Sumatera	INS
2177,00	DSC	B=100Bd S=170Hz	PKP	Dumai R , Sumatera	INS
2177,00	DSC	B=100Bd S=170Hz	PKP	Dumai R , Sumatera	INS
2177,00	DSC	B=100Bd S=170Hz	PKC4	Panjang R , Sumatera	INS
2177,00	DSC	B=100Bd S=170Hz	PKJ28	Sei Kolak Kijang R	INS
2177,00	DSC	B=100Bd S=170Hz	PKB3	Sibolga R , Sumatera	INS

Frequency	Mode	Mode Parameter	Callsign	User	Country
2177,00	DSC	B=100Bd S=170Hz		Benete R , Sumbawa	INS
2177,00	DSC	B=100Bd S=170Hz	PKK	Kupang R , Timor	INS
2177,00	DSC	B=100Bd S=170Hz		Polsih Rescue R	POL
2177,00	DSC	B=100Bd S=170Hz		Coruna R	E
2177,00	DSC	B=100Bd S=170Hz		Valencia R	E
2177,00	DSC	B=100Bd S=170Hz	LGP	Bodoe R	NOR
2177,00	DSC	B=100Bd S=170Hz	LGP	NOR Coastal R North	NOR
2177,00	USB		OXZ	Lyngby R	DNK
2177,50	DSC	B=100Bd S=170Hz	PKN	Balikpapan R , Kalimantan	INS
2182,00	USB			San Juan JRCC	PTR
2182,00	USB			Magadan R	RUS
2182,00	USB		OBC3	Callao MRCC	PRU
2182,00	USB		XFS	Tampico P	MEX
2182,00	USB			Cayewnne MRSC	GUF
2182,00	USB			Cheju MRCC	KOR
2182,00	USB		HLC	Incheon R	KOR
2182,00	USB		HLE	Cheju R	KOR
2182,00	USB		HLC	Incheon MRCC	KOR
2182,00	USB		HSA	Bangkok R	THA
2182,00	USB			Grenada MRSC	GND
2182,00	USB		NMJ	USA CG Juneau	ALS
2182,00	USB		JGD	Kobe R	J
2182,00	USB		JNT	Nagoya R	J
2182,00	USB		JNB	Naha R	J
2182,00	USB		JNN	Shiogama R	J
2182,00	USB		HLK	Kangnung R	KOR
2182,00	USB		HLN	Gunsan R	KOR
2182,00	USB			Mokpo MRCC	KOR
2182,00	SSB			Distress/calling frequency	WW
2182,00	USB			Hodeihdah R	YEM
2182,00	USB		P2M	Port Moresbay R	PNG
2182,00	USB		ZKN	Niue R	NIU
2182,00	USB		OBF4	Mollendo MRSC	PRU
2182,00	USB		XFL	Mazatlan P	MEX
2182,00	USB		HLM	Mogpo R	KOR
2182,00	USB			Busan MRCC	KOR
2182,00	USB			USA CG Kodiak	ALS
2182,00	USB		HLP	Busan R	KOR
2182,00	USB		CWM30	URG SAR net	URG
2182,00	USB		CWC30	La Paloma R	URG
2182,00	USB		XSZ	Dalian R	CHN
2182,00	USB			Chuuk R	FSM
2182,00	USB		PJC	Curacao R	ATN
2182,00	USB		CWS	URG N Montevideo	URG
2182,00	USB		CWF	Punta Carretas R	URG
2182,00	USB		XSP	Shantou R	CHN
2182,00	USB			Pohnpei R	FSM
2182,00	USB			Yap R	FSM
2182,00	USB		6YX	JMC CG Jamaica	JMC
2182,00	USB		OBY2	Paita MRSC	PRU
2182,00	USB		CWC34	Punta del Este R	URG
2182,00	USB		CWC39	Montevideo Trouville R	URG
2182,00	USB		JNJ	Kagoshima R	J
2182,00	USB		JNL	Otura R	J
2182,00	USB		JNE	Hiroshima R	J
2182,00	USB		JNC	Maizuru R	J
2182,00	USB		JNR	Moji R	J
2182,00	USB		JNV	Niigata R	J
2182,00	USB		UFH	Petrapavlovsk-Kamchatskii MRSC	RUS
2182,00	USB		9YL	Trinedad North Post R	TRD
2182,00	USB		UFH	Petrapavlovsk-Kamchatskii R	RUS
2182,00	USB		UFH	Petrapavlovsk-Kamchatskii R	RUS

Frequency	Mode	Mode Parameter	Callsign	User	Country
2182,00	USB		XSL	Fuzhou R	CHN
2182,00	USB			Korror R	PLW
2182,00	USB		H4H	Honiara R	SLM
2182,00	USB		PZN	Paramaribo R	SUR
2182,00	USB		A3A	Nukualofa R	TON
2182,00	USB		FJA	Mahina R	OCE
2182,00	USB		JGC	Yokohama R	J
2182,00	USB		UFL	Vladivostok MRCC	RUS
2182,00	USB			Majuro R	MHL
2182,00	USB		VRC	Hong Kong R	CHN
2182,00	USB		XSV	Tianjin R	CHN
2182,00	USB		NOJ3	USA CG Anchorage , ALS	ALS
2182,00	USB		HLF	Tonghae MRCC	KOR
2182,00	USB		HLU	Ulreung R	KOR
2182,00	USB		HLY	Yeosu R	KOR
2182,00	USB			Yuzhno-Sakhalinsk MRSC	RUS
2182,00	USB			Moule a Chique Lighthouse	LCA
2182,00	USB			Vigie Lighthouse	LCA
2182,00	USB			Saint Vincent and the Grenadines CG Base	VCT
2182,00	USB			Samoa Port Authority	SMA
2182,00	USB			USA CG Cape Hatteras	TUN
2182,00	USB			USA CG Cape Hatteras	USA
2182,00	USB			USA CG Cape May	USA
2182,00	USB			USA CG Charleston	USA
2182,00	USB			USA CG Chincoteague	USA
2182,00	USB			USA CG Fort Macon	USA
2182,00	USB			USA CG Mayport	USA
2182,00	USB			USA CG New Haven	USA
2182,00	USB			USA CG New York	USA
2182,00	USB			USA CG Southwest Harbor	USA
2182,00	USB			USA CG New Orleans JRCC	USA
2182,00	USB			USA CG Honolulu	USA
2182,00	USB			USA CG Astoria	USA
2182,00	USB		XSK	Beihai R	CHN
2182,00	USB			Aasiaat R	GRL
2182,00	USB			Palma R	E
2182,00	USB			Porto Torres R	I
2182,00	USB			Tallin R	EST
2187,50	DSC	B=100Bd S=170Hz		Las Palmas R	E
2187,50	.			DSC distress/safety frequency	WW
2187,50	DSC	B=100Bd S=170Hz		Arkhangelsk R	RUS
2187,50	DSC	B=100Bd S=170Hz		Novorossiysk R	RUS
2187,50	DSC	B=100Bd S=170Hz		Taman R	RUS
2187,50	DSC	B=100Bd S=170Hz		Taganrog R	RUS
2187,50	DSC	B=100Bd S=170Hz	LGP	NOR Coastal T North TX Bodoe	NOR
2187,50	DSC	B=100Bd S=170Hz	LGP	NOR Coastal T North TX Vardoe	NOR
2187,50	USB		DCA	MRCC Bremen Rescue	D
2187,50	DSC	B=100Bd S=170Hz	LGP	NOR Coastal R North TX Sandnessjoen	NOR
2187,50	DSC	B=100Bd S=170Hz	LGP	NOR Coastal R North TX Andenes	NOR
2187,50	DSC	B=100Bd S=170Hz	LGP	NOR Coastal R North TX Tromsoe	NOR
2187,50	DSC	B=100Bd S=170Hz	LGP	NOR Coastal R North TX Hammerfest	NOR
2187,50	DSC	B=100Bd S=170Hz	LGP	NOR Coastal R North TX Berlevag	NOR
2187,50	DSC	B=100Bd S=170Hz	LGP	NOR Coastal R North TX Jan Mayen	NOR
2187,50	DSC	B=100Bd S=170Hz	LGP	NOR Coastal R North TX Bjornoya	NOR
2187,50	DSC	B=100Bd S=170Hz	LGP	NOR Coastal R North TX Svalbard	NOR
2187,50	DSC	B=100Bd S=170Hz	LGQ	NOR Coastal R South TX Farsund	NOR
2187,50	DSC	B=100Bd S=170Hz	LGQ	NOR Coastal R South TX Vigre	NOR
2187,50	DSC	B=100Bd S=170Hz	LGQ	NOR Coastal R South TX Jeloey	NOR
2187,50	DSC	B=100Bd S=170Hz	LGQ	NOR Coastal R South TX Bergen	NOR
2187,50	DSC	B=100Bd S=170Hz	LGQ	NOR Coastal R South TX Floroe	NOR
2187,50	DSC	B=100Bd S=170Hz	LGQ	NOR Coastal R South TX Orlandet	NOR
2189,50	.			SS DSC calling frequency	WW

Frequency	Mode	Mode Parameter	Callsign	User	Country
2191,00	USB			Sao Miguel R	AZR
2191,00	USB			Madeira R	MDR
2191,00	SSB			SS calling frequency	R I
2191,00	SSB		4OB	Bar R	MNE
2191,20	*STANAG 4285*	*B=2400Bd*		*G MIL Crimond*	*G*
2192,00	USB		9PA	Banana R	ZAI
2199,20	*STANAG 4285*	*B=2400Bd*	*LBA*	*NOR N Stavanger*	*NOR*
2200,00	RADAR			MF/HF SA Radar Bear Lake	USA
2201,00	USB		VMC	Charleville M	AUS
2204,00	USB		PBC	NLD N Goeree Island	HOL
2204,00	USB		PBB	HOL N Den Helder	HOL
2207,00	USB		E5R	Rarotonga R	CKH
2207,00	USB		ZLM	Taupo Maritime R	NZL
2207,50	*STANAG 4285*	*B=2400Bd*	*JWT*	*NOR N Stavanger*	*NOR*
2207,50	ALE USB	B=125Bd Ch=8 ChS=250Hz		GRC AF net	GRC
2210,00	SSB		3VB	Bizerte R	TUN
2212,00	PACTOR III	B=100Bd S=200Hz	9Z4DH	Sailmail Chaguaramas	TRD
2216,00	*ALE USB*	*B=125Bd Ch=8 ChS=250Hz*		*G MIL net*	*G*
2219,00	*CIS 12*	*B=1440Bd Ch=12 ChS=200Hz*	*RMP*	*RUS N HQ Kaliningrad*	*RUS*
2220,00	RADAR			MF/HF SA Radar Saskatoon	CAN
2221,00	ALE USB	B=125Bd Ch=8 ChS=250Hz		S FRO net	S
2225,00	USB		OYR	Aasiaat R	GRL
2226,00	SSB			Aberdeen CG MRCC	G
2233,00	MIL 188-110A SER	B=2400Bd		S MIL	S
2241,20	*STANAG 4285*	*B=2400Bd*		*NOR N Oslo*	*NOR*
2242,00	MIL 188-110B SER	B=2400Bd			
2250,00	*ISR N HYBRID MODEM*	*B=2400Bd*	*4XZ*	*ISR N Haifa*	*ISR*
2250,00	USB		OYR	Aasiaat R	GRL
2250,00	ALE USB	B=125Bd Ch=8 ChS=250Hz		DNK N net	DNK
2252,00	USB		LBJ	NOR N Bodoe	NOR
2255,00	SSB			USA Army MARS	
2255,00	ALE USB	B=125Bd Ch=8 ChS=250Hz		S FRO net	S
2256,00	USB		OST	Oostende R	BEL
2259,40	USB		PBC	NLD N Goeree Island	HOL
2259,40	USB		PBB	NLD N Den Helder	HOL
2260,00	USB		VJN	RFDS Cairns	AUS
2261,00	SSB			USCG rescue operation channel	USA
2265,00	USB		OYR	Aasiaat R	GRL
2270,00	USB		XFS	Tampico P	MEX
2274,50	SSB		IGJ	I N Augusta	I
2275,00	USB			G MIL Army Cadets	G
2280,00	USB		VKL	RFDS Port Hedland	AUS
2280,00	USB		VJT	RFDS Carnarvon	AUS
2280,00	USB		VKJ	RFDS Meekathara	AUS
2282,00	SSB		ZAC	Shengjin R	ALB
2282,00	SSB		ZAD	Durres R	ALB
2282,00	SSB		ZAV	Vlore R	ALB
2283,00	LINK 11 SLEW	B=2400Bd	JWT	NOR N Stavanger	NOR
2284,00	USB		HLC	Incheon R	KOR
2284,00	ALE USB	B=125Bd Ch=8 ChS=250Hz		AUS Police NSW	AUS
2285,60	STANAG 4285	B=2400Bd		I N Rome	I
2289,50	USB		ICS	I N La Spezia	I
2289,70	STANAG 4285	B=2400Bd	IDR	I N Rome	I
2295,00	CW		VTP3	IND N Vishakhapatnam	IND
2299,00	USB		HLE	Cheju R	KOR

Frequency	Mode	Mode Parameter	Callsign	User	Country
2304,00	USB		OYR	Aasiaat R	GRL
2306,00	USB			Norhstar Shipping net	G
2311,00	USB			Arklow shipping net	HOL
2320,00	USB		XFS	Tampico P	MEX
2321,20	STANAG 4285	B=2400Bd	SXA	GRC N Piraeus	GRC
2325,00	USB			Arkhangelsk MRSC	RUS
2325,00	CIS 12	B=1440Bd Ch=12 ChS=200Hz	RMP	RUS N HQ Kaliningrad	RUS
2325,00	ALE USB	B=125Bd Ch=8 ChS=250Hz		BEL MIL net	BEL
2336,50	STANAG 4285	B=2400Bd		G MIL Akrotiri	CYP
2348,00	USB			Auckland Tramping Club	NZL
2350,00	USB		IHMM	I N minesweeper MILAZZO	I
2350,00	USB		IATD	I N ship VEGA	I
2350,00	USB		IHMW	I N ship VIAREGGIO	I
2350,00	STANAG 4285	B=2400Bd	IHMW	I N ship VIAREGGIO	I
2352,00	ALE USB	B=125Bd Ch=8 ChS=250Hz		POL MIL net	POL
2352,50	SSB		ASK	Karachi R	PAK
2357,00	USB		HLM	Mogpo R	KOR
2368,00	USB			G AF Air Cadets	G
2376,00	SSB		OST	Oostende R	BEL
2379,70	STANAG 4285	B=2400Bd	PBC	NLD N Goeree Island	HOL
2391,00	USB			Dutch inter ship	HOL
2391,00	STANAG 4285	B=2400Bd	TBB	TUR N Ankara	TUR
2394,00	SSB		7TB	Annaba R	ALG
2395,00	USB		AMBA	Samara ACC	RUS
2395,00	USB		MELODIJA DWA	Kazan ACC	RUS
2400,00	USB		OYR	Aasiaat R	GRL
2400,00	USB		ZAC	Shengjin R	ALB
2400,00	USB		ZAV	Vlore R	ALB
2400,00	RADAR			MF/HF SA Radar Christchurch	NZL
2400,00	OOK			MF/HF SA Radar Syowa Base	ANT
2401,00	STANAG 4285	B=2400Bd		G MIL Akrotiri	CYP
2415,00	USB			G MIL Army Cadets	G
2415,20	STANAG 4285	B=2400Bd	EBA	E N Madrid	E
2415,50	ALE USB	B=125Bd Ch=8 ChS=250Hz		G MIL net	G
2417,50	SITOR , CW	B=100Bd S=170Hz	JNN	Shiogama R	J
2417,50	SITOR , CW	B=100Bd S=170Hz	JNE	Hiroshima R	J
2417,50	SITOR , CW	B=100Bd S=170Hz	JNL	Otura R	J
2417,50	SITOR , CW	B=100Bd S=170Hz	JNJ	Kagoshima R	J
2417,50	SITOR , CW	B=100Bd S=170Hz	JGD	Kobe R	J
2417,50	SITOR , CW	B=100Bd S=170Hz	JNC	Maizuru R	J
2417,50	SITOR , CW	B=100Bd S=170Hz	JNR	Moji R	J
2417,50	SITOR , CW	B=100Bd S=170Hz	JNT	Nagoya R	J
2417,50	SITOR , CW	B=100Bd S=170Hz	JNV	Niigata R	J
2417,50	SITOR , CW	B=100Bd S=170Hz	JNB	Naha R	J
2417,50	SITOR , CW	B=100Bd S=170Hz	JGC	Yokohama R	J
2421,00	CW			J MIL	J
2423,20	STANAG 4285	B=2400Bd		G MIL TX St. Eval	G
2423,20	STANAG 4285	B=2400Bd		G MIL TX Inskip	G
2430,00	RADAR			MF/HF SA Radar Poker Flat	USA
2430,00	RADAR			MF/HF SA Radar Pontianak	INS
2434,70	STANAG 4285	B=2400Bd	IDN	I N Naples	I
2438,00	ALE USB	B=125Bd Ch=8 ChS=250Hz			
2440,00	USB			IRL MIL	IRL
2442,50	MIL 188-110B SER	B=2400Bd		POL MIL	POL
2444,00	NFM			RUS subway net	RUS
2446,00	ALE USB	B=125Bd Ch=8 ChS=250Hz		Guardia di Finanza net	I

Frequency	Mode	Mode Parameter	Callsign	User	Country
2451,50	*ALE USB*	*B=125Bd Ch=8 ChS=250Hz*		*G MIL net*	*G*
2452,00	ALE USB	B=125Bd Ch=8 ChS=250Hz		BEL MIL net	BEL
2457,70	CW		AQP2	PAK N Karachi	PAK
2458,50	CW		AQP2	PAK N Karachi	PAK
2460,00	USB		HLP	Busan R	KOR
2461,50	SITOR A	B=100Bd S=170Hz		IRL N	IRL
2461,50	SITOR A	B=100Bd S=170Hz	0A**	IRL N Dublin	IRL
2461,50	SITOR A	B=100Bd S=170Hz	44**	IRL N	IRL
2461,50	SITOR A	B=100Bd S=170Hz	81**	IRL N	IRL
2461,50	SITOR A	B=100Bd S=162Hz		IRL N	IRL
2461,50	SITOR A	B=100Bd S=170Hz	78**	IRL N	IRL
2461,50	SITOR A	B=100Bd S=170Hz	49**	IRL N	IRL
2464,00	SSB		UFF	Sukhumi R	GEO
2464,00	NFM			RUS subway net	RUS
2470,00	USB		XFL	Mazatlan P	MEX
2474,00	*BAUDOT*	*B=75Bd S=850Hz*	*PBB*	*HOL N Den Helder*	*HOL*
2484,00	SSB		OST	Oostende R	BEL
2490,00	USB			G AF Air Cadets	G
2500,00	CW		BPM	Xi`an TS	CHN
2500,00	CW		WWV	Fort Collins TS , CO	USA
2500,00	CW		WWVH	Kekaha Kauai TS	HWA
2500,00	USB			EGY air defense net	EGY
2507,00	USB		HLN	Gunsan R	KOR
2510,00	USB		LBJ	NOR N Bodoe	NOR
2510,00	USB		JWT	NOR N Stavanger	NOR
2513,00	SSB		5AB	Benghazi R	LBY
2514,00	USB		VCG	Riviere au Renard R	CAN
2514,00	USB		VCM	S. Anthony R	CAN
2514,00	USB		VCP	S. Lawrence R	CAN
2514,00	USB		VFF	Iqaluit R	CAN
2514,00	USB		VOJ	Stephenville R	CAN
2514,00	USB		VOK	Labrador R	CAN
2514,00	USB		VON	S. John`s R	CAN
2514,00	USB		VCS	Halifax R	CAN
2521,50	MIL 188-110A SER	B=2400Bd			
2522,00	SSB		C6N	Nassau R	BAH
2524,00	USB			Tasmanian Maritime R	AUS
2530,00	SSB		KBP	Honolulu R	HWA
2530,00	USB		VCO	Sydney R	CAN
2531,00	ALE USB	B=125Bd Ch=8 ChS=250Hz		BEL MIL net	BEL
2538,00	USB		VCP	S. Lawrence R	CAN
2538,00	USB		VOK	Labrador R	CAN
2538,00	USB		VAR	CAN CG St. John	CAN
2539,50	PACTOR III	B=100Bd S=200Hz	DCA28	MRCC Bremen Rescue	D
2540,00	ALE USB	B=125Bd Ch=8 ChS=250Hz		USA SAC net	USA
2542,50	PACTOR III	B=100Bd S=200Hz	DCA28	MRCC Bremen Rescue	D
2542,50	PACTOR III	B=100Bd S=200Hz	DBAM	SAR vessel ANNELISE KRAMER	D
2545,00	USB			Fort de France MRCC	MRT
2549,70	STANAG 4285	B=2400Bd	XPQ2	DNK N Aarhus	DNK
2554,00	ALE USB	B=125Bd Ch=8 ChS=250Hz		LTU MIL net	LTU
2558,00	USB		VFA	Inuvik R	CAN
2560,00	USB		XFL	Mazatlan P	MEX
2560,50	*STANAG 4285*	*B=2400Bd*		*G MIL TX Crimond*	*G*
2572,50	STANAG 4285	B=2400Bd	CTA	POR N Lisbon	POR
2573,00	USB		PBC	NLD N Goeree Island	HOL
2576,00	ALE USB	B=125Bd Ch=8 ChS=250Hz		BEL MIL net	BEL

Frequency	Mode	Mode Parameter	Callsign	User	Country
2578,00	USB		XSA2	Nanjing R	CHN
2579,00	SSB		5RO	Toliara R	MDG
2579,00	SSB		IPB	Bari R	I
2579,20	STANAG 4285	B=2400Bd	PBB	NLD N TX Goeree Island	HOL
2580,00	BAUDOT	B=75Bd S=850Hz	OSN	BEL N Oostende	BEL
2580,00	CW		RMP	RUS N HQ Kaliningrad	RUS
2582,00	USB			Lisbon R	POR
2582,00	USB		CLT	Habana R	CUB
2582,00	USB		VAR	CAN CG St. John	CAN
2582,00	SSB		C6X2	Marsh Harbour R , Great Abaco Island	BAH
2582,00	USB		VCG	Riviere au Renard R	CAN
2582,00	USB		VCM	S. Anthony R	CAN
2582,00	USB		VCO	Sydney R	CAN
2582,00	USB		VCP	S. Lawrence R	CAN
2582,00	USB		VFF	Iqaluit R	CAN
2582,00	USB		VOJ	Stephenville R	CAN
2582,00	USB		VOK	Labrador R	CAN
2582,00	USB		VON	S. John`s R	CAN
2582,00	USB		ZBR	Bermuda Harbour R	BER
2582,00	USB		VCS	Halifax R	CAN
2583,00	SSB			St. Denis R	REU
2584,80	MIL 188-110B SER	B=2400Bd		S MIL net	S
2585,50	ALE USB	B=125Bd Ch=8 ChS=250Hz		GRC AF net	GRC
2586,00	USB		J2A	Djibouti R	DJI
2586,00	SSB		7TO	Oran R	ALG
2586,00	SSB		CNP	Casablanca R	MRC
2586,00	SSB		TRA	Libreville R	GAB
2586,00	SSB		TUA	Abidjan R	CTI
2586,00	USB		OXZ	Lyngby R TX Ronne R	DNK
2587,60	STANAG 4285	B=2400Bd	FUO	F N Toulon	F
2590,60	STANAG 4285	B=2400Bd	PBB	NLD N TX Goeree Island	HOL
2593,00	SSB		CND	Agadir R	MRC
2594,00	USB			Cape Town R	AFS
2594,00	LINK 11 CLEW	B=2250Bd Ch=16 ChS=330/110/550Hz		TUR MIL	TUR
2595,00	SSB		S7Q	Seychelles R	SEY
2598,00	USB		VOK	Labrador R	CAN
2598,00	USB			Port-aux-Basque R	CAN
2598,00	USB			St. Anthony R	CAN
2598,00	USB			Rivière-au-Renard R	CAN
2598,00	USB		VCP	Placentia R	CAN
2600,00	SSB		IQQ	Mazara del Vallo R	I
2600,00	SSB		FFD	Saint Denis R	REU
2601,00	SSB		A7D	Doha R	QAT
2601,00	SSB		D3E	Luanda R	AGL
2601,20	STANAG 4285	B=2400Bd	FUG	F N Saissac	F
2603,00	USB			RUS Air Defense net Kaliningrad	RUS
2603,50	ALE USB	B=125Bd Ch=8 ChS=250Hz		AUS Police QLD	AUS
2604,00	SSB		A4M	Muscat R	OMA
2605,00	USB		XSH	Basuo R	CHN
2605,00	USB		XSI	Sanya R	CHN
2606,40	STANAG 4285	B=2400Bd	FUO2	F N Toulon	F
2607,00	USB		A9M	Bahrain R	BHR
2607,00	SSB		A4M	Muscat R	OMA
2607,00	SSB		SVK	Kerkyra R	GRC
2608,40	STANAG 4285	B=2400Bd	FUO2	F N Toulon	F
2608,60	STANAG 4285	B=2400Bd	FUO	F N Toulon	F
2610,00	SSB		ESA	Tallin R	EST
2613,70	STANAG 4285	B=2400Bd	DHJ58	D N Glücksburg TX Marlow	D

Frequency	Mode	Mode Parameter	Callsign	User	Country
2613,70	*STANAG 4285*	*B=2400Bd*	*DHJ58*	*D N Glücksburg*	*D*
2615,00	SSB		9AD	Dubrovnik R	HRV
2615,20	STANAG 4285	B=2400Bd	PBB	NLD N Den Helder	HOL
2616,00	USB			Chuuk R	FSM
2616,00	USB			Majuro R	MHL
2618,50	*FAX 120/576*		*GYA*	*G N London TX Inskip*	*G*
2620,00	USB		XSQ	Guangzhou R	CHN
2622,00	SSB			NASA rocket booster recovery	USA
2623,20	*STANAG 4285*	*B=2400Bd*	*DHJ58*	*D N Glücksburg*	*D*
2624,00	SSB		IQX	Trieste R	I
2624,00	USB		SVR	Rhodos R	GRC
2624,00	USB			Genua R	I
2625,00	SSB		9HD	Malta R	MLT
2625,50	CW		TBH3	TUR N Golcuk	TUR
2627,00	USB		XSU	Yantai R	CHN
2628,00	SSB		IQA	Augusta R	I
2628,00	FAX 120/576		VMC	Charleville M	AUS
2629,00	SSB		TAN	Izmir R	TUR
2629,00	SSB		TAI	Iskenderun R	TUR
2630,00	USB			Arkhangelsk MRSC	RUS
2630,50	ALE USB	B=125Bd Ch=8 ChS=250Hz		AUS Police QLD	AUS
2630,50	ALE USB	B=125Bd Ch=8 ChS=250Hz		SAPOL Communication Infrastructure	AUS
2632,00	SSB		IQH	Napoli R	I
2632,00	ALE USB	B=125Bd Ch=8 ChS=250Hz		AUS Police NSW	AUS
2632,00	ALE USB	B=125Bd Ch=8 ChS=250Hz		AUS Police NT	AUS
2635,00	SSB			ship/ship working frequency	R23
2635,00	SSB			Floro R via Orlandet	NOR
2638,00	SSB			ship/ship working frequency	R24
2638,00	ALE LSB	B=125Bd Ch=8 ChS=250Hz		D Red Cross net	D
2638,60	*STANAG 4285*	*B=2400Bd*	*FUE*	*F N Brest*	*F*
2639,00	CW		UGC	St. Petersburg R	RUS
2639,40	USB			Cheju MRCC	KOR
2639,40	USB		HLC	Incheon MRCC	KOR
2639,40	USB			Mokpo MRCC	KOR
2639,40	USB			Busan MRCC	KOR
2639,40	USB		HLF	Tonghae MRCC	KOR
2642,00	SSB		ICB	Genova R	I
2642,00	SSB		LGV	Vardo R	NOR
2643,00	SITOR , CW	B=100Bd S=170Hz	A9M	Bahrain R	BHR
2649,00	SSB		4XA	Elat R	ISR
2649,00	SSB			Floro R	NOR
2649,00	ALE USB	B=125Bd Ch=8 ChS=250Hz		G MIL net	G
2650,00	PACTOR II	B=200Bd S=200Hz	HPPM2	Sailmail Chiriqui	PNR
2650,00	USB		UWH	Berdyansk R	UKR
2655,00	USB		AVROROA	Volgograd ACC	RUS
2655,00	USB		RIDAN	Makhachkala ACC	RUS
2655,00	USB		SHPORA	Mineralnye Vody ACC	USA
2655,00	USB		SHTAT	Sochi ACC	RUS
2655,00	USB		TOM	Stavropol ACC	RUS
2655,00	USB		TOSNO	Nalchik ACC	RUS
2655,00	USB		TRIOL	Elista ACC	RUS
2655,00	USB		TYURIK	Krasnodar ACC	IRL
2655,00	USB		UBEZISCE	Mineralnye Vody Ag3CC	RUS
2656,00	USB		A9M	Bahrain R	BHR
2656,00	USB		VJT	RFDS Carnarvon	AUS
2656,00	SSB		4XA	Elat R	ISR

Frequency	Mode	Mode Parameter	Callsign	User	Country
2656,00	USB		IPA	Ancona R	I
2656,00	SSB			Rogaland R via Vigre	NOR
2656,40	PACTOR II	B=200Bd S=200Hz	KZN508	Sailmail Rockhill , SC	USA
2657,00	USB		CTA	Lisbon R CENCOMAR	POR
2657,00	USB			Madeira R	MDR
2657,00	SSB		CTH	POR N Horta R	AZR
2657,00	SSB		CTN	Leixoes R	POR
2657,00	SSB		CTS	Sagres R	POR
2657,00	USB		CTG	POR N Alges	POR
2658,20	STANAG 4285	B=2400Bd		DNK N Skagen	DNK
2659,00	ALE USB	B=125Bd Ch=8 ChS=250Hz		USA FEMA net	USA
2660,00	USB			Bodo R via Andenes	NOR
2661,40	PACTOR II	B=200Bd S=200Hz	WRD719	Sailmail Palo Alto	USA
2662,00	SSB		YKO	Tartous R	SYR
2663,00	SSB		CNP	Casablanca R	MRC
2663,00	SSB		IPC	Crotone R	I
2667,00	SSB			Rogaland R via Bergen	NOR
2668,00	CIS 12	B=1440Bd Ch=12 ChS=200Hz		RUS MIL Moscow	RUS
2670,00	USB			Falmouth CG MRCC	G
2670,00	USB		NMJ	USA CG Juneau	ALS
2670,00	USB			USA CG Kodiak	ALS
2670,00	SSB		3VT	Tunis R	TUN
2670,00	SSB		TAH	Istanbul R	TUR
2670,00	SSB		UNI	Ventspils R	LVA
2670,00	USB		NOJ3	USA CG Anchorage , ALS	ALS
2670,00	USB			USA CG Cape Hatteras	USA
2670,00	USB			USA CG Cape MAy	USA
2670,00	USB			USA CG Charleston	USA
2670,00	USB			USA CG Chincoteague	USA
2670,00	USB			USA CG Fort Macon	USA
2670,00	USB			USA CG Mayport	USA
2670,00	USB			USA CG New Haven	USA
2670,00	USB			USA CG New York	USA
2670,00	USB			USA CG Southwest Harbor	USA
2670,00	USB			USA CG Astoria	USA
2671,00	CW		JJU22	J MIL Naha	J
2671,00	CW		JJU33	J MIL Yozadake	J
2671,00	CW		JJZ44	J MIL Okinoerabu	J
2671,00	CW		JJU55	J MIL	J
2671,00	CW		JJU66	J MIL Kume Shima	J
2671,00	USB			Oostende R	BEL
2676,00	SSB			Rogaland R via Farsund	NOR
2676,00	ALE USB	B=125Bd Ch=8 ChS=250Hz		AUS Police NSW	AUS
2677,00	USB			CROSS Corsen	F
2677,00	USB			CROSS Etel	F
2677,00	USB			CROSS Gris-Nez	F
2677,00	USB			CROSS Jobourg	F
2677,00	USB			CROSS La Garde	F
2678,00	SSB			NASA ETR day channel	USA
2678,20	STANAG 4285	B=2400Bd	LBA	NOR N TX Stavanger	NOR
2678,20	STANAG 4285	B=2400Bd		NOR N TX Oslo	NOR
2679,60	ALE USB	B=125Bd Ch=8 ChS=250Hz		ARINC Urgent Link net	USA
2680,00	CW		4XZ	ISR N Haifa	ISR
2680,00	USB		IDC	Cagliari R	I
2684,20	STANAG 4285	B=2400Bd		G MIL TX St. Eval	G
2685,00	ALE USB	B=125Bd Ch=8 ChS=250Hz		USA FEMA net	USA
2685,00	SSB		9AS	Split R	HRV

Frequency	Mode	Mode Parameter	Callsign	User	Country
2685,00	CW		JJT44	J MIL Irumagawa	J
2685,00	CW		JJT30	J MIL	J
2685,00	CW		JJT88	J MIL Tokyo	J
2685,00	CW		JJW35	J MIL	J
2685,00	CW		JJV89	J MIL	J
2685,00	CW		JJV40	J MIL	J
2685,00	CW		JJW46	J MIL	J
2685,00	CW		JJS76	J MIL	J
2685,00	CW		JJT99	J MIL	J
2686,00	USB		HSA	Bangkok R	THA
2690,00	USB		PKC	Palembang R , Sumatera	INS
2690,00	USB		PKC4	Panjang R , Sumatera	INS
2690,00	USB		PKD	Surabaya R , Djawa	INS
2690,00	USB		PKD3	Lembar R , Lombok	INS
2690,00	USB		PKD5	Benoa R , Bali	INS
2690,00	USB		PKE	Amboina R , Ceram	INS
2690,00	USB		PKK	Kupang R , Timor	INS
2690,00	USB		PKM	Bitung R , Sulawesi	INS
2690,00	USB		PKM25	Tahuna R	INS
2690,00	USB		PKN	Balikpapan R , Kalimantan	INS
2690,00	USB		PKP38	Batu Ampar R	INS
2690,00	USB		PKY2	Biak R , Irian Jaya	INS
2690,00	USB		PKZ2	Cirebon R , Jawa	INS
2690,00	USB		PKM44	Pantolan R , Sulawesi	INS
2690,00	USB			Benete R , Sumbawa	INS
2691,00	SSB		7TA	Alger R	ALG
2693,00	USB			Lisbon R	POR
2693,00	SSB		TAF	Samsun R	TUR
2693,00	SSB		TAL	Antalya R	TUR
2695,00	USB			Vardo R via Hammerfest	NOR
2695,00	SSB		URL8	Sevastopol R	UKR
2695,00	USB			G N Sea Cadets	G
2700,00	USB		5BA	Cyprus R	CYP
2701,40	PACTOR II	B=200Bd S=200Hz	KUZ533	Sailmail Honululu	HWA
2701,50	*CIS 12*	*B=1440Bd Ch=12 ChS=200Hz*	*RCV*	*RUS N HQ Sevastopol*	*UKR*
2704,00	ALE USB	B=125Bd Ch=8 ChS=250Hz		BEL MIL net	BEL
2705,00	MIL 188-110B SER	B=2400Bd		G MIL	G
2705,00	SSB		TNA	Pointe Noir R	COG
2711,50	USB		CLT	Habana R	CUB
2713,50	ALE LSB	B=125Bd Ch=8 ChS=250Hz		GRC AF net	GRC
2713,50	ALE USB	B=125Bd Ch=8 ChS=250Hz		GRC AF net	GRC
2714,00	USB		SPH	Gdynia R	POL
2718,90	ALE USB	B=125Bd Ch=8 ChS=250Hz		UI Net	
2719,00	SSB		7TO	Oran R	ALG
2719,00	SSB		HSP4	Phuket R	THA
2719,00	SSB		IZN	Porto Torres R	I
2720,00	USB		SPS	Witowo R	POL
2720,00	USB		VNJ	Casey Base , AUS	ANT
2720,00	USB			Davis Base , AUS	ANT
2720,00	USB			Macquarie Island , AUS	ANT
2720,00	USB			Mawson Base , AUS	ANT
2720,80	PACTOR II	B=200Bd S=200Hz	WPTG385	Sailmail Corpus Christi , TX	USA
2721,00	CW		REA4	RUS AF Moscow	RUS
2724,00	USB			Chuuk R	FSM
2724,00	SSB		KUP65	Majuro M	MRL
2724,00	SSB		KUP66	Ponape M	MRL
2724,00	SSB		KUP67	Truk M	MRL

Frequency	Mode	Mode Parameter	Callsign	User	Country
2724,00	USB			Majuro R	MHL
2724,00	USB		TFA	Reykjavik R	ISL
2730,00	USB		SVL	Limnos R	GRC
2730,00	SSB		ZAD	Durres R	ALB
2730,00	SITOR B	B=100Bd S=170Hz	SVO	Olympia R	GRC
2733,00	SSB		SDJ	Stockholm R , Härnösand	S
2735,00	USB		9YL	Trinedad North Post R	TRD
2735,60	ALE USB	B=125Bd Ch=8 ChS=250Hz		ARINC Urgent Link net	USA
2737,00	CW		REA4	RUS AF HQ Moscow	RUS
2737,00	*CW*		*RMP*	*RUS N HQ Kaliningrad*	*RUS*
2737,60	ALE USB	B=125Bd Ch=8 ChS=250Hz			
2738,00	USB		CBA27	Taltal R	CHL
2738,00	USB		CBP31	Puerto Aguirre R	CHL
2738,00	USB		CBP23	Ancud R	CHL
2738,00	USB		CBP56	Baker R	CHL
2738,00	USB		CBP32	Puerto Chacabuco R	CHL
2738,00	USB		CBP24	Chaiten R	CHL
2738,00	SSB			Port au Prince R	HTI
2738,00	USB		CBA	Antofagasta R	CHL
2738,00	USB		CBA2	Arica R	CHL
2738,00	USB		CBA23	Chanaral R	CHL
2738,00	USB		CBP39	Puerto Cisnes R	CHL
2738,00	USB		CBA24	Huasco R	CHL
2738,00	USB		CBA3	Iquique R	CHL
2738,00	USB		CBA4	Coquimbo R	CHL
2738,00	USB		CBA5	Caldera R	CHL
2738,00	USB		CBF	Isla Juan Fernandez R	CHL
2738,00	USB		CBM	Magallanes Zonal R	CHL
2738,00	USB		CBM2	Faro Cabo Raper R	CHL
2738,00	USB		CBM3	Faro Islotes Evangelistas	CHL
2738,00	USB		CBM4	Faro Islote Fairway	CHL
2738,00	USB		CBM5	Punta Delgada R	CHL
2738,00	USB		CBM71	Faro Punta Dungeness	CHL
2738,00	USB		CBM72	Faro Espiritu Santo R	CHL
2738,00	USB		CBN	Wollaston R	CHL
2738,00	USB		CBP2	Castro R	CHL
2738,00	USB		CBP3	Puerto Aysen R	CHL
2738,00	USB		CBP4	Faro Isla Guafo	CHL
2738,00	USB		CBP70	Faro Punta Corona R	CHL
2738,00	USB		CBS	San Pedro R	CHL
2738,00	USB		CBT	Talcahuano Zonal R	CHL
2738,00	USB		CBT2	Faro Cabo Carranza	CHL
2738,00	USB		CBT21	Constitucion R	CHL
2738,00	USB		CBT3	Faro Isla Mocha	CHL
2738,00	SSB		CBT70	Faro Isla Quiriquina	CHL
2738,00	USB		CBV	Playa Ancha Zonal R	CHL
2738,00	USB		CBV22	San Antonio MRSC	CHL
2738,00	USB		CBX	Bahia Felix R	CHL
2738,00	USB		CBY	I. de Pascua R (Easter Iland)	CHL
2738,00	SSB		CBZ22	CHL Bahis Fildes R	ANT
2738,00	USB		CBT24	Coronel R	CHL
2738,00	USB		CBM21	Puerto Eden R	CHL
2738,00	USB		CBM30	Isla Diego Ramirez R	CHL
2738,00	USB		CBT25	Lebu R	CHL
2738,00	USB		CBT26	Los Vilos R	CHL
2738,00	USB		CBA22	Mejillones R	CHL
2738,00	USB		CBP29	Melinka R	CHL
2738,00	USB			Punta Arenas MRCC	CHL
2738,00	USB		CBP28	Quellon R	CHL
2738,00	USB		CBA21	Tocopilla R	CHL

Frequency	Mode	Mode Parameter	Callsign	User	Country
2738,00	USB		CBM24	Puerto Williams MRSC	CHL
2738,00	USB		CBP	Puerto Montt Zonal R	CHL
2740,00	USB		STP	Port Sudan R	SDN
2741,00	USB			Sao Miguel R	AZR
2746,00	USB		HSA	Bangkok R	THA
2748,00	USB		YQI	Constanta R	ROU
2748,00	ALE USB	B=125Bd Ch=8 ChS=250Hz		G MIL net	G
2749,00	USB		CFH	CAN N Halifax VOLMET	CAN
2749,00	USB			Rivière-au-Renard	CAN
2750,00	USB			Sao Miguel R	AZR
2752,00	SSB		9AD	Dubrovnik R	HRV
2752,00	SSB		4OB	Bar R	MNE
2753,00	SSB		PFC	HOL N Kooy	HOL
2756,00	USB		XFS	Tampico P	MEX
2756,20	STANAG 4285	B=2400Bd	EBA	E N Madrid	E
2759,00	PACTOR III	B=100Bd S=200Hz	WQAB964	Sailmail San Diego , CA	USA
2760,00	USB		CLT	Habana R	CUB
2760,00	SSB		TAO	Trabzon R	TUR
2761,00	SSB		OST	Oostende R	BEL
2763,00	USB		CBT4	Valdivia MRSC	CHL
2764,00	SSB			NASA rocket booster recovery	USA
2766,00	STANAG 4285	B=2400Bd		G MIL St. Eval	G
2768,00	SSB		A7D	Doha R	QAT
2768,00	SSB		SAE	Tingstäde R	S
2770,00	USB			CIS aero network night	CIS
2770,00	USB			G MIL Army Cadets	G
2771,50	CIS 12	B=1440Bd Ch=12 ChS=200Hz		RUS MIL Moscow	RUS
2773,20	STANAG 4285	B=2400Bd		G MIL TX Inskip	G
2775,00	SSB		7TA	Alger R	ALG
2776,00	CW		RCV	RUS N HQ Sevastopol	UKR
2780,00	USB			Lisbon R	POR
2780,00	BAUDOT	B=75Bd S=850Hz	FUC	F N Cherbourg	F
2780,00	RADAR			MF/HF SA Radar Tromso	NOR
2781,60	ALE USB	B=125Bd Ch=8 ChS=250Hz			
2783,00	USB			Cape Town R	AFS
2783,00	SSB		V5W	Walvis Bay R	NMB
2783,00	USB		XSJ	Zhanjiang R	CHN
2784,00	ALE USB	B=125Bd Ch=8 ChS=250Hz		G MIL net	G
2789,00	USB		IDF	Messina R	I
2789,00	BAUDOT	B=75Bd S=850Hz	FUE	F N Brest	F
2789,00	STANAG 4285	B=2400Bd	FUE	F N Brest	F
2789,00	USB			Cagliari R	I
2792,00	USB		VJB	RFDS Derby	AUS
2792,00	USB		VJQ	RFDS Kalgoorlie	AUS
2794,40	PACTOR III	B=100Bd S=200Hz	WHV382	Sailmail Friday Habor , WA	USA
2796,50	MIL 188-110B SER	B=2400Bd		S N Karlskrona	S
2799,00	USB		SVH	Iraklion Kritis R	GRC
2800,00	USB			Mauritius MRCC	MAU
2800,00	HAARP waveform			HAARP	USA
2800,40	PACTOR II	B=200Bd S=200Hz	WHV861	Sailmail San Luis Obispo , CA	USA
2803,00	ALE USB	B=125Bd Ch=8 ChS=250Hz		POL MIL NET	POL
2805,00	ALE USB	B=125Bd Ch=8 ChS=250Hz		USA AF net	
2805,00	USB			Taganrog R	RUS
2805,00	USB		XFL	Mazatlan P	MEX
2805,00	4-DPSK	B=250Bd		DGPS Noord Ost Polder	HOL
2807,00	CW		REA4	RUS AF Moscow	RUS

Frequency	Mode	Mode Parameter	Callsign	User	Country
2807,80	PACTOR II	B=200Bd S=200Hz	WPUC469	Sailmail South Daytona , FL	USA
2808,50	ALE USB	B=125Bd Ch=8 ChS=250Hz		USA FBI net	USA
2810,00	USB			Madeira R	MDR
2810,00	LINK 11 CLEW	B=2250Bd Ch=16 ChS=330/110/550Hz			
2810,00	SSB		OHG	Helsinki MRSC	FNL
2810,00	CW			RUS HFDF net	RUS
2810,00	STANAG 5511	B=2400Bd			
2810,20	*STANAG 4285*	*B=2400Bd*	*DHJ58*	*D N Glücksburg*	*D*
2815,00	2FSK	B=600Bd S=850Hz		G MIL St. Eval	G
2815,00	*STANAG 4481*	*B=75Bd S=850Hz*		*G MIL St. Eval*	*G*
2816,00	SSB		5AB	Benghazi R	LBY
2818,00	USB		PZN	Paramaribo R	SUR
2820,00	SSB		TAM	Mersin R	TUR
2821,00	SSB		HSP5	Phuket R	THA
2824,00	PACTOR	B=100Bd S=200Hz	VZX	Sailmail Darawank	AUS
2828,50	PACTOR III	B=100Bd S=200Hz	CEV773	Sailmail Los Lagos	CHL
2830,00	USB		SVK	Kerkyra R	GRC
2833,00	CW		4XZ	ISR N Haifa	ISR
2833,20	*STANAG 4285*	*B=2400Bd*	*DHJ58*	*D N Glücksburg TX Marlow*	*D*
2833,20	*STANAG 4285*	*B=2400Bd*	*DHJ58*	*D N Glücksburg TX Neuharlingersiel*	*D*
2834,00	PACTOR	B=200Bd S=170Hz		UNHCR	
2834,00	4-DPSK	B=250Bd		DGPS	G
2836,00	USB		HLK	Kangnung R	KOR
2842,20	*STANAG 4285*	*B=2400Bd*	*CFH*	*CAN N TX Halifax*	*CAN*
2843,00	USB			Madeira R	MDR
2847,00	*STANAG 4285/5066*	*B=2400Bd*	*FUE*	*F N Brest*	*F*
2848,50	USB			G AF Air Cadets	G
2850,00	USB			EGY N net	EGY
2851,00	SSB			MWARA AFI	AFI
2851,00	SSB			RDARA 11B	
2851,00	SSB			RDARA 13E	
2851,00	SSB			RDARA 13F	
2851,00	SSB			RDARA 14	
2851,00	SSB			RDARA 2A	
2851,00	SSB			RDARA 3B	
2851,00	USB			Barnaul ACC	RUS
2851,00	SSB			RDARA 3C	
2851,00	SSB			RDARA 9C	
2851,00	USB			Irkutsk ACC	RUS
2851,00	USB			Khanty-Mansiysk ACC	RUS
2851,00	USB			Kirensk ACC	RUS
2851,00	USB			Kalpashevo ACC	RUS
2851,00	USB			Krasnoyarsk ACC	RUS
2851,00	USB			Novosibirsk ACC	RUS
2851,00	USB			Podkamennaya ACC	RUS
2851,00	USB			Surgut ACC	RUS
2851,00	USB			Yeniseysk ACC	RUS
2851,00	USB		BOING SEATTLE	Boing Seattle	USA
2851,00	USB		BOING EVERETT	Boing Seattle	USA
2854,00	SSB			Canarias ACC , Las Palmas	CNR
2854,00	SSB			MWARA SAT	SAT
2854,00	SSB			RDARA 10B	
2854,00	SSB			RDARA 3A	
2854,00	SSB			RDARA 3B	
2854,00	SSB			RDARA 6E	
2854,00	SSB			Sal ACC	CPV

Frequency	Mode	Mode Parameter	Callsign	User	Country
2854,00	USB			Recife ACC	B
2854,00	USB			Rio de Janeiro ACC	B
2854,00	USB			Paramaribo ACC	SUR
2854,00	USB			Johannesburg ACC	AFS
2854,00	USB			Manaus ACC	B
2854,00	USB			Dakar ACC	SEN
2854,00	USB			Cayenne ACC	GUF
2857,00	SSB			RDARA 13J	
2857,00	SSB			RDARA 2B	
2857,00	SSB			RDARA 2C	
2857,00	SSB			RDARA 6B	
2857,00	USB			Airbus Toulouse Technique	F
2860,00	SSB			RDARA 10B	
2860,00	SSB			RDARA 12E	
2860,00	SSB			RDARA 12J	
2860,00	SSB			RDARA 13I	
2860,00	SSB			RDARA 1B	
2860,00	SSB			RDARA 3A	
2860,00	SSB			RDARA 3C	
2860,00	SSB			RDARA 9B	
2860,00	SSB			VOLMET AFI	AFI
2860,00	CW		4XZ	ISR N Haifa	ISR
2863,00	SSB			Hongkong VOLMET	HKG
2863,00	SSB			RDARA 13C	
2863,00	SSB			RDARA 13J	
2863,00	SSB			RDARA 13K	
2863,00	SSB			RDARA 2A	
2863,00	SSB			RDARA 2C	
2863,00	SSB			RDARA 7B	
2863,00	SSB			Tokyo VOLMET	J
2863,00	SSB			VOLMET PAC	PAC
2863,00	SSB		KVM70	Honululu VOLMET	HWA
2866,00	SSB			RDARA 10A	
2866,00	SSB			RDARA 2C	
2866,00	SSB			RDARA 3C	
2866,00	SSB			RDARA 4B	
2866,00	SSB			RDARA 6D	
2869,00	SSB		UNNN	Novosibirsk VOLMET	RUS
2869,00	USB			San Francisco ACC , CA	USA
2869,00	SSB			MWARA CEP	CEP
2869,00	SSB			RDARA 10E	
2869,00	SSB			RDARA 13C	
2869,00	SSB			RDARA 14G	
2869,00	SSB			RDARA 2A	
2869,00	SSB			RDARA 2B	
2869,00	SSB			RDARA 3A	
2869,00	SSB			RDARA 6G	
2869,00	SSB		RQCI	Samara VOLMET	RUS
2869,00	SSB		UBB2	Syktyvka VOLMET	RUS
2869,00	SSB		RVPE	Tyumen VOLMET	RUS
2872,00	SSB			Iceland ACC , Reyklavik	ISL
2872,00	SSB			MWARA NAT	NAT
2872,00	SSB			RDARA 13G	
2872,00	SSB			RDARA 3B	
2872,00	SSB			RDARA 6A	
2872,00	SSB			RDARA 6E	
2872,00	*USB*		*EIP*	*Shanwik ACC*	*IRL*
2872,00	SSB		VFG	Gander ACC	CAN
2875,00	SSB			RDARA 10A	
2875,00	SSB			RDARA 12G	
2875,00	SSB			RDARA 2A	
2875,00	SSB			RDARA 2B	

Frequency	Mode	Mode Parameter	Callsign	User	Country
2875,00	SSB			RDARA 3A	
2875,00	SSB			RDARA 6G	
2878,00	SSB			MWARA AFI	AFI
2878,00	SSB			RDARA 11B	
2878,00	SSB			RDARA 13I	
2878,00	SSB			RDARA 13J	
2878,00	SSB			RDARA 14	
2878,00	SSB			RDARA 3B	
2878,00	SSB			RDARA 3C	
2878,00	USB			Maiduguri ACC	NIG
2878,00	USB			Maroua ACC	CME
2878,00	USB			N djamena ACC	TCD
2878,00	USB			Niamey ACC	NGR
2878,00	USB			Niamtougou ACC	TGO
2878,00	USB			Port Gentil ACC	GAB
2878,00	USB			Pointe Noire ACC	COG
2878,00	USB			Accra ACC	GHA
2878,00	USB			Bangui ACC	CAF
2878,00	USB			Sao Tome ACC	CAF
2878,00	USB			Windhoek ACC	NMB
2878,00	USB			Yaounde ACC	CME
2878,00	USB			Doula ACC	CME
2878,00	USB			Entebbe ACC	UGA
2878,00	USB			Garouae ACC	CME
2878,00	USB			Goma ACC	COG
2878,00	USB			Harare ACC	ZWE
2878,00	USB			Kano ACC	NIG
2878,00	USB			Kinshasa ACC	COG
2878,00	USB			Kisangani ACC	COG
2878,00	USB			Lagos ACC	NIG
2878,00	USB			Libreville ACC	GAB
2878,00	USB			Luanda ACC	AGL
2878,00	USB			Lubumbashi ACC	COG
2878,00	USB			Lusaka ACC	ZMB
2878,00	USB			Port Gentil ACC	GAB
2878,00	USB			Roberts ACC	LBR
2878,00	USB			Franceville ACC	GAB
2881,00	SSB			RDARA 1B	
2881,00	SSB			RDARA 2A	
2881,00	SSB			RDARA 2B	
2881,00	SSB			RDARA 3A	
2881,00	SSB			RDARA 6C	
2881,00	SSB			VOLMET SAM	SAM
2881,00	USB		LWB	Ezeiza VOLMET	ARG
2884,00	SSB			RDARA 2C	
2884,00	SSB			RDARA 3B	
2884,00	SSB			RDARA 6D	
2886,50	MIL 188-110B	B=2400Bd		S N Karlskrona	S
2887,00	USB			Barranquilla ACC	CLM
2887,00	USB			Merida ACC	MEX
2887,00	SSB			New York ACC , NY	USA
2887,00	SSB			Boyeros ACC	CUB
2887,00	SSB			MWARA CAR	CAR
2887,00	SSB			Piarco ACC	TRD
2887,00	SSB			RDARA 13I	
2887,00	SSB			RDARA 14C	
2887,00	SSB			RDARA 2A	
2887,00	SSB			RDARA 2B	
2887,00	SSB			RDARA 3A	
2887,00	SSB			RDARA 7E	
2887,00	USB			Panama ACC	PNR
2889,00	USB		LWL	Comodoro VOLMET	ARG

Frequency	Mode	Mode Parameter	Callsign	User	Country
2890,00	USB			PanAM R	USA
2890,00	SSB			RDARA 13J	
2890,00	SSB			RDARA 1B	
2890,00	SSB			RDARA 6G	
2890,00	USB		EIP	Shanwik ACC	IRL
2893,00	SSB			RDARA 10D	
2893,00	SSB			RDARA 12F	
2893,00	SSB			RDARA 2C	
2893,00	SSB			RDARA 3	
2893,00	SSB			RDARA 4B	
2896,00	SSB			RDARA 2A	
2896,00	SSB			RDARA 2B	
2896,00	SSB			RDARA 3A	
2896,00	SSB			RDARA 6G	
2899,00	SSB			Iceland ACC , Reyklavik	ISL
2899,00	SSB			MWARA NAT	NAT
2899,00	SSB		KEA5	New York ACC	USA
2899,00	SSB			RDARA 13H	
2899,00	SSB			RDARA 5D	
2899,00	SSB			RDARA 6G	
2899,00	SSB		CSY	Santa Maria ACC	AZR
2899,00	USB		EIP	Shanwik ACC	IRL
2899,00	SSB		VFG	Gander ACC	CAN
2900,00	RADAR			MF/HF SA Radar Scott Base	ANT
2902,00	SSB			RDARA 12J	
2902,00	SSB			RDARA 2B	
2902,00	SSB			RDARA 2C	
2902,00	SSB			RDARA 3B	
2902,00	SSB			RDARA 6G	
2905,00	SSB			RDARA 3A	
2905,00	SSB			RDARA 3C	
2905,00	SSB			RDARA 5C	
2905,00	SSB			RDARA 9B	
2905,00	SSB			VOLMET NAT	NAT
2908,00	SSB			RDARA 13M	
2908,00	SSB			RDARA 2B	
2908,00	SSB			RDARA 2C	
2908,00	SSB			RDARA 3B	
2911,00	SSB			RDARA 10A	
2911,00	SSB			RDARA 3A	
2911,00	SSB			RDARA 5B	
2911,00	SSB			RDARA 6G	
2914,00	SSB			RDARA 13D	
2914,00	SSB			RDARA 2B	
2914,00	SSB			RDARA 2C	
2914,00	SSB			RDARA 3B	
2917,00	SSB			RDARA 2A	
2917,00	SSB			RDARA 6E	
2917,00	SSB			RDARA 6G	
2920,00	SSB			RDARA 12C	
2920,00	SSB			RDARA 13J	
2920,00	SSB			RDARA 2B	
2920,00	SSB			RDARA 2C	
2920,00	SSB			RDARA 6B	
2920,00	AM			Tundra net	RUS
2923,00	SSB			RDARA 3A	
2923,00	SSB			RDARA 6A	
2926,00	SSB			RDARA 10C	
2926,00	SSB			RDARA 12J	
2926,00	SSB			RDARA 2A	
2926,00	SSB			RDARA 2C	
2926,00	SSB			RDARA 4A	

Frequency	Mode	Mode Parameter	Callsign	User	Country
2926,00	SSB			RDARA 6F	
2929,00	SSB			RDARA 2B	
2929,00	SSB			RDARA 9B	
2932,00	USB			San Francisco ACC , CA	USA
2932,00	SSB			MWARA NP	NP
2932,00	SSB			RDARA 13K	
2932,00	SSB			RDARA 2A	
2932,00	SSB			RDARA 2C	
2932,00	SSB			RDARA 3A	
2932,00	SSB			Tokyo ACC	J
2935,00	SSB			MWARA SAT	SAT
2935,00	SSB			RDARA 10D	
2935,00	SSB			RDARA 3	
2938,00	SSB			RDARA 2	
2938,00	SSB			RDARA 6G	
2941,00	SSB			RDARA 2A	
2941,00	SSB			RDARA 6F	
2941,00	USB		RLAP	Rostov VOLMET	RUS
2941,00	SSB		UHD	St. Petersburg VOLMET	RUS
2941,00	HFDL	B=1800Bd	H10**	Muan	KOR
2944,00	SSB			MWARA MID	MID
2944,00	SSB			MWARA SAM	SAM
2944,00	SSB			RDARA 10A	
2944,00	SSB			RDARA 10E	
2944,00	SSB			RDARA 14G	
2944,00	SSB			Santa Cruz ACC	BOL
2944,00	USB			Bogota ACC	CLM
2944,00	HFDL	B=1800Bd	H09**	Utqiagvik , AK	ALS
2944,00	USB			Maiquetia ACC	VEN
2944,00	USB			Lima ACC	PRU
2944,00	USB			Quito ACC	EQA
2944,00	USB			Asuncion ACC	PRG
2944,00	USB			Buenos Aires ACC	ARG
2944,00	USB			Cordoba ACC	ARG
2944,00	USB			Easter Island ACC	CHL
2944,00	USB			La Paz ACC	CHL
2944,00	USB			Puerto Montt ACC	CHL
2944,00	USB			Punta Arenas ACC	CHL
2944,00	USB			Santiago ACC	CHL
2944,00	USB			Talara ACC	PRU
2944,00	USB			Ushuaia ACC	ARG
2944,00	USB			Aktyubinsk ACC	KAZ
2944,00	USB			Almaty ACC	KAZ
2944,00	USB			Bishkek ACC	KGZ
2944,00	USB			Dushanbe ACC	TJK
2944,00	USB			Kuybyshev ACC	RUS
2944,00	USB			Kzyl-Orda ACC	KAZ
2944,00	USB			Moscow ACC	RUS
2944,00	USB			Samarkhand ACC	UZB
2944,00	USB			Tashkent ACC	UZB
2944,00	USB			Uralsk ACC	KAZ
2944,00	USB			Yerevan ACC	ARM
2947,00	SSB			RDARA 6A	
2950,00	SSB			RDARA 10F	
2950,00	SSB			RDARA 12A	
2950,00	SSB			RDARA 14A	
2950,00	SSB			RDARA 14D	
2950,00	SSB			RDARA 2	
2950,00	SSB			RDARA 3C	
2950,00	SSB			RDARA 7C	
2950,00	SSB			VOLMET CAR	CAR
2953,00	SSB			RDARA 4A	

Frequency	Mode	Mode Parameter	Callsign	User	Country
2953,00	SSB			RDARA 6G	
2956,00	SSB			RDARA 10A	
2956,00	SSB			RDARA 12E	
2956,00	SSB			RDARA 12F	
2956,00	SSB			RDARA 12G	
2956,00	SSB			RDARA 12H	
2956,00	SSB			RDARA 13F	
2956,00	SSB			RDARA 6C	
2956,00	SSB			RDARA 7F	
2956,00	SSB			VOLMET MID	MID
2959,00	SSB			RDARA 3A	
2962,00	SSB			MWARA NAT	NAT
2962,00	SSB		KEA5	New York ACC	USA
2962,00	SSB			RDARA 6G	
2962,00	USB			Santa Maria ACC	AZR
2965,00	SSB		HSD	Bangkok VOLMET	THA
2965,00	USB		AWB	Bombay VOLMET	IND
2965,00	SSB			Calcutta VOLMET	IND
2965,00	SSB			Karachi VOLMET	PAK
2965,00	SSB			RDARA 10C	
2965,00	SSB			RDARA 12F	
2965,00	SSB			RDARA 12J	
2965,00	SSB			RDARA 13H	
2965,00	SSB			RDARA 1E	
2965,00	SSB			RDARA 7B	
2965,00	SSB			AUS VOLMET Brisbane	AUS
2965,00	SSB			VOLMET SEA	SEA
2968,00	SSB			RDARA 3B	
2968,00	SSB			RDARA 5B	
2968,00	SSB			RDARA 6G	
2971,00	USB			Bodo ACC	NOR
2971,00	SSB			Iceland ACC , Reykjavik	ISL
2971,00	SSB			MWARA NAT	NAT
2971,00	SSB			RDARA 13G	
2971,00	SSB			RDARA 5D	
2971,00	SSB			RDARA 6G	
2971,00	USB		EIP	Shanwik ACC	IRL
2971,00	SSB		VFG	Gander ACC	CAN
2974,00	SSB			RDARA 1D	
2974,00	SSB			RDARA 3C	
2977,00	SSB			RDARA 13M	
2977,00	SSB			RDARA 1C	
2977,00	SSB			RDARA 6G	
2978,00	USB		LVE	Cordoba VOLMET	ARG
2980,00	SSB			RDARA 12C	
2980,00	SSB			RDARA 1D	
2980,00	SSB			RDARA 3C	
2980,00	AM			Tundra net	RUS
2983,00	SSB			RDARA 13D	
2983,00	SSB			RDARA 1C	
2983,00	SSB			RDARA 6G	
2986,00	SSB			RDARA 13N	
2986,00	SSB			RDARA 3C	
2986,00	SSB			RDARA 5A	
2989,00	SSB			RDARA 1D	
2989,00	SSB			RDARA 6G	
2989,00	USB			GRC aero net	GRC
2989,00	USB			Naxos R	GRC
2989,00	USB			Kalamata R	GRC
2989,00	USB			Kassos R	GRC
2989,00	USB			Ioannina R	GRC
2989,00	USB			Kavala R	GRC

Frequency	Mode	Mode Parameter	Callsign	User	Country
2989,00	USB			Iraklion R	GRC
2989,00	USB			Paros R	GRC
2989,00	USB			Samos R	GRC
2989,00	USB			Limnos R	GRC
2989,00	USB			Preveza R	GRC
2989,00	USB			Skiathos R	GRC
2989,00	USB			Patras R	GRC
2989,00	USB			Kastoria R	GRC
2989,00	USB			Astypalaia R	GRC
2989,00	USB			Chios R	GRC
2989,00	USB			Rhodes R	GRC
2989,00	USB			Alexandroupoli R	GRC
2989,00	USB			Zakynthos R	GRC
2989,00	USB			Athens R	GRC
2989,00	USB			Kozani R	GRC
2989,00	USB			Ikaria R	GRC
2989,00	USB			Corfu R	GRC
2989,00	USB			Kefalonia R	GRC
2989,00	USB			Kithira R	GRC
2989,00	USB			Kos R	GRC
2989,00	USB			Leros R	GRC
2989,00	USB			Thessaloniki R	GRC
2989,00	USB			Mykonos R	GRC
2989,00	USB			Mytilene R	GRC
2989,00	USB			Samos R	GRC
2989,00	USB			Santorini R	GRC
2989,00	USB			Skyros R	GRC
2989,00	USB			Chania R	GRC
2991,50	AM			Tundra net	RUS
2992,00	USB			Aden ACC	YEM
2992,00	USB			Ankara ACC	TUR
2992,00	USB			Cairo ACC	EGY
2992,00	USB			Odessa ACC	UKR
2992,00	USB			Jeddah ACC	ARS
2992,00	USB			Sanaa ACC	YEM
2992,00	USB			Manama ACC	BHR
2992,00	USB			Simferopol ACC	UKR
2992,00	USB			Tehran ACC	IRN
2992,00	USB			Tbilisi ACC	GEO
2992,00	USB			Yerevan ACC	ARM
2992,00	SSB			Amman ACC	JOR
2992,00	SSB			Beirut ACC	LBN
2992,00	SSB			Damascus ACC	SYR
2992,00	SSB			Kuwait ACC	KWT
2992,00	SSB			MWARA MID	MID
2992,00	SSB			RDARA 10A	
2992,00	SSB			RDARA 10E	
2992,00	SSB			RDARA 13C	
2992,00	HFDL	B=1800Bd	H09**	Utqiagvik , AK	ALS
2995,00	SSB			RDARA 6G	
2998,00	USB			Manila ACC	PHL
2998,00	SSB			MWARA CWP	CWP
2998,00	SSB			RDARA 12E	
2998,00	SSB			RDARA 12F	
2998,00	SSB			RDARA 12G	
2998,00	SSB			RDARA 12H	
2998,00	SSB			RDARA 13F	
2998,00	SSB			RDARA 7D	
2998,00	SSB			Tokyo ACC	J
2998,00	SSB			VOLMET EUR	EUR
2998,00	USB			Hong Kong ACC	HKG
2998,00	USB			Naha ACC	J

Frequency	Mode	Mode Parameter	Callsign	User	Country
2998,00	USB			Port Moresby ACC	PNG
2998,00	USB			San Francisco ACC , CA	USA
2998,00	USB			Seoul ACC	KOR
2998,00	USB			Taipai ACC	TWN
2998,00	HFDL	B=1800Bd	H07**	Shannon	IRL
3000,00	ALE USB	B=125Bd Ch=8 ChS=250Hz		LBY GMRA net	LBY
3000,00	*STANAG 4285*	*B=2400Bd*	*JWT*	*NOR N Stavanger*	*NOR*
3000,00	USB			EGY N net	EGY
3001,00	SSB			RDARA 6A	
3001,00	SSB			RDARA 6E	
3004,00	SSB			Irkutsk ACC	RUS
3004,00	SSB			Khabarovsk ACC	RUS
3004,00	SSB			MWARA NCA	NCA
3004,00	SSB			Pyongyang ACC	KRE
3004,00	SSB			RDARA 11B	
3004,00	SSB			RDARA 13K	
3004,00	SSB			Ulaanbaatar ACC	MNG
3004,00	USB			Chita ACC	RUS
3004,00	USB			Chulman ACC	RUS
3004,00	USB			Ekimchan ACC	RUS
3004,00	USB			Kirensk ACC	RUS
3004,00	USB		BOING EVERETT	Boing Seattle	USA
3004,00	USB		BOING SEATTLE	Boing Seattle	USA
3005,00	AM			Tundra net	RUS
3007,00	SSB			W II	WW
3007,00	SSB			W III	WW
3007,00	HFDL	B=1800Bd	H09**	Utqiagvik , AK	ALS
3010,00	SSB			W I	WW
3010,00	SSB			W IV	WW
3012,00	RUS 4TONE	Ch=4		RUS MIL	RUS
3012,00	USB			RUS MIL Smolensk area	RUS
3013,00	SSB			W II	WW
3013,00	SSB			W V	WW
3016,00	USB			Canarias ACC , Las Palmas	CNR
3016,00	SSB			MWARA EA	EA
3016,00	SSB			MWARA NAT-A	NAT
3016,00	USB		KEA5	New York ACC	USA
3016,00	SSB			RDARA 13G	
3016,00	SSB			RDARA 9D	
3016,00	USB		CSY	Santa Maria ACC	AZR
3016,00	USB		EIP	Shanwik ACC	IRL
3016,00	USB		VFG	Gander ACC	CAN
3016,00	USB			Paramaribo ACC	SUR
3016,00	HFDL	B=1800Bd	H08**	Johannesburg	AFS
3016,00	USB			Beijing ACC	CHN
3016,00	USB			Guangzhou ACC	CHN
3016,00	USB			Hailar ACC	CHN
3016,00	USB			Irkutsk ACC	RUS
3016,00	USB			Jinan ACC	CHN
3016,00	USB			Kunming ACC	CHN
3016,00	USB			Lanzhou ACC	CHN
3016,00	USB			Pyongyang ACC	KRE
3016,00	USB			Shanghai ACC	CHN
3016,00	USB			Shenyangi ACC	CHN
3016,00	USB			Taegu ACC	KOR
3016,00	USB			Ulaanbaatar ACC	MNG
3016,00	USB			Urumqi ACC	CHN
3016,00	USB			Wuhan ACC	CHN
3016,00	USB			Zhengzhou ACC	CHN

Frequency	Mode	Mode Parameter	Callsign	User	Country
3016,00	USB			Piarco ACC	TRD
3017,00	AM			Tundra net	RUS
3019,00	SSB			Moscow ACC	RUS
3019,00	SSB			MWARA NCA-1	NCA
3019,00	SSB			RDARA 11B	
3019,00	SSB			RDARA 13K	
3019,00	USB			RUS	
3019,00	USB			RUS	
3019,00	USB			RUS	
3019,00	USB			RUS	
3023,00	USB		VNJ	Casey Base , AUS	ANT
3023,00	USB			AUS Davis Base	ANT
3023,00	USB		LOK	ARG N Laurie Island , Orcadas	ANT
3023,00	USB		CWM30	URG SAR net	URG
3023,00	USB			Macquarie Island , AUS	ANT
3023,00	USB			Mawson Base , AUS	ANT
3023,00	SSB			Distress/calling frequency	WW
3023,00	SSB			DNK AF Karup	DNK
3023,00	SSB			NOR AF Stavanger	NOR
3023,00	SSB		ONY77	BEL AF Koksijde	BEL
3023,00	USB			Wellington Mountain Radio Service	NZL
3026,00	USB			Kinloss Rescue	G
3027,00	AM			Tundra net	RUS
3034,00	MIL 188-110A SER	B=2400Bd		S MIL	S
3034,00	MIL 188-110A SER	B=2400Bd		S MIL	S
3035,00	USB		LBJ	NOR N Bodoe	NOR
3042,00	VFT: M/S 1,0	B=50Bd S=170Hz Ch=2 ChS=1000Hz			
3044,00	USB		FUN	F N Lorient	F
3047,00	USB		CHR	CAN AF Trenton	CAN
3057,00	AM			Tundra net	RUS
3059,00	ALE USB	B=125Bd Ch=8 ChS=250Hz		USA AF net	
3062,00	AM			Tundra net	RUS
3063,00	BAUDOT	B=50Bd S=1000Hz	NAV12**		
3068,00	ALE USB	B=125Bd Ch=8 ChS=250Hz		USA AF net	
3069,00	AM			Tundra net	RUS
3079,00	AM			Tundra net	RUS
3079,50	ALE USB	B=125Bd Ch=8 ChS=250Hz			
3081,00	SSB		CUW2	USAF Lajes	POR
3085,00	AM			RUS Arctic net	RUS
3086,00	AM			Tundra net	RUS
3088,00	LINK 11 CLEW	B=2250Bd Ch=16 ChS=330/110/550Hz			
3090,00	AM			Tundra net	RUS
3092,00	USB		CHR	CAN AF Trenton	CAN
3095,00	SSB			Dharan ACC	ARS
3095,00	SSB			Riyadh ACC	ARS
3095,00	ALE USB	B=125Bd Ch=8 ChS=250Hz		G MIL net	G
3097,00	AM			Tundra net	RUS
3100,00	AM			Tundra net	RUS
3107,00	USB		DHM91	D AF Münster	D
3107,00	USB		DHO78	D AF Kalkar	D
3107,00	USB		DHJ83	D AF Gatow	D
3110,00	AM			Tundra net	RUS
3113,00	SSB			USA AF Strategic Command	USA
3113,00	ALE USB	B=125Bd Ch=8 ChS=250Hz		USA AF net	
3116,00	USB			Kirensk VOLMET	RUS

Frequency	Mode	Mode Parameter	Callsign	User	Country
3116,00	SSB			RAF Boulmer	G
3116,00	SSB			RAF Buchan	G
3116,00	SSB			RAF Portreath	G
3116,00	USB			Irkutsk VOLMET	RUS
3120,00					
3121,00	AM			Tundra net	RUS
3122,00	SSB			USCG Boston , MA	USA
3122,00	SSB			USCG Charleston , SC	USA
3125,00	USB		OYR	Aasiaat R	GRL
3125,00	SSB			RAF Boulmer	G
3128,00	AM			Tundra net	RUS
3130,50	ALE USB	B=125Bd Ch=8 ChS=250Hz			
3133,00	AM			Tundra net	RUS
3134,00	ALE USB	B=125Bd Ch=8 ChS=250Hz		LTU F net	
3136,00	AM			Tundra net	RUS
3137,00	SSB			Göteborg Rescue	S
3137,00	ALE USB	B=125Bd Ch=8 ChS=250Hz		USA AF net	
3141,00	ALE USB	B=125Bd Ch=8 ChS=250Hz		I N net	I
3143,00	SSB		AGA2	USAF Hickham AFB , Honolulu	HWA
3145,00	AM			Tundra net	RUS
3151,00	AM			Tundra net	RUS
3152,00	CW		HMA	Pyonyang M	KRE
3153,00	AM			Tundra net	RUS
3154,00	AM			Tundra net	RUS
3159,00	CW		9MR5	MLA N Johor Baharu	MLA
3159,00	BAUDOT	B=50Bd S=850Hz	9MR5	MLA N Johor Baharu	MLA
3160,00	AM			Tundra net	RUS
3160,00	MIL 188-110B SER	B=2400Bd			
3161,00	CW		RGT77	RUS MIL Moscow, RVSN	RUS
3162,00	USB			Saint Helena R	SHN
3162,50	CIS 500 FSK BURST	B=500Bd S=1000Hz		RUS N	RUS
3163,50	MIL 188-110B SER	B=2400Bd			
3165,00	USB		9YL	Trinedad North Post R	TRD
3167,00	CW			RUS N Kaliningrad	RUS
3167,00	CW		R**	RUS N Ustinov	RUS
3167,00	CW		P**	RUS N Kaliningrad	RUS
3167,00	AM			Tundra net	RUS
3168,50	CW		L**	RUS N St. Petersburg	RUS
3170,00	RADAR			MF/HF SA Radar Saura	NOR
3172,00	CIS ARQ	B=100Bd S=500Hz		RUS PTT	RUS
3173,00	AM			Tundra net	RUS
3173,00	CIS OFDM 45	Ch=45 ChS=40Hz		RUS diplo	RUS
3175,00	USB		VNJ	Casey Base , AUS	ANT
3175,00	USB			AUS Davis Base	ANT
3175,00	USB			Macquarie Island , AUS	ANT
3175,00	USB			Mawson Base , AUS	ANT
3175,00	USB		VMD750	Austravel Safety Net	AUS
3176,00	CIS OFDM 60	B=2400Bd Ch=60 ChS=44,5Hz		RUS MIL Savolenka	RUS
3178,00	CW		RGT77	RUS MIL Moscow, RVSN	RUS
3180,00	USB		PKA	Sabang R , We	INS
3180,00	USB		PKB	Belawan R , Sumatera	INS
3180,00	USB		PKB3	Sibolga R , Sumatera	INS
3180,00	USB		PKE5	Ternate R , Maluku	INS
3180,00	USB		PKF3	Kendari R	INS
3180,00	USB		PKG	Bandjarmasin R , Kalimantan	INS
3180,00	USB		PKO	Tarakan R , Kalimantan	INS

Frequency	Mode	Mode Parameter	Callsign	User	Country
3180,00	USB		PKP	Dumai R , Sumatera	INS
3180,00	USB		PKP2	Telukbayur R , Sumatera	INS
3180,00	USB		PKR	Semarang R , Djawa	INS
3180,00	USB		PKR3	Cilacap R , Jawa	INS
3180,00	USB		PKS	Pontanak R , Kalimantan	INS
3180,00	USB		PKY23	Fak Fak R	INS
3180,00	USB		PKY3	Manokwari R , Irian Jaya	INS
3180,00	USB		PKY4	Sorong R , Irian Jaya	INS
3180,00	USB		PKY5	Merauke R , Irian Jaya	INS
3180,00	USB		PKZ2	Cirebon R , Jawa	INS
3180,00	USB		PKZ34	Cigading R , Jawa	INS
3180,00	USB		PNK	Jayapura R , Irian Jaya	INS
3180,00	USB		PKE37	Sanana R , Kep Sula	INS
3180,00	USB		PKJ28	Sei Kolak Kijang R	INS
3180,00	RADAR			MF/HF SA Radar Juliusruh	D
3181,00	CW		RMP	RUS N HQ Kaliningrad	RUS
3185,00	CW			RUS Air Defense net	RUS
3187,00	SSB			NASA rocket booster recovery	USA
3188,50	USB			POL Mil net	POL
3188,50	ALE USB	B=125Bd Ch=8 ChS=250Hz		POL net	POL
3188,50	R&S MODEM	B=2400Bd		POL MSW Poznan	POL
3188,50	R&S MODEM	B=2400Bd		POL MSW Warszawa	POL
3192,00	*CW*		*RMP*	*RUS N HQ Kaliningrad*	*RUS*
3192,20	*STANAG 4285*	*B=2400Bd*		*DNK N TX Frederickshavn*	*DNK*
3195,00	CW		JJT88	J MIL Tokyo	J
3195,00	CW		JJT20	J MIL	J
3195,00	CW		JJT55	J MIL Fuchu	J
3195,00	CW		JJS21	J MIL Misawa	J
3195,00	CW		JJV56	J MIL Hamamatsu	J
3195,00	CW		JJZ37	J MIL Kasuga	J
3195,00	CW		JJT44	J MIL Irumagawa	J
3195,00	CW		JJU22	J MIL Naha	J
3200,00	HAARP waveform			HAARP	USA
3200,00	USB			EGY N net	EGY
3204,00	ALE USB	B=125Bd Ch=8 ChS=250Hz		USA Civil Air Patrol net	USA
3207,00	CW			RUS HFDF net	RUS
3210,00	CW		JJT88	J MIL Tokyo	J
3210,00	CW		JJX25	J MIL Miho AB	J
3210,00	CW		JJX36	J MIL Bofu	J
3210,00	CW		JJZ59	J MIL Ashiya Fukuoka	J
3210,00	CW		JJZ48	J MIL Tsuiki	J
3210,00	CW		JJZ26	J MIL Nyutabaru	J
3213,00	AM			Tundra net	RUS
3216,50	ALE USB	B=125Bd Ch=8 ChS=250Hz		GRC AF net	GRC
3221,20	STANAG 4285	B=2400Bd		I N Naples	I
3222,00	CW		RCV	RUS N HQ Sevastopol	UKR
3225,00	ALE USB	B=125Bd Ch=8 ChS=250Hz		ALB MIL net	ALB
3226,00	4-DPSK	B=250Bd		DGPS	DNK
3226,00	MIL 188-110B SER	B=2400Bd		G MIL	
3231,00	SSB			USA AF MARS	
3235,00	*STANAG 4285*	*B=2400Bd*		*TUR N TX Ankara*	*TUR*
3235,50	CW		RFE76	RUS N	RUS
3235,50	CW		RFF78	RUS N	RUS
3236,00	USB			G AF Air Cadets	G
3236,50	MIL 188-110B SER	B=2400Bd		G MIL	
3237,00	SSB			USA Army MARS	
3238,50	SSB			USN Plead Control Long Beach , OR	USA

Frequency	Mode	Mode Parameter	Callsign	User	Country
3239,00	ALE USB	B=125Bd Ch=8 ChS=250Hz		CHL MOI net	CHL
3240,00	ALE USB	B=125Bd Ch=8 ChS=250Hz		ARINC Urgent Link net	USA
3242,00	SSB			USA Army MARS	
3242,00	CW			RUS MIL net	RUS
3243,00	USB			RUS MIL	RUS
3245,00	USB			CIS aero network night	CIS
3245,00	SSB			USA Army MARS	
3245,00	2FSK	B=96Bd S=500Hz			
3247,00	CW			RUS Air Defense net	RUS
3247,20	*STANAG 4285*	*B=2400Bd*	*DHJ58*	*D N Glücksburg TX Neuharlingersiel*	*D*
3249,20	*STANAG 4285*	*B=2400Bd*	*JWT*	*NOR N Stavanger*	*NOR*
3250,00	HAARP waveform			HAARP	ALS
3252,00	ALE USB	B=125Bd Ch=8 ChS=250Hz		AUS Police NSW	AUS
3253,00	FAX 120/576		VFF	CAN CG Iqaluit M	CAN
3254,00	USB			POL mil net	POL
3254,00	ALE USB	B=125Bd Ch=8 ChS=250Hz		LTU MIL net	LTU
3254,00	CW		RJY50	RUS N Pacific fleet	RUS
3254,50	ALE USB	B=125Bd Ch=8 ChS=250Hz		G MIL net	G
3255,00	USB			CIS aero network night	CIS
3255,00	SKYOFDM	B=64Bd Ch=28 ChS=86Hz			
3255,00	FSK		P**	RUS N Kaliningrad	RUS
3255,00	USB			Ashgabat Asgabat ACC	TKM
3255,00	USB		ASSISTENT	Mosdok ACC	RUS
3255,50	SSB			USA Army MARS	
3256,00	CW		P**	RUS N Kaliningrad	RUS
3260,00	ALE USB	B=125Bd Ch=8 ChS=250Hz			
3261,00	USB		ZLJG	Wellington Mountain Radio Service	NZL
3262,00	CW		RMP	RUS N HQ Kaliningrad	RUS
3262,50	ALE USB	B=125Bd Ch=8 ChS=250Hz			D
3266,50	CW			RUS Operational Strategic Command Air Defense net	RUS
3270,00	USB		KAPEL	Omsk ACC	RUS
3270,00	USB		SARATOV	Saratov ACC	RUS
3272,00	SSB			USA Army MARS	
3275,50	ALE USB	B=125Bd Ch=8 ChS=250Hz		USA SAC net	USA
3276,00	USB		OYR	Aasiaat R	GRL
3280,00	USB		OYR	Aasiaat R	GRL
3280,00	*MIL 188-110B SER*	*B=2400Bd*		*G MIL St. Eval*	*G*
3282,00	*T600*	*B=50Bd S=200Hz*	*RMP*	*RUS N Kaliningrad*	*RUS*
3282,00	*ALE USB*	*B=125Bd Ch=8 ChS=250Hz*		*F N net*	*F*
3286,00	*STANAG 4285*	*B=2400Bd*	*JWT*	*NOR N Stavanger*	*NOR*
3287,50	*STANAG 4285*	*B=2400Bd*		*G MIL TX Inskip*	*G*
3289,00	SSB			USA Army MARS	
3291,00	CW		P**	RUS N Kaliningrad	RUS
3291,00	ALE USB	B=125Bd Ch=8 ChS=250Hz		DNK AF	DNK
3292,00	SSB			USA AF MARS	
3295,00	CIS 14	B=96Bd S=500Hz			
3295,00	ALE USB	B=125Bd Ch=8 ChS=250Hz		S FRO net	S
3298,00	USB			UKR Air Defence net	UKR
3299,00	SSB			USA AF MARS	

Frequency	Mode	Mode Parameter	Callsign	User	Country
3300,00	CIS ARQ	B=100Bd S=500Hz		RUS PTT	RUS
3302,00	ALE USB	B=125Bd Ch=8 ChS=250Hz		LTU MIL net	LTU
3305,00	ALE USB	B=125Bd Ch=8 ChS=250Hz		NLD MIL net	NLD
3306,00	USB			G AF Air Cadets	G
3306,50	*STANAG 4285*	*B=2400Bd*		*I N Naples*	*I*
3308,00	SSB			USA AF MARS	
3308,00	USB		AFA5HI	USA AF MARS	USA
3308,00	USB		AFA5ML	USA AF MARS	USA
3308,00	USB		AFA5HF	USA AF MARS	USA
3308,00	USB		AFA5HX	USA AF MARS	USA
3308,00	USB		AFA5JX	USA AF MARS	USA
3308,00	USB		AFA5KK	USA AF MARS	USA
3308,00	USB		AFA5SH	USA AF MARS	USA
3308,00	USB		AFA5WE	USA AF MARS	USA
3310,00	SSB		ESA	Tallin R	EST
3310,00	USB		UTT	Odessa R	UKR
3310,00	CW		RMAE	RUS N	RUS
3312,20	*STANAG 4539*	*B=2400Bd*		*USA AF Croughton*	*G*
3313,00	RUS TACTICAL DATALINK	B=1200Bd S=800Hz Ch=1200		RUS MIL	RUS
3314,00	CW			RUS Operational Strategic Command Air Defense net	RUS
3314,00	CW		RAL2	RUS N HQ Astrakhan	RUS
3315,00	CW		RCV	RUS N HQ Sevastopol	UKR
3315,00	ALE USB	B=125Bd Ch=8 ChS=250Hz			
3318,20	*STANAG 4539*	*B=2400Bd*		*Croughton*	*G*
3318,30	ALE USB	B=125Bd Ch=8 ChS=250Hz		G MIL net	G
3320,00	CW		REA	RUS AF Moscow	RUS
3322,00	CW			RUS Operational Strategic Command Air Defense net	RUS
3322,00	CW		RCV	RUS N HQ Sevastopol	UKR
3322,70	ALE USB	B=125Bd Ch=8 ChS=250Hz		CHN N net	CHN
3325,00	CW		RMP	RUS N HQ Kaliningrad	RUS
3326,00	ALE USB	B=125Bd Ch=8 ChS=250Hz		G MIL net	G
3327,00	CW		P**	RUS N Kaliningrad	RUS
3327,20	*STANAG 4539*	*B=2400Bd*		*Croughton*	*G*
3330,00	CW		CHU	Ottawa TS	CAN
3332,00	USB			LBY CG net	LBY
3333,00	USB			EGY N net	EGY
3335,00	ALE USB	B=125Bd Ch=8 ChS=250Hz		S FRO net	S
3336,00	ALE USB	B=125Bd Ch=8 ChS=250Hz		ALG SONATRECH net	ALG
3336,00	USB			Auckland Tramping Club	NZL
3337,00	CW		RJP54	RUS N ship YENISEY	RUS
3337,00	CW		RMGZ	RUS N ship EPRON	RUS
3338,00	CW		L**	RUS N St. Petersburg	RUS
3340,00	F1B MORSE		UZC2		UKR
3341,00	ALE USB	B=125Bd Ch=8 ChS=250Hz		Bonneville power administrattion net	USA
3341,00	ALE USB	B=125Bd Ch=8 ChS=250Hz		USA FEMA net	USA
3342,00	USB		AAR2AO	USA AF MARS	USA
3342,00	USB		AARTC	US Army MARS station in USA	USA
3343,00	USB			G AF Air Cadets	G
3345,00	USB		ZLJG	Wellington Mountain Radio Service	NZL
3346,50	SSB			USA Army MARS	

Frequency	Mode	Mode Parameter	Callsign	User	Country
3348,00	CW			RUS MIL net	RUS
3350,00	ALE USB	B=125Bd Ch=8 ChS=250Hz			
3350,00	CW		RJC60	RUS N	RUS
3353,00	CW		REL5	RUS N ship KOYDA	RUS
3353,00	CW		RMUW	RUS N ship SHAKTER	RUS
3354,00	ALE USB	B=125Bd Ch=8 ChS=250Hz			
3356,00	ALE USB	B=125Bd Ch=8 ChS=250Hz		SVK MIL net	SVK
3356,00	MIL 188-110A SER	B=2400Bd		SVK MIL Presov	SVK
3356,00	CW		RGR99	RUS N	RUS
3356,00	CW		RGR89	RUS N	RUS
3356,00	CW		RGR90	RUS N	RUS
3356,00	CW		RGR91	RUS N	RUS
3356,00	CW		RGR92	RUS N	RUS
3356,00	CW		RGR94	RUS N	RUS
3356,00	CW		RGR96	RUS N	RUS
3356,00	CW		RGR98	RUS N	RUS
3356,00	CW		RGR95	RUS N	RUS
3356,00	MIL 188-110A SER	B=2400Bd		SVK MIL Velkaida	SVK
3363,00	CW			RUS MIL	RUS
3365,00	CW		RAL46	RUS N ship	RUS
3368,50	*STANAG 4285*	*B=2400Bd*		*G MIL Inskip*	*G*
3370,00	*CIS 12*	*B=1440Bd Ch=12 ChS=200Hz*		*Warszawa*	*POL*
3372,00	CW		RJD99	RUS N Severodvinsk	RUS
3372,00	USB			Hawks Bay Mountain Radio Service	NZL
3375,30	STANAG 4538	B=2400Bd			
3379,00	CW		RCV	RUS N HQ Sevastopol	UKR
3382,00	CIS 50-500 BAUDOT	B=50Bd S=500Hz		RUS	RUS
3384,50	*STANAG 4285*	*B=2400Bd*		*G MIL Crimond*	*G*
3388,00	ALE USB	B=125Bd Ch=8 ChS=250Hz		USA FEMA net	USA
3390,00	USB			EGY N net	EGY
3391,00	CIS 50-500 BAUDOT	B=50Bd S=500Hz		RUS	RUS
3394,00	ALE USB	B=125Bd Ch=8 ChS=250Hz		D Red Cross net	D
3395,50	CW		RCV	RUS N HQ Sevastopol	UKR
3395,50	CW		RCV	RUS N HQ Sevastopol	UKR
3395,50	CW		RIR96	RUS N landing ship AZOV	RUS
3395,50	CW		RFH77	RUS N ship LADNYY	RUS
3398,00	USB			POL MIL net	POL
3398,50	CW		RJD99	RUS N Severodvinsk	RUS
3399,00	USB			EGY N net	EGY
3401,00	SSB			RDARA 12C	
3401,00	SSB			RDARA 13K	
3401,00	SSB			RDARA 2B	
3401,00	SSB			RDARA 2C	
3401,00	SSB			RDARA 3B	
3401,00	SSB			RDARA 9B	
3404,00	SSB			RDARA 10B	
3404,00	SSB			RDARA 3A	
3404,00	SSB			RDARA 3C	
3404,00	SSB			RDARA 9C	
3404,00	SSB			RDARA 9D	
3404,00	SSB			VOLMET AFI	AFI
3407,00	SSB			Aktyubinsk VOLMET	KAZ
3407,00	SSB			RDARA 12D	
3407,00	SSB			RDARA 14D	
3407,00	SSB			RDARA 2B	

Frequency	Mode	Mode Parameter	Callsign	User	Country
3407,00	SSB			RDARA 2C	
3407,00	SSB			RDARA 3B	
3407,00	SSB			RDARA 7C	
3407,00	USB		RDFG	Tashkent VOLMET	UZB
3410,00	SSB			RDARA 11B	
3410,00	SSB			RDARA 13J	
3410,00	SSB			RDARA 1D	
3410,00	SSB			RDARA 3C	
3413,00	USB			San Francisco ACC , CA	USA
3413,00	SSB			MWARA CEP	CEP
3413,00	SSB			RDARA 13C	
3413,00	SSB			RDARA 14A	
3413,00	SSB			RDARA 14E	
3413,00	SSB			RDARA 3B	
3413,00	SSB			RDARA 6G	
3413,00	SSB			VOLMET EUR	EUR
3413,00	*USB*		*EIP*	*Shannon VOLMET*	*IRL*
3416,00	SSB			Kaunas ACC	LTU
3416,00	SSB			RDARA 1D	
3416,00	SSB			RDARA 2A	
3416,00	SSB			RDARA 2B	
3416,00	SSB			RDARA 3A	
3416,00	SSB			RDARA 6D	
3416,00	SSB			Tallinn ACC	EST
3416,00	SSB			Vilnius ACC	LTU
3418,00	USB		VNJ	Casey Base , AUS	ANT
3418,00	USB			AUS Davis Base	ANT
3418,00	USB			Macquarie Island , AUS	ANT
3418,00	USB			Mawson Base , AUS	ANT
3419,00	USB			Gao ACC	MLI
3419,00	USB			Timimoun ACC	ALG
3419,00	SSB			Alger ACC	ALG
3419,00	SSB			Kano ACC	NIG
3419,00	SSB			MWARA AFI	AFI
3419,00	SSB			Niamey ACC	NGR
3419,00	SSB			RDARA 10D	
3419,00	SSB			RDARA 12J	
3419,00	SSB			RDARA 13I	
3419,00	SSB			RDARA 3B	
3419,00	SSB			RDARA 3C	
3419,00	SSB			RDARA 9B	
3419,00	SSB			Tamanrasset ACC	ALG
3419,00	SSB			Tripoli ACC	LBY
3419,00	SSB			Tunis ACC	TUN
3419,00	USB			N Djamena ACC	TCD
3422,00	USB			RUS areo network night	
3422,00	SSB			RDARA 2A	
3422,00	SSB			RDARA 2B	
3422,00	SSB			RDARA 3A	
3422,00	SSB			RDARA 6G	
3423,00	CW		KYW4		
3425,00	SSB			MWARA AFI	AFI
3425,00	SSB			RDARA 10D	
3425,00	SSB			RDARA 13D	
3425,00	SSB			RDARA 3B	
3425,00	SSB			RDARA 3C	
3425,00	SSB			RDARA 9B	
3428,00	SSB			RDARA 11B	
3428,00	SSB			RDARA 13J	
3428,00	SSB			RDARA 2B	
3428,00	SSB			RDARA 2C	
3431,00	SSB			RDARA 3A	

Frequency	Mode	Mode Parameter	Callsign	User	Country
3431,00	SSB			RDARA 3B	
3431,00	SSB			RDARA 5B	
3431,00	SSB			RDARA 6G	
3434,00	SSB			RDARA 11B	
3434,00	SSB			RDARA 13G	
3434,00	SSB			RDARA 2A	
3434,00	SSB			RDARA 2C	
3434,00	SSB			RDARA 6F	
3437,00	SSB			RDARA 13M	
3437,00	SSB			RDARA 3B	
3437,00	SSB			RDARA 4A	
3437,00	SSB			RDARA 6G	
3440,00	SSB			RDARA 12	
3440,00	SSB			RDARA 2A	
3440,00	SSB			RDARA 2C	
3440,00	SSB			RDARA 6F	
3440,00	USB			IRL MIL	IRL
3443,00	SSB			RDARA 11B	
3443,00	SSB			RDARA 13N	
3443,00	SSB			RDARA 3A	
3443,00	SSB			RDARA 3B	
3443,00	SSB			RDARA 4B	
3443,00	SSB			RDARA 6E	
3443,00	USB		BOING SEATTLE	Boing Seattle	USA
3443,00	USB		BOING EVERETT	Boing Seattle	USA
3446,00	SSB			RDARA 10E	
3446,00	SSB			RDARA 13F	
3446,00	SSB			RDARA 14	
3446,00	SSB			RDARA 1D	
3446,00	SSB			RDARA 6G	
3446,00	USB		EIP	Shanwik ACC	IRL
3449,00	SSB			RDARA 10A	
3449,00	SSB			RDARA 13M	
3449,00	SSB			RDARA 2B	
3449,00	SSB			RDARA 2C	
3449,00	SSB			RDARA 6G	
3451,50	SITOR	B=100Bd S=170Hz	0A**	IRL N Dublin	IRL
3451,50	SITOR A	B=100Bd S=170Hz		IRL N	IRL
3452,00	USB			Brasilia ACC	B
3452,00	USB			Cayenne ACC	GUF
3452,00	USB			Manaus ACC	B
3452,00	USB			Paramaribo ACC	SUR
3452,00	USB			Rio de Janerio ACC	B
3452,00	USB			Bamako ACC	MLI
3452,00	USB			Bangui ACC	CAF
3452,00	USB			BissauACC	GNB
3452,00	USB			Bouake ACC	CTI
3452,00	USB			Casablanca ACC	MRC
3452,00	SSB			Canarias ACC , Las Palmas	CNR
3452,00	SSB			Dakar ACC	SEN
3452,00	SSB			MWARA SAT	SAT
3452,00	SSB			RDARA 14C	
3452,00	SSB			RDARA 3A	
3452,00	SSB			RDARA 5A	
3452,00	SSB			RDARA 5C	
3452,00	SSB			Recife ACC	B
3452,00	SSB			Sal ACC	CPV
3452,00	USB			Conakry ACC	GUI
3452,00	USB			Canarias ACC	E
3452,00	USB			Dakar ACC	SEN

Frequency	Mode	Mode Parameter	Callsign	User	Country
3452,00	USB			Freetown ACC	SRL
3452,00	USB			Johannesburg ACC	AFS
3452,00	USB			Kano ACC	NIG
3452,00	USB			Niamey ACC	NGR
3452,00	USB			Nouadhibou ACC	MTN
3452,00	USB			Nouakchott ACC	MTN
3452,00	USB			Ouagadougou ACC	BFA
3452,00	USB			Roberts ACC	LBR
3452,00	USB			San Francisco ACC , CA	USA
3455,00	USB			New York ACC	USA
3455,00	SSB			Boyeros ACC	CUB
3455,00	SSB			MWARA CAR	CAR
3455,00	SSB			MWARA CWP	CWP
3455,00	SSB			RDARA 13H	
3455,00	SSB			RDARA 2A	
3455,00	SSB			RDARA 2C	
3455,00	SSB			RDARA 7B	
3455,00	USB			Barranquilla ACC	CLM
3455,00	USB			Cayenne ACC	GUF
3455,00	USB			Georgetown ACC	GUY
3455,00	USB			Maiquetia ACC	VEN
3455,00	HFDL	B=1800Bd	H07**	Shannon	IRL
3455,00	USB			Panama ACC	PNR
3455,00	USB			Paramaribo ACC	SUR
3455,00	USB			Piarco ACC	TRD
3455,00	USB			San Francisco ACC , CA	USA
3458,00	SSB			RDARA 10D	
3458,00	SSB			RDARA 13C	
3458,00	SSB			RDARA 13J	
3458,00	SSB			RDARA 13K	
3458,00	SSB			RDARA 1B	
3458,00	SSB			VOLMET SEA	SEA
3458,00	USB		BSQ	Bejing VOLMET	CHN
3461,00	SSB			RDARA 12E	
3461,00	SSB			RDARA 12F	
3461,00	SSB			RDARA 12G	
3461,00	SSB			RDARA 12H	
3461,00	SSB			RDARA 14	
3461,00	SSB			RDARA 7F	
3461,00	SSB			RDARA 9C	
3461,00	SSB			VOLMET NCA	NCA
3464,00	SSB			Northsea Helicopter Esbjerg	DNG
3464,00	SSB			RDARA 12C	
3464,00	SSB			RDARA 13K	
3464,00	SSB			RDARA 1C	
3464,00	SSB			RDARA 6G	
3467,00	SSB			Aden ACC	YEM
3467,00	SSB			Ashkabad ACC	TKM
3467,00	SSB			Asmara ACC	ERT
3467,00	SSB			Mumbai ACC	IND
3467,00	SSB			Cairo ACC	EGY
3467,00	SSB			Dushanbe ACC	TJK
3467,00	SSB			Kabul ACC	AFG
3467,00	SSB			Karachi ACC	PAK
3467,00	SSB			Khartoum ACC	SDN
3467,00	SSB			Lahore ACC	PAK
3467,00	SSB			Mogadishu ACC	SOM
3467,00	SSB			MWARA AFI	AFI
3467,00	SSB			MWARA MID	MID
3467,00	SSB			MWARA SP	SP
3467,00	SSB			Nadi ACC	
3467,00	SSB			New Delhi ACC	IND

Frequency	Mode	Mode Parameter	Callsign	User	Country
3467,00	SSB			RDARA 10B	
3467,00	SSB			RDARA 13D	
3467,00	SSB		ETD3	Addis Abbeba ACC	ETH
3467,00	USB			Bahrain ACC	BHR
3467,00	USB			Benghazi ACC	LBY
3467,00	USB			Bujumbura ACC	BDI
3467,00	USB			Comoros ACC	COM
3467,00	USB			Dar es Salaam ACC	TZA
3467,00	USB			Entebbe ACC	UGA
3467,00	USB			Hargeisa ACC	SOM
3467,00	USB			Djibouti ACC	DJI
3467,00	USB			Jeddah ACC	ARS
3467,00	USB			Kigali ACC	RRW
3467,00	USB			Kisimayu ACC	SOM
3467,00	USB			Male ACC	MLD
3467,00	USB			Nairobi ACC	KEN
3467,00	USB			Port Sudan ACC	SDN
3467,00	USB			Sanaa ACC	YEM
3467,00	USB			Seychelles ACC	SEY
3467,00	USB			Tripoli ACC	LBY
3467,00	USB			Abadan ACC	IRN
3467,00	USB			Almaty ACC	KAZ
3467,00	USB			Bishkek ACC	KGZ
3467,00	USB			Mumbai ACC	IND
3467,00	USB			Kathmandu ACC	NPL
3467,00	USB			Kuwait ACC	KWT
3467,00	USB			Muscat ACC	OMA
3467,00	USB			Odessa ACC	UKR
3467,00	USB			Tehran ACC	IRN
3467,00	USB			Tbilisi ACC	GEO
3467,00	USB			Urumqi ACC	CHN
3467,00	USB			Yerevan ACC	ARM
3467,00	USB			Auckland ACC	AUS
3467,00	USB			Brisbane ACC	AUS
3467,00	USB			Pascua ACC	CHL
3467,00	USB			Port Vila ACC	VUT
3467,00	USB			Rarotonga ACC	CKH
3467,00	USB			San Francisco ACC , CA	USA
3467,00	USB			Wallis ACC	WAL
3467,00	USB			Tahiti ACC	OCE
3470,00	SSB			Calcutta ACC	IND
3470,00	SSB			Colombo ACC	CLN
3470,00	SSB			Kunming ACC	CHN
3470,00	SSB			Madras ACC	IND
3470,00	SSB			Male ACC	MLD
3470,00	SSB			MWARA SEA	SEA
3470,00	SSB			RDARA 10A	
3470,00	SSB			RDARA 13G	
3470,00	SSB			RDARA 1C	
3470,00	SSB			Yangon ACC	BRM
3470,00	SSB			Kuala Lumpur ACC	MLA
3470,00	USB			Bali ACC	IND
3470,00	USB			Bangkok ACC	THA
3470,00	USB			Dhaka ACC	BGD
3470,00	USB			Guangzhou ACC	CHN
3470,00	USB			Jakarta ACC	INS
3470,00	USB			Kathmandu ACC	NPL
3470,00	USB			Singapore ACC	SNG
3470,00	USB			Brisbane ACC	AUS
3470,00	USB			Ujung Pandang ACC	INS
3473,00	SSB			MWARA MID	MID
3473,00	SSB		ETD3	Northsea Helicopter Amsterdam	HOL

Frequency	Mode	Mode Parameter	Callsign	User	Country
3473,00	SSB			RDARA 10E	
3473,00	SSB			RDARA 13C	
3473,00	SSB			RDARA 1B	
3473,00	SSB			RDARA 6C	
3476,00	SSB			Anatananarivo ACC	MDG
3476,00	SSB			Beira ACC	MOZ
3476,00	SSB			Cocos Island	ICO
3476,00	SSB			Mauritius ACC , Plaisance	MAU
3476,00	SSB			Moroni ACC	COM
3476,00	SSB			MWARA INO	INO
3476,00	SSB			MWARA NAT	NAT
3476,00	SSB			Perth ACC	AUS
3476,00	SSB			RDARA 13F	
3476,00	SSB			RDARA 9B	
3476,00	SSB			Seychelles ACC , Mahe	SEY
3476,00	SSB		VFG	Gander ACC	CAN
3476,00	USB		EIP	Shanwik ACC	IRL
3476,00	USB			Mumbai ACC	IND
3476,00	USB			Brisbane ACC	AUS
3476,00	USB			Colombo ACC	CLN
3476,00	USB			Dar es Salaam ACC	TZA
3476,00	USB			Harare ACC	ZWE
3476,00	USB			Jeddah ACC	ARS
3476,00	USB			Kigali ACC	RRW
3476,00	USB			Lilongwe ACC	MWI
3476,00	USB			Lusaka ACC	ZMB
3476,00	USB			Madras ACC	IND
3476,00	USB			Mahajanga ACC	MDG
3476,00	USB			Male ACC	MLD
3476,00	USB			Nairobi ACC	KEN
3476,00	USB			St. Denis ACC	REU
3476,00	USB			Toamasina ACC	MDG
3479,00	SSB			Belem ACC	B
3479,00	SSB			Brasilia ACC	B
3479,00	SSB			Rio de Janeiro ACC	B
3479,00	SSB			Iquitos ACC	PRU
3479,00	SSB			Maiquetia ACC , Caracas	VEN
3479,00	SSB			Montevideo ACC	URG
3479,00	SSB			MWARA EUR	EUR
3479,00	SSB			MWARA SAM	SAM
3479,00	SSB			Piarco ACC	TRD
3479,00	SSB			Porto Velho ACC	B
3479,00	SSB			RDARA 14	
3479,00	SSB			RDARA 6A	
3479,00	SSB			RDARA 6B	
3479,00	USB			Bogota ACC	CLM
3479,00	USB			Leticia ACC	CLM
3479,00	USB			Manaus ACC	B
3479,00	USB			Maiquetia ACC	VEN
3479,00	USB			Cayenne ACC	GUF
3479,00	USB			Georgetown ACC	GUY
3479,00	USB			Paramaribo ACC	SUR
3479,00	USB			Recife ACC	B
3479,00	USB			Asuncion ACC	PRG
3479,00	USB			Buenos Aires ACC	ARG
3479,00	USB			Campo Grande ACC	B
3479,00	USB			La Paz ACC	MEX
3479,00	USB			Montevideo ACC	URG
3479,00	USB			Lima ACC	PRU
3479,00	USB			Porto Alegre ACC	B
3479,00	USB			Salvador ACC	B
3479,00	USB			Santa Cruz ACC	CHL

Frequency	Mode	Mode Parameter	Callsign	User	Country
3479,00	USB			Arkhangelsk ACC	RUS
3479,00	USB			Beirut ACC	LBN
3479,00	USB			Berlin ACC	D
3479,00	USB			Kiev ACC	UKR
3479,00	USB			Lvov ACC	UKR
3479,00	USB			Minsk ACC	BLR
3479,00	USB			Moscow ACC	RUS
3479,00	USB			Murmansk ACC	RUS
3479,00	USB			Odessa ACC	UKR
3479,00	USB			Riga ACC	LVA
3479,00	USB			Simferopol ACC	RUS
3479,00	USB			Sofial ACC	BUL
3479,00	USB			St. Petersburg ACC	RUS
3479,00	USB			Syktyvkar ACC	RUS
3479,00	USB			Tunis ACC	TUN
3479,00	USB			Velikiye ACC	RUS
3479,00	USB			Vilnius ACC	LTU
3479,00	USB			Vologda ACC	RUS
3482,00	SSB			RDARA 5D	
3482,00	SSB			RDARA 6G	
3485,00	SSB			MWARA EA	EA
3485,00	SSB			MWARA SEA	SEA
3485,00	SSB		WSY70	New York VOLMET	USA
3485,00	SSB			RDARA 13H	
3485,00	SSB			RDARA 7E	
3485,00	SSB			VOLMET NAT	NAT
3485,00	SSB		VFG	Gander VOLMET	CAN
3485,00	USB			Guangzhou ACC	CHN
3485,00	USB			Irkutsk ACC	RUS
3485,00	USB			Pyongyang ACC	KRE
3485,00	USB			Ulaanbaatar ACC	MNG
3485,00	USB			Bali ACC	INS
3485,00	USB			Bangkok ACC	THA
3485,00	USB			Hanoi ACC	VTN
3485,00	USB			Ho Chi Minh ACC	VTN
3485,00	USB			Hong Kong ACC	HKG
3485,00	USB			Jakarta ACC	INS
3485,00	USB			Kuala Lumpur ACC	MLA
3485,00	USB			Kota Kinabalu ACC	MLA
3485,00	USB			Manila ACC	PHL
3485,00	USB			Seoul ACC	KOR
3485,00	USB			Singapore ACC	SNG
3485,00	USB			Tokyo ACC	J
3485,00	USB			Vientianne ACC	LAO
3488,00	SSB			Heliport Beccles	G
3488,00	SSB			RDARA 10B	
3488,00	SSB			RDARA 12E	
3488,00	SSB			RDARA 12F	
3488,00	SSB			RDARA 12G	
3488,00	SSB			RDARA 12H	
3488,00	SSB			RDARA 14B	
3488,00	SSB			RDARA 14F	
3488,00	SSB			RDARA 1B	
3488,00	SSB			RDARA 5B	
3488,00	SSB			RDARA 6B	
3490,00	SSB		YKM7	Lattakia R	SYR
3491,00	SSB			MWARA EA	EA
3491,00	SSB			RDARA 10C	
3491,00	SSB			RDARA 13E	
3491,00	SSB			RDARA 1E	
3491,00	SSB			RDARA 4A	
3494,00	SSB			W II	WW

Frequency	Mode	Mode Parameter	Callsign	User	Country
3494,00	USB		SDJ	Stockholm R	S
3494,00	USB			San Francisco ACC , CA	USA
3497,00	SSB			W II	WW
3497,00	HFDL	B=1800Bd	H09**	Utqiagvik , AK	ALS
3503,50	*ALE USB*	*B=125Bd Ch=8 ChS=250Hz*		*G MIL net*	*G*
3518,50	CW		W1AW	ARRL morse training	USA
3522,00	CIS 75-250	B=75Bd S=250Hz		RUS N Moscow	RUS
3527,00	ALE USB	B=125Bd Ch=8 ChS=250Hz		HFLINK network	
3531,00	CW		REA4	RUS AF HQ Moscow	RUS
3532,00	LINK 11 CLEW	B=2250Bd Ch=16 ChS=330/110/550Hz			
3552,00	*STANAG 4285*	*B=2400Bd*		*G MIL Akrotiri*	*CYP*
3553,70	STANAG 4285	B=2400Bd		TUR N	TUR
3554,00	CW			RUS MIL net	RUS
3558,00	ROS USB	B=1Bd Ch=144 ChS=15,625Hz		ROS frequency	
3568,00	CW		V**	RUS N Khiva	RUS
3568,60	WSPR	B=1,46Bd Ch=4 ChS=1,46Hz		WSPR net	
3570,00	CW			RUS N	RUS
3570,00	JT65	B=2,69Bd Ch=65 ChS=2,69Hz		JT65 channel	
3572,00	JT9	B=1,736Bd Ch=9 ChS=1,736Hz		JT9 channel	
3573,00	FT8	B=5,86Bd Ch=8 ChS=5,86Hz		FT8 channel	
3575,00	FT4	B=5,86Bd Ch=8 ChS=5,86Hz		FT4 channel	
3578,00	JS8	B=6,25Bd Ch=8 ChS=6,25Hz		JS8 user	
3579,00	CW		DK0WCY	Aurora Beacon	D
3580,00	ALE USB	B=125Bd Ch=8 ChS=250Hz		JOR net	JOR
3580,00	*STANAG 4285*	*B=2400Bd*		*NOR N Oslo*	*NOR*
3580,00	PSK31	B=31Bd		PSK31 channel	
3582,50	MFSK			MFSK modes user	
3585,00	FAX 120/576		HLL2	Seoul M	KOR
3586,00	*STANAG 4285*	*B=2400Bd*		*TUR MIL Ankara*	*TUR*
3587,50	CW		JJT88	J MIL Tokyo	J
3587,50	CW		JJS21	J MIL Misawa	J
3587,50	CW		JJR20	J MIL Chitose	J
3587,50	CW		JJT77	J MIL	J
3587,50	CW		JJT22	J MIL Kisarazu	J
3587,50	CW		JJT46	J MIL	J
3590,00	BAUDOT	B=45,45Bd S=170Hz		RTTY user	
3593,70	CW		D**	RUS N Sevastopol	UKR
3593,70	CW		L**	RUS N St. Petersburg	RUS
3593,80	CW		P**	RUS N Kaliningrad	RUS
3593,90	CW		S**	RUS N Severomorsk	RUS
3594,00	CW		A**	RUS N Astrakhan	RUS
3594,50	CW		OK0EU	CZE QRP beacon net	CZE
3595,00	VARA	Ch=1		VARA chat freqeuency	
3595,00	USB			CIS aero network night	CIS
3595,00	LONGCHAT	B=40Bd Ch=1		LongChat net	
3596,00	ALE USB	B=125Bd Ch=8 ChS=250Hz		HFLINK network	
3600,00	CW		OK0EN	Kam. Zehrovice	TCH
3600,00	LSB			Wireless Institute Civil Emergency Service WICEN	AUS

Frequency	Mode	Mode Parameter	Callsign	User	Country
3600,50	ALE USB	B=125Bd Ch=8 ChS=250Hz		HFLINK network	
3605,00	USB		UUT	Odessa R	UKR
3606,00	MIL 188-110B SER	B=2400Bd	TP21**	ALG F	ALG
3606,00	ARD9800 MODEM	B=2250Bd Ch=36 ChS=62,5Hz		AOR Digital Voice net	USA
3610,00	HARRIS AVS	Ch=24 ChS=112,5Hz			
3610,00	ROBUST PACKET RADIO	B=200Bd Ch=8 ChS=60Hz		Robust Packet Radio net, HF-APRS	
3613,00	SSB		SVK	Kerkyra R	GRC
3621,00	FAX 120/576		JMH	Tokyo M	J
3624,00	SSB		TMC88	ICRC Paris	F
3624,00	SSB		YKM7	Lattakia R	SYR
3628,00	SSB			Floro R via Orlandet	NOR
3628,80	STANAG 4538	B=2400Bd			
3630,00	SSB		SVR	Rhodos R	GRC
3631,00	SSB			Rogaland R via Bergen	NOR
3640,00	ALE USB	B=125Bd Ch=8 ChS=250Hz		G Mil net	G
3643,00	FREEDV	B=1900Bd Ch=14			
3644,30	STANAG 4538	B=2400Bd			
3645,00	SSB			Floro R	NOR
3648,00	SSB		TAI	Iskenderun R	TUR
3650,50	CW		JJT44	J MIL Irumagawa	J
3650,50	CW		JJT88	J MIL Tokyo	J
3650,50	CW		JJV34	J MIL Miho AB	J
3650,50	CW		JJV56	J MIL Hamamatsu	J
3650,50	CW		JJT66	J MIL Kumagaya	J
3650,50	CW		JJW24	J MIL Nara	J
3652,00	USB			Vardo R via Hammerfest	NOR
3655,00	USB			CIS aero network night	CIS
3656,00	SSB		4XA	Elat R	ISR
3657,00	CW		V**	RUS N Khiva	RUS
3660,00	USB			G N Sea Cadets	G
3670,00	SSB		9AD	Dubrovnik R	HRV
3673,00	SSB		A7D	Doha R	QAT
3673,00	USB		PBK	NLD CG Ijmuiden	NLD
3678,00	USB			G AF Air Cadets	G
3685,00	CW		SXA	GRC N Piraeus	GRC
3691,00	CW		JJT88	J MIL Tokyo	J
3691,00	CW		JJT20	J MIL	J
3691,00	CW		JJT55	J MIL Fuchu	J
3691,00	CW		JJS32	J MIL Matsushima	J
3691,00	CW		JJT33	J MIL Hyakurigahara	J
3691,00	CW		JJV23	J MIL Komaki	J
3691,00	CW		JJV90	J MIL Komatsu	J
3691,00	CW		JJT44	J MIL Irumagawa	J
3692,00	MIL 188-110B SER	B=2400Bd	BT31**	ALG F	ALG
3692,00	MIL 188-110B SER	B=2400Bd	FS30**	ALG F	ALG
3692,50	CW		RJD56	RUS N HQ Murmansk	RUS
3699,50	CW		P**	RUS N Kaliningrad	RUS
3710,00	MIL 188-110A SER	B=2400Bd		S MIL Oland	S
3715,00	USB			G AF Air Cadets	G
3715,30	STANAG 4538	B=2400Bd			
3733,00	ALE USB	B=125Bd Ch=8 ChS=250Hz		POL net	POL
3740,00	USB		LZW	Varna R	BUL
3742,00	SSB		A4M	Muscat R	OMA
3742,30	STANAG 4538	B=2400Bd			
3745,00	USB		A9M	Bahrain R	BHR
3745,00	SSB		A4M	Muscat R	OMA

Frequency	Mode	Mode Parameter	Callsign	User	Country
3745,00	HFPAGER	B=5,86Bd Ch=18		RUS HFPager net	RUS
3746,00	CW		RMP	RUS N HQ Kaliningrad	RUS
3747,00	CW		RAL2	RUS N HQ Astrakhan	RUS
3752,00	USB			G AF Air Cadets	G
3752,00	ALE USB	B=125Bd Ch=8 ChS=250Hz		AUS Police NSW	AUS
3752,00	ALE USB	B=125Bd Ch=8 ChS=250Hz		AUS Police QLD	AUS
3752,00	ALE USB	B=125Bd Ch=8 ChS=250Hz		SAPOL Communication Infrastructure	AUS
3755,00	CW		RTS	RUS N Magadan	RUS
3756,00	CW			RUS MIL Rostov	RUS
3756,00	USB			RUS MIL Rostov-na-Donu	RUS
3760,00	LSB			Ham radio emergency frequency	R1
3760,00	USB			HFoZ/Radtel net	AUS
3764,00	BAUDOT	B=75Bd S=850Hz	PBB	HOL N Den Helder	HOL
3764,20	BAUDOT	B=75Bd S=850Hz	PBH	HOL N Den Helder	HOL
3765,00	SSB		4XA	Elat R	ISR
3770,00	CW			RUS Operational Strategic Command Air Defense net	RUS
3771,00	USB			RUS MIL Rostov-na-Donu	RUS
3772,00	CW		P**	RUS N Kaliningrad	RUS
3773,00	USB		VMS469	Reids Radiodata net	AUS
3775,00	HFPAGER	B=5,86Bd Ch=18		RUS HFPager net	RUS
3782,00	*BAUDOT*	*B=75Bd S=850Hz*	*CTP*	*NATO Lisbon*	*POR*
3790,00	USB			ARM MIL net	ARM
3791,00	ALE USB	B=125Bd Ch=8 ChS=250Hz		HFLINK network	
3792,00	CW		RFH61	RUS N ship PEREKOP	RUS
3793,00	USB		SVL	Limnos R	GRC
3797,00	CW		RCV	RUS N HQ Sevastopol	UKR
3799,00	CW		EYH66		
3800,00	RADAR			MF/HF SA Radar Bear Lake	USA
3801,50	PACTOR	B=100Bd S=200Hz	DEK88	DRK	D
3802,50	USB			E Civil Protection net REMER	E
3803,00	PACTOR III	B=100Bd S=200Hz	DEK27	DRK Hannover	D
3803,00	PACTOR III	B=100Bd S=200Hz	DEK77	DRK Nienburg	D
3803,20	*STANAG 4285*	*B=2400Bd*		*G MIL St. Eval*	*G*
3803,50	*STANAG 4285*	*B=2400Bd*		*G MIL Inskip*	*G*
3805,20	STANAG 4285	B=2400Bd	DHJ58	D N Glücksburg	D
3808,50	CW		RJD69	RUS F	RUS
3810,00	CW		HD2IOA	Guayaquil TS	EQA
3812,00	ISR N HYBRID MODEM	B=2400Bd	4XZ	ISR N Haifa	ISR
3815,00	USB			CIS aero network night	CIS
3815,00	ISR N HYBRID MODEM	B=2400Bd	4XZ	ISR N Haifa	ISR
3818,00	PACTOR III	B=100Bd S=200Hz	DEK24	DRK Münster	D
3818,00	PACTOR III	B=100Bd S=200Hz	DEK2811	DRK Mücke-Merlan	D
3818,20	*STANAG 4285*	*B=2400Bd*	*JXU*	*NOR N TX Bodo*	*NOR*
3820,00	ALE USB	B=125Bd Ch=8 ChS=250Hz		SAPOL Communication Infrastructure	AUS
3820,20	STANAG 4285	B=2400Bd	DHJ58	D N Glücksburg	D
3824,00	ALE USB	B=125Bd Ch=8 ChS=250Hz		AUS Police WA	AUS
3825,00	*CIS 46-500*	*B=46Bd S=500Hz*		*UKR MIL Lvov*	*UKR*
3828,00	3FSK			RUS MIL Rostov-on-Don	RUS
3837,00	CW		P**	RUS N Kaliningrad	RUS
3846,00	*LINK 11 SLEW*	*B=2400Bd*		*Prishtine*	*SRB*
3848,00	ALE USB	B=125Bd Ch=8 ChS=250Hz		AUS Police QLD	AUS
3850,00	USB			G MIL Army Cadets	G
3850,00	USB			RUS MIL net	RUS

Frequency	Mode	Mode Parameter	Callsign	User	Country
3852,00	BAUDOT	B=75Bd S=850Hz	OSN	BEL N Oostende	BEL
3852,50	CW		JJZ37	J MIL Kasuga	J
3852,50	CW		JJZ82	J MIL	J
3852,50	CW		JJX58	J MIL	J
3852,50	CW		JJX47	J MIL	J
3852,50	CW		JJZ71	J MIL Nyutabaru	J
3852,50	CW		JJZ50	J MIL	J
3852,50	CW		JJZ93	J MIL	J
3852,50	CW		JJZ60	J MIL	J
3855,00	*FAX 120/288/576*		*DDH3*	*DWD TX Pinneberg*	*D*
3855,20	*STANAG 4285*	*B=2400Bd*	*JXU*	*NOR N TX Bodo*	*NOR*
3861,00	T600	B=50Bd S=250Hz	RDL	RUS N Moscow	RUS
3862,20	*STANAG 4285*	*B=2400Bd*		*NOR N TX Oslo*	*NOR*
3865,00	USB			G MIL Army Cadets	G
3865,00	*ALE USB*	*B=125Bd Ch=8 ChS=250Hz*		*ISR AF net*	*ISR*
3868,20	*STANAG 4285*	*B=2400Bd*		*NOR N TX Nordmaela*	*NOR*
3868,70	STANAG 4285	B=2400Bd	DRAC	D N Frigate Hessen F 221	D
3870,30	STANAG 4538	B=2400Bd			
3873,00	ALE USB	B=125Bd Ch=8 ChS=250Hz		POL MIL net	POL
3875,00	CW		RJD99	RUS N Severodvinsk	RUS
3875,50	RUS TACTICAL DATALINK	B=1200Bd S=800Hz Ch=1200		RUS MIL	RUS
3877,50	CW		RBGR	RUS N ship	RUS
3881,00	CW		FAV22	F F Mont-Valerien	F
3882,00	CW		RGT77	RUS MIL Moscow, RVSN	RUS
3885,00	SSB		FJY4	St.Paul et Amsterdam R	AMS
3885,00	USB			HFoZ/Radtel net	AUS
3888,00	USB			EGY N net	EGY
3890,50	USB		VKE237	AUS HF Radio Club	AUS
3891,00	CW		RJD99	RUS N Severodvinsk	RUS
3891,20	*STANAG 4285*	*B=2400Bd*	*DHJ58*	*D N Glücksburg TX Marlow*	*D*
3891,20	*STANAG 4285*	*B=2400Bd*	*DHJ58*	*D N TX Glücksburg*	*D*
3900,00	HFDL	B=1800Bd	H03**	Reykjavik	ISL
3901,40	MIL 188-110B SER	B=2400Bd	CM10**	ALG F	ALG
3903,00	ALE USB	B=125Bd Ch=8 ChS=250Hz		F AF net	F
3911,00	RFSM8000	B=2400Bd			
3923,00	LSB			Wyoming Cowboy net	USA
3923,00	LSB			Tar Heel emergency net , NC	USA
3924,00	USB			Plymouth Air Safety	G
3927,00	USB		0A**	IRL N AF Dublin	IRL
3932,00	CW		RFFN	RUS MIL	RUS
3932,00	CW		RFFCH	RUS MIL	RUS
3932,00	CW		RFFR	RUS MIL	RUS
3932,00	CW		RFFP	RUS MIL	F
3932,00	CW		RFFO	RUS MIL	RUS
3947,00	CW			RUS Operational Strategic Command Air Defense net	RUS
3952,00	CW		RMP	RUS N HQ Kaliningrad	RUS
3968,00	CW			CHN MIL net	CHN
3995,00	USB		VMS469	Reids Radiodata net	AUS
3995,00	SSB		VKS737	AUS 4WD net	AUS
3996,00	ALE USB	B=125Bd Ch=8 ChS=250Hz		HFLINK network	
4000,00	SSB			USA Army MARS	
4000,00	SSB			SS Ch 1	
4000,20	STANAG 4285	B=2400Bd	DHJ58	D N Glücksburg TX Neuharlingersiel	D
4003,00	USB			Bahamas weather net	BAH
4003,00	ALE USB	B=125Bd Ch=8 ChS=250Hz		SVK MIL net	SVK
4003,00	MIL 188-110A SER	B=2400Bd		SVK MIL Nitra	SVK

Frequency	Mode	Mode Parameter	Callsign	User	Country
4003,00	MIL 188-110A SER	B=2400Bd		SVK MIL Presov	SVK
4003,00	MIL 188-110A SER	B=2400Bd		SVK MIL Sliac	SVK
4005,00	STANAG 4481	B=50Bd S=850Hz	NAU	USA N San Juan	PTR
4005,00	STANAG 4481	B=50Bd S=850Hz	NSS	USA N TX Davidsonville , MD	USA
4005,00	ALE USB	B=125Bd Ch=8 ChS=250Hz			
4007,00	USB			US Army MARS	USA
4007,00	STANAG 4481	B=50Bd S=850Hz	NSS	USA N TX Davidsonville , MD	USA
4007,00	CW		RCV	RUS N HQ Sevastopol	UKR
4009,00	USB		EAL	Las Palmas R , Haria , Lanzarote	CNR
4009,00	USB			Bilbao R via Machichaco	E
4009,00	USB			Coruna R	E
4009,00	USB			Malaga R via Tarfifa	E
4009,00	USB			Valencia R via Cabo de la Nao	E
4010,00	USB		VKJ	RFDS Meekathara	AUS
4010,00	USB		VNZ	RFDS Pt. Augusta	AUS
4011,00	USB			USA Army MARS	
4011,20	PACTOR I	B=100Bd S=170Hz	DEK27	DRK Hannover	D
4013,00	ALE LSB	B=125Bd Ch=8 ChS=250Hz		D Red Cross net	D
4013,50	SSB			US Army MARS	
4014,00	CW			RUS N calling frequency	
4015,00	PACTOR I	B=100Bd S=170Hz	DEK3110	DRK München mobile	D
4015,00	ALE USB	B=125Bd Ch=8 ChS=250Hz		D Red Cross net	D
4017,00	SSB			US AF MARS	
4018,00	USB		EAL	Las Palmas R , Fuencaliente , Gran Canaria	CNR
4018,00	USB			Coruna R	E
4018,00	USB			Malaga R via Tarfifa	E
4018,00	USB			Valencia R via Cabo de la Nao	E
4018,00	USB			USA Army MARS	
4018,50	CIS 1200	B=1200Bd		RUS	RUS
4020,00	CW		RDL	RUS N Moscow	RUS
4020,00	USB			USA AF MARS	
4020,00	ALE USB	B=125Bd Ch=8 ChS=250Hz		JOR net	JOR
4020,00	UI BURST	Ch=1		RUS	RUS
4020,70	STANAG 4285	B=2400Bd	IDR	I N TX Rome	I
4022,00	USB			USA AF MARS	
4022,20	STANAG 4539	B=2400Bd		Croughton	G
4024,00	SSB			USA Army MARS	
4025,00	USB			USA AF MARS	
4027,00	USB			USA Army MARS	USA
4028,20	STANAG 4539	B=2400Bd		Croughton	G
4030,00	USB		VKL	RFDS Port Hedland	AUS
4030,00	USB			USA Army MARS	
4030,00	ALE USB	B=125Bd Ch=8 ChS=250Hz		TUN MIL net	TUN
4031,00	ALE USB	B=125Bd Ch=8 ChS=250Hz		SUI MIL net	SUI
4031,00	CW		P**	RUS N Kaliningrad	RUS
4031,00	CW		V**	RUS N Khiva	RUS
4033,00	STANAG 4285	B=2400Bd	IGDD	I N ship GORGONA	I
4033,00	USB		IABC	I N supply ship VESUVIO	I
4033,00	USB		IHMC	I N minesweeper CROTONE	I
4033,00	USB		IHMV	I N minesweeper VIESTE	I
4033,00	STANAG 4285	B=2400Bd	IHMV	I N minesweeper VIESTE	I
4033,00	STANAG 4285	B=2400Bd	IABC	I N supply ship VESUVIO	I
4033,00	STANAG 4285	B=2400Bd	IHMC	I N minesweeper CROTONE	I
4033,00	USB			USA Army MARS	USA
4033,00	USB		IHMW	I N ship VIAREGGIO	I

Frequency	Mode	Mode Parameter	Callsign	User	Country
4035,00	ALE USB	B=125Bd Ch=8 ChS=250Hz		TUN MIL net	TUN
4036,00	USB			USA Army/AF MARS	
4036,70	ALE USB	B=125Bd Ch=8 ChS=250Hz		F MIL	F
4037,20	*STANAG 4539*	*B=2400Bd*		*Croughton*	*G*
4038,50	USB			USA Army MARS	
4040,00	USB		VNJ	Casey Base , AUS	ANT
4040,00	USB			Davis Base , AUS	ANT
4040,00	USB			Macquarie Island , AUS	ANT
4040,00	USB			Mawson Base , AUS	ANT
4041,00	USB			USA Army MARS	
4043,00	CW		P**	RUS N Kaliningrad	RUS
4043,00	CIS 1200	B=1200Bd		RUS	RUS
4043,50	SSB			USA Army MARS	
4045,00	CW			RUS Operational Strategic Command Air Defense net	RUS
4045,00	USB		VJT	RFDS Carnarvon	AUS
4045,00	USB			CIS aero network night	CIS
4046,00	CW		YDS	S NG	S
4046,00	CW		EHV	S NG	S
4046,00	ALE USB	B=125Bd Ch=8 ChS=250Hz		S MIL net	S
4047,50	ALE USB	B=125Bd Ch=8 ChS=250Hz		G MIL net	G
4048,00	CW		RJS	RUS N HQ Vladivostok	RUS
4049,00	PACTOR III	B=100Bd S=200Hz	OL1A	CZE MIL Prague	CZE
4049,00	CW		RMP	RUS N HQ Kaliningrad	RUS
4051,00	CW		RAL2	RUS N HQ Astrakhan	RUS
4051,00	CW		P**	RUS N Kaliningrad	RUS
4052,00	ALE USB	B=125Bd Ch=8 ChS=250Hz			
4055,00	CW		RCV	RUS N HQ Sevastopol	UKR
4055,00	ALE USB	B=125Bd Ch=8 ChS=250Hz		AUS Police NSW	AUS
4055,20	*STANAG 4285*	*B=2400Bd*	*DHJ58*	*D N Glücksburg TX Marlow*	*D*
4055,20	*STANAG 4285*	*B=2400Bd*	*DHJ58*	*D N Glücksburg TX Neuharlingersiel*	*D*
4059,50	UDB	Ch=1	DRDF	D N submarine U36	D
4065,00	SSB			SS Ch 401	
4066,10	USB			Falkland R	FLK
4068,00	SSB			SS Ch 402	
4068,20	STANAG 4285	B=2400Bd		G N Akrotiri	CYP
4071,00	SSB			SS Ch 403	
4071,50	CW			RUS air defence net	RUS
4074,00	USB			RUS Operational Strategic Command Air Defense net	RUS
4074,00	SSB			SS Ch 404	
4075,00	PACTOR II	B=200Bd S=200Hz	HPPM1	Sailmail Chiriqui	PNR
4075,00	PACTOR III	B=100Bd S=200Hz	HPPM3	Sailmail Panama	PNR
4077,00	SSB			SS Ch 405	
4078,50	CW		P**	RUS N Kaliningrad	RUS
4079,00	CW		RMP	RUS N HQ Kaliningrad	RUS
4079,00	*CIS 12*	*B=1440Bd Ch=12 ChS=200Hz*		*RUS ship*	
4080,00	SSB			SS Ch 406	
4080,00	*HYBRID MODEM*	*B=2400Bd*		*ISR*	
4081,20	STANAG 4285	B=2400Bd	DHJ58	D N Glücksburg	D
4082,00	USB			RUS Operational Strategic Command Air Defense net	RUS
4083,00	SSB			SS Ch 407	
4085,50	USB			USA Army MARS	
4086,00	ALE USB	B=125Bd Ch=8 ChS=250Hz		POL F net	

Frequency	Mode	Mode Parameter	Callsign	User	Country
4089,00	SSB			SS Ch 409	
4092,00	SSB			SS Ch 410	
4095,00	USB		DRDA	D N submarine U31	D
4095,00	USB		DRDD	D N submarine U34	D
4095,00	SSB			SS Ch 411	
4098,00	SSB			SS Ch 412	
4100,20	*STANAG 4285*	*B=2400Bd*	*PBC*	*NLD N Goeree Island*	*HOL*
4101,00	SSB			SS Ch 413	
4101,00	SITOR B	B=100Bd S=170Hz	SVO	Olympia R	GRC
4102,00	OFDM	B=1560Bd Ch=26 ChS=75Hz			
4102,80	CW		W**		USA
4103,80	CW			RUS HFDF net	RUS
4104,00	SSB			SS Ch 414	
4104,00	OFDM	B=1700Bd Ch=34 ChS=62,5Hz			
4105,00	CW		XSG	Shanghai R	CHN
4107,00	SSB			SS Ch 415	
4109,00	CIS OFDM 60	B=2400Bd Ch=60 ChS=44,5Hz			RUS
4110,00	SSB			SS Ch 416	
4113,00	SSB			SS Ch 417	
4115,60	USB			I F	I
4116,00	SSB			SS Ch 418	
4119,00	SSB			SS Ch 419	
4122,00	SSB			SS Ch 420	
4125,00	USB		ZBR	Bermuda Harbour R	BER
4125,00	USB		NMF	USA CG Boston , MA	USA
4125,00	USB			Cape Town R	AFS
4125,00	USB		VNJ	Casey Base , AUS	ANT
4125,00	USB			Bethel R	ALS
4125,00	USB			AUS Davis Base	ANT
4125,00	USB		HCY	Ayora R , Santa Cruz	EQA
4125,00	USB			King Salmon R	ALS
4125,00	USB		KWL38	Kodiak R	ALS
4125,00	USB			USA CG Kodiak	ALS
4125,00	USB		JNA	J CG HQ Tokyo	J
4125,00	USB		XFL	Mazatlan P	MEX
4125,00	USB		CWC30	La Paloma R	URG
4125,00	USB		CWS	URG N Montevideo	URG
4125,00	USB		ZLM	Taupo Maritime R	NZL
4125,00	USB			Nome R	ALS
4125,00	USB		CWF	Punta Carretas R	URG
4125,00	USB			Macquarie Island , AUS	ANT
4125,00	USB		CWC34	Punta del Este R	URG
4125,00	USB		CWC39	Montevideo Trouville R	URG
4125,00	USB			Mawson Base , AUS	ANT
4125,00	USB			Yakutat R	ALS
4125,00	SSB			calling frequency	WW
4125,00	SSB			SS Ch 421	
4125,00	USB		P2M	Port Moresbay R	PNG
4125,00	USB		PPR	Rio R	B
4125,00	USB		PZN	Paramaribo R	SUR
4125,00	SSB		V5W	Walvis Bay R	NMB
4125,00	USB		VAE	Tofino R	CAN
4125,00	USB		YJM	Port Vila R	VUT
4125,00	USB		ZKN	Niue R	NIU
4125,00	USB		E5R	Rarotonga R	CKH
4125,00	SSB		D3E	Luanda R	AGL
4125,00	USB		VFF	Iqaluit R	CAN
4125,00	USB		A3A	Nukualofa R	TON
4125,00	USB		VRC	Hong Kong R	CHN

Frequency	Mode	Mode Parameter	Callsign	User	Country
4125,00	USB		NMA	USA CG Miami , FL	USA
4125,00	USB			Anette R	ALS
4125,00	USB			Samoa Port Authority	SMA
4125,00	USB			Borrow R	ALS
4125,00	USB			Cold Bay R	ALS
4125,00	USB			USA CG Honolulu	HWA
4125,00	USB		PKG	Bandjarmasin R , Kalimantan	INS
4125,00	USB			USA CG Point Reyes	USA
4125,00	USB		PKY3	Manokwari R , Irian Jaya	INS
4125,00	USB		PKY%	Biak R , Irian Jaya	INS
4125,00	USB		PKY5	Merauke R , Irian Jaya	INS
4125,00	USB		PKM	Bitung R , Sulawesi	INS
4125,00	USB		PKF3	Kendari R	INS
4125,00	USB		NHG	USARP Palmer Station	ANT
4125,00	USB		UGC	St. Petersburg R	RUS
4125,00	USB		PKA	Sabang R , We	INS
4125,00	ALE USB	B=125Bd Ch=8 ChS=250Hz		ARINC Urgent Link net	USA
4128,00	SSB			SS Ch 422	
4131,00	SSB			SS Ch 423	
4131,00	PACTOR III	B=100Bd S=200Hz	PYB45	Sao Paulo	B
4132,00	MIL 188.110A SER	B=2400Bd		SUI MIL	SUI
4132,00	ALE USB	B=125Bd Ch=8 ChS=250Hz		SUI MIL	SUI
4132,50	*STANAG 4285*	*B=2400Bd*	*IDR*	*I N TX Roma*	*I*
4134,00	USB		L3A	ARG N Comodoro Rivadavia	ARG
4134,00	SSB			SS Ch 424	
4134,00	ALE USB	B=125Bd Ch=8 ChS=250Hz		AUS Police NSW	AUS
4134,50	ALE 3G	B=2400Bd		UI MIL net	
4135,00	ALE USB	B=125Bd Ch=8 ChS=250Hz		SUI MIL	SUI
4137,00	ALE USB	B=125Bd Ch=8 ChS=250Hz		SUI MIL net	SUI
4137,00	SSB			SS Ch 425	
4141,70	*STANAG 4481*	*B=75Bd S=850Hz*		*G MIL Inskip*	*G*
4141,70	*STANAG 4481*	*B=75Bd S=850Hz*		*G MIL St. Eval*	*G*
4143,00	ALE USB	B=125Bd Ch=8 ChS=250Hz		SUI MIL net	SUI
4143,00	SSB			SS Ch 427	
4143,00	USB			Berdyansk R	UKR
4143,00	USB		UUO	Kerch R	RUS
4143,60	USB		YJM	Port Vila R	VUT
4145,00	USB			TUR MIL net	TUR
4145,00	STANAG 4285	B=2400Bd		TUR MIL net	TUR
4146,00	USB			Tasmanian Maritime R	AUS
4146,00	USB		SXE	Aspropyrgos Attikis R	GRC
4146,00	USB		VIA	Adelaide R	AUS
4146,00	USB		L2T	ARG N Mar del Plata	ARG
4146,00	USB		CBP23	Ancud R	CHL
4146,00	USB		CBT2	Faro Cabo Carranza	CHL
4146,00	USB		CBP24	Chaiten R	CHL
4146,00	USB		CBA23	Chanaral R	CHL
4146,00	USB		CBP39	Puerto Cisnes R	CHL
4146,00	USB		CBT21	Constitucion R	CHL
4146,00	SSB			simplex frequency ship - ship	
4146,00	USB		CBA	Antofagasta R	CHL
4146,00	USB		CBA2	Arica R	CHL
4146,00	USB		CBA3	Iquique R	CHL
4146,00	USB		CBA4	Coquimbo R	CHL
4146,00	USB		CBA5	Caldera R	CHL
4146,00	USB		CBF	Isla Juan Fernandez R	CHL

Frequency	Mode	Mode Parameter	Callsign	User	Country
4146,00	USB		CBM	Magallanes Zonal R	CHL
4146,00	USB		CBM2	Faro Cabo Raper R	CHL
4146,00	USB		CBM3	Faro Islotes Evangelistas	CHL
4146,00	USB		CBM4	Faro Islote Fairway	CHL
4146,00	USB		CBM5	Punta Delgada R	CHL
4146,00	USB		CBM71	Faro Punta Dungeness	CHL
4146,00	USB		CBM72	Faro Espiritu Santo R	CHL
4146,00	USB		CBN	Wollaston R	CHL
4146,00	USB		CBP	Puerto Montt Zonal R	CHL
4146,00	USB		CBP2	Castro R	CHL
4146,00	USB		CBP3	Puerto Aysen R	CHL
4146,00	USB		CBP4	Faro Isla Guafo	CHL
4146,00	USB		CBS	San Pedro R	CHL
4146,00	USB		CBT	Talcahuano Zonal R	CHL
4146,00	USB		CBV	Playa Ancha Zonal R	CHL
4146,00	USB		CBV22	San Antonio MRSC	CHL
4146,00	USB		CBX	Bahia Felix R	CHL
4146,00	USB		CBY	I. de Pascua R (Easter Iland)	CHL
4146,00	USB		ZLM	Taupo Maritime R	NZL
4146,00	USB		CBM21	Puerto Eden R	CHL
4146,00	USB		CBA24	Huasco R	CHL
4146,00	USB		CBT3	Faro Isla Mocha R	CHL
4146,00	USB		CBT26	Los Vilos R	CHL
4146,00	USB		CBA22	Mejillones R	CHL
4146,00	USB		CBP29	Melinka R	CHL
4146,00	USB		CBP70	Faro Punta Corona R	CHL
4146,00	USB		CBP28	Quellon R	CHL
4146,00	USB		CBA27	Taltal R	CHL
4146,00	USB		CBA21	Tocopilla R	CHL
4146,00	USB		CBM24	Puerto Williams MRSC	CHL
4146,00	USB		WQCN860	Ocean-Pro R Naples , FL	USA
4149,00	USB		VMW	Wiluna M	AUS
4149,00	USB		ZOE	Tristan da Cunha R	TRC
4149,00	USB		E5R	Rarotonga R	CKH
4149,00	SSB			simplex frequency ship - ship	
4149,00	USB		UKW	Kertch R	UKR
4149,00	USB		WPE	Crowly Marine , Jacksonville	USA
4150,70	ALE USB	B=125Bd Ch=8 ChS=250Hz		TUN N net	TUN
4153,00	J PSK	B=1500Bd		J PSK	J
4153,20	USB		PBB	NLD N Den Helder	HOL
4158,00	SITOR A	B=100Bd S=170Hz		J CG Tokyo	J
4158,00	SKYOFDM	B=64Bd Ch=22 ChS=86Hz		Helsinki	FIN
4162,00	PACTOR II	B=200Bd S=200Hz	VZX	Sailmail Darawank	AUS
4162,00	USB			POL MIL net	POL
4163,20	*STANAG 4285*	*B=2400Bd*		*G MIL Crimond*	*G*
4168,00	PACTOR III	B=100Bd S=200Hz	ZKN2SM	Sailmail Niue	NIU
4170,00	FAX 120/576		XSG	Shanghai R	CHN
4171,50	USB		PBC	NLD N Goeree Island	HOL
4171,50	USB		PBB	NLD N Den Helder	HOL
4172,50	SITOR , CW	B=100Bd S=170Hz		SS Ch 1	
4173,00	SITOR , CW	B=100Bd S=170Hz		SS Ch 2	
4173,50	SITOR , CW	B=100Bd S=170Hz		SS Ch 3	
4174,00	SITOR , CW	B=100Bd S=170Hz		SS Ch 4	
4174,50	SITOR , CW	B=100Bd S=170Hz		SS Ch 5	
4175,00	SITOR , CW	B=100Bd S=170Hz		SS Ch 6	
4175,00	SITOR A	B=100Bd S=170Hz	ZLM	Taupo Maritime R	NZL
4175,50	SITOR , CW	B=100Bd S=170Hz		SS Ch 7	
4176,00	SITOR , CW	B=100Bd S=170Hz		SS Ch 8	
4176,50	SITOR , CW	B=100Bd S=170Hz		SS Ch 9	
4177,00	SITOR , CW	B=100Bd S=170Hz		SS Ch 10	

Frequency	Mode	Mode Parameter	Callsign	User	Country
4177,40	SITOR , CW	B=100Bd S=170Hz		Cheju MRCC	KOR
4177,50	SITOR , CW	B=100Bd S=170Hz	HLC	Incheon MRCC	KOR
4177,50	SITOR , CW	B=100Bd S=170Hz		Mokpo MRCC	KOR
4177,50	SITOR , CW	B=100Bd S=170Hz	VRC	Hong Kong R	CHN
4177,50	SITOR , CW	B=100Bd S=170Hz		Busan MRCC	KOR
4177,50	SITOR B	B=100Bd S=170Hz	ZLM	Taupo Maritime R	NZL
4177,50	SITOR , CW	B=100Bd S=170Hz	JNA	J CG HQ Tokyo	J
4177,50	.			Distress/safety frequency	WW
4177,50	SITOR , CW	B=100Bd S=170Hz	HLF	Tonghae MRCC	KOR
4177,50	SITOR , CW	B=100Bd S=170Hz	PKD	Surabaya R , Djawa	INS
4177,50	SITOR , CW	B=100Bd S=170Hz	PNK	Jayapura R , Irian Jaya	INS
4177,50	SITOR , CW	B=100Bd S=170Hz	PKM	Bitung R , Sulawesi	INS
4177,50	SITOR , CW	B=100Bd S=170Hz	PKP	Dumai R , Sumatera	INS
4178,00	SITOR , CW	B=100Bd S=170Hz		SS Ch 12	
4178,50	SITOR , CW	B=100Bd S=170Hz		SS Ch 13	
4179,00	SITOR , CW	B=100Bd S=170Hz		SS Ch 14	
4179,00	CW		REA4	RUS AF HQ Moscow	RUS
4179,50	SITOR , CW	B=100Bd S=170Hz		SS Ch 15	
4180,00	SITOR , CW	B=100Bd S=170Hz		SS Ch 16	
4180,50	SITOR , CW	B=100Bd S=170Hz		SS Ch 17	
4180,50	ALE USB	B=125Bd Ch=8 ChS=250Hz			
4181,00	SITOR , CW	B=100Bd S=170Hz		SS Ch 18	
4181,50	SITOR , CW	B=100Bd S=170Hz		SS Ch 19	
4182,00	CW			SS calling frequency Ch 1	
4182,00	*USB*		*T***	*RUS MIL Naro-Fominsk*	*RUS*
4182,50	CW			SS calling frequency Ch 2	
4183,00	CW			SS calling frequency Ch 3	
4183,50	CW			SS calling frequency Ch 4	
4184,00	CW			SS calling frequency Ch 5	
4184,50	CW			SS calling frequency Ch 6	
4185,00	CW			SS calling frequency Ch 7	
4185,50	CW			SS calling frequency Ch 8	
4186,00	CW			SS calling frequency Ch 9	
4186,50	CW			SS calling frequency Ch 10	
4190,70	ALE USB	B=125Bd Ch=8 ChS=250Hz		TUN N net	TUN
4191,00	USB			POL MIL net	POL
4195,00	CW		RGR35	RUS N	RUS
4197,00	CIS 75-250	B=75Bd S=250Hz		RUS N Moscow	RUS
4199,75	FAX 120/576		XSQ	Guangzhou R	CHN
4200,00	CW			RUS HFDF net	RUS
4200,00	USB			EGY N net	EGY
4202,50	.			SS Ch 1	
4203,00	.			SS Ch 2	
4203,20	*STANAG 4285*	*B=2400Bd*	*PBB*	*NLD N Den Helder*	*HOL*
4203,50	.			SS Ch 3	
4204,00	.			SS Ch 4	
4204,50	.			SS Ch 5	
4205,00	.			SS Ch 6	
4205,50	.			SS Ch 7	
4206,00	.			SS Ch 8	
4206,50	PACTOR III	B=100Bd S=200Hz	FOHXM	Sailmail Manihi Atoll	OCE
4207,00	.			SS Ch 10	
4207,50	DSC	B=100Bd S=170Hz	JNA	J CG HQ Tokyo	J
4207,50	USB		VFF	Iqaluit R	CAN
4207,50	.			DSC distress/safety frequency	WW
4209,50	SITOR B	B=100Bd S=170Hz	SVH	Irakleio R	GRC
4209,50	SITOR B	B=100Bd S=170Hz	SUZ	Serapeum R	ÈGY
4209,50	SITOR B	B=100Bd S=170Hz	XVN	Nha Trang R	VTN
4209,50	SITOR B	B=100Bd S=170Hz	XVG	Hai Phong R	VTN
4209,50	SITOR B	B=100Bd S=170Hz	XSG	Shanghai R	CHN

Frequency	Mode	Mode Parameter	Callsign	User	Country
4209,50	SITOR B	B=100Bd S=170Hz	XSV	Tianjin R	CHN
4209,50	SITOR B	B=100Bd S=170Hz	TAH	Istanbul R	TUR
4209,50	SITOR B	B=100Bd S=170Hz	TAF	Samsun R	TUR
4209,50	SITOR B	B=100Bd S=170Hz	TAL	Antalya R	TUR
4209,50	SITOR B	B=100Bd S=170Hz	NRV	USA CG Guam	GUM
4209,50	SITOR B	B=100Bd S=170Hz	XSQ	Guangzhou R	CHN
4210,00	SITOR B	B=100Bd S=170Hz		MSI frequency	WW
4210,50	.			CS Ch 1	
4210,50	SITOR , CW	B=100Bd S=170Hz	A9M	Bahrain R	BHR
4210,50	SITOR , CW	B=100Bd S=170Hz	UFL	Vladivostok R	RUS
4211,00	.			CS Ch 2	
4211,00	SITOR , CW	B=100Bd S=170Hz	JCS	Choshi R	J
4211,50	.			CS Ch 3	
4212,00	.			CS Ch 4	
4212,00	SITOR , CW	B=100Bd S=170Hz	XSQ	Guangzhou R	CHN
4212,00	SITOR , CW	B=100Bd S=170Hz	UGC	St. Petersburg R	RUS
4212,50	SITOR , CW	B=100Bd S=170Hz		Cheju MRCC	KOR
4212,50	SITOR , CW	B=100Bd S=170Hz	HLC	Incheon MRCC	KOR
4212,50	SITOR , CW	B=100Bd S=170Hz		Mokpo MRCC	KOR
4212,50	SITOR , CW	B=100Bd S=170Hz		Busan MRCC	KOR
4212,50	.			CS Ch 5	
4212,50	SITOR , CW	B=100Bd S=170Hz	WOO	Ocean Gate R , NJ	USA
4212,50	SITOR , CW	B=100Bd S=170Hz	HLF	Tonghae MRCC	KOR
4212,50	SITOR , CW	B=100Bd S=170Hz	XSV	Tianjin R	CHN
4213,00	SITOR , CW	B=100Bd S=170Hz	UFM3	Nevelsk R	RUS
4213,00	.			CS Ch 6	
4213,00	SITOR , CW	B=100Bd S=170Hz	SVO25	Olympia R	GRC
4213,50	.			CS Ch 7	
4214,00	.			CS Ch 8	
4214,50	SITOR , CW	B=100Bd S=170Hz	UDK2	Murmansk R	RUS
4214,50	.			CS Ch 9	
4214,50	SITOR , CW	B=100Bd S=170Hz	SVO26	Olympia R	GRC
4215,00	.			CS Ch 10	
4215,00	SITOR , CW	B=100Bd S=170Hz	XSG	Shanghai R	CHN
4215,50	BAUDOT	B=75Bd S=850Hz	IDR2	I N Rome	I
4215,50	.			CS Ch 12	
4216,00	.			CS Ch 13	
4216,00	SITOR , CW	B=100Bd S=170Hz	TAH	Istanbul R	TUR
4216,00	SITOR , CW	B=100Bd S=170Hz	SVO2	Olympic R	GRC
4216,50	SITOR , CW	B=100Bd S=170Hz	JNA	J CG HQ Tokyo	J
4216,50	.			CS Ch 14	
4216,50	SITOR , CW	B=100Bd S=170Hz	XSA2	Nanjing R	CHN
4217,00	.			CS Ch 15	
4217,00	SITOR , CW	B=100Bd S=170Hz	UFH	Petrapavlovsk-Kamchatskii R	RUS
4217,50	.			CS Ch 16	
4217,70	ALE USB	B=125Bd Ch=8 ChS=250Hz		TUN N net	TUN
4218,00	SITOR , CW	B=100Bd S=170Hz	EQI	Abbas R	IRN
4218,00	.			CS Ch 17	
4218,00	SITOR , CW	B=100Bd S=170Hz	4PB	Colombo R	CLN
4218,00	SITOR , CW	B=100Bd S=170Hz	UDB2	Kholmsk R	
4218,50	.			CS Ch 18	
4219,00	SITOR , CW	B=100Bd S=170Hz	XSQ	Guangzhou R	CHN
4219,00	.			CS Ch 19	
4219,00	SITOR , CW	B=100Bd S=170Hz	4PB	Colombo R	CLN
4219,00	SITOR , CW	B=100Bd S=170Hz	TAH	Istanbul R	TUR
4219,50	DSC	B=100Bd S=170Hz	JNA	J CG HQ Tokyo	J
4219,50	DSC	B=100Bd S=170Hz	PKX	Jakarta R , Jawa	INS
4220,00	ALE USB	B=125Bd Ch=8 ChS=250Hz		G MIL net	G
4220,00	SITOR , CW	B=100Bd S=170Hz	UFL	Vladivostok R	RUS

Frequency	Mode	Mode Parameter	Callsign	User	Country
4220,00	RUS TACTICAL DATALINK	B=1200Bd S=800Hz Ch=1200		RUS MIL	RUS
4220,50	CW		C6N	Nassau R	BAH
4221,00	CW		ODR9	Beyrouth R	LBN
4222,00	CW		RMP	RUS N HQ Kaliningrad	RUS
4222,00	CW		UNU3	RUS N CHMS Kaliningrad	RUS
4222,00	USB			EGY N net	EGY
4224,60	ALE 3G	B=2400Bd		UI MIL net	
4225,20	*STANAG 4285*	*B=2400Bd*	*IDR*	*I N Rome*	*I*
4228,00	MIL 188-110A SER	B=2400Bd		S MIL	S
4228,00	FAX 120/576		CBV	Valpariso R	CHL
4228,00	*STANAG 4285*	*B=2400Bd*	*FUO*	*F N Toulon*	*F*
4228,20	*STANAG 4285*	*B=2400Bd*	*PBB*	*NLD N TX Goeree Island*	*HOL*
4231,00	J PSK	B=1500Bd		J MIL	J
4232,00	STANAG 4285	B=2400Bd	FUF	F N Fort de France	MRT
4233,00	CW		A4M	Muscat R	OMA
4233,00	STANAG 4481	B=50Bd S=850Hz	FUF	F N Fort de France	MRT
4233,00	STANAG 4481	B=50Bd S=850Hz	NPM	USA N Lualualei	HWA
4234,00	ALE USB	B=125Bd Ch=8 ChS=250Hz		ARINC Urgent Link net	USA
4235,00	FAX 120/576		NMF	USCG Boston , MA	USA
4236,00	USB		USI	Kherson R	UKR
4238,00	CW		VTP4	IND N Vishakhapatnam	IND
4238,00	CW		PKD	Surabaya R , Djawa	INS
4238,00	CW		PKG	Bandjarmasin R , Kalimantan	INS
4238,40	STANAG 4285	B=2400Bd	FUE	F N Brest	F
4239,50	ALE USB	B=125Bd Ch=8 ChS=250Hz		G MIL net	G
4239,50	CW		RFK95	RUS N	RUS
4243,00	CHN 4+4	B=75Bd Ch=8 ChS=300/450Hz		CHN N	CHN
4244,00	CW		RJP42	RUS N ship	RUS
4245,00	MHF-50 MODEM	B=54,3Bd Ch=32 ChS=64,5Hz	ZSJ	AFS N Capetown	AFS
4247,00	CW		KPH	San Francisco R , CA (Bolinas/Port Reyes)	USA
4248,50	MIL 188-110A SER	B=2400Bd		AUS MHFCS North West Cape	AUS
4250,00	USB			Mauritius MRCC	MAU
4250,00	HAARP waveform			HAARP	USA
4251,00	MIL 188-110B 39TONE	B=2400Bd Ch=39 ChS=56,25Hz			
4252,00	PACTOR II	B=200Bd S=200Hz	PWX33	B N Brasilia	B
4258,00	T600	B=50Bd S=200Hz		RUS N Murmansk	RUS
4258,50	ALE USB	B=125Bd Ch=8 ChS=250Hz		G MIL net	G
4259,00	CW		XSG	Shanghai R	CHN
4259,20	*STANAG 4285*	*B=2400Bd*	*FUG*	*F N Saissac*	*F*
4261,20	STANAG 4285	B=2400Bd	EBA	E N Madrid	E
4263,20	*STANAG 4285*	*B=2400Bd*		*G F Inskip*	*G*
4268,00	CW		VTG4	IND N MUmbai	IND
4271,00	FAX 120/576		CFH	CAN F Halifax	CAN
4271,00	BAUDOT	B=75Bd S=630Hz	CFH	CAN F Halifax	CAN
4271,00	STANAG 4285	B=2400Bd	FUJ	F N Noumea	NCL
4273,00	CW		RFH82	RUS N ship POVORINO	RUS
4273,00	CW		RJE65	RUS N HQ Novorossiysk	RUS
4275,00	CW		RGN90	RUS N	RUS
4277,20	*STANAG 4285*	*B=2400Bd*	*FPI*	*F N St. Assise*	*F*
4279,70	*STANAG 4285*	*B=2400Bd*	*IDR*	*I N TX Rome*	*I*
4280,00	BAUDOT	B=75Bd S=850Hz	PBB	NLD N Den Helder	HOL
4280,00	CIS 500 FSK BURST	B=500Bd S=1000Hz		RUS N	RUS
4281,00	CW		AQK2	PAK N Karachi	PAK
4282,00	CW		CWM	Montevideo Armada R	URG
4282,00	CW		RGR99	RUS N	RUS

Frequency	Mode	Mode Parameter	Callsign	User	Country
4282,00	CW		RGR90	RUS N	RUS
4282,00	CW		RGR88	RUS N	RUS
4282,00	CW		RGR96	RUS N	RUS
4282,00	CW		RGR92	RUS N	RUS
4282,00	CW		RGR91	RUS N	RUS
4282,00	CW		RGR89	RUS N	RUS
4282,00	CW		RGR95	RUS N	RUS
4282,00	CW		RGR87	RUS N	RUS
4283,00	CW		XSV	Tianjin R	CHN
4283,00	CHN 4+4	B=75Bd Ch=8 ChS=300/450Hz	XSV70	CHN Great Wall Base	ANT
4284,20	STANAG 4285	B=2400Bd	CKN	CAN MIL Aldergrove	CAN
4288,60	*STANAG 4285*	*B=2400Bd*	*FUG*	*F N Saissac*	*F*
4289,00	STANAG 4481	B=50Bd S=850Hz	NPN	USA N Guam	GUM
4290,00	PACTOR IV	B=100Bd S=200Hz	DBAA	SAR vessel ALFRIED KRUPP	D
4291,00	USB			Bilbao R via Madrid	E
4291,00	J PSK	B=1500Bd		J MIL	J
4294,00	J OFDM 30+2	B=1500Bd Ch=32 ChS=25Hz		J N broadcast	J
4294,50	J 16 TONE QPSK	B=1200Bd Ch=16 ChS=100Hz			J
4295,00	CW		PKF	Makassar R , Sulawesi	INS
4295,00	CW		PKC	Palembang R , Sumatera	INS
4295,00	CW		PKE	Amboina R , Ceram	INS
4295,00	*STANAG 4285*	*B=2400Bd*	*FUE*	*F N Brest*	*F*
4295,70	PACTOR II FEC	B=100Bd S=200Hz			
4298,00	FAX 120/576		NOJ	USA CG Kodiak , ALS	ALS
4299,20	*STANAG 4285*	*B=2400Bd*		*NOR N TX Stavanger*	*NOR*
4300,00	SSTV			SURA TX	RUS
4305,00	SITOR , CW	B=100Bd S=170Hz	XSZ	Dalian R	CHN
4306,00	ALE USB	B=125Bd Ch=8 ChS=250Hz		USA FEMA net	USA
4306,00	*STANAG 4285*	*B=2400Bd*	*CTA*	*POR N Oeiras*	*POR*
4310,00	USB			RUS MIL Smolensk area	RUS
4310,20	*STANAG 4285*	*B=2400Bd*	*PBB*	*NLD N TX Goeree Island*	*HOL*
4311,20	STANAG 4481	B=75Bd S=850Hz	NAU	USA N San Juan	PTR
4314,20	*STANAG 4285*	*B=2400Bd*	*PBC*	*NLD N Goeree Island*	*HOL*
4314,20	*STANAG 4285*	*B=2400Bd*		*G MIL TX Inskip*	*G*
4316,00	FAX 120/576		JJC	Tokyo M	J
4316,00	USB		NMG	USA CG New Orleans , LA	USA
4317,90	FAX 120/576		NMG	USA CG New Orleans	USA
4321,00	MIL 188-110A SER	B=2400Bd		S MIL	S
4321,00	USB			CHN air defense	CHN
4322,00	FAX 120/576		CBM	Magallanes Zonal R	CHL
4322,50	CW		RFK95	RUS N	RUS
4325,00	*STANAG 4285*	*B=2400Bd*	*FUG*	*F N Saissac*	*F*
4325,00	CW		R**	RUS N Ustinov	RUS
4325,00	CW		T**	RUS MIL Naro-Fominsk	RUS
4328,40	STANAG 4285	B=2400Bd	FUE	F N Brest	F
4331,00	CW		4XZ	ISR N Haifa	ISR
4338,00	STANAG 4285	B=2400Bd	FUG	F N Saissac	F
4338,00	*F N FSK*	*B=50Bd S=850Hz Ch=1*	*FUG*	*F N Saissac*	*F*
4338,00	USB			BGD N net	BGD
4340,00	CW			RUS N calling frequency	RUS
4343,00	RUS TACTICAL DATALINK	B=1200Bd S=800Hz Ch=1200		RUS MIL	RUS
4345,00	ALE USB	B=125Bd Ch=8 ChS=250Hz		POL net	POL
4345,70	*STANAG 4285*	*B=2400Bd*	*CFH*	*CAN N Halifax*	*CAN*
4346,00	FAX 120/576		NMC	USA CG Pt. Reyes , CA	USA
4347,50	*ALE USB*	*B=125Bd Ch=8 ChS=250Hz*		*G MIL net*	*G*

Frequency	Mode	Mode Parameter	Callsign	User	Country
4348,00	ALE USB	B=125Bd Ch=8 ChS=250Hz		PAK N net	
4349,00	CW		JCS	Choshi R	J
4349,00	USB		CBT4	Valdivia MRSC	CHL
4349,50	CHN 4+4	B=75Bd Ch=8 ChS=300/450Hz		CHN MIL	
4350,00	FMCW			Angel Island, CA	USA
4350,00	FMCW			Cabo Rojo Lighthouse at Cabo Rojo, Puerto Rico	USA
4350,00	FMCW			Punta Tuna Lighthouse, Maunabo, Puerto Rico	USA
4351,00	USB		SXE	Aspropyrgos Attikis R	GRC
4351,00	USB			Peiraias CG R	GRC
4351,00	USB			Tiksi MRSC	RUS
4351,00	CW		XSZ	Dalian R	CHN
4351,00	SSB			CS Ch 428	
4354,00	USB		L2T	ARG N Mar del Plata	ARG
4354,00	USB		UUT	Odessa R	UKR
4354,00	USB		JNA	J CG HQ Tokyo	J
4354,00	USB		L2A	ARG N Mar del Plata	ARG
4354,00	CW		ZRQ	AFS N Capetown	AFS
4354,00	SSB			CS Ch 429	
4354,00	*CIS 12*	*B=1440Bd Ch=12 ChS=200Hz*		*RUS MIL Tartus*	*SYR*
4357,00	SSB			Bodo R via Svalbard	NOR
4357,00	USB		STP	Port Sudan R	SDN
4357,00	USB		SDJ	Stockholm R	S
4357,00	USB		CBV	Valpariso R	CHL
4357,00	CW		REA4	RUS AF Moscow	RUS
4357,00	USB		CWF	Punta Carretas R	URG
4357,00	DSC	B=100Bd S=170Hz	HLS	Seoul R	KOR
4357,00	SSB			CS Ch 401	
4357,00	SSB		V5W	Walvis Bay R	NMB
4357,00	SSB		ZPA	Asuncion R	PRG
4357,00	USB		PKN	Balikpapan R , Kalimantan	INS
4357,00	USB		PKP	Dumai R , Sumatera	INS
4357,00	USB		UFL	Vladivostok R	RUS
4357,40	USB		PNK	Jayapura R , Irian Jaya	INS
4360,00	USB		EQJ	Chahbahar R	IRN
4360,00	USB		LSD836	Argentina R	ARG
4360,00	SSB			CS Ch 402	
4361,00	STANAG 4529	B=1200Bd		G	
4363,00	USB			Cape Town R	AFS
4363,00	CW		USI	Kherson R	UKR
4363,00	USB		VFF	Iqaluit R	CAN
4363,00	SSB			CS Ch 403	
4363,00	USB		3AC	Monaco R	MCO
4363,00	SSB		4PB	Colombo R	CLN
4363,00	USB		VFA	Inuvik R	CAN
4363,60	SSB		4PB	Colombo R	CLN
4363,60	SSB		ETC	Assab R	ETH
4364,00	CHN 4+4	B=75Bd Ch=8 ChS=300/450Hz	XSV85	CHN N	CHN
4365,00	STANAG 4285	B=2400Bd	TBB	TUR N Incirlik	TUR
4366,00	USB		EQI	Abbas R	IRN
4366,00	USB		CBV	Valpariso R	CHL
4366,00	SSB		EQM	Bushehr R	IRN
4366,00	SSB			CS Ch 404	
4366,00	SSB		A4M	Muscat R	OMA
4366,00	USB		PPR	Rio R	B
4369,00	USB		EQO	Now Shar R	IRN
4369,00	USB			Cape Town R	AFS

Frequency	Mode	Mode Parameter	Callsign	User	Country
4369,00	USB		XSL	Fuzhou R	CHN
4369,00	USB		CBV	Valpariso R	CHL
4369,00	SSB			CS Ch 405	
4369,00	SSB		EQM	Bushehr R	IRN
4369,00	USB		XSG	Shanghai R	CHN
4369,00	USB		PPR	Rio R	B
4372,00	USB		LSD836	Argentina R	ARG
4372,00	SSB			CS Ch 406	
4372,00	USB		3DP	Suva R	FJI
4372,00	SSB		4PB	Colombo R	CLN
4372,00	USB		5BA	Cyprus R	CYP
4372,00	USB			BLR MIL net	BLR
4373,00	MIL 188-110A SER	B=2400Bd		S MIL	S
4375,00	USB		EQJ	Chahbahar R	IRN
4375,00	USB			Cape Town R	AFS
4375,00	SSB			CS Ch 407	
4375,00	USB		VOK	Labrador R	CAN
4375,00	ALE USB	B=125Bd Ch=8 ChS=250Hz		AUS Police NSW	AUS
4376,50	CW		RAG43	RUS N ship	RUS
4376,50	CW		RMLZ	RUS N ship M. RUDNITSKIY	RUS
4376,50	CW		RJI92	RUS N ship INGURI	RUS
4378,00	USB		EQL	Anzali R	IRN
4378,00	SSB			CS Ch 408	
4378,00	USB		PKE	Amboina R , Ceram	INS
4379,00	CW		REA4	RUS AF Moscow	RUS
4379,10	USB		XFL	Mazatlan P	MEX
4379,50	USB		PKD	Surabaya R , Djawa	INS
4380,00	USB			RUS Air Defense net Kaliningrad	RUS
4381,00	USB		OYR	Aasiat R	GRL
4381,00	USB		P2M	Port Moresbay R	PNG
4381,00	USB		XSU	Yantai R	CHN
4381,00	SSB			CS Ch 409	
4381,00	USB		PPR	Rio R	B
4382,00	*STANAG 4285*	*B=2400Bd*		*TUR N TX Izmir*	*TUR*
4384,00	USB		EQC	Amirabad R	IRN
4384,00	USB		ZBR	Bermuda Harbour R	BER
4384,00	USB		XST	Qingdao R	CHN
4384,00	USB		LSD836	Argentina R	ARG
4384,00	SSB			CS Ch 410	
4384,00	SSB		4PB	Colombo R	CLN
4384,00	SSB		S7Q	Seychelles R	SEY
4384,00	USB		XSQ	Guangzhou R	CHN
4385,30	USB		YJM	Port Vila R	VUT
4387,00	USB		EQC	Amirabad R	IRN
4387,00	USB			San Lorenzo R	EQA
4387,00	USB		HCY	Ayora R , Santa Cruz	EQA
4387,00	USB		LSD836	Argentina R	ARG
4387,00	USB			Cristobal R	EQA
4387,00	USB			Floreana R , Santa Maria	EQA
4387,00	SSB			CS Ch 411	
4387,00	SSB		OSU21	Oostende R	BEL
4387,00	USB		PKA	Sabang R , We	INS
4387,00	USB		PKG	Bandjarmasin R , Kalimantan	INS
4387,00	SSB		TAH	Istanbul R	TUR
4387,00	USB		HCG	Guayaquil R	EQA
4387,00	USB			Seymour R , Baltra	EQA
4387,00	USB			Villamil R , Isabela	EQA
4390,00	USB		UFH	Petrapavlovsk-Kamchatskii R	RUS
4390,00	SSB			CS Ch 412	
4390,00	PACTOR III	B=100Bd S=200Hz	DBAG	SAR vessel VORMANN JANTZEN	D
4390,00	PACTOR III	B=100Bd S=200Hz	DBAK	SAR Vessel HARRO KOEBKE	D

Frequency	Mode	Mode Parameter	Callsign	User	Country
4390,00	PACTOR III	B=100Bd S=200Hz	DBAT	SAR vessel HANS HACKMACK	D
4390,00	PACTOR III	B=100Bd S=200Hz	DCA28	MRCC Bremen Rescue	D
4390,00	PACTOR III	B=100Bd S=200Hz	DBAD	SAR vessel ARKONA	D
4391,00	CW			RUS Operational Strategic Command Air Defense net	RUS
4393,00	USB		A9M	Bahrain R	BHR
4393,00	USB			Lisbon R	POR
4393,00	USB		HCY	Ayora R , Santa Cruz	EQA
4393,00	USB		LSD836	Argentina R	ARG
4393,00	SSB			CS Ch 413	
4393,00	SSB		D4A	S. Vincente de Cabo R	CPV
4393,00	SSB		SVO21	Olympia R	GRC
4393,00	SSB		USO6	Izmail R	UKR
4393,00	PACTOR III	B=100Bd S=200Hz	DBAS	SAR vessel BREMEN	D
4393,00	PACTOR III	B=100Bd S=200Hz	DBAH	SAR vessel ARKONA	D
4393,00	PACTOR III	B=100Bd S=200Hz	DBBR	SAR vessel THEO FISCHER	D
4394,00	CW		RMP	RUS N HQ Kaliningrad	RUS
4395,00	RUS TACTICAL DATALINK	B=1200Bd S=800Hz Ch=1200		RUS MIL	RUS
4396,00	USB			Saint Helena R	SHN
4396,00	SSB			CS Ch 414	
4396,00	USB		PKF	Makassar R , Sulawesi	INS
4396,00	MIL 188-110A SER	B=2400Bd		S MIL	S
4396,00	ALE USB	B=125Bd Ch=8 ChS=250Hz		TUR Disaster and Emergency net	TUR
4397,00	CW		RGZ58	RUS N	RUS
4397,60	BPSK	B=1200Bd			
4397,70	USB		PKF	Makassar R , Sulawesi	INS
4398,60	STANAG 4539	B=2400Bd		UI MIL user	
4399,00	USB		EQI	Abbas R	IRN
4399,00	USB		YQI	Constanta R	ROU
4399,00	USB			Arkhangelsk MRSC	RUS
4399,00	SSB			CS Ch 415	
4399,00	SSB		SVO22	Olympia R	GRC
4399,00	USB		XSV	Tianjin R	CHN
4401,70	*STANAG 4285*	*B=2400Bd*	*EBA*	*E N Madrid*	*E*
4402,00	USB		EQL	Anzali R	IRN
4402,00	USB		3BM	Mauritius R	MAU
4402,00	USB		EQN	Khomeini R	IRN
4402,00	SSB			CS Ch 416	
4402,00	USB		PPR	Rio R	B
4405,00	USB		UFM3	Nevelsk R	RUS
4405,00	USB		LSD836	Argentina R	ARG
4405,00	USB		P2M	Port Moresbay R	PNG
4405,00	SSB			CS Ch 417	
4405,00	SSB		TAH	Istanbul R	TUR
4405,00	USB			CIS aero network night	CIS
4405,00	USB		KLB	Seattle R , WA	USA
4408,00	MIL 188-110A SER	B=2400Bd		S MIL	S
4408,00	USB		LZW	Varna R	BUL
4408,00	USB		J2A	Djibouti R	DJI
4408,00	USB		SUH	Alexandria R	EGY
4408,00	USB		LSD836	Argentina R	ARG
4408,00	SSB			CS Ch 418	
4408,00	USB		PKF3	Kendari R	INS
4408,00	USB		PKY5	Merauke R , Irian Jaya	INS
4408,00	USB		PKY3	Manokwari R , Irian Jaya	INS
4408,00	USB		PKM	Bitung R , Sulawesi	INS
4411,00	USB		XSR	Haikou R	CHN
4411,00	USB		CBV	Valpariso R	CHL
4411,00	SSB			CS Ch 419	
4411,00	SSB		TUA	Abidjan R	CTI

Frequency	Mode	Mode Parameter	Callsign	User	Country
4414,00	USB		EQI	Abbas R	IRN
4414,00	USB		EQO	Now Shar R	IRN
4414,00	USB		SDJ	Stockholm R	S
4414,00	USB		TAH	Istanbul R	TUR
4414,00	USB		9MG	Penang R	MLA
4414,00	SSB			CS Ch 420	
4416,00	FAX 120/576		VCO	CAN CG Sydney	CAN
4416,00	USB		CBV22	San Antonio MRSC	CHL
4417,00	SSB		V5W	Walvis Bay R	NMB
4417,00	USB			Cape Town R	AFS
4417,00	PACTOR III	B=100Bd S=200Hz	WHX	Droop Mountain , WV	USA
4417,00	USB		CBP23	Ancud R	CHL
4417,00	USB		CBT2	Faro Cabo Carranza	CHL
4417,00	USB		CBA23	Chanaral R	CHL
4417,00	USB		9MG	Penang R	MLA
4417,00	USB		CBP39	Puerto Cisnes R	CHL
4417,00	USB		CBT21	Constitucion R	CHL
4417,00	USB		CBT24	Coronel R	CHL
4417,00	USB		CBA24	Huasco R	CHL
4417,00	USB		CBM21	Puerto Eden R	CHL
4417,00	USB		CBT3	Faro Isla Mocha R	CHL
4417,00	USB		CBT26	Los Vilos R	CHL
4417,00	USB		CBA22	Mejillones R	CHL
4417,00	USB		CBP29	Melinka R	CHL
4417,00	USB		CBP70	Faro Punta Corona R	CHL
4417,00	USB		CBP28	Quellon R	CHL
4417,00	USB		CBA27	Taltal R	CHL
4417,00	USB		CBA21	Tocopilla R	CHL
4417,00	USB		CBT4	Valdivia MRSC	CHL
4417,00	USB		CBM24	Puerto Williams MRSC	CHL
4417,00	USB		CBV	Valpariso R	CHL
4417,00	SSB			CS Ch 421	
4417,00	SSB		A4M	Muscat R	OMA
4417,00	USB		CBA	Antofagasta R	CHL
4417,00	USB		CBA2	Arica R	CHL
4417,00	USB		CBA3	Iquique R	CHL
4417,00	USB		CBA4	Coquimbo R	CHL
4417,00	USB		CBA5	Caldera R	CHL
4417,00	USB		CBF	Isla Juan Fernandez R	CHL
4417,00	USB		CBM	Magallanes Zonal R	CHL
4417,00	USB		CBM2	Faro Cabo Raper R	CHL
4417,00	USB		CBM3	Faro Islotes Evangelistas	CHL
4417,00	USB		CBM4	Faro Islote Fairway	CHL
4417,00	USB		CBM5	Punta Delgada R	CHL
4417,00	USB		CBM71	Faro Punta Dungeness	CHL
4417,00	USB		CBM72	Faro Espiritu Santo R	CHL
4417,00	USB		CBN	Wollaston R	CHL
4417,00	USB		CBP	Puerto Montt Zonal R	CHL
4417,00	USB		CBP2	Castro R	CHL
4417,00	USB		CBP3	Puerto Aysen R	CHL
4417,00	USB		CBP4	Faro Isla Guafo	CHL
4417,00	USB		CBS	San Pedro R	CHL
4417,00	USB		CBT	Talcahuano Zonal R	CHL
4417,00	USB		CBX	Bahia Felix R	CHL
4417,00	USB		CBY	I. de Pascua R (Easter Iland)	CHL
4417,00	SSB		CBZ22	CHL Bahis Fildes R	ANT
4417,00	SSB		D3E	Luanda R	AGL
4419,00	CW			RUS MIL	RUS
4419,40	USB		XFL	Mazatlan P	MEX
4419,40	USB		PKI2	Jakarta R	INS
4419,40	USB		PKC5	Pangkal Balam R	INS
4419,70	STANAG 4285	B=2400Bd		G MIL TX Inskip	G

Frequency	Mode	Mode Parameter	Callsign	User	Country
4419,70	*STANAG 4285*	*B=2400Bd*		*G MIL St. Eval*	*G*
4420,00	USB		ESA	Tallin R	EST
4420,00	USB		CNP	Casablanca R	MRC
4420,00	SSB			CS Ch 422	
4421,00	HDR MODEM	B=2400Bd			
4422,70	STANAG 4285	B=2400Bd		G MIL Inskip	G
4423,00	SSB			CS Ch 423	
4423,00	USB		PKC2	Plaju R , Sumatera	INS
4423,20	*STANAG 4285*	*B=2400Bd*	*DHJ58*	*D N TX Neuharlingersiel*	*D*
4423,60	PACTOR III	B=100Bd S=200Hz	WNU	Austin , TX	USA
4425,00	CW		RIT	RUS N HQ Severomorsk	RUS
4425,60	USB		PKC5	Pangkal Balam R	INS
4425,60	USB		PKI2	Jakarta R	INS
4426,00	USB		VMC	Charleville M	AUS
4426,00	USB		SDJ	Stockholm R	S
4426,00	USB		HSA	Bangkok R	THA
4426,00	USB		L2A	ARG N Buenos Aires	ARG
4426,00	SSB			CS Ch 424	
4426,00	SSB		SVO23	Olympia R	GRC
4426,00	USB		XSA2	Nanjing R	CHN
4426,00	USB		NMF	USA CG Boston , MA	USA
4426,00	USB		NMN	USA CG Portsmouth , VA	USA
4426,00	USB		NMA	USA CG Miami , FL	USA
4426,00	USB			USA CG Honolulu	HWA
4426,00	USB		NMC	USA CG Pt. Reyes , CA	USA
4426,00	ALE USB	B=125Bd Ch=8 ChS=250Hz		AUS Police NSW	AUS
4429,00	USB		CBV	Valpariso R	CHL
4429,00	SSB			CS Ch 425	
4429,00	SSB		SVO24	Olympia R	GRC
4429,00	PACTOR III	B=100Bd S=200Hz	PYB45	Sao Paulo	B
4432,00	USB			Madeira R	MDR
4432,00	SSB			CS Ch 426	
4432,00	SSB		D4A	S. Vincente de Cabo R	CPV
4433,00	CW		RIT	RUS N HQ Severomorsk	RUS
4433,70	STANAG 4481	B=75Bd S=850Hz	RBDECR**	G N CINCFLEET CTF	G
4434,60	STANAG 4539	B=2400Bd		UI MIL user	
4434,90	SSB		ODR5	Beyrouth R	LBN
4435,00	USB			Sao Miguel R	AZR
4435,00	USB		CBV	Valpariso R	CHL
4435,00	SSB			CS Ch 427	
4435,00	MIL 188-110B SER	B=2400Bd		TUN	
4435,20	ALE 3G	B=2400Bd		UI MIL net	
4438,10	PACTOR III	B=100Bd S=200Hz	KDS	Ontario	CAN
4438,50	USB			USA Army MARS	
4439,50	PACTOR III	B=200Bd S=200Hz	WGM	Hollywood , FL	USA
4440,00	*T600*	*B=50Bd S=200Hz*		*RUS N Moscow*	*RUS*
4442,00	ALE USB	B=125Bd Ch=8 ChS=250Hz		Public health net , TX	USA
4443,00	CW		WH2XWF	Traveling Ionospheric Disturbance Detector	USA
4443,00	ALE USB	B=125Bd Ch=8 ChS=250Hz		S MIL net	S
4443,50	PACTOR II FEC	B=100Bd S=200Hz			
4445,00	USB			USA Army MARS	USA
4446,00	LINK 11 CLEW	B=2250Bd Ch=16 ChS=330/110/550Hz	DHJ58	D N Glücksburg	D
4447,00	USB		XVD	Hue R	VTN
4447,00	CW			RUS MIL net	RUS
4448,50	USB			USA AF MARS	J
4449,00	CW			RUS MIL	RUS
4450,00	FMCW			Pt. Arguello, CA	USA

Frequency	Mode	Mode Parameter	Callsign	User	Country
4450,00	FMCW			Big Creek, CA	USA
4450,00	FMCW			Aransas National Wildlife Refuge	USA
4450,00	FMCW			Matagorda Bay Nature Park	USA
4450,00	FMCW			Marthas Vineyard, MA	USA
4450,00	FMCW			Nantucket Island, MA	USA
4450,00	FMCW			Nauset, MA	USA
4450,00	FMCW			Padre Island, TX	USA
4450,00	FMCW			Rollover Pass, TX	USA
4450,00	FMCW			Surfside, TX	USA
4450,50	FMCW			Bonilla Island, BC, Canada	USA
4451,00	CW		RGT77	RUS MIL Moscow, RVSN	RUS
4454,00	ALE USB	B=125Bd Ch=8 ChS=250Hz		POL MIL net	POL
4454,00	MIL 188-110B SER	B=2400Bd		POL MIL	POL
4457,00	ALE LSB	B=125Bd Ch=8 ChS=250Hz			
4457,00	STANAG 4197	B=1800Bd Ch=16/39 ChS=112/56Hz			
4460,00	ALE USB	B=125Bd Ch=8 ChS=250Hz		F N net	F
4461,30	PACTOR III	B=100Bd S=200Hz	WHX	Droop Mountain , WV	USA
4463,00	FMCW			Crissy Field, CA	USA
4463,00	FMCW			Diablo Canyon Long Range, CA	USA
4463,00	FMCW			The Exploratorium Museum, CA	USA
4463,00	FMCW			Finisterre	E
4463,00	FMCW			Silleiro	E
4463,00	FMCW			Vilan	E
4463,00	FMCW			Prior	E
4463,20	STANAG 4285	B=2400Bd	DHJ58	D N Glücksburg TX Marlow	D
4464,00	PACTOR III	B=100Bd S=200Hz	KLT	Austin , TX	USA
4464,00	CODAR			Biscarosse	F
4467,00	CW			RUS MIL	RUS
4467,00	USB			POL MIL net	POL
4469,00	LINK 11 CLEW	B=2250Bd Ch=16 ChS=330/110/550Hz	DHJ58	D N TX Glücksburg	D
4470,00	FMCW			Point San Luis, CA	USA
4471,00	CW			RUS MIL net	RUS
4472,00	CW		RBHM	RUS N ship	RUS
4475,50	FMCW			Manhattan Beach, OR	USA
4476,00	CW		P**	RUS N Kaliningrad	RUS
4477,00	ALE USB	B=125Bd Ch=8 ChS=250Hz		USA Civil Air Patrol net	USA
4478,00	CW		REH85	RUS N	F
4480,00	FMCW			Montara Sanitary District, Montara, CA	USA
4480,00	ALE USB	B=125Bd Ch=8 ChS=250Hz		F N net	F
4480,00	FMCW			Amagansett, New York	USA
4480,00	FMCW			Hempstead, NY	USA
4480,00	FMCW			Sandy Hook	USA
4480,00	FMCW			Loveladies	USA
4480,00	FMCW			Moriches, NY	USA
4480,00	FMCW			Wildwood	USA
4480,50	CIS 12	B=1440Bd Ch=12 ChS=200Hz	RCV	RUS N HQ Sevastopol	UKR
4481,00	FAX 120/576		SVO	Olympia R	GRC
4483,00	USB			Sydney Hobart Race	AUS
4487,80	STANAG 4539	B=2400Bd		UI MIL user	
4488,20	ALE 3G	B=2400Bd		UI MIL net	
4488,50	MIL 188-110B SER	B=2400Bd			
4489,00	STANAG 4285	B=2400Bd		G MIL TX Inskip	G
4489,50	CW		RMBN	RUS N ship	RUS

Frequency	Mode	Mode Parameter	Callsign	User	Country
4489,50	CW		RMTM	RUS N ship	RUS
4489,50	CW		UCXA7	RUS N ship	RUS
4489,50	CW		RMXK	RUS N ship	RUS
4490,00	FMCW			Chevron Pipeline Facility, LA	USA
4490,00	ALE USB	B=125Bd Ch=8 ChS=250Hz		USA DA net	USA
4490,00	FMCW			Southwest Pass, LA	USA
4490,00	ALE USB	B=125Bd Ch=8 ChS=250Hz		USA SHARES net	USA
4491,00	SSB			Missile Range Barking Sands	HWA
4494,00	CW			RUS Operational Strategic Command Air Defense net	RUS
4496,00	CW			RUS air defence net	RUS
4498,00	J OFDM 30+2	B=1500Bd Ch=32 ChS=25Hz		J N broadcast	J
4498,60	J 16 TONE QPSK	B=1200Bd Ch=16 ChS=100Hz			J
4500,00	*BAUDOT*	*B=75Bd S=850Hz*	*OSN*	*BEL N Oostende*	*BEL*
4500,00	USB		SVO	Olympia R	GRC
4500,00	USB			EGY N net	EGY
4502,00	CIS OFDM 120	B=2559Bd Ch=120 ChS=25Hz		RUS diplo Moscow	RUS
4505,00	*STANAG 4285*	*B=2400Bd*		*G MIL TX Inskip*	*G*
4506,70	2FSK	B=300Bd S=850Hz		G MIL Crimond	G
4510,00	SSB			NASA rocket booster recovery	USA
4510,00	USB			TKMS test frequency	D
4510,00	ALE USB	B=125Bd Ch=8 ChS=250Hz		G MIL net	G
4510,00	CODAN	B=2400Bd Ch=16 ChS=112,5Hz	36011		
4510,00	CODAN	B=2400Bd Ch=16 ChS=112,5Hz	1986		
4513,00	MIL 188-110A SER	B=2400Bd		S MIL	S
4513,00	FMCW			Ponce Yacht and Fishing Club at Ponce, Puerto Rico	USA
4514,00	ALE USB	B=125Bd Ch=8 ChS=250Hz		PAK N net	PAK
4517,00	USB			USA AF MARS	J
4517,00	*CW*		*RIT*	*RUS N HQ Severomorsk*	*RUS*
4520,00	ALE USB	B=125Bd Ch=8 ChS=250Hz		Queensland Police net	AUS
4521,00	CW		REA4	RUS AF Moscow	RUS
4521,00	USB		RFK99	RUS N	RUS
4522,00	MIL 188-110A SER	B=2400Bd		S MIL	S
4522,00	*STANAG 4285*	*B=2400Bd*		*G MIL TX Inskip*	*G*
4523,00	CW		RFN69	RUS N	RUS
4523,00	CW		RFN33	RUS N	RUS
4523,30	STANAG 4538	B=2400Bd			
4525,00	FMCW			Matxitxako	E
4525,00	FMCW			Higer	E
4529,10	ALE 3G	B=2400Bd		UI MIL net	
4530,00	ALE USB	B=125Bd Ch=8 ChS=250Hz		S FRO net	S
4534,00	FMCW			Henderson Beach State Park, FL	USA
4537,00	FMCW			Core Banks, NC	USA
4537,00	FMCW			Duck, NC	USA
4537,00	FMCW			Little Island Park, VA	USA
4537,50	ALE USB	B=125Bd Ch=8 ChS=250Hz		USA SAC net	USA
4538,00	STANAG 4538	B=2400Bd			
4539,70	STANAG 4481	B=75Bd S=850Hz		G MIL Crimond	G
4539,70	*STANAG 4285*	*B=2400Bd*	*IDR*	*I N TX Rome*	*I*
4540,00	USB		VNJ	Casey Base , AUS	ANT

Frequency	Mode	Mode Parameter	Callsign	User	Country
4540,00	USB			Davis Base , AUS	ANT
4540,00	USB			Macquarie Island , AUS	ANT
4540,00	USB			Mawson Base , AUS	ANT
4540,00	ALE USB	B=125Bd Ch=8 ChS=250Hz		SNG N net	SNG
4540,00	CIS ARQ	B=100Bd S=500Hz	RTW54	RUS PTT	RUS
4540,00	CIS 75-200	B=75Bd S=200Hz		RUS N ship	
4541,00	USB			POL MIL net	POL
4544,00	CW			RUS HFDF net	RUS
4550,00	FMCW			Pillar Point Air Force Station, CA	USA
4550,00	*ALE USB*	*B=125Bd Ch=8 ChS=250Hz*		*JOR net*	*JOR*
4550,00	*CIS 12*	*B=1440Bd Ch=12 ChS=200Hz*		*RUS MIL Orscha*	*BLR*
4551,00	ALE USB	B=125Bd Ch=8 ChS=250Hz		S FRO net	S
4553,60	ALE USB	B=125Bd Ch=8 ChS=250Hz		USA diplo net	
4556,00	CW		RJD99	RUS N Severodvinsk	RUS
4557,00	USB			USA AF MARS	
4557,00	ALE USB	B=125Bd Ch=8 ChS=250Hz		AUS Police NSW	AUS
4557,00	ALE USB	B=125Bd Ch=8 ChS=250Hz		AUS Police QLD	AUS
4557,00	ALE USB	B=125Bd Ch=8 ChS=250Hz		AUS Police WA	AUS
4557,00	ALE USB	B=125Bd Ch=8 ChS=250Hz		SAPOL Communication Infrastructure	AUS
4557,70	CW		D**	RUS N Sevastopol	UKR
4557,80	CW		S**	RUS N Severomorsk	RUS
4557,90	CW		P**	RUS N Kaliningrad	RUS
4558,00	CW		C**	RUS N Moscow	RUS
4558,10	CW		A**	RUS N Astrakhan	RUS
4559,00	CW			RUS HFDF net	RUS
4560,00	CIS 14	B=96Bd S=500Hz			
4560,00	ALE USB	B=125Bd Ch=8 ChS=250Hz		AUS Police NSW	AUS
4560,00	ALE USB	B=125Bd Ch=8 ChS=250Hz		AUS Police NT	AUS
4560,00	ALE USB	B=125Bd Ch=8 ChS=250Hz		AUS Police QLD	AUS
4560,00	ALE USB	B=125Bd Ch=8 ChS=250Hz		AUS Police WA	AUS
4560,00	SITOR , CW	B=100Bd S=170Hz	TAH	Istanbul R	TUR
4563,00	CW		RAL2	RUS N HQ Astrakhan	RUS
4563,20	*STANAG 4285*	*B=2400Bd*		*DNK N TX Frederikshavn*	*DNK*
4563,20	*STANAG 4285*	*B=2400Bd*	*DHJ58*	*D N TX Neuharlingersiel*	*D*
4564,00	CW		RGT77	RUS MIL Moscow, RVSN	RUS
4568,60	ALE 3G	B=2400Bd		UI MIL net	
4570,00	CIS ARQ	B=100Bd S=500Hz	RZT76	RUS PTT	RUS
4573,00	ALE USB	B=125Bd Ch=8 ChS=250Hz		F N net	F
4575,00	FMCW			Ragged Point, CA	USA
4575,00	FMCW			Assateague, MD	USA
4575,00	FMCW			Cape Hatteras, NC	USA
4576,20	*STANAG 4285*	*B=2400Bd*	*IDN*	*I N TX Napoli*	*I*
4580,00	CW		RBHV	RUS N ship NARYAN MAR	RUS
4580,00	CW		RMJZ	RUS N ship PELYM	CLN
4580,00	CW		RMB81	RUS N ship ARAGON	RUS
4580,00	CW		RBC89	RUS N ship GS-260	RUS
4580,00	FMCW			Orange Beach State Park, AL	USA
4580,50	CW		RBE99	RUS N ship GS-192	RUS
4580,50	CW		RAS82	RUS N ship GS-405	RUS

Frequency	Mode	Mode Parameter	Callsign	User	Country
4580,50	CW		RBHV	RUS N ship NARYAN MAR	RUS
4582,00	CW			Kaliningrad	RUS
4583,00	*BAUDOT*	*B=50Bd S=425Hz*	*DDK2*	*DWD TX Pinneberg*	*D*
4584,00	KOR 2FSK	B=1200Bd S=850Hz Ch=1		KOR MIL Busan	KOR
4586,00	CW		RSVN	RUS MIL net	RUS
4586,00	USB			Auckland Tramping Club	NZL
4588,00	ALE LSB	B=125Bd Ch=8 ChS=250Hz		D Red Cross net	D
4589,70	PACTOR III	B=100Bd S=200Hz	DEK27	DRK Hannover	D
4590,00	USB			USA AF MARS	
4590,00	*STANAG 4285*	*B=2400Bd*	*6WW*	*F N Dakar*	*SEN*
4590,00	*CIS 50-500*	*B=50Bd S=500Hz*		*UKR F Lvov*	*UKR*
4590,00	CHN 4+4	B=75Bd Ch=8 ChS=300/450Hz		CHN MIL	
4593,00	ALE USB	B=125Bd Ch=8 ChS=250Hz		ARS AF net	ARS
4593,20	*STANAG 4285*	*B=2400Bd*		*G MIL TX Inskip*	*G*
4593,50	USB			USA AF MARS	
4600,00	ALE USB	B=125Bd Ch=8 ChS=250Hz		G MIL net	G
4603,00	ALE USB	B=125Bd Ch=8 ChS=250Hz		USA FEMA net	USA
4604,00	CW		RIT	RUS N HQ Severomorsk	RUS
4605,00	CW		P**	RUS N Kaliningrad	RUS
4607,00	MIL 188-110C APP D	B=2400Bd		UKR MIL	UKR
4610,00	FAX 120/576		GYA	G N London	G
4614,00	USB			BGD N net	BGD
4620,00	FMCW			Point Sur Long Range, CA	USA
4620,00	USB		MKL	G F AMCC Northwood	G
4620,00	FMCW			Romberg Tiburon Center, CA	USA
4621,00	CW		4XZ	ISR N Haifa	ISR
4621,00	CW		RGT77	RUS MIL Moscow, RVSN	RUS
4625,00	*AM*		*UVB76*	*RUS MIL Naro-Fominsk*	*RUS*
4625,00	ALE USB	B=125Bd Ch=8 ChS=250Hz		ALG net	ALG
4626,00	KOR 2FSK	B=1200Bd S=850Hz Ch=1		KOR MIL Busan	KOR
4627,00	CW			RUS Operational Strategic Command Air Defense net	RUS
4627,00	ALE USB	B=125Bd Ch=8 ChS=250Hz		S MIL net	S
4632,00	CW			RUS air defence net	RUS
4632,00	CHN 4+4	B=75Bd Ch=8 ChS=300/450Hz		CHN MIL	
4643,20	*STANAG 4285*	*B=2400Bd*	*FUG*	*F N Saissac*	*F*
4645,00	USB			Tallinn VOMLET	EST
4645,00	BAUDOT	B=400Bd S=850Hz			
4645,00	CHP200 ALE	B=250Bd ChS=170Hz			
4645,00	*STANAG 4481*	*B=150Bd S=850Hz*	*FUG*	*F N Saissac*	*F*
4647,00	FMCW			Sausalito-Marin City Sanitary District, CA	USA
4647,00	FMCW			San Diego, Scripps Long-range	USA
4649,00	CHN 4+4	B=75Bd Ch=8 ChS=300/450Hz		CHN MIL	CHN
4649,00	CW		REA4	RUS AF HQ Moscow	RUS
4649,49	CW		OKVA	IAP CAS Beacon Vackov	CZE
4649,50	CW		OKDL	IAP CAS Beacon Dlouha Louka	CZE
4649,50	CW		OKPV	IAP CAS Beacon Panska Ves	CZE
4649,50	CW		OKPR	IAP CAS Beacon Pruhonice	CZE
4649,51	CW		OKTR	IAP CAS Beacon Trebon	CZE
4651,00	USB		EIP	Shanwik ACC	IRL

Frequency	Mode	Mode Parameter	Callsign	User	Country
4651,00	SSB			Antofagasta ACC	CHL
4651,00	SSB			Iqaluit ACC	CAN
4651,00	SSB			RDARA 10B	
4651,00	SSB			RDARA 10E	
4651,00	SSB			RDARA 13E	
4651,00	SSB			RDARA 13F	
4651,00	SSB			RDARA 1D	
4651,00	SSB			RDARA 6C	
4651,00	SSB			RDARA 6G	
4651,00	SSB			Resolute Bay ACC	CAN
4651,00	SSB			Santiago ACC	CHL
4651,00	SSB		VFG	Gander ACC	CAN
4651,00	CW		S	RUS AF marker	RUS
4654,00	SSB			W I	WW
4654,00	SSB			W II	WW
4654,00	HFDL	B=1800Bd	H09**	Utqiagvik , AK	ALS
4656,00	USB			Salehard ACC	RUS
4656,00	USB			Nadym ACC	RUS
4656,00	USB			Novy Urengoy ACC	RUS
4656,00	USB			Naryan-Mar ACC	RUS
4657,00	SSB			Anatananarivo ACC	MDG
4657,00	SSB			Beira ACC	MOZ
4657,00	SSB			Moroni ACC	COM
4657,00	SSB			MWARA AFI	AFI
4657,00	SSB			MWARA CEP	CEP
4657,00	SSB			RDARA 13H	
4657,00	SSB			RDARA 2A	
4657,00	SSB			RDARA 2C	
4657,00	SSB			RDARA 3B	
4657,00	SSB			RDARA 6A	
4657,00	SSB			RDARA 6E	
4657,00	USB		LWL	Comodoro VOLMET	ARG
4660,00	SSB			RDARA 10C	
4660,00	SSB			RDARA 13D	
4660,00	SSB			RDARA 13M	
4660,00	SSB			RDARA 2B	
4660,00	SSB			RDARA 2C	
4660,00	SSB			RDARA 9B	
4660,00	HFDL	B=1800Bd	H13**	Santa Cruz	BOL
4663,00	SSB		UGEF	Khabarovsk VOLMET	RUS
4663,00	USB		UNNN	Novosibirsk VOLMET	RUS
4663,00	SSB			RDARA 10F	
4663,00	SSB			RDARA 13E	
4663,00	SSB			RDARA 13F	
4663,00	SSB			RDARA 13K	
4663,00	SSB			RDARA 6G	
4663,00	USB		RDFG	Tashkent VOLMET	UZB
4663,00	SSB			VOLMET NCA	NCA
4666,00	USB			Hong Kong ACC	HKG
4666,00	USB			Manila ACC	PHL
4666,00	USB			Naha ACC	J
4666,00	USB			Port Moresby ACC	PNG
4666,00	SSB			MWARA CWP	CWP
4666,00	SSB			RDARA 10D	
4666,00	SSB			RDARA 10B	
4666,00	SSB			RDARA 10E	
4666,00	SSB			RDARA 1C	
4666,00	USB			San Francisco ACC , CA	USA
4666,00	USB			Seoul ACC	KOR
4666,00	USB			Taipei ACC	TWN
4666,00	USB			Barranquilla ACC	CLM
4666,00	USB			Bogota ACC	CLM

Frequency	Mode	Mode Parameter	Callsign	User	Country
4666,00	USB			Tokyo ACC	J
4666,00	USB			Maiquetia ACC	VEN
4666,00	USB			Lima ACC	PRU
4666,00	USB			Quito ACC	EQA
4666,00	USB			Antofagasta ACC	CHL
4666,00	USB			Asuncion ACC	PRG
4666,00	USB			Buenos Aires ACC	ARG
4666,00	USB			Cordoba ACC	ARG
4666,00	USB			Easter Island ACC	CHL
4666,00	USB			La Paz ACC	BOL
4666,00	USB			Puerto Montt ACC	CHL
4666,00	USB			Punta Arenas ACC	CHL
4666,00	USB			Santiago ACC	CHL
4666,00	USB			Talara ACC	PRU
4666,00	USB			Ushuaia ACC	ARG
4666,00	USB			Aden ACC	YEM
4666,00	USB			Amman ACC	JOR
4666,00	USB			Ankara ACC	TUR
4666,00	USB			Beirut ACC	LBN
4666,00	USB			Cairo ACC	EGY
4666,00	USB			Damascus ACC	SYR
4666,00	USB			Jeddah ACC	ARS
4666,00	USB			Kuwait ACC	KWT
4666,00	USB			Manama ACC	BHR
4666,00	USB			Odessa ACC	UKR
4666,00	USB			Sanaa ACC	YEM
4666,00	USB			Simferopol ACC	UKR
4666,00	USB			Tehran ACC	IRN
4666,00	USB			Yerevan ACC	ARM
4666,00	USB			Aktyubinsk ACC	KAZ
4666,00	USB			Almaty ACC	KAZ
4666,00	USB			Bishkek ACC	KGZ
4666,00	USB			Dushanbe ACC	TJK
4666,00	USB			Kuybyshev ACC	RUS
4666,00	USB			Kzyl-Orda ACC	KAZ
4666,00	USB			Moscow ACC	RUS
4666,00	USB			Samarkhand ACC	UZB
4666,00	USB			Tashkent ACC	UZB
4666,00	USB			Uralsk ACC	KAZ
4666,00	USB			Yerevan ACC	ARM
4669,00	SSB			Aktyubinsk ACC	KAZ
4669,00	SSB			Antofagasta ACC	CHL
4669,00	SSB			Aralsk ACC	KAZ
4669,00	SSB			Ashkabad ACC	TKM
4669,00	SSB			Asuncion ACC	PRG
4669,00	SSB			Baltra ACC	GPG
4669,00	SSB			Guayaquil ACC	EQA
4669,00	SSB			Kyzl-Orda ACC	RUS
4669,00	SSB			Lima ACC	PRU
4669,00	SSB			MWARA MID	MID
4669,00	SSB			MWARA SAM	SAM
4669,00	SSB			Puerto Montt ACC	CHL
4669,00	SSB			Punta Arenas ACC	CHL
4669,00	SSB			Quito ACC	EQA
4669,00	SSB			RDARA 10C	
4669,00	SSB			RDARA 10D	
4669,00	SSB			RDARA 6G	
4669,00	SSB			Santa Cruz ACC	BOL
4669,00	SSB			Santiago ACC	CHL
4669,00	SSB			Tashkent ACC	UZB
4669,00	SSB			Uralsk ACC	KAZ
4671,00	CW			RUS AF net	RUS

Frequency	Mode	Mode Parameter	Callsign	User	Country
4672,00	SSB			RDARA 11B	
4672,00	SSB			RDARA 13K	
4672,00	SSB			RDARA 2A	
4672,00	SSB			RDARA 2B	
4672,00	SSB			RDARA 3A	
4672,00	SSB			RDARA 4A	
4672,00	SSB			RDARA 6G	
4672,50	ALE USB	B=125Bd Ch=8 ChS=250Hz		IRQ MIL net	IRQ
4675,00	SSB			Bodoe ACC	NOR
4675,00	SSB			Iceland ACC , Reykjavik	ISL
4675,00	SSB			MWARA NAT	NAT
4675,00	SSB			RDARA 13G	
4675,00	SSB			RDARA 6A	
4675,00	SSB			RDARA 6E	
4675,00	SSB			RDARA 9C	
4675,00	USB		EIP	Shanwik ACC	IRL
4675,00	SSB		VFG	Gander ACC	CAN
4675,00	USB		LVR	Resistenca VOLMET	ARG
4678,00	USB			Barnaul ACC	RUS
4678,00	USB		VNJ	Casey Base , AUS	ANT
4678,00	USB			AUS Davis Base	ANT
4678,00	USB			Macquarie Island , AUS	ANT
4678,00	USB			Mawson Base , AUS	ANT
4678,00	USB			Irkutsk ACC	RUS
4678,00	USB			Kirensk ACC	RUS
4678,00	SSB			MWARA NCA	NCA
4678,00	SSB			RDARA 10D	
4678,00	SSB			RDARA 13I	
4678,00	SSB			RDARA 14A	
4678,00	SSB			RDARA 14G	
4678,00	USB			Kolpashevo ACC	RUS
4678,00	USB			Krasnoyarsk ACC	RUS
4678,00	USB			Novosibirsk ACC	RUS
4678,00	USB			Podkamennaya ACC	RUS
4678,00	USB			Surgut ACC	RUS
4678,00	USB			Yeniseysk ACC	RUS
4681,00	ALE USB	B=125Bd Ch=8 ChS=250Hz		ISR AF net	ISR
4681,00	SSB			RDARA 10B	
4681,00	SSB			RDARA 12E	
4681,00	SSB			RDARA 2B	
4681,00	SSB			RDARA 2C	
4681,00	SSB			RDARA 3B	
4681,00	HFDL	B=1800Bd	H08**	Johannesburg	AFS
4684,00	SSB			RDARA 10E	
4684,00	SSB			RDARA 13J	
4684,00	SSB			RDARA 14B	
4684,00	SSB			RDARA 14C	
4684,00	SSB			RDARA 3A	
4684,00	SSB			RDARA 3C	
4686,00	STANAG 4285	B=2400Bd		G F Inskip	G
4687,00	SSB			W I	WW
4687,00	SSB			W II	WW
4687,00	SSB			W III	WW
4687,00	HFDL	B=1800Bd	H09**	Barrow , AK	ALS
4690,00	SSB			RDARA 10B	
4690,00	SSB			RDARA 13M	
4690,00	SSB			RDARA 2A	
4690,00	SSB			RDARA 2B	
4690,00	SSB			RDARA 3A	
4690,00	SSB			RDARA 6G	

Frequency	Mode	Mode Parameter	Callsign	User	Country
4693,00	SSB			RDARA 10B	
4693,00	SSB			RDARA 12C	
4693,00	SSB			RDARA 13I	
4693,00	SSB			RDARA 14D	
4693,00	SSB			RDARA 2B	
4693,00	SSB			RDARA 2C	
4693,00	SSB			RDARA 3	
4696,00	SSB			RDARA 10	
4696,00	SSB			RDARA 13J	
4696,00	SSB			RDARA 2	
4696,00	SSB			RDARA 6G	
4696,00	SSB			RDARA 9	
4700,00	ALE USB	B=125Bd Ch=8 ChS=250Hz	OPE**	MOD Warszawa	POL
4703,00	USB		CHR	CAN AF Trenton	CAN
4703,50	LINK 11 LSB	B=2250Bd Ch=16 ChS=330/110/550Hz			
4706,00	*ALE USB*	*B=125Bd Ch=8 ChS=250Hz*		*G MIL net*	*G*
4707,00	SSB			G AF Natishead	G
4712,00	SSB		SHAPORA	Rostov ACC	RUS
4712,00	SSB			Sochi ACC	RUS
4712,00	SSB			Tbilisi ACC	GEO
4712,00	SSB			Uralsk ACC	KAZ
4712,00	USB			Saratov ACC	RUS
4712,00	USB			Ufa ACC	RUS
4713,00	STANAG 4481	B=75Bd S=850Hz			
4716,00	ALE USB	B=125Bd Ch=8 ChS=250Hz		I AF net	I
4718,00	USB			Kinloss Rescue	G
4721,00	ALE USB	B=125Bd Ch=8 ChS=250Hz		USA AF net	
4721,00	USB		IDR	I N Rome	I
4724,00	USB			USA AF Sigonella	I
4724,00	USB			USA AF Lajes AFB	AZR
4724,00	USB			USA AF Ascension AFB	ASC
4724,00	USB		AFS	USA AF Offutt AFB , Omaha , NE	USA
4724,00	USB		AIE	USA AF Andersen AFB	GUM
4724,00	SSB			USA AF Yokota AFB	J
4724,00	SSB		AKA	USA AF Elmendorf AFB , Anchorage	ALS
4724,00	ALE USB	B=125Bd Ch=8 ChS=250Hz		USA AF net	USA
4724,00	USB			USA AF Croughton	G
4724,00	USB			USA AF Guam	GUM
4724,00	USB			USA AF Hickam , HWA	USA
4724,00	USB			USA AF Puerto Rico	PTR
4724,00	USB		AFA3	USA AF Andrews AFB , Camp Springs , MD	USA
4726,00	ALE USB	B=125Bd Ch=8 ChS=250Hz		SNG N net	SNG
4726,00	STANAG 4481	B=50Bd S=850Hz	NPG	USA N Dixon , CA	USA
4727,00	USB			F Navy Air Reporting Control Net	F
4732,00	*STANAG 4481*	*B=75Bd S=850Hz*	*MKL*	*G N Inskip*	*G*
4733,00	ALE USB	B=125Bd Ch=8 ChS=250Hz		G MIL net	G
4735,00	CW			RUS air defense net	RUS
4735,00	CW			BUL MIL Bankya	BUL
4735,00	USB		0A**	IRL N AF Dublin	IRL
4739,00	USB		DHJ83	D AF Gatow	D
4739,00	USB		DHO78	D AF Kalkar	D
4739,00	USB		DHM91	D AF Münster	D
4740,00	LSB			EGY air defense net	EGY
4740,00	BAUDOT	B=50Bd S=300Hz		SRB MIL net	SRB

Frequency	Mode	Mode Parameter	Callsign	User	Country
4741,00	STANAG 4481	B=50Bd S=850Hz	NAU	USA N San Juan	PTR
4742,00	USB	Ch=1	GFW	G AF Akrotiri VOLMET	CYP
4742,00	USB			USA AF Diega Garcia	DGA
4742,20	STANAG 4539 HDR	B=2400Bd		UI MIL net	
4745,00	ALE USB	B=125Bd Ch=8 ChS=250Hz		F AF net	F
4745,00	*STANAG 4285*	*B=2400Bd*	*TBB*	*TUR N Ankara*	*TUR*
4747,00	STANAG 4481	B=75Bd S=850Hz	NAU	USA N San Juan	PTR
4750,00	FMCW			Catalina Island, CA	USA
4750,00	FMCW			Block Island	USA
4750,00	FMCW			Singing River, MS	USA
4751,00	ALE USB	B=125Bd Ch=8 ChS=250Hz		E Guadia Civil net	E
4755,00	USB			CIS aero network night	CIS
4755,00	SSB			USA AF MARS	
4757,00	ALE USB	B=125Bd Ch=8 ChS=250Hz		Public health net , TX	USA
4765,00	KOR 2FSK	B=1200Bd S=850Hz Ch=1		KOR MIL Busan	KOR
4766,00	CW		RMP	RUS N HQ Kaliningrad	RUS
4766,00	CW		RKA80	RUS N ship VASSILY TATISCHEV	RUS
4770,00	LINK 11 CLEW	B=2250Bd Ch=16 ChS=330/110/550Hz			
4770,00	USB			McMurdo Ops	
4770,00	USB			RUS MIL Smolensk	RUS
4775,00	SKYOFDM	B=64Bd Ch=28 ChS=86Hz			
4780,00	ALE USB	B=125Bd Ch=8 ChS=250Hz		USA FEMA net	USA
4780,00	ALE USB	B=125Bd Ch=8 ChS=250Hz		AUS Police QLD	AUS
4782,20	*STANAG 4539*	*B=2400Bd*		*Croughton*	*G*
4785,00	FMCW			Bodega Marine Laboratory Long Range, CA	USA
4785,00	FMCW			Cape Blanco, OR	USA
4785,00	FMCW			Loomis Lake, OR	USA
4785,00	FMCW			Point Arena Field Station, CA	USA
4785,00	FMCW			Point St. George, CA	USA
4785,00	FMCW			Torrance Beach, CA	USA
4785,00	FMCW			Camp Pendleton, CA	USA
4785,00	FMCW			San Clemente Is, CA	USA
4785,00	FMCW			Shelter Cove, CA	USA
4785,00	FMCW			Trinidad, CA	USA
4785,00	FMCW			Winchester Bay, OR	USA
4785,00	FMCW			Yaquina Head Long, OR	USA
4788,20	*STANAG 4539*	*B=2400Bd*		*Croughton*	*G*
4788,30	ALE USB	B=125Bd Ch=8 ChS=250Hz		G MIL net	G
4789,00	MIL 188-110B 39TONE	B=2400Bd Ch=39 ChS=56,25Hz		CHN F	CHN
4789,50	USB			USA Army MARS	
4791,00	CW			RUS Air Defense net	RUS
4792,00	CW		WROZ722	CODAR Oregon State University	USA
4797,20	*STANAG 4539*	*B=2400Bd*		*Croughton*	*G*
4800,00	FMCW			Pt. Arguello, CA	USA
4800,00	FMCW			San Mateo Point, CA	USA
4800,00	FMCW			Wainwright, AK	USA
4802,50	USB			POL MIL net	POL
4804,40	ALE 3G	B=2400Bd		UI MIL net	
4804,40	STANAG 4539	B=2400Bd		UI MIL user	
4805,00	PACTOR II	B=200Bd S=200Hz	XJN714	Sailmail Lunenburg , NS	CAN
4807,00	CW		RIT	RUS N HQ Severomorsk	RUS

Frequency	Mode	Mode Parameter	Callsign	User	Country
4810,00	CW		RMAE	RUS N	RUS
4810,10	STANAG 4285	B=2400Bd	FUF	F N Fort de France	MRT
4812,50	CW		RDL	RUS N Moscow	RUS
4816,60	STANAG 4539	B=2400Bd		UI MIL user	
4817,90	STANAG 4539	B=2400Bd		UI MIL user	
4818,00	CW		RDL	RUS N Moscow	RUS
4818,00	LSB			CHN air defense	CHN
4820,00	FMCW			Seaside, OR	USA
4820,00	FMCW			Cape St. Mary, NS, CA	USA
4820,80	STANAG 4539	B=2400Bd		UI MIL user	
4820,80	ALE 3G	B=2400Bd		UI MIL net	
4823,00	CIS 75-250	B=75Bd S=250Hz		RUS N Moscow	RUS
4825,50	USB			POL MIL net	POL
4828,00	CW		P**	RUS N Kaliningrad	RUS
4830,00	*CIS 12*	*B=1440Bd Ch=12 ChS=200Hz*		*RUS MIL Moscow*	*RUS*
4835,00	CW		REA4	RUS AF Moscow	RUS
4841,00	CIS OFDM 120	B=2559Bd Ch=120 ChS=25Hz		RUS diplo Moscow	RUS
4844,50	CHN 4+4	B=75Bd Ch=8 ChS=300/450Hz		CHN MIL	
4845,00	ALE USB	B=125Bd Ch=8 ChS=250Hz		G MIL net	G
4848,00	USB			POL MIL net	POL
4851,00	CW		RIR98	RUS N	RUS
4855,00	CW		RIT	RUS N HQ Severomorsk	RUS
4855,00	HFPAGER	B=5,86Bd Ch=18		RUS HFPager net	RUS
4862,00	USB		VMD750	Austravel Safety Net	AUS
4862,00	ALE 3G	B=2400Bd		UI MIL net	
4862,00	STANAG 4539	B=2400Bd		UI MIL user	
4865,50	CW			RUS Air Defense net	RUS
4867,00	BAUDOT	B=200Bd S=720Hz			
4870,00	CIS 12	B=1440Bd Ch=12 ChS=200Hz	RCV	RUS N HQ Sevastopol	UKR
4872,00	USB			USA AF MARS	
4872,00	CW			RUS MIL net	RUS
4875,00	USB			USA AF MARS	USA
4880,00	ALE USB	B=125Bd Ch=8 ChS=250Hz		SVK MIL net	SVK
4880,00	MIL 188-110A SER	B=2400Bd		SVK MIL Trencin (ZASKIS)	SVK
4880,00	MIL 188-110A SER	B=2400Bd		SVK MIL Sliac	SVK
4880,00	CIS 1200	B=1200Bd		RUS	RUS
4882,00	ALE USB	B=125Bd Ch=8 ChS=250Hz		USA FEMA net	USA
4882,50	ALE USB	B=125Bd Ch=8 ChS=250Hz			D
4886,00	CW		RMP	RUS N HQ Kaliningrad	RUS
4888,00	STANAG 4197	B=1800Bd Ch=16/39 ChS=112/56Hz		EGY N net	EGY
4888,00	USB			EGY N net	EGY
4890,00	USB			POL Mil net	POL
4890,00	ALE USB	B=125Bd Ch=8 ChS=250Hz		SVK AF net	SVK
4899,00	CW		P**	RUS N Kaliningrad	RUS
4900,00	FMCW			Marathon, FL	USA
4900,00	FMCW			Naples, FL	USA
4900,00	FMCW			Reddington, FL	USA
4900,00	FMCW			Venice, FL	USA
4904,00	CW			RUS Operational Strategic Command Air Defense net	RUS
4905,00	STANAG 4481	B=50Bd S=850Hz	NPG	USA N Dixon , CA	USA
4905,00	STANAG 4481	B=50Bd S=850Hz	NAU	USA N San Juan	PTR

Frequency	Mode	Mode Parameter	Callsign	User	Country
4908,00	CW			RUS MIL net	RUS
4910,50	ALE USB	B=125Bd Ch=8 ChS=250Hz		GRC AF net	GRC
4912,50	ALE USB	B=125Bd Ch=8 ChS=250Hz		USA SAC net	USA
4920,00	USB			G MIL Army Cadets	G
4922,50	USB			G MIL Army Cadets	G
4925,00	ALE USB	B=125Bd Ch=8 ChS=250Hz		GRC MOI net	GRC
4925,00	USB			G AF Air Cadets	G
4930,00	USB			RUS MIL Smolensk	RUS
4930,00	*STANAG 4285*	*B=2400Bd*	*JWT*	*NOR N TX Stavanger*	*NOR*
4940,50	CW		RAI62	RUS N	RUS
4940,50	CW		RIQ88	RUS N	RUS
4940,50	CW		RAS92	RUS N	RUS
4940,50	CW		RAI63	RUS N	RUS
4947,00	CIS 12	B=1440Bd Ch=12 ChS=200Hz	RCV	RUS N HQ Sevastopol	UKR
4948,00	CW		RGN90	RUS N	RUS
4948,00	CW		RHO62	RUS N	RUS
4948,00	CW		RMP	RUS N HQ Kaliningrad	RUS
4948,00	STANAG 4538	B=2400Bd			
4948,00	CIS 12	B=1440Bd Ch=12 ChS=200Hz	RMP	RUS N HQ Kaliningrad	RUS
4950,00	*LINK 11 ISB*	*B=2250Bd Ch=16 ChS=330/110/550Hz*	*JWT*	*NOR N Stavanger*	*NOR*
4950,00	ALE USB	B=125Bd Ch=8 ChS=250Hz		USA AF net	
4951,00	CW			RUS air defence net	RUS
4951,50	CW			RUS Operational Strategic Command Air Defense net	RUS
4954,00	CIS 1200	B=1200Bd		RUS	RUS
4954,10	*CIS 12*	*B=1440Bd Ch=12 ChS=200Hz*		*RUS MIL Samara*	*RUS*
4955,00	BAUDOT	B=75Bd S=300Hz		SRB F net	SRB
4955,00	USB			G MIL Army Cadets	G
4957,00	CW		RMP	RUS N HQ Kaliningrad	RUS
4960,00	ALE USB	B=125Bd Ch=8 ChS=250Hz		UKR MIL net	UKR
4961,00	CW			RUS Operational Strategic Command Air Defense net	RUS
4963,00	ALE USB	B=125Bd Ch=8 ChS=250Hz		F N net	F
4965,20	*STANAG 4285*	*B=2400Bd*	*IDR*	*I N TX Rome*	*I*
4969,00	SSB			RDARA 10	
4970,00	CIS ARQ	B=100Bd S=500Hz	RFT6	RUS PTT Voronezeh	RUS
4977,00	CW			RUS HFDF net	RUS
4979,00	CW		RMW2	RUS N	RUS
4979,00	CW		RAL2	RUS N HQ Astrakhan	RUS
4979,00	CW		RHQ2	RUS N	RUS
4979,00	CW		RBL75	RUS N	RUS
4979,00	CW		RLO2	RUS N	RUS
4979,70	*STANAG 4285*	*B=2400Bd*	*IDR*	*I N TX Rome*	*I*
4980,00	USB		VJJ	RFDS Charleville	AUS
4983,00	T600	B=50Bd S=200Hz		RUS N Sveromorsk	UKR
4985,00	STANAG 4481	B=50Bd S=850Hz	NAU	USA N San Juan	PTR
4987,00	STANAG 4481	B=50Bd S=850Hz	NSS	USA N TX Davidsonville , MD	USA
4988,00	CW			RUS HFDF net	RUS
4988,00	ISR N HYBRID MODEM	B=2400Bd	4XZ	ISR N Haifa	ISR
4991,00	ALE USB	B=125Bd Ch=8 ChS=250Hz		USA FBI net	USA

Frequency	Mode	Mode Parameter	Callsign	User	Country
4992,00	MCPSK 14CH	Ch=14 ChS=3000Hz		Lvov	UKR
4996,00	*CW*		*RWM*	*Moscow TS*	*RUS*
4998,00	AM		EBC	San Fernando TS	E
5000,00	AM		IAM	Rome TS	I
5000,00	AM		LOL	Buenos Aires TS	ARG
5000,00	AM		HD2IOA	Guayaquil TS	EQA
5000,00	AM		HLA	Taejon TS	KOR
5000,00	AM		YVTO	Caracas TS	VEN
5000,00	CW		WWV	Fort Collins TS , CO	USA
5000,00	CW		WWVH	Kekaha Kauai TS	HWA
5000,00	CW , AM		BPM	Xi`an TS	CHN
5000,00	CW , AM		BSF	Chung-Li TS	TWN
5000,00	STANAG 4285	B=2400Bd	FUV	F N Djibouti	DJI

5000 – 10000 kHz

Frequency	Mode	Mode Parameter	Callsign	User	Country
5000,00	AM		IAM	Rome TS	I
5000,00	AM		LOL	Buenos Aires TS	ARG
5000,00	AM		HD2IOA	Guayaquil TS	EQA
5000,00	AM		HLA	Taejon TS	KOR
5000,00	AM		YVTO	Caracas TS	VEN
5000,00	CW		WWV	Fort Collins TS , CO	USA
5000,00	CW		WWVH	Kekaha Kauai TS	HWA
5000,00	CW , AM		BPM	Xi`an TS	CHN
5000,00	CW , AM		BSF	Chung-Li TS	TWN
5000,00	STANAG 4285	B=2400Bd	FUV	F N Djibouti	DJI
5005,00	PACTOR III	B=100Bd S=200Hz	9Z4DH	Sailmail Chaguaramas	TRD
5006,00	CW		RHQ33	RUS N	RUS
5006,00	ALE USB	B=125Bd Ch=8 ChS=250Hz		USA Civil Air Patrol net	USA
5006,00	CW		JG2XA	Experimental station Tokyo	J
5006,00	T600	B=50Bd S=100Hz	RMP	RUS N HQ Kaliningrad	RUS
5007,00	CW		V**	RUS N Khiva	RUS
5008,00	CIS 1200	B=1200Bd		RUS	RUS
5015,00	CIS ARQ	B=100Bd S=500Hz		RUS PTT	RUS
5015,60	PACTOR III	B=100Bd S=200Hz	KDS	Ontario	CAN
5016,30	ALE USB	B=125Bd Ch=8 ChS=250Hz		F	F
5018,00	CW		RJD23		RUS
5019,00	CW		RJD99	RUS N Severodvinsk	RUS
5019,00	MIL 188-110B 39TONE	B=2400Bd Ch=39 ChS=56,25Hz		CHN MIL	CHN
5020,90	ALE 3G	B=2400Bd		UI MIL net	
5020,90	STANAG 4539 HDR	B=2400Bd		UI MIL net	
5023,00	CW		REA4	RUS AF Moscow	RUS
5023,50	T600	B=50Bd S=200Hz	RIT	RUS N HQ Severomorsk	RUS
5024,00	*LINK 11 CLEW*	*B=2250Bd Ch=16 ChS=330/110/550Hz*		*F MIL La Rochelle*	*F*
5028,00	BAUDOT	B=75Bd S=850Hz			
5029,40	ALE 3G	B=2400Bd		UI MIL net	
5035,00	ALE USB	B=125Bd Ch=8 ChS=250Hz		USA AF net	
5036,00	ALE USB	B=125Bd Ch=8 ChS=250Hz		ALG oilfield net	
5036,00	ALE USB	B=125Bd Ch=8 ChS=250Hz		SVK AF net	SVK

Frequency	Mode	Mode Parameter	Callsign	User	Country
5040,00	MIL 188-110A SER	B=2400Bd		POL MIL	POL
5040,00	ALE USB	B=125Bd Ch=8 ChS=250Hz		POL MIL net	POL
5044,50	CIS OFDM 93	B=2400Bd Ch=93 ChS=31,25Hz		CIS	RUS
5050,00	*STANAG 4285*	*B=2400Bd*	*PBB*	*NLD N Den Helder*	*HOL*
5052,00	2FSK	B=300Bd S=850Hz		area Sardinia	F
5056,00	LINK 11 CLEW	B=2250Bd Ch=16 ChS=330/110/550Hz		GRC MIL	GRC
5058,50	ALE USB	B=125Bd Ch=8 ChS=250Hz		USA FBI net	USA
5058,70	PACTOR III	B=200Bd S=200Hz	WGM	Hollywood , FL	USA
5060,00	ALE USB	B=125Bd Ch=8 ChS=250Hz		EST MIL net	EST
5062,00	CW		LZQ76	BUL N	BUL
5064,00	CW		9MB	MLA N Georgtown , Penang Island	MLA
5064,40	ALE 3G	B=2400Bd		UI MIL net	
5065,50	ALE LSB	B=125Bd Ch=8 ChS=250Hz		GRC AF net	GRC
5065,50	ALE USB	B=125Bd Ch=8 ChS=250Hz		GRC AF net	GRC
5065,50	HARRIS AVS	Ch=24 ChS=112,5Hz		GRC AF	GRC
5065,60	ALE 3G	B=2400Bd		UI MIL net	
5066,00	CW		RMPV	RUS N ship ALEKSANDR OTRAKOVSKIY	RUS
5066,00	CW		RHI99	RUS N ship ALEXANDER SHABALIN	RUS
5066,70	SITOR A	B=100Bd S=170Hz		EGY E Berne	SUI
5070,50	*STANAG 4285*	*B=2400Bd*		*DNK MIL TX Frederikshavn*	*DNK*
5071,00	LINK 11 CLEW	B=2250Bd Ch=16 ChS=330/110/550Hz			I
5075,00	ALE USB	B=125Bd Ch=8 ChS=250Hz		MTN net	MTN
5075,50	ALE USB	B=125Bd Ch=8 ChS=250Hz		GRC AF net	GRC
5076,00	BPSK	B=1000Bd ChS= Hz			
5077,00	USB		LBJ	NOR N Bodoe	NOR
5077,00	USB		LBO	NOR N	NOR
5077,20	*STANAG 4285*	*B=2400Bd*		*NOR N TX Oslo*	*NOR*
5078,50	USB			USA Army MARS	USA
5080,00	SSB		ZHF33	BAS Signy Base , G	SOK
5080,00	CIS ARQ	B=100Bd S=500Hz		RUS PTT	RUS
5081,50	SSB			USN Plead Control Long Beach , OR	USA
5082,00	USB		PEGASO	E Combat Air Command Torrejon de Ardoz	E
5085,80	PACTOR II	B=200Bd S=200Hz	VZX	Sailmail Darawank	AUS
5086,00	CW		RGO86	RUS N	RUS
5086,00	CW		RGR70	RUS N	RUS
5086,00	*CW*		*RMP*	*RUS N HQ Kaliningrad*	*RUS*
5088,00	USB			G AF Air Cadets	G
5088,60	ALE USB	B=125Bd Ch=8 ChS=250Hz		ARINC Urgent Link net	USA
5090,00	ALE USB	B=125Bd Ch=8 ChS=250Hz			
5093,00	CW		RGT77	RUS MIL Moscow RVSN	RUS
5093,00	CW		RGT77	RUS MIL Moscow, RVSN	RUS
5094,00	ALE USB	B=125Bd Ch=8 ChS=250Hz		TUR Disaster and Emergency net	TUR
5095,00	USB			POL MIL net	POL
5095,20	STANAG 4285	B=2400Bd	CFH	CAN N Halifax	CAN

Frequency	Mode	Mode Parameter	Callsign	User	Country
5099,00	*LINK 11 CLEW*	*B=2250Bd Ch=16 ChS=330/110/550Hz*		*NOR N Bergen*	*NOR*
5100,00	FAX 120/576		VMC	Charleville M	AUS
5104,00	CW		RMP	RUS N HQ Kaliningrad	RUS
5105,00	USB			HFoZ/Radtel net	AUS
5105,00	ALE USB	B=125Bd Ch=8 ChS=250Hz		HFoZ/Radtel net	AUS
5105,00	CIS 12	B=1440Bd Ch=12 ChS=200Hz		RUS MIL	RUS
5105,80	ALE 3G	B=2400Bd		UI MIL net	
5105,90	STANAG 4539	B=2400Bd		UI MIL user	
5110,00	USB		VJN	RFDS Cairns	AUS
5110,00	SSB			USA AF Strategic Command	USA
5110,00	USB		VJI	RFDS Mount Isa	AUS
5111,00	*T600*	*B=50Bd S=250Hz*	*RMP*	*RUS N HQ Kaliningrad*	*RUS*
5111,00	CW		P**	RUS N Kaliningrad	RUS
5112,00	USB			USA Army MARS	USA
5114,00	THALES SKYMASTER ALE	B=125Bd Ch=8 ChS=250Hz			
5115,00	USB			LBY CG net	LBY
5115,00	SSB			USA Army MARS	
5116,10	2FSK	B=200Bd S=200Hz			
5120,00	RUS TACTICAL DATALINK	B=1200Bd S=800Hz Ch=1200		RUS MIL	RUS
5120,00	ALE USB	B=125Bd Ch=8 ChS=250Hz		BIH MIL net	BIH
5120,00	MIL 188-110A SER	B=2400Bd		BIH MIL net	BIH
5120,00	MIL 188-110B 39TONE	B=2400Bd Ch=39 ChS=56,25Hz		BIH MIL net	BIH
5120,00	ALE USB	B=125Bd Ch=8 ChS=250Hz		SRB Mil net	SRB
5120,00	STANAG 4481	B=75Bd S=850Hz	NSY	USA N Niscemi , Sicily	I
5121,50	ALE 3G	B=2400Bd		UI MIL net	
5122,00	*LINK 11 CLEW*	*B=2250Bd Ch=16 ChS=330/110/550Hz*		*F F Nancy*	*F*
5122,00	CW		RGT77	RUS MIL Moscow, RVSN	RUS
5122,00	*STANAG 4285*	*B=2400Bd*	*TBB*	*TUR N Ankara*	*TUR*
5123,00	ALE LSB	B=125Bd Ch=8 ChS=250Hz		TWN N net	TWN
5123,00	ALE USB	B=125Bd Ch=8 ChS=250Hz		ISR AF net	ISR
5124,00	RUS TACTICAL DATALINK	B=1200Bd S=800Hz Ch=1200		RUS MIL Kaliningrad	RUS
5125,50	ALE LSB	B=125Bd Ch=8 ChS=250Hz		GRC AF net	GRC
5125,50	HARRIS AVS	Ch=24 ChS=112,5Hz		GRC AF	GRC
5126,00	ALE USB	B=125Bd Ch=8 ChS=250Hz		USA Army net	
5126,00	USB			EGY N	EGY
5126,00	STANAG 4197	B=1800Bd Ch=16/39 ChS=112/56Hz		EGY N net	EGY
5127,00	USB		VMD750	Austravel Safety Net	AUS
5129,00	CW		RIR98	RUS N repair ship PM-56 Armur Class	RUS
5133,00	USB			RUS AF net	RUS
5135,00	CW		RFV99	RUS N	RUS
5135,00	ALE USB	B=125Bd Ch=8 ChS=250Hz		USA Red Cross net	USA
5135,00	ALE USB	B=125Bd Ch=8 ChS=250Hz		USA NG net	USA

Frequency	Mode	Mode Parameter	Callsign	User	Country
5135,00	USB		VKE237	AUS HF Radio Club	AUS
5138,80	ALE 3G	B=2400Bd		UI MIL net	
5140,00	ALE USB	B=125Bd Ch=8 ChS=250Hz		USA Red Cross net	USA
5145,00	USB		VJN	RFDS Cairns	AUS
5145,00	USB			CIS AMS net	
5145,00	USB		AMBARCI K	Aktobe ACC	KAZ
5145,00	USB		AMBA	Samara ACC	RUS
5145,00	USB		ASSISTEN T	Mosdok ACC	RUS
5145,00	USB		ASKHABA D	Ashkabad ACC	TKM
5145,00	USB		SARDINA	Baku ACC	AZE
5145,00	USB		AMBARCH IK	Aktyubinsk ACC	KAZ
5145,00	STANAG 4539	B=2400Bd		UI MIL user	
5146,00	*MIL 188-110A SER*	*B=2400Bd*		*S MIL Karlskrona*	*S*
5146,00	T600	B=50Bd S=200Hz	RDL	RUS N HQ Moscow	RUS
5146,00	ALE USB	B=125Bd Ch=8 ChS=250Hz		ARINC Urgent Link net	USA
5149,70	*STANAG 4285*	*B=2400Bd*	*IDR*	*I N TX Rome*	*I*
5150,00	CW		VTK3	IND N Tuticorin	IND
5152,00	*CIS 12*	*B=1440Bd Ch=12 ChS=200Hz*		*RUS MIL Wjasma*	*RUS*
5153,60	CW		D**	RUS N Sevastopol	UKR
5153,80	CW		P**	RUS N Kaliningrad	RUS
5154,00	CW		C**	RUS N Moscow	RUS
5154,00	CW		R**	RUS N Ustinov	RUS
5154,00	CW		S**	RUS N Severomorsk	RUS
5154,10	CW		A**	RUS N Astrakhan	
5154,40	CW		M**	RUS N Magadan	RUS
5155,00	*STANAG 4285*	*B=2400Bd*	*TBB*	*TUR N TX Incirlik*	*TUR*
5156,80	CW		L**	RUS N St. Petersburg	RUS
5158,00	*CIS 50-1000*	*B=50Bd S=1000Hz*	REA4	*RUS AF HQ Moscow*	*RUS*
5160,20	*STANAG 4285*	*B=2400Bd*	PBB	*NLD N TX Goeree Island*	*HOL*
5161,00	CW		RGZ59	RUS N	RUS
5165,00	CIS ARQ	B=100Bd S=500Hz		RUS PTT	RUS
5169,00	CIS 75-250	B=75Bd S=250Hz		RUS N Moscow	RUS
5174,00	*STANAG 4285*	*B=2400Bd*		*G MIL TX Inskip*	*G*
5176,00	MFSK16	B=2400Bd Ch=16 ChS=125Hz			
5178,00	*T600*	*B=50Bd S=250Hz*		*RUS N Murmansk*	*RUS*
5178,80	ALE 3G	B=2400Bd		UI MIL net	
5178,80	STANAG 4539 HDR	B=2400Bd		UI MIL net	
5179,00	MFSK20	B=10Bd Ch=20 ChS=40Hz			
5179,00	CW		RMP	RUS N HQ Kaliningrad	RUS
5180,00	SSB			NASA tracking ships	USA
5180,00	ALE USB	B=125Bd Ch=8 ChS=250Hz		MRC F net	MRC
5180,00	ALE USB	B=125Bd Ch=8 ChS=250Hz		AUS Police QLD	AUS
5180,00	ALE USB	B=125Bd Ch=8 ChS=250Hz		AUS Police NSW	AUS
5180,00	ALE USB	B=125Bd Ch=8 ChS=250Hz		AUS Police WA	AUS
5180,00	ALE USB	B=125Bd Ch=8 ChS=250Hz		SAPOL Communication Infrastructure	AUS
5181,00	USB			US AF MARS	
5182,00	BAUDOT	B=50Bd S=300Hz		SRB MIL net	SRB
5183,00	MIL 188-110A SER	B=2400Bd		AUS MHFCS North West Cape	AUS

Frequency	Mode	Mode Parameter	Callsign	User	Country
5185,00	CIS ARQ	B=100Bd S=500Hz	RYF2	RUS PTT	RUS
5185,00	CIS ARQ	B=100Bd S=500Hz	RBW	RUS PTT	RUS
5187,00	SSB			NASA tracking ships	USA
5189,00	CIS 12	B=1440Bd Ch=12 ChS=200Hz	RCV	RUS N HQ Sevastopol	UKR
5189,50	STANAG 4481	B=75Bd S=850Hz		F N St Assise	F
5190,00	SSB			NASA ETR day channel	USA
5190,00	RS-ARQ	B=228,5Bd S=170Hz		CZE F	CZE
5190,00	CIS 12	B=1440Bd Ch=12 ChS=200Hz		RUS MIL St. Petersburg	RUS
5192,00	ALE USB	B=125Bd Ch=8 ChS=250Hz		USA emergency net	USA
5194,00	ALE USB	B=125Bd Ch=8 ChS=250Hz		AUT MIL net	AUT
5194,00	MIL 188-110B 39TONE	B=2400Bd Ch=39 ChS=56,25Hz		AUT MIL Wien-Breitensee	AUT
5195,00	*CW*		*DRA5*	*Aurora Beacon Scheggerott*	*D*
5195,00	PSK31	B=31Bd	DRA5	Aurora Beacon Scheggerott	D
5195,00	BAUDOT	B=45,45Bd S=170Hz	DRA5	Aurora Beacon Scheggerott	D
5196,00	STANAG 4285	B=2400Bd		DNK N Frederikshavn	DNK
5196,00	STANAG 4539	B=2400Bd		UI MIL user	
5198,00	CW			RUS Operational Strategic Command Air Defense net	RUS
5198,10	ALE 3G	B=2400Bd		UI MIL net	
5200,00	HAARP WAVEFORM			HAARP	USA
5200,00	USB			EGY N net	EGY
5200,00	STANAG 4197	B=1800Bd Ch=16/39 ChS=112/56Hz		EGY N net	EGY
5201,00	CW			RUS Operational Strategic Command Air Defense net	RUS
5202,00	CIS 12	B=1440Bd Ch=12 ChS=200Hz	RMP	RUS N HQ Kaliningrad	RUS
5202,00	USB			USA Army MARS	USA
5203,00	STANAG 4197	B=1800Bd Ch=16/39 ChS=112/56Hz			
5203,00	T600	B=50Bd S=500Hz		RUS N Moscow	RUS
5205,00	USB			Chuuk R	FSM
5205,00	USB			Pohnpei R	FSM
5205,00	USB			Yap R	FSM
5205,00	USB			Korror R	PLW
5205,00	USB			Majuro R	MHL
5205,00	USB			USA Army MARS	USA
5206,00	USB			Plymouth Air Safety	G
5209,00	ALE 3G	B=2400Bd		UI MIL net	
5210,30	ALE USB	B=125Bd Ch=8 ChS=250Hz		F MIL net	F
5212,00	PACTOR II	B=200Bd S=200Hz	V8V2222	Sailmail Brunai Bay	BRU
5212,00	PACTOR III	B=200Bd S=200Hz	RC01	Sailmail Maputo	MOZ
5212,00	ALE USB	B=125Bd Ch=8 ChS=250Hz		USA FEMA net	USA
5213,00	CW		P**	RUS N Kaliningrad	RUS
5213,00	CW		RMP	RUS N HQ Kaliningrad	RUS
5214,00	USB			USA Army MARS	USA
5215,60	*STANAG 4285*	*B=2400Bd*	*FUO2*	*F N Toulon*	*F*
5220,00	CIS ARQ	B=100Bd S=500Hz		RUS PTT	RUS
5221,50	CW			RUS Operational Strategic Command Air Defense net	RUS
5222,00	USB			EGY N net	EGY

Frequency	Mode	Mode Parameter	Callsign	User	Country
5224,00	CW		RCV	RUS N HQ Sevastopol	UKR
5225,00	CIS ARQ	B=100Bd S=500Hz	RVR39	RUS PTT	RUS
5227,00	SSTV			SURA TX	RUS
5227,70	*STANAG 4285*	*B=2400Bd*	*JXU*	*NOR N TX Bodo*	*NOR*
5227,70	*STANAG 4285*	*B=2400Bd*		*NOR N TX Oslo*	*NOR*
5232,00	CW		RMP	RUS N HQ Kaliningrad	RUS
5235,00	FMCW			San Clemente Island North, CA	USA
5235,70	*STANAG 4285*	*B=2400Bd*	*IDN*	*I N Napoli*	*I*
5240,00	*STANAG 4285*	*B=2400Bd*		*G N Crimond*	*G*
5242,00	MIL 188-110A SER	B=2400Bd		S MIL	S
5242,00	T600	B=50Bd S=200Hz		RUS N Moscow	RUS
5243,50	CHN 4+4	B=75Bd Ch=8 ChS=300/450Hz		CHN MIL	CHN
5245,00	USB			G AF Air Cadets	G
5246,00	MIL 188-110B SER	B=2400Bd		S MIL	S
5250,00	ALE USB	B=125Bd Ch=8 ChS=250Hz		ALG MIL net	ALG
5251,00	USB			CHN air defense	CHN
5254,00	ALE USB	B=125Bd Ch=8 ChS=250Hz		ALG net	ALG
5254,00	USB			Auckland Tramping Club	NZL
5254,00	ALE 3G	B=2400Bd		UI MIL net	
5255,00	ALE USB	B=125Bd Ch=8 ChS=250Hz		S FRO net	S
5259,60	ALE 3G	B=2400Bd		UI MIL net	
5259,60	STANAG 4539	B=2400Bd		UI MIL user	
5261,40	PACTOR I	B=100Bd S=200Hz		FRONET net	S
5264,00	ISR VFT	B=75Bd Ch=5		ISR	
5264,50	USB			POL MIL net	POL
5266,00	CW		RMP	RUS N HQ Kaliningrad	RUS
5266,00	CW		RMED	RUS N	RUS
5266,50	PACTOR III	B=100Bd S=200Hz	CEV773	Sailmail Los Lagos	CHL
5266,70	*STANAG 4285*	*B=2400Bd*	*PBB*	*NLD N TX Goeree Island*	*HOL*
5268,00	*T600*	*B=50Bd S=200Hz*	*RMP*	*RUS N Kaliningrad*	*RUS*
5269,00	ALE USB	B=125Bd Ch=8 ChS=250Hz		ISR AF net	ISR
5270,00	ALE USB	B=125Bd Ch=8 ChS=250Hz		AZE emergency net	AZE
5270,00	USB		VMD750	Austravel Safety Net	AUS
5271,00	USB			Thanh Hoa R	TUN
5274,00	USB			DNK MIL Daneborg Station	GRL
5274,00	USB		XVY	Phu Yen R	VTN
5275,00	MIL 188-110A SER	B=2400Bd		SUI MIL	SUI
5275,00	*STANAG 4481*	*B=75Bd S=850Hz*		*G MIL St. Eval*	*G*
5275,00	ALE USB	B=125Bd Ch=8 ChS=250Hz		SUI MIL net	SUI
5276,00	ALE USB	B=125Bd Ch=8 ChS=250Hz			
5280,00	ALE USB	B=125Bd Ch=8 ChS=250Hz		SUI MIL net	SUI
5280,00	CW			RUS HFDF net	RUS
5281,00	ALE USB	B=125Bd Ch=8 ChS=250Hz		SVK MIL net	SVK
5282,00	CIS ARQ	B=100Bd S=500Hz		RUS PTT	RUS
5282,80	ALE 3G	B=2400Bd		UI MIL net	
5286,50	ALE 3G	B=2400Bd		UI MIL net	
5287,00	*ALE USB*	*B=125Bd Ch=8 ChS=250Hz*		*S MIL net*	*S*
5287,20	WSPR	B=1,46Bd Ch=4 ChS=1,46Hz		WSPR net	
5289,50	STANAG 4481	B=50Bd S=850Hz	NSS	USA N TX Davidsonville , MD	USA
5290,00	JT9	Ch=1	GB3ORK	Orkney Isles	G

Frequency	Mode	Mode Parameter	Callsign	User	Country
5290,00	CW		GB3RAL	Rutherford Appleton Laboratories , Oxfordshire	G
5290,00	CW		GB3WES	Cumbria	G
5290,00	CW		GB3ORK	Orkney Isles	G
5293,00	CW		REA4	RUS AF Moscow	RUS
5293,00	USB			USA Army MARS	USA
5293,00	CW		RMP	RUS N HQ Kaliningrad	RUS
5295,00	ALE USB	B=125Bd Ch=8 ChS=250Hz		G MIL net	G
5295,50	ALE USB	B=125Bd Ch=8 ChS=250Hz		POL MIL net	
5299,00	USB			Auckland Tramping Club	NZL
5300,00	USB		VJB	RFDS Derby	AUS
5300,00	USB		VKL	RFDS Port Hedland	AUS
5300,00	USB		VKJ	RFDS Meekathara	AUS
5300,00	USB		VJQ	RFDS Kalgoorlie	AUS
5300,00	*STANAG 4285*	*B=2400Bd*	*TBO*	*TUR N Izmir*	*TUR*
5300,50	CW		RJP24	RUS N	RUS
5301,00	USB			G RAF Air Cadets	G
5302,60	ALE 3G	B=2400Bd		UI MIL net	
5303,00	CW		V**	RUS N Khiva	RUS
5303,50	USB			USA Army MARS	USA
5304,00	USB			G RAF Air Cadets	G
5305,50	*T600*	*B=50Bd S=250Hz*	*RMP*	*RUS N Kaliningrad*	*RUS*
5306,00	STANAG 4481	B=50Bd S=850Hz	NKW	USA N Diego Garcia	DGA
5309,70	*STANAG 4285*	*B=2400Bd*	*EBA*	*E N Madrid*	*E*
5310,00	RUS TACTICAL DATALINK	B=1200Bd S=800Hz Ch=800		RUS MIL	RUS
5312,00	CW			RUS Strategic AF	RUS
5312,00	CW		RCV	RUS N HQ Sevastopol	UKR
5312,00	CIS 12	B=1440Bd Ch=12 ChS=200Hz	RCV	RUS N HQ Sevastopol	UKR
5312,20	*STANAG 4285*	*B=2400Bd*		*G MIL Inskip*	*G*
5312,30	STANAG 4539	B=2400Bd		UI MIL user	
5312,30	ALE 3G	B=2400Bd		UI MIL net	
5313,00	CW			RUS air defence net	RUS
5313,00	CIS 12	B=1440Bd Ch=12 ChS=200Hz	RCV	RUS N HQ Sevastopol	UKR
5315,00	CIS 12	B=1440Bd Ch=12 ChS=200Hz		RUS MIL St. Petersburg	RUS
5316,00	ALE USB	B=125Bd Ch=8 ChS=250Hz		HRV NPRD net	HRV
5316,00	ALE USB	B=125Bd Ch=8 ChS=250Hz		SVK AF net	SVK
5317,00	USB			G RAF Air Cadets	G
5317,50	R&S MODEM	B=2400Bd		POL MSW Warszawa	POL
5317,50	R&S MODEM	B=2400Bd		POL MSW Bydgoszcz	POL
5318,00	SSB			ARG F net	ARG
5320,00	ALE USB	B=125Bd Ch=8 ChS=250Hz		AZE emergency net	AZE
5320,00	USB			G RAF Air Cadets	G
5320,00	*STANAG 4285*	*B=2400Bd*		*G MIL Akrotiri*	*CYP*
5322,00	CIS 12	B=1440Bd Ch=12 ChS=200Hz		RUS N ship	
5322,50	USB			USA Army MARS	USA
5323,80	ALE 3G	B=2400Bd		UI MIL net	
5325,00	RUS TACTICAL DATALINK	B=1200Bd S=800Hz Ch=800		RUS MIL	RUS
5325,00	CIS ARQ	B=100Bd S=500Hz	RND79	RUS PTT Moscow	RUS
5325,60	ALE 3G	B=2400Bd		UI MIL net	
5327,50	CHN 4+4	B=75Bd Ch=8 ChS=300/450Hz		CHN MIL	
5330,00	USB			G MIL Army Cadets	G

Frequency	Mode	Mode Parameter	Callsign	User	Country
5332,00	M/S 1.0	B=200Bd S=850Hz			
5332,00	*STANAG 4285*	*B=2400Bd*	*TBB*	*TUR N TX Ankara*	*TUR*
5334,30	ALE 3G	B=2400Bd		UI MIL net	
5335,00	ALE USB	B=125Bd Ch=8 ChS=250Hz		S FRO net	S
5335,00	USB			G RAF Air Cadets	Gg
5336,00	FAX 90/120/576		RBW41	Murmansk M	RUS
5338,00	HC-265			ALG AF	ALG
5340,00	STANAG 4481	B=75Bd S=850Hz	NAR	USN Key West , FL	USA
5340,00	STANAG 4481	B=75Bd S=850Hz	NAU	USA N San Juan	PTR
5340,00	STANAG 4481	B=50Bd S=850Hz	NAU	USA N San Juan	PTR
5342,00	CW		V**	RUS N Khiva	RUS
5343,00	*CW*		*RIT*	*RUS N HQ Severomorsk*	*RUS*
5343,00	CW		RHQ33	RUS N ship PRIAZOVYE	RUS
5343,00	CW		RCV	RUS N HQ Sevastopol	UKR
5345,00	USB			G MIL Army Cadets	G
5345,00	STANAG 4539	B=2400Bd		UI MIL user	
5350,00	SSB			NASA launch support aircraft	USA
5351,00	*CIS 12*	*B=1440Bd Ch=12 ChS=200Hz*		*RUS MIL Orscha*	*BLR*
5352,00	MIL 188-110B SER	B=2400Bd		E MOI	E
5354,00	ROBUST PACKET RADIO	B=200Bd Ch=8 ChS=60Hz		Robust Packet Radio net, HF-APRS	
5354,00	USB			G RAF Air Cadets	G
5354,50	ALE USB	B=125Bd Ch=8 ChS=250Hz		HFLINK network	
5355,00	VARA	Ch=1		VARA chat freqeuency	
5357,00	FT8	B=5,86Bd Ch=8 ChS=5,86Hz		FT8 channel	
5357,00	ALE USB	B=125Bd Ch=8 ChS=250Hz		HFLINK network	
5358,50	USB		LBJ	NOR F Bodo	NOR
5360,00	USB		VKL	RFDS Port Hedland	AUS
5360,00	USB		VJT	RFDS Carnarvon	AUS
5360,00	USB		VKJ	RFDS Meekathara	AUS
5360,00	USB		VJQ	RFDS Kalgoorlie	AUS
5362,20	*STANAG 4285*	*B=2400Bd*	*OVG*	*DNK N Frederickshavn*	*DNK*
5363,00	USB			G RAF Air Cadets	G
5364,20	ALE 3G	B=2400Bd		UI MIL net	
5364,70	WSPR	B=1,46Bd Ch=4 ChS=1,46Hz		WSPR net	
5365,00	ALE USB	B=125Bd Ch=8 ChS=250Hz		Queensland Deptartment of Community Safety	AUS
5366,70	ALE 3G	B=2400Bd		UI MIL net	
5366,70	STANAG 4539 HDR	B=2400Bd		UI MIL net	
5367,00	ROS USB	B=1Bd Ch=144 ChS=15,625Hz		ROS frequency	
5368,00	SSB			ARG F net	ARG
5369,50	CHN 4+4	B=75Bd Ch=8 ChS=300/450Hz		CHN MIL	CHN
5370,00	CIS ARQ	B=100Bd S=500Hz		RUS PTT	RUS
5371,00	THALES ROBUST WAVEFORM	B=1000Bd Ch=8 ChS=250Hz		MRC MIL net	MRC
5371,50	ALE USB	B=125Bd Ch=8 ChS=250Hz		HFLINK network	
5371,50	ALE USB	B=125Bd Ch=8 ChS=250Hz		USA Civil Air Patrol net	USA
5372,00	CIS 12	B=1440Bd Ch=12 ChS=200Hz	RCV	RUS N HQ Sevastopol	UKR
5376,00	CW			RUS MIL net	RUS
5377,30	ALE 3G	B=2400Bd		UI MIL net	
5377,80	ALE 3G	B=2400Bd		UI MIL net	

Frequency	Mode	Mode Parameter	Callsign	User	Country
5378,00	ALE USB	B=125Bd Ch=8 ChS=250Hz		USA FEMA net	USA
5379,00	USB			G RAF Air Cadets	G
5379,30	ALE 3G	B=2400Bd		UI MIL net	
5379,60	ALE3G	B=2400Bd			
5381,80	STANAG 4285	B=2400Bd	OUA15	DNK N Stevns	DNK
5382,00	CW		RJD99	RUS N Severodvinsk	RUS
5384,00	ALE USB	B=125Bd Ch=8 ChS=250Hz		LTU MIL net	LTU
5384,00	ALE USB	B=125Bd Ch=8 ChS=250Hz		POL MIL net	POL
5385,00	USB			USA Army MARS	USA
5387,50	ALE USB	B=125Bd Ch=8 ChS=250Hz		NLD MIL net	HOL
5388,50	ALE USB	B=125Bd Ch=8 ChS=250Hz		USA FBI net	USA
5389,00	CW		RGT77	RUS MIL Moscow, RVSN	RUS
5389,60	ALE 3G	B=2400Bd		UI MIL net	
5389,60	STANAG 4539 HDR	B=2400Bd		UI MIL net	
5390,80	ALE 3G	B=2400Bd		UI MIL net	
5391,00	USB		LBJ	NOR N Bodoe	NOR
5394,00	CHN 4+4	B=75Bd Ch=8 ChS=300/450Hz		CHN MIL	CHN
5394,00	CW			MNG MIL net	MNG
5394,00	USB			USA Army MARS	USA
5394,50	USB		AAR0QC	US Army MARS	
5394,50	USB		AAR0XV	USA Army MARS	
5394,50	USB		AAR0MB	USA Army MARS	
5394,50	USB		AAR0AR	USA Army MARS	
5395,00	USB			G RAF Air Cadets	G
5396,50	ALE USB	B=125Bd Ch=8 ChS=250Hz		USA SAC net	USA
5398,00	T600	B=50Bd S=200Hz	RMP	RUS N Kaliningrad	RUS
5398,50	USB			G RAF Air Cadets	G
5399,00	USB			USA Army MARS	USA
5399,80	ALE 3G	B=2400Bd		UI MIL net	
5400,00	USB		VNJ	Casey Base , AUS	ANT
5400,00	USB			AUS Davis Base	ANT
5400,00	USB			Macquarie Island , AUS	ANT
5400,00	USB			Mawson Base , AUS	ANT
5400,00	HAARP WAVEFORM			HAARP	ALS
5400,00	MIL 188-110B SER	B=2400Bd	DRINI**	ALB F	ALB
5400,70	LINK 11 CLEW	B=2250Bd Ch=16 ChS=330/110/550Hz		G N Plymouth	G
5401,00	RUS TACTICAL DATALINK	B=1200Bd S=800Hz Ch=1200		RUS MIL	RUS
5401,00	ALE USB	B=125Bd Ch=8 ChS=250Hz		POL Net	POL
5402,00	ALE USB	B=125Bd Ch=8 ChS=250Hz		USA FEMA net	USA
5403,50	USB			G RAF Air Cadets	G
5405,00	ALE USB	B=125Bd Ch=8 ChS=250Hz		ARS AF net	ARS
5406,00	MAHRS	B=2400Bd	DHJ58	D N Glücksburg TX Rostock	D
5407,00	ALE USB	B=125Bd Ch=8 ChS=250Hz		AZE emergency net	AZE
5407,00	CW		RGT77	RUS MIL Moscow, RVSN	RUS
5407,00	ALE USB	B=125Bd Ch=8 ChS=250Hz			
5408,30	FSK2	B=200Bd S=400Hz			

Frequency	Mode	Mode Parameter	Callsign	User	Country
5410,00	ALE LSB	B=125Bd Ch=8 ChS=250Hz		ALG gas & oil net Sonatrach	ALG
5410,00	USB		VJD	RFDS Alice Springs	AUS
5413,00	USB			BGD N net	BGD
5414,30	ALE 3G	B=2400Bd		UI MIL net	
5417,00	STANAG 4285	B=2400Bd		G MIL	G
5418,00	ALE USB	B=125Bd Ch=8 ChS=250Hz		ALG Sonatrach net	ALG
5418,40	ALE USB	B=125Bd Ch=8 ChS=250Hz		USA N net	USA
5419,00	ALE ISB	B=125Bd Ch=8 ChS=125Hz		ARS AF net	
5420,00	USB			MMR MIL net	MMR
5421,00	ALE USB	B=125Bd Ch=8 ChS=250Hz		G MIL net	G
5423,00	CW			RUS HFDF net	RUS
5427,50	STANAG 4285	B=2400Bd	JXU	NOR N TX Bodo	NOR
5429,00	USB			RUS AF net	RUS
5431,00	CW			UKR MIL net	UKR
5434,00	ALE USB	B=125Bd Ch=8 ChS=250Hz		USA National Guard net	USA
5435,00	USB		MKL	G AF London VOLMET	G
5437,00	ALE USB	B=125Bd Ch=8 ChS=250Hz		F N net	F
5437,00	MIL 188-110A SER	B=2400Bd		F N	F
5437,00	MIL 188-110A SER	B=2400Bd		POL MIL Katowice	POL
5437,00	ALE USB	B=125Bd Ch=8 ChS=250Hz		SVK MIL net	SVK
5438,00	T600	B=50Bd S=200Hz		RUS N Murmansk	RUS
5440,00	ALE USB	B=125Bd Ch=8 ChS=250Hz		MTN net	MTN
5440,00	LSB			CHN air defense	CHN
5440,00	CIS ARQ	B=100Bd S=500Hz	RWD59	RUS PTT Moscow	RUS
5441,00	CW		RCV	RUS N HQ Sevastopol	UKR
5441,00	STANAG 4285	B=2400Bd		G MIL TX Inskip	G
5441,00	STANAG 4285	B=2400Bd		G MIL St. Eval	G
5442,00	CHN 4+4	B=75Bd Ch=8 ChS=300/450Hz		CHN MIL	
5443,00	CW		RCB	RUS AF	RUS
5443,00	CW		RIT	RUS N HQ Severomorsk	RUS
5444,00	ALE USB	B=125Bd Ch=8 ChS=250Hz		D net	D
5448,00	USB			RUS MIL Rostov-na-Donu	RUS
5450,00	USB		MKL	G AF VOLMET Inskip	G
5451,00	SSB			RDARA 10F	
5451,00	SSB			RDARA 11B	
5451,00	SSB			RDARA 12F	
5451,00	SSB			RDARA 12H	
5451,00	SSB			RDARA 13I	
5451,00	SSB			RDARA 13J	
5451,00	HFDL	B=1800Bd	H16**	Agana	GUM
5451,00	USB		BOING EVERETT	Boing Seattle	USA
5451,00	USB		BOING SEATTLE	Boing Seattle	USA
5451,50	SYSTEME 3000	B=2400Bd			
5452,00	RUS MFSK32 OFDM60	B=40Bd Ch=16		MFA Moscow	RUS
5452,00	ALE USB	B=125Bd Ch=8 ChS=250Hz		USA MIL net	
5452,00	CW		USY47		
5454,00	SSB			RDARA 10	
5454,00	SSB			RDARA 12E	

Frequency	Mode	Mode Parameter	Callsign	User	Country
5454,00	SSB			RDARA 12F	
5454,00	SSB			RDARA 13F	
5454,00	SSB			RDARA 13J	
5454,00	STANAG 4481	B=75Bd S=850Hz		POR N Lisbon TX Azores	POR
5455,00	USB		VMS469	Reids Radiodata net	AUS
5455,00	SSB		VKS737	AUS 4WD net	AUS
5456,00	USB		RMP	RUS N HQ Kaliningrad	RUS
5457,00	SSB			RDARA 10C	
5457,00	SSB			RDARA 13N	
5457,00	*MIL 188-110B SER*	*B=2400Bd*		*S MIL Karlskrona*	*S*
5460,00	SSB			RDARA 10B	
5460,00	SSB			RDARA 10E	
5460,00	SSB			RDARA 12C	
5460,00	SSB			RDARA 13D	
5460,00	ALE USB	B=125Bd Ch=8 ChS=250Hz		AZE emergency net	AZE
5463,00	SSB			RDARA 11B	
5463,00	SSB			RDARA 13H	
5463,00	SSB			RDARA 13K	
5463,00	SSB			RDARA 13M	
5464,50	ALE 3G	B=2400Bd		UI MIL net	
5464,50	STANAG 4539	B=2400Bd		UI MIL user	
5465,80	CW		R**	RUS N Ustinov	RUS
5466,00	SSB			RDARA 10B	
5466,00	SSB			RDARA 13I	
5466,00	ALE USB	B=125Bd Ch=8 ChS=250Hz		G MIL net	G
5466,00	ALE 3G	B=2400Bd		UI MIL net	
5466,90	STANAG 4539	B=2400Bd		UI MIL user	
5467,00	ALE USB	B=125Bd Ch=8 ChS=250Hz		ALG MIL net	ALG
5469,00	SSB			RDARA 11B	
5469,00	SSB			RDARA 13G	
5470,00	SYSTEME 3000 BURST	B=2400Bd		Prague	CZE
5471,00	CW			RUS MOI net	RUS
5472,00	SSB			RDARA 10D	
5472,00	SSB			RDARA 10A	
5472,00	SSB			RDARA 13H	
5472,00	*LINK 11 CLEW*	*B=2250Bd Ch=16 ChS=330/110/550Hz*		*G MIL Upavon*	*G*
5472,00	CW		V**	RUS N Khiva	RUS
5472,00	CHN 4+4	B=75Bd Ch=8 ChS=300/450Hz		CHN MIL	
5473,00	CW		RJD99	RUS N Severodvinsk	RUS
5473,00	3FSK			RUS MIL Rostov-on-Don	RUS
5474,00	LINK 11 SLEW	B=2400Bd			
5475,00	USB			Salta VOLMET	ARG
5475,00	USB			Cordoba VOLMET	ARG
5475,00	SSB			RDARA 10D	
5475,00	SSB			RDARA 10A	
5475,00	SSB			RDARA 12E	
5475,00	SSB			RDARA 12F	
5475,00	SSB			RDARA 13G	
5476,00	ALE USB	B=125Bd Ch=8 ChS=250Hz		GRC MOI net	GRC
5476,00	LINK 22	B=2250Bd			
5477,00	CW			RUS HFDF net	RUS
5478,00	ALE USB	B=125Bd Ch=8 ChS=250Hz		ROU diplo net	ROU
5479,00	CW			CHN MIL net	CHN

Frequency	Mode	Mode Parameter	Callsign	User	Country
5480,00	ALE USB	B=125Bd Ch=8 ChS=250Hz		AZE emergency net	AZE
5481,00	SSB			RDARA 10C	
5481,00	SSB			RDARA 10E	
5481,00	SSB			RDARA 12E	
5481,00	SSB			RDARA 12F	
5481,00	SSB			RDARA 12J	
5481,00	SSB			RDARA 13E	
5481,00	SSB			RDARA 14D	
5481,00	SSB			RDARA 14G	
5481,00	SSB			RDARA 2A	
5481,00	SSB			RDARA 2C	
5481,00	SSB			RDARA 4B	
5481,00	SSB			RDARA 6G	
5481,00	SSB			RDARA 7D	
5484,00	SSB			RDARA 10D	
5484,00	SSB			RDARA 10A	
5484,00	SSB			RDARA 12C	
5484,00	SSB			RDARA 12G	
5484,00	SSB			RDARA 13H	
5484,00	SSB			RDARA 1B	
5484,00	SSB			RDARA 3A	
5484,00	SSB			RDARA 3C	
5484,00	SSB			RDARA 6A	
5484,00	SSB			RDARA 9B	
5487,00	SSB			RDARA 10C	
5487,00	SSB			RDARA 12E	
5487,00	SSB			RDARA 2C	
5487,00	SSB			RDARA 6G	
5490,00	SSB			RDARA 10D	
5490,00	SSB			RDARA 10A	
5490,00	SSB			RDARA 12C	
5490,00	SSB			RDARA 13C	
5490,00	SSB			RDARA 2A	
5490,00	SSB			RDARA 2B	
5490,00	SSB			RDARA 3A	
5490,00	SSB			RDARA 6D	
5493,00	USB			Bangui ACC	CAF
5493,00	USB			Douala ACC	CME
5493,00	USB			Entebbe ACC	UGA
5493,00	USB			FrancevilleACC	GAB
5493,00	SSB			Accra ACC	GHA
5493,00	SSB			Kano ACC	NIG
5493,00	USB			Garoua ACC	CME
5493,00	USB			Goma ACC	COG
5493,00	USB			Harare ACC	ZWE
5493,00	USB			Lubumbashi ACC	COG
5493,00	USB			Maroua ACC	CME
5493,00	USB			Pointe Noire ACC	COG
5493,00	USB			Port Gentil ACC	GAB
5493,00	SSB			Kinshasa ACC	ZAI
5493,00	SSB			Kisangani ACC	ZAI
5493,00	SSB			Lagos ACC	NIG
5493,00	SSB			Luanda ACC	AGL
5493,00	SSB			Lusaka ACC	ZMB
5493,00	SSB			MWARA AFI	AFI
5493,00	SSB			Niamey ACC	NGR
5493,00	SSB			N`djamena ACC	TCD
5493,00	SSB			RDARA 3B	
5493,00	SSB			RDARA 6G	
5493,00	USB			Roberts ACC	LBR
5493,00	USB			Sao Tome ACC	CAF

Frequency	Mode	Mode Parameter	Callsign	User	Country
5493,00	USB			Windhoek ACC	NMB
5493,00	USB			Yaounde ACC	CME
5496,00	SSB			RDARA 10D	
5496,00	SSB			RDARA 10A	
5496,00	SSB			RDARA 12C	
5496,00	SSB			RDARA 12J	
5496,00	SSB			RDARA 13I	
5496,00	SSB			RDARA 2A	
5496,00	SSB			RDARA 2B	
5496,00	SSB			RDARA 3A	
5496,00	SSB			RDARA 6F	
5496,00	MIL 188-110B SER	B=2400Bd		South Sweden	S
5499,00	SSB			Antananarivo VOLMET	MDG
5499,00	SSB		TNL	Brazzaville VOLMET	COG
5499,00	SSB			RDARA 3B	
5499,00	SSB			RDARA 6G	
5499,00	SSB			VOLMET AFI	AFI
5502,00	SSB			RDARA 10C	
5502,00	SSB			RDARA 12C	
5502,00	SSB			RDARA 13M	
5502,00	SSB			RDARA 2A	
5502,00	SSB			RDARA 2B	
5502,00	SSB			RDARA 3A	
5502,00	SSB			RDARA 6B	
5502,00	HFDL	B=1800Bd	H10**	Muan	KOR
5504,00	CW			RUS HFDF net	RUS
5505,00	SSB			RDARA 3B	
5505,00	SSB			RDARA 6G	
5505,00	SSB			VOLMET EUR	EUR
5505,00	USB		EIP	Shannon VOLMET	IRL
5507,00	MIL 188-110A SER	B=2400Bd		S MIL Stockholm	S
5507,30	STANAG 4539	B=2400Bd		UI MIL user	
5508,00	HFDL	B=1800Bd	H01**	San Francisco , CA	USA
5508,00	SSB			Bogota ACC	CLM
5508,00	SSB			Cucuta ACC	CLM
5508,00	SSB			Medellin ACC	CLM
5508,00	SSB			RDARA 11B	
5508,00	SSB			RDARA 12F	
5508,00	SSB			RDARA 13N	
5508,00	SSB			RDARA 2B	
5508,00	SSB			RDARA 2C	
5508,00	SSB			RDARA 6F	
5508,00	SSB			RDARA 7	
5508,00	SSB			RDARA 9B	
5511,00	USB			PanAM R	USA
5511,00	SSB			RDARA 3A	
5511,00	SSB			RDARA 5B	
5511,00	SSB			RDARA 6G	
5512,00	*ISR N HYBRID MODEM*	*B=2400Bd*	*4XZ*	*ISR N Haifa*	*ISR*
5514,00	SSB			RDARA 11B	
5514,00	SSB			RDARA 13C	
5514,00	SSB			RDARA 2C	
5514,00	SSB			RDARA 3B	
5514,00	SSB			RDARA 3C	
5514,00	SSB			RDARA 6E	
5514,00	HFDL	B=1800Bd	H02**	Molokai , HI	HWA
5514,00	CW			RUS MIL net	RUS
5517,00	SSB			Cairo ACC	EGY
5517,00	USB			Sanaa ACC	YEM
5517,00	SSB			Tripoli ACC	LBY
5517,00	SSB			RDARA 3A	

Frequency	Mode	Mode Parameter	Callsign	User	Country
5517,00	SSB			RDARA 6G	
5517,00	USB			Addis Ababa ACC	ETH
5517,00	USB			Aden ACC	YEM
5517,00	USB			Asmara ACC	ERI
5517,00	USB			Bahrain ACC	BHR
5517,00	USB			Benghazi ACC	LBY
5517,00	USB			Mumbai ACC	IND
5517,00	USB			Bujumbura ACC	BDI
5517,00	USB			Comoros ACC	COM
5517,00	USB			Dar es Salaam ACC	TZA
5517,00	USB			Entebbe ACC	UGA
5517,00	USB			Hargeisa ACC	SOM
5517,00	USB			Djibouti ACC	DJI
5517,00	USB			Jeddah ACC	ARS
5517,00	USB			Khartoum ACC	SDN
5517,00	USB			Kigali ACC	RRW
5517,00	USB			Kisimayu ACC	SOM
5517,00	USB			Male ACC	MLD
5517,00	USB			Nairobi ACC	KEN
5517,00	USB			Mogadishu ACC	SOM
5517,00	USB			Port Sudan ACC	SDN
5517,00	USB			Seychelles ACC	SEY
5520,00	SSB			Boyeros ACC	CUB
5520,00	SSB			MWARA CAR	CAR
5520,00	SSB		KEA5	New York ACC	USA
5520,00	SSB			Panama ACC	PNR
5520,00	SSB			RDARA 2B	
5520,00	SSB			RDARA 2C	
5520,00	SSB			RDARA 6D	
5520,00	SSB			RDARA 7E	
5520,00	USB			Barranquilla ACC	CLM
5520,00	USB			Cayenne ACC	GUF
5520,00	USB			Georgetown ACC	GUY
5520,00	USB			Maiquetia ACC	VEN
5520,00	USB			Paramaribo ACC	SUR
5520,00	USB			Piarco ACC	TRD
5523,00	SSB			RDARA 11B	
5523,00	SSB			RDARA 12G	
5523,00	SSB			RDARA 13I	
5523,00	SSB			RDARA 2A	
5523,00	SSB			RDARA 6G	
5523,00	SSB			RDARA 9B	
5526,00	SSB			Asuncion ACC	PRG
5526,00	SSB			AUS HF Network SE region	AUS
5526,00	SSB			Belem ACC	B
5526,00	SSB			Bogota ACC	CLM
5526,00	SSB			Brasilia ACC	B
5526,00	SSB			Cayenne ACC	GUF
5526,00	SSB			La Paz ACC	BOL
5526,00	SSB			Leticia ACC	CLM
5526,00	SSB			Maiquetia ACC , Caracas	VEN
5526,00	SSB			Manaus ACC	B
5526,00	SSB			Montevideo ACC	URG
5526,00	SSB			MWARA SAM	SAM
5526,00	SSB			Paramaribo ACC	SUR
5526,00	SSB			Piarco ACC	TRD
5526,00	SSB			Porto Allegro ACC	B
5526,00	SSB			Porto Velho ACC	B
5526,00	SSB			RDARA 10F	
5526,00	SSB			RDARA 14	
5526,00	SSB			RDARA 2B	
5526,00	SSB			RDARA 2C	

Frequency	Mode	Mode Parameter	Callsign	User	Country
5526,00	SSB			RDARA 3B	
5526,00	SSB			RDARA 5D	
5526,00	SSB			RDARA 6E	
5526,00	SSB			Rio de Janeiro ACC	B
5526,00	SSB			Santa Cruz ACC	BOL
5526,00	USB			Iquitos ACC	PRU
5526,00	USB			Georgetown ACC	CYM
5526,00	USB			Recife ACC	B
5526,00	USB			Brasilia ACC	B
5526,00	USB			Buenos Aires ACC	ARG
5526,00	USB			Campo Grande ACC	B
5526,00	USB			Lima ACC	PRU
5526,00	USB			Salvador ACC	B
5529,00	SSB			Boyeros ACC	CUB
5529,00	SSB			W I	WW
5529,00	SSB			W II	WW
5529,00	USB			IBERIA Madrid ACC	E
5529,00	HFDL	B=1800Bd	H09**	Utqiagvik , AK	ALS
5529,00	HFDL	B=1800Bd	H08**	Johannesburg	AFS
5531,00	USB			EGY N net	EGY
5531,00	STANAG 4197	B=1800Bd Ch=16/39 ChS=112/56Hz		EGY N net	EGY
5532,00	SSB			W I	WW
5532,00	SSB			W V	WW
5535,00	SSB			Lima ACC	PRU
5535,00	SSB			W I	WW
5535,00	SSB			W IV	WW
5535,00	ALE USB	B=125Bd Ch=8 ChS=250Hz		S FRO net	S
5538,00	SSB			W II	WW
5538,00	SSB			W V	WW
5538,00	HFDL	B=1800Bd	H09**	Utqiagvik , AK	ALS
5540,00	SSB			RDARA 3B	
5540,00	ALE USB	B=125Bd Ch=8 ChS=250Hz			
5541,00	USB		SDJ	Stockholm R	S
5541,00	SSB			W I	WW
5541,00	SSB			W IV	WW
5541,00	SSB			World ACCways Oakland	USA
5542,00	STANAG 4285	B=2400Bd		G MIL Inskip	G
5544,00	SSB			W II	WW
5544,00	SSB			W V	WW
5544,00	HFDL	B=1800Bd	H09**	Utqiagvik , AK	ALS
5544,00	HFDL	B=1800Bd	H15**	Al Muharraq	BHR
5547,00	USB			San Francisco ACC , CA	USA
5547,00	SSB			MWARA CEP	CEP
5547,00	SSB			RDARA 13H	
5547,00	SSB			RDARA 13K	
5547,00	SSB			RDARA 2A	
5547,00	SSB			RDARA 4A	
5547,00	SSB			RDARA 6G	
5547,00	SSB			RDARA 7F	
5547,00	HFDL	B=1800Bd	H07**	Shannon	IRL
5550,00	USB			Barranquilla ACC	CLM
5550,00	SSB			Boyeros ACC	CUB
5550,00	SSB			Merida ACC	MEX
5550,00	SSB			MWARA CAR	CAR
5550,00	SSB		KEA5	New York ACC , NY	USA
5550,00	SSB			Panama ACC	PNR
5550,00	SSB			Piarco ACC	TRD
5550,00	SSB			RDARA 14G	

Frequency	Mode	Mode Parameter	Callsign	User	Country
5550,00	SSB			RDARA 2B	
5550,00	SSB			RDARA 2C	
5550,00	SSB			RDARA 3B	
5550,00	SSB			RDARA 5D	
5550,00	SSB			RDARA 6C	
5550,00	SSB			RDARA 6E	
5550,00	SSB			San Jose ACC	CTR
5550,00	ALE USB	B=125Bd Ch=8 ChS=250Hz		UN MINUSCA net	
5551,00	*ALE USB*	*B=125Bd Ch=8 ChS=250Hz*		*TUR Disaster and Emergency net*	*TUR*
5553,00	SSB			RDARA 10B	
5553,00	SSB			RDARA 13C	
5553,00	SSB			RDARA 6G	
5555,00	USB			EGY CG net	EGY
5555,00	CHN 4+4	B=75Bd Ch=8 ChS=300/450Hz		CHN MIL	CHN
5555,00	*STANAG 4285*	*B=2400Bd*		*G MIL Akrotiri*	*CYP*
5559,00	SSB			MWARA SP	SP
5559,00	SSB			RDARA 10E	
5559,00	SSB			RDARA 12G	
5559,00	SSB			RDARA 13J	
5559,00	SSB			RDARA 2A	
5559,00	SSB			RDARA 4A	
5559,00	SSB			RDARA 6G	
5560,00	ALE USB	B=125Bd Ch=8 ChS=250Hz		AZE emergency net	AZE
5560,00	USB		ALFUR	Tyumen ACC	RUS
5560,00	USB		BEREZNY	Chanty-Mansijsk ACC	RUS
5560,00	USB		BRUSNIC HKA	Noyabrsk ACC	RUS
5562,00	SSB			NOAA Hurricane Center Miami , FL	USA
5562,00	SSB			RDARA 10C	
5562,00	SSB			RDARA 12D	
5562,00	SSB			RDARA 13D	
5562,00	SSB			RDARA 2C	
5562,00	SSB			RDARA 3B	
5562,00	SSB			RDARA 3C	
5565,00	USB			Cayenne ACC	GUF
5565,00	USB			Manaus ACC	B
5565,00	USB			Paramaribo ACC	SUR
5565,00	USB			Rio de Janerio ACC	B
5565,00	SSB			Bissau ACC	GNB
5565,00	SSB			Canarias ACC , Las Palmas	CNR
5565,00	SSB			Dakar ACC	SEN
5565,00	SSB			Johannesburg ACC	AFS
5565,00	SSB			MWARA SAT	SAT
5565,00	SSB			RDARA 10A	
5565,00	SSB			RDARA 6G	
5565,00	SSB			RDARA 9B	
5565,00	SSB			Recife ACC	B
5565,00	SSB			Sal ACC	CPV
5565,00	ALE USB	B=125Bd Ch=8 ChS=250Hz		UI net	
5565,00	BAUDOT	B=50Bd S=100Hz	7X6R**	RUS	RUS
5566,00	MIL 188-110A	B=2400Bd		EGY N net	EGY
5566,00	USB			EGY N net	EGY
5568,00	SSB			RADAR 12	
5568,00	SSB			RDARA 10B	
5568,00	SSB			RDARA 13J	
5568,00	SSB			RDARA 1B	
5568,00	SSB			RDARA 3A	

Frequency	Mode	Mode Parameter	Callsign	User	Country
5568,00	SSB			RDARA 3C	
5568,00	SSB			RDARA 5B	
5568,00	SSB			RDARA 6D	
5568,00	SSB			RDARA 7F	
5568,00	USB		SHPORA	Rostov ACC	RUS
5568,00	USB		YURIST	Vladivostok ACC	RUS
5568,00	USB		RIDAN	Makhachkala ACC	RUS
5568,00	USB		TJURIK	Krasnodar ACC	RUS
5568,00	USB		TRIOL	Elista ACC	RUS
5568,00	USB		KAZACOK	Nizhny-Novgorod ACC	RUS
5568,00	USB		TOM	Stavropol ACC	RUS
5568,00	USB		STAVOK	Vladikavkaz ACC	RUS
5568,00	USB		TOSNO	Nalchik ACC	RUS
5568,00	USB		SERIOSHKA	Astrakhan ACC	S
5568,00	USB		AVROROA	Volgograd ACC	RUS
5568,00	USB		ASSISTENT	Mosdok ACC	RUS
5568,00	USB		SHTAT	Sochi ACC	RUS
5568,00	USB		UZBEZISCHE	Mineralny Vody ACC	RUS
5568,00	USB		ATLANTIDA TRI	Maikop ACC	RUS
5571,00	SSB			RDARA 11B	
5571,00	SSB			RDARA 13C	
5571,00	SSB			RDARA 6G	
5571,60	PACTOR III	B=100Bd S=200Hz	WHX	Droop Mountain , WV	USA
5574,00	SSB			MWARA CEP	CEP
5574,00	SSB			RDARA 13G	
5574,00	SSB			RDARA 2B	
5574,00	SSB			RDARA 2C	
5574,00	SSB			RDARA 4B	
5574,00	SSB			RDARA 6D	
5574,00	USB			San Francisco ACC , CA	USA
5577,00	SSB			RDARA 10E	
5577,00	SSB			RDARA 13C	
5577,00	SSB			RDARA 13J	
5577,00	SSB			RDARA 13K	
5577,00	SSB			RDARA 1C	
5577,00	SSB			RDARA 5A	
5577,00	SSB			RDARA 6G	
5577,00	SSB			RDARA 7B	
5580,00	SSB			RDARA 14G	
5580,00	SSB			RDARA 3A	
5580,00	SSB			RDARA 3B	
5580,00	SSB			RDARA 6A	
5580,00	SSB			RDARA 6C	
5580,00	SSB			VOLMET CAR	CAR
5580,00	USB			ARM MIL net	ARM
5581,00	ALE USB	B=125Bd Ch=8 ChS=250Hz		ISR AF net	ISR
5581,30	STANAG 4539	B=2400Bd		UI MIL user	
5583,00	SSB			RDARA 10B	
5583,00	SSB			RDARA 12E	
5583,00	SSB			RDARA 12F	
5583,00	SSB			RDARA 12H	
5583,00	SSB			RDARA 13E	
5583,00	SSB			RDARA 13F	
5583,00	SSB			RDARA 1E	
5583,00	SSB			RDARA 5A	
5583,00	SSB			RDARA 5C	
5583,00	SSB			RDARA 6G	

Frequency	Mode	Mode Parameter	Callsign	User	Country
5583,00	SSB			RDARA 7B	
5583,00	SSB			RDARA 9	
5583,00	HFDL	B=1800Bd	H05**	Auckland	NZL
5586,00	SSB			RDARA 10D	
5586,00	SSB			RDARA 2C	
5586,00	SSB			RDARA 3C	
5589,00	HFDL	B=1800Bd	H11**	Albrook	PNR
5589,00	SSB			RDARA 12C	
5589,00	SSB			VOLMET MID	MID
5592,00	SSB			RDARA 6G	
5592,00	SSB			RDARA 7C	
5592,00	SSB			RDARA 9D	
5592,00	SSB			VOLMET NAT	NAT
5595,00	SSB			RDARA 10C	
5595,00	SSB			RDARA 12E	
5595,00	SSB			RDARA 1C	
5595,00	SSB			RDARA 2B	
5595,00	SSB			RDARA 6B	
5596,00	USB			Workuta ACC	RUS
5596,00	USB			Arkhangelsk ACC	RUS
5596,00	USB			Syktyvkar ACC	RUS
5596,00	USB			Amderma ACC	RUS
5596,00	USB			Nirjan-Mar ACC	RUS
5596,00	USB			Pecora ACC	RUS
5596,00	USB			Ukhta ACC	RUS
5596,00	USB			Vorkuta ACC	RUS
5596,00	USB			Kotlas ACC	RUS
5596,00	USB			Kargopol ACC	RUS
5596,00	USB			Leshukonskoye ACC	RUS
5596,00	USB			RUS areo network day	
5598,00	USB			Paramaribo ACC	SUR
5598,00	SSB			Canarias ACC , Las Palmas	CNR
5598,00	SSB			MWARA NAT	NAT
5598,00	SSB		KEA5	New York ACC	USA
5598,00	SSB			Piarco ACC	TRD
5598,00	SSB			RDARA 6G	
5598,00	SSB		CSY	Santa Maria ACC	AZR
5598,00	USB		EIP	Shanwik ACC	IRL
5598,00	SSB		VFG	Gander ACC	CAN
5600,00	USB			EGY N net	EGY
5601,00	SSB			RDARA 3A	
5601,00	SSB			RDARA 3B	
5601,00	SSB			RDARA 6A	
5601,00	SSB			VOLMET Asuncion	ASC
5601,00	SSB			VOLMET SAM	SAM
5601,00	USB		LWB	Ezeiza VOLMET	ARG
5604,00	SSB			RDARA 10	
5604,00	SSB			RDARA 12A	
5604,00	SSB			RDARA 12E	
5604,00	SSB			RDARA 12F	
5604,00	SSB			RDARA 13E	
5604,00	SSB			RDARA 13F	
5604,00	SSB			RDARA 13K	
5604,00	SSB			RDARA 14	
5604,00	SSB			RDARA 2A	
5604,00	SSB			RDARA 2C	
5604,00	SSB			RDARA 4B	
5604,00	SSB			RDARA 6G	
5607,00	SSB			RDARA 2B	
5610,00	SSB			RDARA 6G	
5610,00	ALE USB	B=125Bd Ch=8 ChS=250Hz			

Frequency	Mode	Mode Parameter	Callsign	User	Country
5613,00	SSB			RDARA 12C	
5613,00	SSB			RDARA 2B	
5616,00	SSB			Iceland ACC , Reykjavik	ISL
5616,00	SSB			MWARA NAT	NAT
5616,00	SSB		KEA5	New York ACC	USA
5616,00	SSB			RDARA 6G	
5616,00	SSB		CSY	Santa Maria ACC	AZR
5616,00	USB		EIP	Shanwik ACC	IRL
5616,00	SSB		VFG	Gander ACC	CAN
5619,00	SSB			RDARA 12J	
5619,00	SSB			RDARA 2B	
5620,00	CW			RUS Strategic AF net	RUS
5620,00	BPSK	B=2000Bd			
5622,00	SSB			RDARA 1D	
5622,00	SSB			RDARA 6G	
5622,00	HFDL	B=1800Bd	H14**	Krasnoyarsk	RUS
5623,00	USB			EGY N net	EGY
5625,00	SSB			RDARA 10D	
5625,00	SSB			RDARA 3A	
5625,00	SSB			RDARA 5B	
5625,00	SSB			RDARA 6B	
5627,00	MIL 188-110B SER	B=2400Bd		ROU F	ROU
5628,00	USB			San Francisco ACC , CA	USA
5628,00	SSB			MWARA NP	NP
5628,00	SSB			RDARA 1D	
5628,00	SSB			RDARA 6G	
5628,00	SSB			Tokyo ACC	J
5631,00	SSB			RDARA 10A	
5631,00	SSB			RDARA 6D	
5634,00	USB			Mumbai ACC	IND
5634,00	USB			Colombo ACC	CLN
5634,00	USB			Dar es Salaam ACC	TZA
5634,00	USB			Brisbane ACC	AUS
5634,00	USB			Harare ACC	ZWE
5634,00	USB			Jeddah ACC	ARS
5634,00	USB			Kigali ACC	RRW
5634,00	USB			Lilongwe ACC	MWI
5634,00	USB			Madras ACC	IND
5634,00	USB			Mahajanga ACC	MDG
5634,00	USB			Male ACC	MLD
5634,00	USB			Nairobi ACC	KEN
5634,00	USB			St. Denis ACC	REU
5634,00	USB			Toamasina ACC	MDG
5634,00	USB			Beira ACC	MOZ
5634,00	SSB			Antananarivo ACC	MDG
5634,00	USB			Cocos Island ACC	ICO
5634,00	SSB			Lusaka ACC	ZMB
5634,00	SSB			Mauritius ACC , Plaisance	MAU
5634,00	SSB			Moroni ACC	COM
5634,00	SSB			MWARA INO	INO
5634,00	SSB			Perth ACC	AUS
5634,00	SSB			RDARA 6G	
5634,00	SSB			Seychelles ACC , Mahe	SEY
5637,00	SSB			RDARA 1D	
5637,00	SSB			RDARA 3C	
5637,00	USB			GRC aero net	GRC
5637,00	USB			Naxos R	GRC
5637,00	USB			Kalamata R	GRC
5637,00	USB			Kassos R	GRC
5637,00	USB			Kavala R	GRC
5637,00	USB			Iraklion R	GRC
5637,00	USB			Paros R	GRC

Frequency	Mode	Mode Parameter	Callsign	User	Country
5637,00	USB			Samos R	GRC
5637,00	USB			Limnos R	GRC
5637,00	USB			Preveza R	GRC
5637,00	USB			Skiathos R	GRC
5637,00	USB			Patras R	GRC
5637,00	USB			Kastoria R	GRC
5637,00	USB			Astypalaia R	GRC
5637,00	USB			Chios R	GRC
5637,00	USB			Alexandroupoli R	GRC
5637,00	USB			Zakynthos R	GRC
5637,00	USB			Athens R	GRC
5637,00	USB			Kozani R	GRC
5637,00	USB			Ikaria R	GRC
5637,00	USB			Corfu R	GRC
5637,00	USB			Kefalonia R	GRC
5637,00	USB			Kithira R	GRC
5637,00	USB			Kos R	GRC
5637,00	USB			Leros R	GRC
5637,00	USB			Thessaloniki R	GRC
5637,00	USB			Mykonos R	GRC
5637,00	USB			Mytilene R	GRC
5637,00	USB			Samos R	GRC
5637,00	USB			Santorini R	GRC
5637,00	USB			Skyros R	GRC
5637,00	USB			Chania R	GRC
5640,00	SSB			RDARA 6G	
5643,00	USB			Brisbane ACC	AUS
5643,00	USB			Pascua ACC	CHL
5643,00	USB			Port Vila ACC	VUT
5643,00	USB			Rarotonga ACC	CKH
5643,00	USB			San Francisco ACC , CA	USA
5643,00	USB			Tahiti ACC	OCE
5643,00	USB			Wallis ACC	WAL
5643,00	USB			Auckland ACC	NZL
5643,00	SSB			MWARA SP	SP
5643,00	SSB			Nandi ACC	FJI
5643,00	SSB			RDARA 3C	
5646,00	USB			Ivdel ACC	RUS
5646,00	USB			Khanty-Mansiysk ACC	RUS
5646,00	USB			Moscow ACC	RUS
5646,00	USB			Syktyvkar ACC	RUS
5646,00	USB			Vologda ACC	RUS
5646,00	SSB			MWARA NCA	NCA
5646,00	SSB			RDARA 12G	
5649,00	USB			Guangzhou ACC	CHN
5649,00	USB			Irkutsk ACC	RUS
5649,00	USB			Pyongyang ACC	KRE
5649,00	USB			Ulaanbaatar ACC	MNG
5649,00	USB			Bali ACC	INS
5649,00	USB			Bangkok ACC	THA
5649,00	SSB			Iceland ACC , Reykjavik	ISL
5649,00	SSB			MWARA NAT	NAT
5649,00	SSB			MWARA SEA	SEA
5649,00	USB		EIP	Shanwik ACC	IRL
5649,00	SSB		VFG	Gander ACC	CAN
5649,00	USB			Hanoi ACC	VTN
5649,00	USB			Ho Chi Minh ACC	VTN
5649,00	USB			Jakarta ACC	INS
5649,00	USB			Kuala Lumpur ACC	MLA
5649,00	USB			Kota Kinabalu ACC	MLA
5649,00	USB			Manila ACC	PHL
5649,00	USB			Seoul ACC	KOR

Frequency	Mode	Mode Parameter	Callsign	User	Country
5649,00	USB			Singapore ACC	SNG
5649,00	USB			Tokyo ACC	J
5649,00	USB			Vientianne ACC	LAO
5652,00	SSB			Alger ACC	ALG
5652,00	SSB			Kano ACC	NIG
5652,00	SSB			MWARA AFI	AFI
5652,00	SSB			MWARA CWP	CWP
5652,00	SSB			Niamey ACC	NIG
5652,00	SSB			N`djamena ACC	TCD
5652,00	SSB			Tamanrasset ACC	ALG
5652,00	SSB			Tripoli ACC	LBY
5652,00	SSB			Tunis ACC	TUN
5652,00	USB			Gao ACC	MLI
5652,00	USB			Timimoun ACC	ALG
5652,00	HFDL	B=1800Bd	H04**	Riverhead , NY	USA
5652,00	USB			San Francisco ACC , CA	USA
5655,00	SSB			Bangkok ACC	THA
5655,00	SSB			Hanoi ACC	VTN
5655,00	SSB			Ho Chi Min Ville ACC	VTH
5655,00	SSB			Honkong ACC	HKG
5655,00	SSB			Manila ACC	PHL
5655,00	SSB			MWARA EA	EA
5655,00	SSB			MWARA SEA	SEA
5655,00	SSB			Singapore ACC	SNG
5655,00	SSB			Vientiane ACC	LAO
5655,00	HFDL	B=1800Bd	H06**	Hat Yai	THA
5655,00	USB			Guangzhou ACC	CHN
5655,00	USB			Irkutsk ACC	RUS
5655,00	USB			Pyongyang ACC	KRE
5655,00	USB			Ulaanbaatar ACC	MNG
5655,00	USB			Bali ACC	INS
5655,00	USB			Jakarta ACC	INS
5655,00	USB			Kuala Lumpur ACC	MLA
5655,00	USB			Kota Kinabalu ACC	MLA
5655,00	USB			Seoul ACC	KOR
5655,00	USB			Tokyo ACC	J
5655,40	STANAG 4539	B=2400Bd		UI MIL user	
5656,00	CCIR 493-4	B=100Bd S=170Hz			
5658,00	SSB			Bishpek ACC	KGZ
5658,00	USB			Abadan ACC	IRN
5658,00	USB			Almaty ACC	KAZ
5658,00	USB			Ashkabad ACC	TKM
5658,00	USB			Kuwait ACC	KWT
5658,00	SSB			Mumbai ACC	IND
5658,00	SSB			Dushanbe ACC	TJK
5658,00	SSB			Kabul ACC	AFG
5658,00	SSB			Karachi ACC	PAK
5658,00	SSB			Kathmandu ACC	NPL
5658,00	SSB			Lahore ACC	PAK
5658,00	SSB			Muscat ACC	OMA
5658,00	SSB			MWARA AFI	AFI
5658,00	SSB			MWARA MID	MID
5658,00	SSB			New Delhi ACC	IND
5658,00	SSB			Teheran ACC	IRN
5658,00	SSB			Urumchi ACC	CHN
5658,00	USB			Male ACC	MLD
5658,00	USB			Odessa ACC	UKR
5658,00	USB			Tbilisi ACC	GEO
5658,00	USB			Yerevan ACC	ARM
5661,00	USB			Vologda ACC	RUS
5661,00	SSB			MWARA CWP	CWP
5661,00	SSB			MWARA EUR	EUR

Frequency	Mode	Mode Parameter	Callsign	User	Country
5661,00	USB			Arkhangelsk ACC	RUS
5661,00	USB			Beirut ACC	LBN
5661,00	USB			Berlin ACC	D
5661,00	USB			Kiev ACC	UKR
5661,00	USB			Lvov ACC	UKR
5661,00	USB			MinskACC	BLR
5661,00	USB			Moscow ACC	RUS
5661,00	USB			Murmansk ACC	RUS
5661,00	USB			Odessa ACC	UKR
5661,00	USB			Riga ACC	LVA
5661,00	USB			Simferopol ACC	UKR
5661,00	USB			Sofia ACC	BUL
5661,00	USB			St. Petersburg ACC	RUS
5661,00	USB			Syktyvkar ACC	RUS
5661,00	USB			Tunis ACC	TUN
5661,00	USB			Velikiye ACC	RUS
5661,00	USB			Vilnius ACC	LTU
5664,00	SSB			Irkutsk ACC	RUS
5664,00	SSB			Khabarovsk ACC	RUS
5664,00	SSB			MWARA NCA	NCA
5664,00	SSB			Pyongyang ACC	KRE
5664,00	SSB			Ulan Bator ACC	MNG
5664,00	SSB			Ulan Ude ACC	RUS
5664,00	USB			Chita ACC	RUS
5664,00	USB			Chulman ACC	RUS
5664,00	USB			Ekimchan ACC	RUS
5664,00	USB			Kirensk ACC	RUS
5666,00	USB			EGY N net	EGY
5667,00	SSB			Aden ACC	YEM
5667,00	SSB			Amman ACC	JOR
5667,00	SSB			Ankara ACC	TUR
5667,00	SSB			Beirut ACC	LBN
5667,00	SSB			Damascus ACC	SYR
5667,00	SSB			Kuwait ACC	KWT
5667,00	SSB			MWARA MID	MID
5667,00	SSB			Teheran ACC	IRN
5667,00	USB			Sanaa ACC	YEM
5667,00	USB			Cairo ACC	EGY
5667,00	USB			Jeddah ACC	ARS
5667,00	USB			Manama ACC	BHR
5667,00	USB			Odessa ACC	UKR
5667,00	USB			Simferropol ACC	UKR
5667,00	USB			Tbilisi ACC	GEO
5667,00	USB			Yerevan ACC	ARM
5670,00	USB			Bali ACC	INS
5670,00	SSB			Kuala Lumpur ACC	MLA
5670,00	SSB			Colombo ACC	CLN
5670,00	SSB			Madras ACC	IND
5670,00	SSB			MWARA EA	EA
5670,00	USB			Bangkok ACC	THA
5670,00	USB			Calcutta ACC	IND
5670,00	USB			Dhaka ACC	BGD
5670,00	USB			Guangzhou ACC	CHN
5670,00	USB			Jakarta ACC	INS
5670,00	USB			Kathmandu ACC	NPL
5670,00	USB			Kunming ACC	CHN
5670,00	USB			Male ACC	MLD
5670,00	USB			Singapore ACC	SNG
5670,00	USB			Yangon ACC	BRM
5673,00	LSB			Beijing VOLMET	CHN
5673,00	SSB			VOLMET SEA	SEA
5673,00	SSB			Guangzhou VOLMET	CHN

Frequency	Mode	Mode Parameter	Callsign	User	Country
5676,00	SSB			VOLMET NCA	NCA
5676,50	STANAG 4285	B=2400Bd	TBB	TUR N Ankara	TUR
5677,00	USB			Sam Francisco ACC , CA	USA
5677,00	USB			Tokyo ACC	J
5680,00	USB		DHN66	NATO Geilenkirchen	D
5680,00	USB		LOK	ARG N Laurie Island , Orcadas	ANT
5680,00	USB		CWM30	URG SAR net	URG
5680,00	SSB			NASA launch support ships	USA
5680,00	SSB			Bodoe Rescue	NOR
5680,00	SSB			D N Glücksburg	D
5680,00	SSB			Distress/calling frequency	WW
5680,00	SSB			DNK AF Karup	DNK
5680,00	SSB			DNK AF Rönne	DNK
5680,00	SSB			DNK AF Skagen	DNK
5680,00	SSB			G AF Finningley	G
5680,00	SSB			Gotland Rescue	S
5680,00	SSB			Ijmuiden Rescue	HOL
5680,00	SSB			Karup Recue	DNK
5680,00	USB			Kinloss Rescue	G
5680,00	SSB			Koksijde Rescue	DNK
5680,00	SSB			Longyear Rescue	SVB
5680,00	SSB			NOR AF Stavanger	NOR
5680,00	SSB			Riga Rescue	LVA
5680,00	SSB			S AF Stockholm	S
5680,00	SSB		ONY77	BEL AF Koksiyde	BEL
5680,00	SSB		SAE	Tingstaedte Air	S
5680,00	USB			Wellington Mountain Radio Service	NZL
5684,00	SSB			USA AF Strategic Command	USA
5684,00	LSB		INSPIRA**	MLA N ship	MLA
5686,50	ALE USB	B=125Bd Ch=8 ChS=250Hz			
5687,00	USB		DHJ83	D AF Gatow	D
5687,00	USB		DHM91	D AF Münster	D
5687,00	USB		DHO78	D AF Kalkar	D
5689,00	RUS TACTICAL DATALINK	B=1200Bd S=800Hz Ch=1200		RUS MIL	RUS
5690,00	ALE USB	B=125Bd Ch=8 ChS=250Hz		USA FAA net	USA
5690,00	USB		0A**	IRL N AF Dublin	IRL
5691,00	SSB		UGEF	Kirensk VOLMET	RUS
5691,00	SSB			Irkutsk VOLMET	RUS
5696,00	SSB			G AF Culdrose	G
5696,00	SSB			USCG New Orleans , LO	USA
5696,00	SSB			USCG Guantanamo Bay	GTB
5696,00	SSB			USCG Kodiak	ALS
5696,00	SSB			USCG Miami , FL	USA
5696,00	SSB			USCG San Francisco , CA	USA
5696,00	SSB			USCG San Juan	PTR
5698,00	USB			Vologda ACC	RUS
5700,00	SSB			USA AF Strategic Command	USA
5702,00	ALE USB	B=125Bd Ch=8 ChS=250Hz		USA AF net	
5708,00	SSB			USAF Lajes	POR
5711,00	USB			Cape Canaveral AFS , FL	USA
5713,00	ALE USB	B=125Bd Ch=8 ChS=250Hz			
5714,00	BAUDOT	B=50Bd S=300Hz		SRB MIL net	SRB
5715,00	STANAG 4481	B=50Bd S=850Hz	NAU	USA N San Juan	PTR
5715,20	*STANAG 4285*	*B=2400Bd*		*G MIL TX Inskip*	*G*
5717,00	USB		CHR	CAN AF Trenton	CAN
5717,00	SSB		CJX	CAN AF St. John`s	CAN

Frequency	Mode	Mode Parameter	Callsign	User	Country
5718,90	ALE USB	B=125Bd Ch=8 ChS=250Hz			
5720,00	HFDL	B=1800Bd	H03**	Reykjavik	ISL
5723,00	ALE USB	B=125Bd Ch=8 ChS=250Hz		SVK Mil net	SVK
5726,00	USB		NPX	USARP Amundson-ScottSouth Pole Station	ANT
5728,50	*SITOR B*	*B=100Bd S=170Hz*	*HBM46*	*SUI MIL Berne*	*SUI*
5729,30	STANAG 4539	B=2400Bd		UI MIL user	
5730,00	ALE USB	B=125Bd Ch=8 ChS=250Hz		CHL SENAPRED net	CHL
5731,20	*STANAG 4285*	*B=2400Bd*		*G MIL TX St. Eval*	*G*
5733,00	USB			Bali ACC	INS
5733,00	USB			Brisbane ACC	AUS
5733,00	USB			Jakarta ACC	INS
5733,00	USB			Male ACC	MLD
5733,00	USB			Singapore ACC	SNG
5733,00	USB			Ujung Pandang ACC	INS
5733,00	USB			MWARA SEA	
5733,50	CW		PWZW	RUS Strategic Command Aerospace Defense	RUS
5734,00	USB			HFoZ/Radtel net	AUS
5735,00	PACTOR II	B=200Bd S=200Hz	HPPM1	Sailmail Chiriqui	PNR
5735,00	PACTOR III	B=100Bd S=200Hz	HPPM3	Sailmail Panama	PNR
5736,00	MIL 188-110B APP C	B=2400Bd			
5740,00	LINK 22	B=2250Bd			
5740,00	PACTOR III	B=100Bd S=200Hz	WQAB964	Sailmail San Diego , CA	USA
5742,20	*STANAG 4539*	*B=2400Bd*		*Croughton*	*G*
5744,00	USB			HFoZ/Radtel net	AUS
5744,00	USB		VMS469	Reids Radiodata net	AUS
5745,00	MFSK32	B=31,25Bd Ch=16 ChS=39,375Hz		VOA Radiogram , TX Edward R. Murrow , NC	USA
5745,00	MFSK8	B=7,8125Bd Ch=32 ChS=9,875Hz		VOA Radiogram , TX Edward R. Murrow , NC	USA
5746,00	CW			RUS NHQ Vladivostok	RUS
5748,20	*STANAG 4539*	*B=2400Bd*		*Croughton*	*G*
5748,30	ALE USB	B=125Bd Ch=8 ChS=250Hz		G MIL net	G
5748,60	ALE USB	B=125Bd Ch=8 ChS=250Hz		USA diplo net	
5749,90	MIL 188.110A SER	B=2400Bd		S MIL	S
5749,90	ALE 3G	B=2400Bd		S MIL	S
5750,00	USB			GRC CG	GRC
5750,00	HAARP WAVEFORM			HAARP	USA
5750,00	ALE USB	B=125Bd Ch=8 ChS=250Hz		JOR net	JOR
5750,00	ALE USB	B=125Bd Ch=8 ChS=250Hz		EGY net	EGY
5751,00	ISR N HYBRID MODEM	B=2400Bd	4XZ	ISR N Haifa	ISR
5751,00	*CIS 12*	*B=1440Bd Ch=12 ChS=200Hz*		*RUS MIL Smolensk*	*RUS*
5752,00	CW			RUS Operational Strategic Command Air Defense net	RUS
5752,50	PACTOR	B=200Bd S=170Hz	HCHRVME**	UNHCR	HRV
5752,50	PACTOR	B=200Bd S=170Hz	HCHRVZA**	UNHCR Zagreb	HRV
5753,00	*CW*		*RIT*	*RUS N HQ Severomorsk*	*RUS*
5755,00	ALE USB	B=125Bd Ch=8 ChS=250Hz		F Nnet	
5755,00	FAX 120/576		VMW	Wiluna M	AUS

Frequency	Mode	Mode Parameter	Callsign	User	Country
5757,20	*STANAG 4539*	*B=2400Bd*		*Croughton*	*G*
5757,80	ALE 3G	B=2400Bd		UI MIL net	
5758,70	*STANAG 4285*	*B=2400Bd*	*FUO*	*F N Toulon*	*F*
5760,00	ALE USB	B=125Bd Ch=8 ChS=250Hz		USA SHARES net	USA
5760,00	USB			USA Army MARS	USA
5760,00	ALE USB	B=125Bd Ch=8 ChS=250Hz		SVK AF net	SVK
5763,00	USB			TUR F net	TUR
5763,00	USB			USA Army MARS	USA
5763,00	USB			TUR N	TUR
5766,00	USB			USA Army MARS	USA
5768,00	CW			CHN air defense net	CHN
5770,00	BAUDOT	B=50Bd S=300Hz		SRB MIL net	SRB
5770,00	CW		RWI2	RUS MIL	RUS
5771,00	ALE USB	B=125Bd Ch=8 ChS=250Hz		POL MIL net	POL
5773,10	STANAG 4538	B=2400Bd			
5775,00	CW		RCV	RUS N HQ Sevastopol	UKR
5776,00	ALE USB	B=125Bd Ch=8 ChS=250Hz		POL MIL net	POL
5777,00	ALE USB	B=125Bd Ch=8 ChS=250Hz		LTU MIL net	LTU
5777,50	ALE 3G	B=2400Bd		UI MIL net	
5778,50	ALE USB	B=125Bd Ch=8 ChS=250Hz		USA AF net	USA
5779,40	ALE 3G	B=2400Bd		UI MIL net	
5780,00	CW			RUS MIL	RUS
5781,00	ALE USB	B=125Bd Ch=8 ChS=250Hz		AZE emergency net	AZE
5782,00	ALE USB	B=125Bd Ch=8 ChS=250Hz		SVK AF net	SVK
5784,00	ALE USB	B=125Bd Ch=8 ChS=250Hz			
5784,00	MIL 188-110A SER	B=2400Bd			
5785,00	ALE USB	B=125Bd Ch=8 ChS=250Hz		I Guardia di Finanza net	I
5785,50	CW			RUS MIL net	RUS
5786,00	ALE USB	B=125Bd Ch=8 ChS=250Hz		R&S test net	D
5786,00	ALE USB	B=125Bd Ch=8 ChS=250Hz		AUT MIL net	AUT
5787,00	BAUDOT	B=200Bd S=500Hz		RUS INTEL	RUS
5787,50	ALE USB	B=125Bd Ch=8 ChS=250Hz		POL MIL net	POL
5787,50	MIL 188-110B SER	B=2400Bd		POL MIL	POL
5788,90	ALE 3G	B=2400Bd		UI MIL net	
5789,00	ALE USB	B=125Bd Ch=8 ChS=250Hz		EST F net	EST
5790,00	CIS 1200	B=1200Bd		RUS	RUS
5790,00	*STANAG 4285*	*B=2400Bd*	*TBB*	*TUR N Izmir*	*TUR*
5791,50	ALE USB	B=125Bd Ch=8 ChS=250Hz		TUR Ministry of Health net	TUR
5796,80	ALE USB	B=125Bd Ch=8 ChS=250Hz		IRQ ISOF net	IRQ
5797,00	CW		RAL2	RUS N HQ Astrakhan	RUS
5798,60	ALE 3G	B=2400Bd		UI MIL net	
5798,60	STANAG 4539	B=2400Bd		UI MIL user	
5800,00	HAARP waveform			HAARP	ALS
5800,00	*BPSK*	*B=1284Bd*		*Sevastopol*	*UKR*
5800,00	CW		RYF2	RUS PTT	RUS
5803,00	USB		VFA	Inuvik R	CAN
5805,00	CW		RMP	RUS N HQ Kaliningrad	RUS

Frequency	Mode	Mode Parameter	Callsign	User	Country
5805,00	CW		RJQ86	RUS N	RUS
5807,80	ALE 3G	B=2400Bd		UI MIL net	
5807,80	STANAG 4539	B=2400Bd		UI MIL user	
5808,00	ALE USB	B=125Bd Ch=8 ChS=250Hz		S MIL net	S
5808,00	CW		RCB	RUS AF	RUS
5808,00	ALE USB	B=125Bd Ch=8 ChS=250Hz		G MIL net	G
5808,80	PACTOR III	B=100Bd S=200Hz	WPTG385	Sailmail Corpus Christi , TX	USA
5809,20	*STANAG 4285*	*B=2400Bd*		*NOR N TX Stavanger*	*NOR*
5810,00	SSB			NASA ETR night channel	USA
5810,00	ALE USB	B=125Bd Ch=8 ChS=250Hz		AZE emergency net	AZE
5810,00	USB			BGD N net	BGD
5811,70	PACTOR II	B=100Bd S=200Hz	STAT151**	MFA Tunis	TUN
5813,00	*STANAG 4285*	*B=2400Bd*		*G MIL TX St. Eval*	*G*
5815,00	ALE USB	B=125Bd Ch=8 ChS=250Hz		POL MIL net	POL
5817,00	CW			RUS HFDF net	RUS
5821,00	ALE USB	B=125Bd Ch=8 ChS=250Hz		USA FEMA net	USA
5821,00	ALE USB	B=125Bd Ch=8 ChS=250Hz		CHL SENAPRED net	CHL
5821,00	CHP200 ALE	B=250Bd ChS=170Hz			
5823,00	CW		RAL2	RUS N HQ Astrakhan	RUS
5825,00	CW			RUS HFDF net	RUS
5826,00	SSB			USA AF Strategic Command	USA
5828,00	SSTV			SURA TX	RUS
5830,00	PACTOR III	B=100Bd S=200Hz	WHV382	Sailmail Friday Habor , WA	USA
5835,00	USB		VNJ	Casey Base , AUS	ANT
5835,00	USB			AUS Davis Base	ANT
5835,00	USB			Macquarie Island , AUS	ANT
5835,00	USB			Mawson Base , AUS	ANT
5835,00	CW			RUS Strategic AF net	RUS
5836,00	PACTOR II	B=200Bd S=200Hz	KUZ533	Sailmail Honululu	HWA
5836,80	ALE 3G	B=2400Bd		UI MIL net	
5838,00	CW		RIT	RUS N HQ Severomorsk	RUS
5838,00	CW			RUS MIL	RUS
5838,60	ALE3G	B=2400Bd			
5840,00	CW			RUS HFDF net	RUS
5843,00	USB			EGY N net	EGY
5844,70	ALE 3G	B=2400Bd		UI MIL net	
5846,50	SITOR B	B=100Bd S=170Hz	HBM46	SUI MIL N Geneva	SUI
5847,50	ALE USB	B=125Bd Ch=8 ChS=250Hz		GRC AF net	GRC
5848,20	*STANAG 4285*	*B=2400Bd*		*S MIL Goeteborg*	*S*
5848,20	*STANAG 4285*	*B=2400Bd*		*DNK N TX Frederikshavn*	*DNK*
5850,75	BAUDOT	B=50Bd S=340Hz		SRB MIL net	SRB
5853,00	CW		REA4	RUS AF Moscow	RUS
5853,00	CIS 50-1000	B=50Bd S=1000Hz	REA4	RUS AF HQ Moscow	RUS
5853,00	CW		REA4	RUS AF HQ Moscow	RUS
5855,00	CW		RCV	RUS N HQ Sevastopol	UKR
5855,00	CIS ARQ	B=100Bd S=500Hz		RUS PTT	RUS
5857,50	FAX 120/576		HLL2	Seoul M	KOR
5857,50	ALE USB	B=125Bd Ch=8 ChS=250Hz		GRC AF net	GRC
5859,00	ALE USB	B=125Bd Ch=8 ChS=250Hz			
5859,40	PACTOR II	B=200Bd S=200Hz	WPTG385	Sailmail Corpus Christi , TX	USA
5860,00	ALE USB	B=125Bd Ch=8 ChS=250Hz		USA FAA net	USA
5861,40	PACTOR II	B=200Bd S=200Hz	WHV861	Sailmail San Luis Obispo , CA	USA

Frequency	Mode	Mode Parameter	Callsign	User	Country
5861,70	*CW*		*P***	*RUS N Kaliningrad*	*RUS*
5862,00	MFSK32	B=31,25Bd Ch=16 ChS=39,375Hz	AFA0SN	USA MARS	USA
5862,00	*T600*	*B=50Bd S=200Hz*	*RMP*	*RUS N HQ Kaliningrad*	*RUS*
5864,50	CW		RMP	RUS N HQ Kaliningrad	RUS
5867,00	CIS OFDM 121	B=2541Bd Ch=121 ChS=25Hz		RUS E Lisbon	POR
5870,00	PACTOR II	B=200Bd S=200Hz	HPPM2	Sailmail Chiriqui	PNR
5870,70	*STANAG 4285*	*B=2400Bd*	*IDR*	*I N TX Rome*	*I*
5871,50	ALE 3G	B=2400Bd		UI MIL net	
5873,00	CW			RUS Operational Strategic Command Air Defense net	RUS
5874,00	CW			RUS MIL	RUS
5874,50	ALE 3G	B=2400Bd		UI MIL net	
5876,00	CIS 50-500	B=50Bd S=500Hz		RUS	RUS
5876,40	PACTOR II	B=200Bd S=200Hz	KZN508	Sailmail Rockhill , SC	USA
5881,00	*CW*		*RMP*	*RUS N HQ Kaliningrad*	*RUS*
5881,40	PACTOR III	B=100Bd S=200Hz	WQLI952	Sailmail Watsonville , CA	USA
5881,40	PACTOR II	B=200Bd S=200Hz	WRD719	Sailmail Palo Alto	USA
5882,00	CW			RUS MIL net	RUS
5883,70	ALE 3G	B=2400Bd		UI MIL net	
5884,00	ALE USB	B=125Bd Ch=8 ChS=250Hz		F N net	F
5885,00	ALE USB	B=125Bd Ch=8 ChS=250Hz		AZE emergency net	AZE
5885,00	ALE USB	B=125Bd Ch=8 ChS=250Hz			
5887,00	CW			RUS Air Defense net	RUS
5888,00	CW			RUS N net	RUS
5890,00	*T600*	*B=50Bd S=200Hz*		*RUS N Murmansk*	*RUS*
5892,00	ALE USB	B=125Bd Ch=8 ChS=250Hz		VEN N net	VEN
5894,00	CW		RMP	RUS N HQ Kaliningrad	RUS
5895,00	CIS ARQ	B=100Bd S=500Hz		RUS PTT	RUS
5895,50	CW		LLE	Bergen-Kringkaster R	NOR
5897,40	PACTOR II	B=200Bd S=200Hz	WPUC469	Sailmail South Daytona , FL	USA
5900,00	ALE 3G	B=2400Bd		UI MIL net	
5901,00	ALE USB	B=125Bd Ch=8 ChS=250Hz		USA Department of Agriculture net	USA
5902,00	ALE USB	B=125Bd Ch=8 ChS=250Hz		AZE emergency net	AZE
5905,00	ISR N HYBRID MODEM	B=2400Bd	4XZ	ISR N Haifa	ISR
5906,00	USB			Paramaribo ACC	SUR
5909,50	ALE USB	B=125Bd Ch=8 ChS=250Hz		USA COTHEN net	USA
5915,00	ALE USB	B=125Bd Ch=8 ChS=250Hz		AUS Police NT	AUS
5915,00	ALE USB	B=125Bd Ch=8 ChS=250Hz		AUS Police QLD	AUS
5916,00	CW		RCV	RUS N HQ Sevastopol	UKR
5919,00	*CIS 12*	*B=1440Bd Ch=12 ChS=200Hz*	*RMP*	*RUS N HQ Kaliningrad*	*RUS*
5920,00	CW			RUS air defence net	RUS
5921,00	CW			RUS MOI net	RUS
5940,00	CIS 12	B=1440Bd Ch=12 ChS=200Hz		RUS MIL Babrusk	BLR
5945,00	ALE USB	B=125Bd Ch=8 ChS=250Hz		SAPOL Communication Infrastructure	AUS
5948,00	CW		RAL2	RUS N HQ Astrakhan	RUS
5961,00	ALE USB	B=125Bd Ch=8 ChS=250Hz		USA FEMA net	USA

Frequency	Mode	Mode Parameter	Callsign	User	Country
5971,00	ALE USB	B=125Bd Ch=8 ChS=250Hz		TUR Disaster and Emergency net	TUR
5985,00	CW		RGZ59	RUS landing ship NIKOLA FILCHENKOV	RUS
5986,00	CIS 12	B=1440Bd Ch=12 ChS=200Hz		Area of Slupsk	POL
6000,00	*STANAG 4285*	*B=2400Bd*	*TBB*	*TUR N Ankara*	*TUR*
6007,50	CW		RJD56	RUS N HQ Murmansk	RUS
6021,50	CW		RJD99	RUS N Severodvinsk	RUS
6043,00	CW		RCV	RUS N HQ Sevastopol	UKR
6043,00	CW		RCIV	RUS N ship	RUS
6049,00	ALE USB	B=125Bd Ch=8 ChS=250Hz		USA FEMA net	USA
6050,00	*STANAG 4285*	*B=2400Bd*	*PBC*	*NLD N Goeree Island*	*HOL*
6059,00	CIS 50-1000	B=50Bd S=1000Hz	REA4	RUS AF HQ Moscow	RUS
6059,00	CW		RMP	RUS N HQ Kaliningrad	RUS
6087,00	CW		REA4	RUS AF Moscow	RUS
6087,00	CW		REA4	RUS AF Moscow	RUS
6106,00	ALE USB	B=125Bd Ch=8 ChS=250Hz		USA FEMA net	USA
6110,00	CW		RMP	RUS N HQ Kaliningrad	RUS
6113,00	CW		RJD99	RUS N Severodvinsk	RUS
6114,00	STANAG 4481	B=75Bd S=850Hz		POR N Lisbon TX Azores	POR
6123,00	RUS TACTICAL DATALINK	B=1200Bd S=800Hz Ch=1200		RUS MIL	RUS
6151,00	ALE USB	B=125Bd Ch=8 ChS=250Hz		USA FEMA net	USA
6183,00	T600	B=50Bd S=200Hz	RCV	RUS N HQ Sevastopol	UKR
6199,00	F1B MORSE	S=200Hz	RIT	RUS N HQ Severomorsk	RUS
6200,00	LINK 22	B=2250Bd			
6200,00	SSB			SS Ch 601	
6200,00	SITOR B	B=100Bd S=170Hz	SVO	Olympia R	GRC
6202,00	LINK 11 SLEW	B=2400Bd			
6203,00	SSB			SS Ch 602	
6203,30	ALE 3G	B=2400Bd		UI MIL net	
6205,00	ALE USB	B=125Bd Ch=8 ChS=250Hz		MLT N net	MLT
6206,00	SSB			SS Ch 603	
6206,00	ALE USB	B=125Bd Ch=8 ChS=250Hz		AUS Police NSW	AUS
6208,40	ALE USB	B=125Bd Ch=8 ChS=250Hz		F	F
6209,00	ALE USB	B=125Bd Ch=8 ChS=250Hz			
6209,00	SSB			SS Ch 604	
6210,00	CIS 12	B=1440Bd Ch=12 ChS=200Hz		RUS MIL Orscha	BLR
6210,20	STANAG 4285	B=2400Bd	IDN	I N Napoli	I
6212,00	SSB			SS Ch 605	
6215,00	USB		NMF	USA CG Boston , MA	USA
6215,00	USB			Cape Town R	AFS
6215,00	USB		VNJ	Casey Base , AUS	ANT
6215,00	USB			AUS coast net	AUS
6215,00	USB			AUS Davis Base	ANT
6215,00	USB			USA CG Kodiak	ALS
6215,00	USB		CWC30	La Paloma R	URG
6215,00	USB		CWS	URG N Montevideo	URG
6215,00	USB		HCY	Ayora R , Santa Cruz	EQA
6215,00	USB		CWF	Punta Carretas R	URG
6215,00	USB			Macquarie Island , AUS	ANT
6215,00	USB		ZLM	Taupo Maritime R	NZL
6215,00	USB		CWC34	Punta del Este R	URG
6215,00	USB		CWC39	Montevideo Trouville R	URG
6215,00	USB		HSA	Bangkok R	THA

Frequency	Mode	Mode Parameter	Callsign	User	Country
6215,00	USB		JNA	J CG HQ Tokyo	J
6215,00	USB			Mawson Base , AUS	ANT
6215,00	USB		VFF	Iqaluit R	CAN
6215,00	USB		H4H	Honiara R	SLM
6215,00	USB		A3A	Nukualofa R	TON
6215,00	USB		VRC	Hong Kong R	CHN
6215,00	USB		NMA	USA CG Miami , FL	USA
6215,00	USB		3DZ	Suva RCC	FJI
6215,00	USB		PKZ34	Cigading R , Jawa	INS
6215,00	.			SS calling frequency	WW
6215,00	SSB			SS Ch 606	
6215,00	USB		PKA	Sabang R , We	INS
6215,00	USB		PKC	Palembang R , Sumatera	INS
6215,00	USB		PKD	Surabaya R , Djawa	INS
6215,00	USB		PKD3	Lembar R , Lombok	INS
6215,00	USB		PKE5	Ternate R , Maluku	INS
6215,00	USB		PKF3	Kendari R	INS
6215,00	USB		PKK	Kupang R , Timor	INS
6215,00	USB		PKM25	Tahuna R	INS
6215,00	USB			Samoa Port Authority	SMA
6215,00	USB		PKO	Pertamina Merak R , Jawa	INS
6215,00	USB		PKP	Dumai R , Sumatera	INS
6215,00	USB		PKP38	Batu Ampar R	INS
6215,00	USB		PKR3	Cilacap R , Jawa	INS
6215,00	USB		PKY2	Biak R , Irian Jaya	INS
6215,00	USB		PKY23	Fak Fak R	INS
6215,00	USB		PKY3	Manokwari R , Irian Jaya	INS
6215,00	USB		PKY4	Sorong R , Irian Jaya	INS
6215,00	USB		PKY5	Merauke R , Irian Jaya	INS
6215,00	USB		P2M	Port Moresbay R	PNG
6215,00	USB		PNK	Jayapura R , Irian Jaya	INS
6215,00	USB		3DP	Suva R	FJI
6215,00	USB			USA CG Honolulu	HWA
6215,00	USB		PKG	Bandjarmasin R , Kalimantan	INS
6215,00	USB			USA CG Point Reyes	USA
6215,00	USB		PKE37	Sanana R , Kep Sula	INS
6215,00	USB		PKM	Bitung R , Sulawesi	INS
6215,00	USB		PKM44	Pantolan R , Sulawesi	INS
6215,00	USB		PKJ28	Sei Kolak Kijang R	INS
6215,00	USB			Benete R , Sumbawa	INS
6215,50	USB		PKB	Belawan R , Sumatera	INS
6215,50	USB		PKB3	Sibolga R , Sumatera	INS
6215,50	USB		PKC4	Panjang R , Sumatera	INS
6215,50	USB		PKD5	Benoa R , Bali	INS
6215,50	USB		PKE	Amboina R , Ceram	INS
6215,50	USB		PKF	Makassar R , Sulawesi	INS
6215,50	USB		PKN	Balikpapan R , Kalimantan	INS
6215,50	USB		PKP2	Telukbayur R , Sumatera	INS
6215,50	USB		PKS	Pontanak R , Kalimantan	INS
6215,50	USB		PKX	Jakarta R , Jawa	INS
6215,50	USB		PKZ2	Cirebon R , Jawa	INS
6215,50	SSB		T3C	Tarawa R	KIR
6215,50	USB		YJM	Port Vila R	VUT
6215,50	USB		ZKN	Niue R	NIU
6218,00	SSB			SS Ch 607	
6218,00	CW			RUS MIL	RUS
6218,20	*STANAG 4285*	*B=2400Bd*	*IDR*	*I N Rome, TX Vigna di Valle*	*I*
6218,60	USB		PKX	Jakarta R , Jawa	INS
6218,60	USB		VFA	Inuvik R	CAN
6218,60	USB		YJM	Port Vila R	VUT
6221,00	SSB			SS Ch 608	
6221,50	ALE 3G	B=2400Bd		UI MIL net	

Frequency	Mode	Mode Parameter	Callsign	User	Country
6221,50	STANAG 4539	B=2400Bd		UI MIL user	
6222,40	PACTOR III	B=100Bd S=200Hz	FOHXM	Sailmail Manihi Atoll	OCE
6222,60	*LINK 11 CLEW*	*B=2250Bd Ch=16 ChS=330/110/550Hz*		*F N Bordeaux*	*F*
6223,20	*STANAG 4285*	*B=2400Bd*	*IDR*	*I N Rome, TX Vigna di Valle*	*I*
6223,60	ALE USB	B=125Bd Ch=8 ChS=250Hz			
6224,00	USB		PKS	Pontanak R , Kalimantan	INS
6224,00	SSB			simplex frequency ship - ship	
6224,00	USB		PKY2	Biak R , Irian Jaya	INS
6224,00	USB		ZLM	Taupo Maritime R	NZL
6224,00	USB		PKE37	Sanana R , Kep Sula	INS
6224,00	USB		XVK	Kien Giang R	VTN
6224,00	PACTOR III	B=100Bd S=200Hz	PYB45	Sao Paulo	B
6224,00	USB		WQCN860	Ocean-Pro R Naples , FL	USA
6224,00	ALE USB	B=125Bd Ch=8 ChS=250Hz		ISR AF net	ISR
6227,00	USB		VIA	Adelaide R	AUS
6227,00	USB		XVU	Can Tho R	VTN
6227,00	USB		H4H	Honiara R	SLM
6227,00	USB		PKY4	Sorong R , Irian Jaya	INS
6227,00	USB		PNK	Jayapura R , Irian Jaya	INS
6227,00	SSB			simplex frequency ship - ship	
6227,00	USB		PKY3	Manokwari R , Irian Jaya	INS
6227,00	USB		PKY5	Merauke R , Irian Jaya	INS
6227,00	CW		RIT81	RUS N ship SYZRAN	RUS
6227,60	*STANAG 4285*	*B=2400Bd*	*FUE*	*F N Brest*	*F*
6229,50	CW		RJD99	RUS N Severodvinsk	RUS
6229,60	ALE USB	B=125Bd Ch=8 ChS=250Hz			
6230,00	USB		VMW	Wiluna M	AUS
6230,00	USB		ZOE	Tristan da Cunha R	TRC
6230,00	SSB			simplex frequency ship - ship	
6230,00	USB		XVR	Vung Tau R	VTN
6230,00	PACTOR III	B=100Bd S=200Hz	PYB45	Sao Paulo	B
6230,00	CW			RUS MIL	RUS
6230,00	ALE USB	B=125Bd Ch=8 ChS=250Hz		ARINC Urgent Link net	USA
6232,50	RUS TACTICAL DATALINK	B=1200Bd S=800Hz Ch=1200		RUS MIL Kaliningrad	RUS
6234,00	*STANAG 4285*	*B=2400Bd*	*TBB*	*TUR N TX Izmir*	*TUR*
6234,00	ALE USB	B=125Bd Ch=8 ChS=250Hz		F N net	
6235,00	SITOR A	B=100Bd S=170Hz		J CG Tokyo	J
6236,00	USB		6YX	JMC CG Jamaica	JMC
6236,20	*STANAG 4481*	*B=75Bd S=850Hz*		*G MIL St. Eval*	*G*
6237,50	USB		PBC	NLD N Goeree Island	HOL
6237,50	USB		PBB	NLD N Den Helder	HOL
6240,00	*ALE USB*	*B=125Bd Ch=8 ChS=250Hz*		*JOR net*	*JOR*
6240,30	ALE 3G	B=2400Bd		UI MIL net	
6241,50	PACTOR III	B=100Bd S=200Hz	ZKN2SM	Sailmail Niue	NIU
6242,00	USB		PBC	NLD N Goeree Island	HOL
6242,00	USB		PBB	NLD N Den Helder	HOL
6243,00	ALE USB	B=125Bd Ch=8 ChS=250Hz		USA AF net	
6243,00	CW			RUS HFDF net	RUS
6243,00	*STANAG 4285*	*B=2400Bd*		*G MIL Akrotiri*	*CYP*
6248,60	ALE 3G	B=2400Bd		UI MIL net	
6249,00	PACTOR II FEC	B=100Bd S=200Hz		TUN diplo	
6250,00	J PSK	B=1500Bd		J MIL	J
6250,80	STANAG 4539	B=2400Bd		UI MIL user	

Frequency	Mode	Mode Parameter	Callsign	User	Country
6251,00	MIL 188-110A SER	B=2400Bd		G MIL	G
6251,00	ALE USB	B=125Bd Ch=8 ChS=250Hz		G MIL net	G
6251,00	STANAG 4539	B=2400Bd		UI MIL user	
6255,00	LINK 11 ISB	B=2250Bd Ch=16 ChS=330/110/550Hz	PBB	NLD N Den Helder	HOL
6255,00	CIS 12	B=1440Bd Ch=12 ChS=200Hz		RUS MIL Tartus	SYR
6258,00	CW		AQK2	PAK N Karachi	PAK
6258,00	CW		AQP4	PAK N Karachi	PAK
6259,50	CW		RJD99	RUS N Severodvinsk	RUS
6260,00	ALE LSB	B=125Bd Ch=8 ChS=250Hz		VEN N net	VEN
6262,00	CW		RHC841	RUS N ship EKVATOR	RUS
6262,50	USB		IHCH	I N ship CHIOGGIA	B
6262,50	USB		IHMT	I N ship TERMOLI	I
6262,50	USB		IADP	I N ship LUIGI DURAND DE LA PENNE (DDHM)(D560)	I
6262,50	USB		IAME	I N ship MAESTRALE	I
6262,50	STANAG 4285	B=2400Bd	IAME	I N ship MAESTRALE	I
6262,50	STANAG 4285	B=2400Bd	IAOC	I N corvette DANAIDE	I
6262,50	USB		IAOC	I N corvette DANAIDE	I
6262,50	USB		IATD	I N ship VEGA	I
6262,50	STANAG 4285	B=2400Bd	IATD	I N ship VEGA	I
6262,50	STANAG 4285	B=2400Bd	IHMM	I N minesweeper MILAZZO	I
6262,50	USB		IHMM	I N minesweeper MILAZZO	I
6262,50	USB		IHMV	I N minesweeper VIESTE	I
6262,50	STANAG 4285	B=2400Bd	IHMV	I N minesweeper VIESTE	I
6262,50	USB		IHMW	I N ship VIAREGGIO	I
6262,50	STANAG 4285	B=2400Bd	IHMW	I N ship VIAREGGIO	I
6262,50	STANAG 4285	B=2400Bd Ch=1	IADP	I N ship LUIGI DURAND DE LA PENNE (DDHM)(D560)	I
6262,50	USB		IGDE	I N ship TREMITI	I
6262,50	STANAG 4285	B=2400Bd	IGDE	I N ship TREMITI	I
6262,50	USB		IDR	I N Rome	I
6262,50	STANAG 4285	B=2400Bd	IDR	I N Rome	I
6262,50	USB		IGDL	I N ship	I
6263,00	.			SS Ch 1	
6263,00	CW			RUS HFDF net	RUS
6263,50	.			SS Ch 2	
6264,00	.			SS Ch 3	
6264,50	.			SS Ch 4	
6265,00	.			SS Ch 5	
6265,00	ALE USB	B=125Bd Ch=8 ChS=250Hz		ISR AF net	ISR
6265,50	SITOR A	B=100Bd S=170Hz	ZLM	Taupo Maritime R	NZL
6265,50	.			SS Ch 6	
6266,00	.			SS Ch 7	
6266,50	.			SS Ch 8	
6267,00	.			SS Ch 9	
6267,50	.			SS Ch 10	
6268,00	SITOR B	B=100Bd S=170Hz	ZLM	Taupo Maritime R	NZL
6268,00	SITOR , CW	B=100Bd S=170Hz	JNA	J CG HQ Tokyo	J
6268,00	SITOR , CW	B=100Bd S=170Hz	VRC	Hong Kong R	CHN
6268,00	.			Distress/safety frequency	WW
6268,00	SITOR , CW	B=100Bd S=170Hz	PKD	Surabaya R , Djawa	INS
6268,00	SITOR , CW	B=100Bd S=170Hz	PNK	Jayapura R , Irian Jaya	INS
6268,00	SITOR , CW	B=100Bd S=170Hz	PKP	Dumai R , Sumatera	INS
6268,20	STANAG	B=2400Bd	SXH	GRC N Thessaloniki	GRC
6268,50	.			SS Ch 12	
6268,50	SITOR , CW	B=100Bd S=170Hz	PKM	Bitung R , Sulawesi	INS
6269,00	.			SS Ch 13	

Frequency	Mode	Mode Parameter	Callsign	User	Country
6269,50	.			SS Ch 14	
6270,00	.			SS Ch 15	
6270,50	.			SS Ch 16	
6271,00	.			SS Ch 17	
6271,50	.			SS Ch 18	
6272,00	.			SS Ch 19	
6272,50	.			SS Ch 20	
6273,00	.			SS Ch 21	
6273,50	.			SS Ch 22	
6274,00	.			SS Ch 23	
6274,20	STANAG 4539 HDR	B=2400Bd		UI MIL net	
6274,50	.			SS Ch 24	
6274,80	ALE 3G	B=2400Bd		UI MIL net	
6275,00	.			SS Ch 25	
6275,00	ALE USB	B=125Bd Ch=8 ChS=250Hz		B N net	B
6275,50	.			SS Ch 26	
6276,00	CW			SS calling frequency Ch 1	
6276,00	CW		3XO27	Kamsar R	GUF
6276,00	STANAG 4539 HDR	B=2400Bd		UI MIL net	
6276,50	CW			SS calling frequency Ch 2	
6277,00	CW			SS calling frequency Ch 3	
6277,50	CW			SS calling frequency Ch 4	
6278,00	CW			SS calling frequency Ch 5	
6278,50	CW			SS calling frequency Ch 6	
6279,00	CW			SS calling frequency Ch 7	
6279,50	CW			SS calling frequency Ch 8	
6280,00	BPSK	B=100Bd		RUS MIL	RUS
6280,00	CW			SS calling frequency Ch 9	
6280,50	CW			SS calling frequency Ch 10	
6281,00	.			SS Ch 27	
6281,50	.			SS Ch 28	
6282,00	.			SS Ch 29	
6282,00	RUS TACTICAL DATALINK	B=1200Bd S=800Hz Ch=1200		RUS MIL	RUS
6282,50	.			SS Ch 30	
6283,00	.			SS Ch 31	
6283,50	.			SS Ch 32	
6284,00	.			SS Ch 33	
6284,50	.			SS Ch 34	
6290,00	CW		UCNA7	RUS ship	RUS
6290,00	CW		RMHW	RUS N survey ship ROMUALD MUKLEVICH	RUS
6290,00	CW		RME81	RUS N ship BILBINO MB-119	RUS
6292,00	CODAN CHIRP	B=80Bd Ch=30 ChS=250Hz			
6292,00	CODAN CHIRP	B=80Bd Ch=30 ChS=250Hz			
6292,00	CODAN CHIRP	B=80Bd Ch=30 ChS=250Hz			
6295,20	*STANAG 4285*	*B=2400Bd*	*IDR*	*I N Rome*	*I*
6300,00	*ALE USB*	*B=125Bd Ch=8 ChS=250Hz*		*F Airbus net*	*F*
6300,50	.			SS Ch 1	
6300,50	ALE 3G	B=2400Bd		UI MIL net	
6301,00	.			SS Ch 2	
6301,50	.			SS Ch 3	
6302,00	.			SS Ch 4	
6302,50	.			SS Ch 5	
6303,00				SS Ch 6	
6303,50	.			SS Ch 7	

Frequency	Mode	Mode Parameter	Callsign	User	Country
6304,00	.			SS Ch 8	
6304,50	.			SS Ch 9	
6305,00	PACTOR II	B=200Bd S=200Hz	V8V2222	Sailmail Brunai Bay	BRU
6305,00	.			SS Ch 10	
6305,50	.			SS Ch 11	
6306,00	.			SS Ch 12	
6306,50	.			SS Ch 13	
6307,00	.			SS Ch 14	
6307,00	SYSTEME 3000 ALE	B=125Bd Ch=8 ChS=250Hz			F
6307,50	.			SS Ch 15	
6308,00	.			SS Ch 16	
6308,50	.			SS Ch 17	
6309,00	.			SS Ch 18	
6309,50	.			SS Ch 19	
6310,00	.			SS Ch 20	
6310,00	CW			RUS N calling frequency	RUS
6310,20	ALE USB	B=125Bd Ch=8 ChS=250Hz		TUN MIL net	TUN
6310,50	.			SS Ch 21	
6311,00	.			SS Ch 22	
6311,50	.			SS Ch 23	
6312,00	DSC	B=100Bd S=170Hz	JNA	J CG HQ Tokyo	J
6312,00	USB		VFF	Iqaluit R	CAN
6312,00	.			DSC distress/safety frequency	WW
6312,00	DSC	B=100Bd S=170Hz	NMC	USA CG Pt. Reyes , CA	USA
6312,00	DSC	B=100Bd S=170Hz	UDK2	Murmansk R	RUS
6312,00	DSC	B=100Bd S=170Hz		Chilung Kelung R	TWA
6312,00	DSC	B=100Bd S=170Hz	XVS	Ho Chi Minh R	VTN
6312,00	DSC	B=100Bd S=170Hz	VWB	Mumbai R	IND
6312,00	DSC	B=100Bd S=170Hz	XVN	Nha Trang R	VTN
6312,00	DSC	B=100Bd S=170Hz	XVG	Hai Phong R	VTN
6312,00	DSC	B=100Bd S=170Hz	NPM	CAMSPAC Honolulu R	HWA
6312,50	.			SS DSC frequency	WW
6313,00	.			SS DSC frequency	WW
6313,50	.			SS DSC frequency	WW
6314,00				MSI frequency	WW
6314,00	SITOR B	B=100Bd S=170Hz	NMF	USA CG Boston , MA	USA
6314,50	.			CS Ch 1	
6315,00	SITOR , CW	B=100Bd S=170Hz	UFL	Vladivostok R	RUS
6315,00	.			CS Ch 2	
6315,00	SITOR , CW	B=100Bd S=170Hz	PKM	Bitung R , Sulawesi	INS
6315,50	.			CS Ch 3	
6316,00	SITOR , CW	B=100Bd S=170Hz	PKR	Semarang R , Djawa	INS
6316,00	.			CS Ch 4	
6316,00	SITOR , CW	B=100Bd S=170Hz	XSQ	Guangzhou R	CHN
6316,00	SITOR , CW	B=100Bd S=170Hz	PKD	Surabaya R , Djawa	INS
6316,50	.			CS Ch 5	
6317,00	.			CS Ch 6	
6317,00	SITOR , CW	B=100Bd S=170Hz	PNK	Jayapura R , Irian Jaya	INS
6317,00	SITOR , CW	B=100Bd S=170Hz	PKY4	Sorong R , Irian Jaya	INS
6317,50	.			CS Ch 7	
6317,50	SITOR , CW	B=100Bd S=170Hz	JCS	Choshi R	J
6318,00	.			CS Ch 8	
6318,00	SITOR , CW	B=100Bd S=170Hz	KLB	Seattle R	USA
6318,00	SITOR , CW	B=100Bd S=170Hz	PKP	Dumai R , Sumatera	INS
6318,00	SITOR , CW	B=100Bd S=170Hz	PKK	Kupang R , Timor	INS
6318,50	SITOR , CW	B=100Bd S=170Hz	UFL	Vladivostok R	RUS
6318,50	.			CS Ch 9	
6318,50	SITOR , CW	B=100Bd S=170Hz	PKN	Balikpapan R , Kalimantan	INS
6319,00	.			CS Ch 10	
6319,00	F1B MORSE	S=200Hz	RIT	RUS N HQ Severomorsk	RUS

Frequency	Mode	Mode Parameter	Callsign	User	Country
6319,50	.			CS Ch 12	
6320,00	.			CS Ch 13	
6320,00	MIL 188-110A SER	B=2400Bd		SVK AF net	SVK
6320,00	ALE USB	B=125Bd Ch=8 ChS=250Hz		SVK AF net	SVK
6320,50	SITOR , CW	B=100Bd S=170Hz	SVO33	Olympia R	GRC
6320,50	SITOR , CW	B=100Bd S=170Hz	JNA	J CG HQ Tokyo	J
6320,50	.			CS Ch 14	
6321,00	SITOR , CW	B=100Bd S=170Hz	XSA2	Nanjing R	CHN
6321,00	SITOR , CW	B=100Bd S=170Hz	PKR3	Cilacap R , Jawa	INS
6321,00	.			CS Ch 15	
6321,00	SITOR , CW	B=100Bd S=170Hz	PKJ28	Sei Kolak Kijang R	INS
6321,50	.			CS Ch 16	
6322,00	.			CS Ch 17	
6322,50	SITOR , CW	B=100Bd S=170Hz	UDK2	Murmansk R	RUS
6322,50	.			CS Ch 18	
6323,00	SITOR , CW	B=100Bd S=170Hz	UFM3	Nevelsk R	RUS
6323,00	DSC	B=100Bd S=170Hz	PKZ2	Cirebon R , Jawa	INS
6323,00	.			CS Ch 19	
6323,00	SITOR , CW	B=100Bd S=170Hz	PKP38	Batu Ampar R	INS
6323,00	CW		RGR99	RUS N	RUS
6323,00	CW		RGR99	RUS N	RUS
6323,00	CW		REO4	RUS N	RUS
6323,00	CW		RGR49	RUS N	RUS
6323,00	CW		RGR87	RUS N	RUS
6323,00	CW		RGR88	RUS N	RUS
6323,00	CW		RGR89	RUS N	RUS
6323,00	CW		RGR90	RUS N	RUS
6323,00	CW		RGR91	RUS N	RUS
6323,00	CW		RGR92	RUS N	RUS
6323,00	CW		RGR93	RUS N	RUS
6323,00	CW		RGR94	RUS N	RUS
6323,00	CW		RGR95	RUS N	RUS
6323,00	CW		RGR96	RUS N	RUS
6323,00	CW		RGR97	RUS N	RUS
6323,00	CW		RGR98	RUS N	RUS
6323,50	.			CS Ch 20	
6324,00	CW		RKH80	RUS N ship	RUS
6324,00	CW		RMEV	RUS N ship	RUS
6324,00	.			CS Ch 21	
6324,00	ALE USB	B=125Bd Ch=8 ChS=250Hz		AZE emergency net	AZE
6324,00	SITOR , CW	B=100Bd S=170Hz	PKF3	Kendari R	INS
6324,00	SITOR , CW	B=100Bd S=170Hz	PKC4	Panjang R , Sumatera	INS
6324,00	CW		RMJT	RUS N ship PEREKOP	RUS
6324,00	CW		UCBC5	RUS N ship YAKOV GREBELSKIY	RUS
6324,00	CW		UCQA2	RUS N ship	RUS
6324,50	BAUDOT	B=45,45Bd S=170Hz	KPH	San Francisco R , CA (Bolinas/Port Reyes)	USA
6324,50	.			CS Ch 22	
6324,50	SITOR A	B=100Bd S=170Hz	KPH	San Francisco R , CA (Bolinas/Port Reyes)	USA
6325,00	CW		RGZ59	RUS N	RUS
6325,00	SITOR , CW	B=100Bd S=170Hz	PKE5	Ternate R , Maluku	INS
6325,00	DSC	B=100Bd S=170Hz	PKE5	Ternate R , Maluku	INS
6325,00	.			CS Ch 23	
6325,00	SITOR , CW	B=100Bd S=170Hz	PKS	Pontianak R , Kalimantan	INS
6325,00	SITOR , CW	B=100Bd S=170Hz	PKY5	Merauke R , Irian Jaya	INS
6325,00	SITOR , CW	B=100Bd S=170Hz	PKM44	Pantolan R , Sulawesi	INS
6325,50	.			CS Ch 24	
6326,00	.			CS Ch 25	
6326,00	SITOR , CW	B=100Bd S=170Hz	XSG	Shanghai R	CHN
6326,50				CS Ch 26	

Frequency	Mode	Mode Parameter	Callsign	User	Country
6327,00	.			CS Ch 27	
6327,00	SITOR , CW	B=100Bd S=170Hz	PKY2	Biak R , Irian Jaya	INS
6327,00	SITOR , CW	B=100Bd S=170Hz	PKB3	Sibolga R , Sumatera	INS
6327,50	.			CS Ch 28	
6327,50	SITOR , CW	B=100Bd S=170Hz	UAT	Moscow R	RUS
6328,00	.			CS Ch 29	
6328,00	SITOR , CW	B=100Bd S=170Hz	WOO	Ocean Gate R , NJ	USA
6328,50	.			CS Ch 30	
6328,50	FAX 120/576		RBW	Murmansk M	RUS
6329,00	SITOR , CW	B=100Bd S=170Hz	XSQ	Guangzhou R	CHN
6329,00	DSC	B=100Bd S=170Hz	PKZ34	Cigading R , Jawa	INS
6329,00	.			CS Ch 31	
6329,00	SITOR , CW	B=100Bd S=170Hz	4PB	Colombo R	CLN
6329,00	SITOR , CW	B=100Bd S=170Hz	PKO	Tarakan R , Kalimantan	INS
6329,00	SITOR , CW	B=100Bd S=170Hz	PKD3	Lembar R , Lombok	INS
6329,00	SITOR , CW	B=100Bd S=170Hz	PKY3	Manokwari R , Irian Jaya	INS
6329,50	.			CS Ch 32	
6330,00	SITOR , CW	B=100Bd S=170Hz	PKD5	Benoa R , Bali	INS
6330,00	.			CS Ch 33	
6330,00	CW			RUS N calling frequency	RUS
6330,00	CW		UNU3	RUS N CHMS Kaliningrad	RUS
6330,00	ALE USB	B=125Bd Ch=8 ChS=250Hz		AZE emergency net	AZE
6330,00	SITOR , CW	B=100Bd S=170Hz	PKM25	Tahuna R	INS
6330,00	ALE USB	B=125Bd Ch=8 ChS=250Hz		ISR AF net	ISR
6330,50	PACTOR III	B=200Bd S=200Hz	OSY	Sailmail Brugge	BEL
6330,50	.			CS Ch 34	
6330,70	BAUDOT	B=50Bd S=100Hz	N4O4**	RUS	RUS
6331,00	DSC	B=100Bd S=170Hz	JNA	J CG HQ Tokyo	J
6331,00	DSC	B=100Bd S=170Hz	PKD5	Benoa R , Bali	INS
6331,00	DSC	B=100Bd S=170Hz	PKE5	Ternate R , Maluku	INS
6331,00	DSC	B=100Bd S=170Hz	PKZ34	Cigading R , Jawa	INS
6331,00	DSC	B=100Bd S=170Hz	PKR3	Cilacap R , Jawa	INS
6331,00	DSC	B=100Bd S=170Hz	PKZ2	Cirebon R , Jawa	INS
6331,00	DSC	B=100Bd S=170Hz	PKX	Jakarta R , Jawa	INS
6331,00	DSC	B=100Bd S=170Hz	PKS	Pontianak R , Kalimantan	INS
6331,00	DSC	B=100Bd S=170Hz	PKO	Tarakan R , Kalimantan	INS
6331,00	DSC	B=100Bd S=170Hz	PKE37	Sanana R , Kep Sula	INS
6331,00	DSC	B=100Bd S=170Hz	PKD3	Lembar R , Lombok	INS
6331,00	DSC	B=100Bd S=170Hz	PKY2	Biak R , Irian Jaya	INS
6331,00	DSC	B=100Bd S=170Hz	PKY23	Fak Fak R	INS
6331,00	DSC	B=100Bd S=170Hz	PKY3	Manokwari R , Irian Jaya	INS
6331,00	DSC	B=100Bd S=170Hz	PKY5	Merauke R , Irian Jaya	INS
6331,00	DSC	B=100Bd S=170Hz	PKF3	Kendari R	INS
6331,00	DSC	B=100Bd S=170Hz	PKM44	Pantolan R , Sulawesi	INS
6331,00	DSC	B=100Bd S=170Hz	PKM25	Tahuna R	INS
6331,00	DSC	B=100Bd S=170Hz	PKP38	Batu Ampar R	INS
6331,00	DSC	B=100Bd S=170Hz	PKC4	Panjang R , Sumatera	INS
6331,00	DSC	B=100Bd S=170Hz	PKJ28	Sei Kolak Kijang R	INS
6331,00	DSC	B=100Bd S=170Hz	PKB3	Sibolga R , Sumatera	INS
6331,20	*STANAG 4285*	*B=2400Bd*	*IDR*	*I N Rome*	*I*
6331,50	SITOR , CW	B=100Bd S=170Hz	UFL	Vladivostok R	RUS
6335,20	*STANAG 4285*	*B=2400Bd*		*G MIL St. Eval*	*G*
6335,60	ALE USB	B=125Bd Ch=8 ChS=250Hz		ARINC Urgent Link net	USA
6337,00	CW		PKR	Semarang R , Djawa	INS
6337,00	CW		PKG	Bandjarmasin R , Kalimantan	INS
6337,00	CW		PKP	Dumai R , Sumatera	INS
6337,00	CW		PKY4	Sorong R , Irian Jaya	INS
6339,50	CW		RJD56	RUS N HQ Murmansk	RUS
6340,50	FAX 120/576		NMF	USCG Boston	USA

Frequency	Mode	Mode Parameter	Callsign	User	Country
6342,00	T600	B=50Bd S=250Hz		RUS N Moscow	RUS
6343,00	CW			RUS Operational Strategic Command Air Defense net	RUS
6343,00	CW		PWZW	RUS Strategic Command Aerospace Defense	RUS
6343,00	CW		RGT77	RUS MIL Moscow, RVSN	RUS
6343,00	CW		RMP	RUS N HQ Kaliningrad	RUS
6343,20	STANAG 4285	B=2400Bd	IDR	I N TX Roma	I
6344,00	CW		YJM6	Port Vila R	VUT
6346,00	ALE 3G	B=2400Bd		UI MIL net	
6346,80	PACTOR II FEC	B=100Bd S=200Hz	AQP4	PAK N Karachi	PAK
6348,00	STANAG 4285	B=2400Bd	FUE	F N Brest	F
6348,00	T600	B=50Bd S=200Hz	RMP	RUS N Kaliningrad	RUS
6348,70	ALE 3G	B=2400Bd		UI MIL net	
6353,30	ALE 3G	B=2400Bd		UI MIL net	
6355,00	CW		PKB3	Sibolga R , Sumatera	INS
6355,00	CW		PKC4	Panjang R , Sumatera	INS
6355,00	CW		PKS	Pontanak R , Kalimantan	INS
6357,00	PACTOR II	B=200Bd S=200Hz	VZX	Sailmail Darawank	AUS
6357,20	STANAG 4285	B=2400Bd	PBC	NLD N Goeree Island	HOL
6358,00	CW		REA4	RUS AF Moscow	RUS
6358,50	BAUDOT	B=75Bd S=850Hz	PBC	NLD N Goeree Island	HOL
6358,50	BAUDOT	B=47,3Bd S=1000Hz	PBB	NLD N Den Helder	HOL
6358,50	BAUDOT	B=75Bd S=850Hz	PBB	NLD N Den Helder	HOL
6360,00	RUS 4TONE	Ch=4		RUS MIL	RUS
6360,60	STANAG 4285	B=2400Bd	FUX	F N Le Port	REU
6362,00	CW		RMP	RUS N HQ Kaliningrad	RUS
6363,00	CW		RGT77	RUS MIL Moscow, RVSN	RUS
6365,50	F N FSK	B=50Bd S=850Hz Ch=1	FUG	F N Saissac	F
6368,50	LINK 11 CLEW	B=2250Bd Ch=16 ChS=330/110/550Hz		POL N Gdansk	POL
6369,20	STANAG 4285	B=2400Bd		G MIL TX Inskip	G
6372,00	STANAG 4481	B=75Bd S=850Hz	AJE	USA AF AWS Croughton	G
6374,00	CW			RUS Operational Strategic Command Air Defense net	RUS
6374,00	PACTOR , SITOR	B=100Bd S=200Hz	KHF	Agana R	GUM
6374,00	CW			RUS HFDF net	RUS
6374,80	BAUDOT	B=50Bd S=300Hz		SRB MIL	SRB
6375,70	STANAG 4285	B=2400Bd	EBA	E N Madrid	E
6376,50	SITOR , CW	B=100Bd S=170Hz	PKP	Dumai R , Sumatera	INS
6377,70	STANAG 4481	B=75Bd S=850Hz		G F Gibraltar	GIB
6381,30	ALE 3G	B=2400Bd		UI MIL net	
6382,20	STANAG 4285	B=2400Bd	PBC	NLD N Goeree Island	HOL
6383,00	STANAG 4481	B=75Bd S=850Hz	NSY	USA N Niscemi , Sicily	I
6384,70	STANAG 4285	B=2400Bd	CFH	CAN N Halifax	CAN
6385,00	USB		302**	MMR MIL net	MMR
6385,00	USB		202**	MMR Mil net	MMR
6385,00	USB		402**	MMR MIL net	MMR
6385,00	USB		502**	MMR MIL net	MMR
6385,00	USB		602**	MMR MIL net	MMR
6385,00	USB		606**	MMR MIL net NCS	MMR
6385,00	USB		310**	MMR MIL net	MMR
6385,00	USB		311**	MMR MIL net	MMR
6385,00	USB		312**	MMR MIL net	MMR
6385,00	USB		313**	MMR MIL net	MMR
6385,00	USB		314**	MMR MIL net	MMR
6385,00	USB		315**	MMR MIL net	MMR
6385,00	USB		316**	MMR MIL net	MMR
6385,00	USB		317**	MMR MIL net	MMR
6385,00	USB		318**	MMR MIL net	MMR

Frequency	Mode	Mode Parameter	Callsign	User	Country
6386,00	CW		RGT77	RUS MIL Moscow, RVSN	RUS
6387,00	ALE USB	B=125Bd Ch=8 ChS=250Hz		G MIL net	G
6388,20	*STANAG 4285*	*B=2400Bd*	*IDR*	*I N Vigna di Valle*	*I*
6390,00	CW		AQP4	PAK N Karachi	PAK
6391,00	CW		AQP	PAK N Karachi	PAK
6391,20	*STANAG 4285*	*B=2400Bd*		*Gdansk*	*POL*
6392,00	J OFDM 30+2	B=1500Bd Ch=32 ChS=25Hz		J N broadcast	J
6393,10	ALE USB	B=125Bd Ch=8 ChS=250Hz		F	F
6393,20	*STANAG 4285*	*B=2400Bd*	*CFH*	*CAN N Halifax*	*CAN*
6393,50	BAUDOT	B=50Bd S=170Hz	UDK	Murmansk R	RUS
6396,20	*STANAG 4285*	*B=2400Bd*	*CFH*	*CAN N Halifax*	*CAN*
6396,30	STANAG 4539	B=2400Bd		UI MIL user	
6397,00	STANAG 4285	B=2400Bd	NSS	USA N TX Davidsonville , MD	USA
6398,70	STANAG 4285	B=2400Bd	CKN	CAN MIL Aldergrove	CAN
6402,00	CW			RUS MIL	RUS
6403,00	USB		PKC2	Plaju R , Sumatera	INS
6403,20	STANAG 4285	B=2400Bd	CFH	CAN N Halifax	CAN
6405,00	USB			CIS aero network day	CIS
6405,20	*STANAG 4285*	*B=2400Bd*		*NOR N Oslo*	*NOR*
6405,20	STANAG 4285	B=2400Bd	JWT	NOR N Stavanger	NOR
6406,00	*STANAG 4285*	*B=2400Bd*	*FUE*	*F N Brest*	*F*
6407,00	MHF-50 MODEM	B=54,3Bd Ch=32 ChS=64,5Hz	ZSJ	AFS N Capetown	AFS
6408,00	MFSK	B=10Bd Ch=12 ChS=40Hz		AFS N	AFS
6408,70	*STANAG 4285*	*B=2400Bd*	*IDR*	*I N Rome TX Santa Rosa*	*I*
6411,00	*MAHRS*	*B=2400Bd*	*DHJ58*	*D N Glücksburg TX Staberhuk*	*D*
6411,00	CW		RJD97	RUS N HQ Strelok	RUS
6411,10	GTOR	B=100Bd S=200Hz	SXTDIS	B N	B
6411,10	GTOR	B=100Bd S=200Hz	MANIBA	B N	B
6412,00	ALE USB	B=125Bd Ch=8 ChS=250Hz		B N net	B
6412,70	*STANAG 4285*	*B=2400Bd*		*G MIL Crimond*	*G*
6413,00	BAUDOT	B=45,45Bd S=170Hz	OSN	BEL N Oostende	BEL
6416,50	*ALE USB*	*B=125Bd Ch=8 ChS=250Hz*		*G MIL net*	*G*
6417,00	J PSK	B=1500Bd		J MIL	J
6417,20	*STANAG 4285*	*B=2400Bd*	*PBC*	*NLD N Goeree Island*	*HOL*
6418,00	CW		VTP5	IND N Vishakhapatnam	IND
6418,50	MIL 188-110A SER	B=2400Bd	NWC	USN Exmouth , WA	USA
6418,50	MIL 188-110B SER	B=2400Bd			
6418,50	STANAG 4285	B=2400Bd		AUS MIL Humpty Doo , NT	
6420,00	SITOR	B=100Bd S=170Hz	PWZ33	B N Rio de Janeiro	B
6420,20	STANAG 4285	B=2400Bd	DHJ58	D N Glücksburg TX Marlow	D
6421,00	ALE USB	B=125Bd Ch=8 ChS=250Hz		G MIL net	G
6421,00	CW			RUS Air Defense net	RUS
6423,00	CW		RJS	RUS NHQ Vladivostok	RUS
6423,00	CW		LZL3	Bourgas R	BUL
6423,00	CW			RUS HFDF net	RUS
6423,20	*STANAG 4285*	*B=2400Bd*	*CFH*	*CAN N Halifax*	*CAN*
6423,50	CW		RAS82	RUS N ship GS-405	RUS
6425,30	STANAG 4538	B=2400Bd			
6428,20	STANAG 4285	B=2400Bd	CFH	CAN N Halifax	CAN
6428,50	CW		PKE5	Ternate R , Maluku	INS
6428,50	CW		PKM	Bitung R , Sulawesi	INS
6428,50	LINK 11 CLEW	B=2250Bd Ch=16 ChS=330/110/550Hz			

Frequency	Mode	Mode Parameter	Callsign	User	Country
6429,20	*STANAG 4285*	*B=2400Bd*	*OVG*	*DNK N Frederikshavn*	*DNK*
6429,20	*STANAG 4285*	*B=2400Bd*		*G MIL Crimond*	*G*
6429,80	STANAG 4285	B=2400Bd	FUG	F N Saissac	F
6430,00	SITOR , CW	B=100Bd S=170Hz	UFH	Petrapavlovsk-Kamchatskii R	RUS
6430,00	CW		URA2	Nikolayev R	UKR
6430,00	PACTOR III	B=100Bd S=200Hz	WNU	Austin , TX	USA
6430,00	CW		ESP	Parnu R.	
6430,50	CW		V**	RUS N Khiva	RUS
6431,70	*F N FSK*	*B=50Bd S=850Hz Ch=1*	*FUG*	*F N Saissac*	*F*
6431,80	STANAG 4538	B=2400Bd			
6432,00	USB		USI	Kherson R	UKR
6432,00	ALE 3G	B=2400Bd		UI MIL net	
6436,00	CW		XSG	Shanghai R	CHN
6438,00	CW		RGT77	RUS MIL Moscow, RVSN	RUS
6439,20	*STANAG 4285*	*B=2400Bd*	*PBC*	*NLD N Goeree Island*	*NLD*
6439,20	STANAG 4285	B=2400Bd	OSN	BEL N Oostende	BEL
6445,00	J PSK	B=1500Bd		J MIL	J
6445,50	FAX 90/120/576			Murmansk M	RUS
6447,70	STANAG 4285	B=2400Bd	CFH	CAN N Halifax	CAN
6448,20	*STANAG 4285*	*B=2400Bd*	*IDR*	*I N Rome, TX Vigna di Valle*	*I*
6449,00	CW		4XZ	ISR N Haifa	ISR
6449,00	*CW*		*RIT*	*RUS N HQ Severomorsk*	*RUS*
6449,50	BAUDOT	B=75Bd S=850Hz	PWZ33	B N Rio de Janeiro	B
6450,00	USB			ARM MIL net	ARM
6450,20	SITOR B	B=100Bd S=170Hz	PWZ33	B N Rio de Janeiro	B
6451,50	CW		RJD56	RUS N HQ Murmansk	RUS
6451,50	CW		RJD99	RUS N Severodvinsk	RUS
6452,00	CIS MFSK 64	B=40Bd Ch=64 ChS=46Hz		RUS diplo	RUS
6453,00	MIL 188-110A SER	B=2400Bd		AUS MHFCS North West Cape	AUS
6454,50	ALE USB	B=125Bd Ch=8 ChS=250Hz		F N net	F
6456,00	CW		XSG	Shanghai R	CHN
6456,00	CW		RCV	RUS N HQ Sevastopol	UKR
6456,00	CW		RQN3	RUS N CHMS Sevastopol	UKR
6456,30	*STANAG 4285*	*B=2400Bd*	*EBA*	*E N Madrid*	*E*
6456,50	CW		PKN2	Balikpapan R , Kalimantan	INS
6458,00	CW		RIT	RUS N HQ Severomorsk	RUS
6460,00	USB			CHN air defense	CHN
6461,50	CW		RMP	RUS N HQ Kaliningrad	RUS
6462,00	ALE USB	B=125Bd Ch=8 ChS=250Hz		G MIL net	G
6462,00	STANAG 4285	B=2400Bd	FUM	F N Papeete , Tahiti	OCE
6463,50	CW		RJD56	RUS N HQ Murmansk	RUS
6467,00	CW		VTG5	IND N Mumbai	IND
6468,00	USB			BGD N net	BGD
6468,20	STANAG 4285	B=2400Bd	EBA	E N Madrid	E
6469,00	ALE USB	B=125Bd Ch=8 ChS=250Hz		IRQ N net	IRQ
6470,70	STANAG 4539	B=2400Bd		UI MIL user	
6471,00	USB		SXA24	GRC N Piraeus	GRC
6471,00	USB			POL MIL	POL
6475,20	*STANAG 4285*	*B=2400Bd*		*G MIL Akrotiri*	*CYP*
6477,50	CW		KPH	San Francisco R , CA (Bolinas/Port Reyes)	USA
6480,00	*BPSK*	*B=62,5Bd*		*POL INT Warsaw*	*POL*
6480,00	CW		TBH3	TUR N Golcuk	TUR
6480,00	SSB			RDARA 14	
6481,00	CW		9MB3	MLA N Georgtown , Penang Island	MLA
6483,00	BAUDOT	B=75Bd S=850Hz	PBB	HOL N Den Helder	HOL
6483,00	BAUDOT	B=50Bd S=850Hz	9MB	MLA N Johore Bahru	MLA
6483,00	BAUDOT	B=50Bd S=850Hz	9MR	MLA N Johor Baharu	MLA

Frequency	Mode	Mode Parameter	Callsign	User	Country
6484,00	CW		LZL3	Bourgas R	BUL
6484,00	STANAG 4539	B=2400Bd		UI MIL user	
6487,00	STANAG 4481	B=50Bd S=850Hz	NSS	USA N TX Davidsonville , MD	USA
6487,30	ALE 3G	B=2400Bd		UI MIL net	
6489,00	STANAG 4285	B=2400Bd	NSS	USA N TX Davidsonville , MD	USA
6489,10	ALE3G	B=2400Bd			
6491,00	CW			RUS Operational Strategic Command Air Defense net	RUS
6491,50	CW		PKD5	Benoa R , Bali	INS
6491,50	CW		PKC	Palembang R , Sumatera	INS
6491,50	CW		PKC3	Jambi R , Sumatera	INS
6491,50	CW		PKZ2	Cirebon R , Jawa	INS
6491,60	*STANAG 4285*	*B=2400Bd*	*EBA*	*E N Madrid*	*E*
6492,30	ALE 3G	B=2400Bd		UI MIL net	
6493,00	BAUDOT	B=50Bd S=850Hz	9MR	MLA N Johor Baharu	MLA
6494,50	CW		RCLH	RUS N Ship SIBIRYAKOV	RUS
6494,50	CW		RMOK	RUS N ship	RUS
6494,50	CW		RMPK	RUS SAR vessel NINA SOKOLOVA SB-121	RUS
6494,50	CW		RMJT	RUS N ship PEREKOP	RUS
6494,50	CW		RMTO	RUS N SAR ship ALEKSANDR FROLOV	RUS
6494,50	CW		RMFE	RUS SAR vessel SB-123	RUS
6494,50	CW		RJQ71	RUS N ship	RUS
6496,00	BAUDOT	B=75Bd S=500Hz	CFH	CAN F Halifax	CAN
6498,10	*STANAG 4285*	*B=2400Bd*	*PBB*	*NLD N Den Helder TX Flevo*	*HOL*
6499,90	CW		PKY5	Merauke R , Irian Jaya	INS
6500,00	FMCW			ONERA research center ROSMed project Golfe du Lion	F
6500,00	ALE LSB	B=125Bd Ch=8 ChS=250Hz		ALG police net	ALG
6501,00	FAX 120/576		NMN	USA CG Chesapeake , VA	USA
6501,00	USB		CBV	Valpariso R	CHL
6501,00	USB		XSL	Fuzhou R	CHN
6501,00	USB		NRV	USA CG Guam	GUM
6501,00	USB		PKX	Jakarta R , Jawa	INS
6501,00	USB		NMA	USA CG Miami , FL	USA
6501,00	USB		NMF	USA CG Boston , MA	USA
6501,00	USB		NMN	USA CG Portsmouth , VA	USA
6501,00	USB			USA CG Point Reyes	USA
6501,00	CW		EQN	Khomeini R	IRN
6501,00	SSB			CS Ch 601	
6501,00	SSB		4PB	Colombo R	CLN
6501,00	USB		PKR3	Cilacap R , Jawa	INS
6501,00	USB		PKY4	Sorong R , Irian Jaya	INS
6501,00	SSB		SVO31	Olympia R	GRC
6501,00	SSB		V5W	Walvis Bay R	NMB
6501,00	USB		VFA	Inuvik R	CAN
6501,00	USB		XSG	Shanghai R	CHN
6501,00	USB			USA CG Honolulu	HWA
6501,00	USB		XSJ	Zhanjiang R	CHN
6501,00	USB		NMO	USA CG Honolulu	HWA
6501,00	USB		NOJ	USA CG Kodiak , ALS	ALS
6501,00	CHN 4+4	B=75Bd Ch=8 ChS=300/450Hz		CHN MIL	
6504,00	USB		LZW	Varna R	BUL
6504,00	USB		UFM3	Nevelsk R	RUS
6504,00	USB		EQJ	Chahbahar R	IRN
6504,00	USB			Lisbon R	POR
6504,00	USB			Cape Town R	AFS
6504,00	USB			San Lorenzo R	EQA
6504,00	USB		HCY	Ayora R , Santa Cruz	EQA
6504,00	USB			Cristobal R	EQA

Frequency	Mode	Mode Parameter	Callsign	User	Country
6504,00	USB			Floreana R , Santa Maria	EQA
6504,00	USB			Seymour R , Baltra	EQA
6504,00	USB			Villamil R , Isabela	EQA
6504,00	SSB			CS Ch 602	
6504,00	USB		3DP	Suva R	FJI
6504,00	USB		PKC4	Panjang R , Sumatera	INS
6504,00	USB		PKD3	Lembar R , Lombok	INS
6504,00	USB		PKG	Bandjarmasin R , Kalimantan	INS
6504,00	USB		PKY5	Merauke R , Irian Jaya	INS
6504,00	USB		HCG	Guayaquil R	EQA
6504,00	SSB		V5W	Walvis Bay R	NMB
6504,00					
6504,00	MHF-50 MODEM	B=54,3Bd Ch=32 ChS=64,5Hz	ZSJ	AFS N Capetown	AFS
6506,00	USB		HBM11	SUI MIL	SUI
6506,00	USB		HBM46	SUI MIL Aarau	SUI
6506,40	USB		PKX	Jakarta R , Jawa	INS
6507,00	USB		UDK2	Murmansk R	RUS
6507,00	USB		VMC	Charleville M	AUS
6507,00	USB		EQC	Amirabad R	IRN
6507,00	USB		3BM	Mauritius R	MAU
6507,00	USB		YQI	Constanta R	ROU
6507,00	USB		SDJ	Stockholm R	S
6507,00	CW		USI	Kherson R	UKR
6507,00	USB		VFF	Iqaluit R	CAN
6507,00	USB		HSA	Bangkok R	THA
6507,00	USB		LSD836	Argentina R	ARG
6507,00	USB		ZBR	Bermuda Harbour R	BER
6507,00	USB		UFL	Vladivostok R	RUS
6507,00	USB		XVN	Nha Trang R	VTN
6507,00	SSB			CS Ch 603	
6507,00	USB		5BA	Cyprus R	CYP
6507,00	SSB		SVO32	Olympia R	GRC
6507,00	SSB		TUA	Abidjan R	CTI
6507,00	CW		JFH	Hammajima Fishery R	J
6507,00	ALE USB	B=125Bd Ch=8 ChS=250Hz		AUS Police NSW	AUS
6507,00	CW		ZRQ3	AFS N Capetown	AFS
6508,00	SYSTEME 3000 ALE	B=125Bd Ch=8 ChS=250Hz			F
6510,00	USB		EQL	Anzali R	IRN
6510,00	USB		PKE5	Ternate R , Maluku	INS
6510,00	USB		CBV	Valpariso R	CHL
6510,00	USB		P2M	Port Moresbay R	PNG
6510,00	USB		PKY23	Fak Fak R	INS
6510,00	USB		XVS	Ho Chi Minh R	VTN
6510,00	USB		PKD5	Benoa R , Bali	INS
6510,00	SSB			CS Ch 604	
6510,00	USB		PKB3	Sibolga R , Sumatera	INS
6510,00	USB		PKF3	Kendari R	INS
6510,00	USB		PKG	Bandjarmasin R , Kalimantan	INS
6510,00	USB		PKK	Kupang R , Timor	INS
6510,00	USB		PKM25	Tahuna R	INS
6510,00	USB		PKN	Balikpapan R , Kalimantan	INS
6510,00	USB		PKP38	Batu Ampar R	INS
6510,00	USB		PNK	Jayapura R , Irian Jaya	INS
6510,00	SSB		TAH	Istanbul R	TUR
6510,00	USB		XSQ	Guangzhou R	CHN
6510,00	USB		PKZ34	Cigading R , Jawa	INS
6510,00	USB		PKY3	Manokwari R , Irian Jaya	INS
6510,00	USB		PKM	Bitung R , Sulawesi	INS
6510,00	USB		PKM44	Pantolan R , Sulawesi	INS

Frequency	Mode	Mode Parameter	Callsign	User	Country
6510,00	USB		A4M	Muscat R	OMA
6510,00	USB		PKJ28	Sei Kolak Kijang R	INS
6510,00	MIL 188.110A SER	B=2400Bd		SVK AF Zvolen	SVK
6510,00	ALE USB	B=125Bd Ch=8 ChS=250Hz		SVK AF net	SVK
6510,00	MIL 188.110A SER	B=2400Bd		SVK AF Presov	SVK
6512,50	USB		PKC5	Pangkal Balam R	INS
6512,60	USB		PKI2	Jakarta R	INS
6513,00	USB		EQO	Now Shar R	IRN
6513,00	USB		STP	Port Sudan R	SDN
6513,00	USB		SDJ	Stockholm R	S
6513,00	DSC	B=100Bd S=170Hz	HLS	Seoul R	KOR
6513,00	USB		VOK	Labrador R	CAN
6513,00	USB		HCY	Ayora R , Santa Cruz	EQA
6513,00	USB		PKO	Tarakan R , Kalimantan	INS
6513,00	CW		EQM	Bushehr R	IRN
6513,00	USB		XVR	Vung Tau R	VTN
6513,00	SSB			CS Ch 605	
6513,00	USB		PKA	Sabang R , We	INS
6513,00	USB		PKP2	Telukbayur R , Sumatera	INS
6513,00	SSB		SUH	Alexandria R	EGY
6513,00	SSB		USO6	Izmail R	UKR
6513,00	SSB		ETC	Assab R	ETH
6513,00	USB		PKS	Pontanak R , Kalimantan	INS
6513,00	USB			Benete R , Sumbawa	INS
6515,00	USB			Mediterranean Cruiser Net , MedNet	
6515,70	USB		PKB	Belawan R , Sumatera	INS
6516,00	USB			Sydney Hobart Race	AUS
6516,00	USB		CBV	Valpariso R	CHL
6516,00	SSB			CS Ch 606	
6516,00	SSB		4PB	Colombo R	CLN
6516,00	USB		CBM	Magallanes Zonal R	CHL
6516,00	USB			Southbound Eveneing net	
6519,00	USB		LSD836	Argentina R	ARG
6519,00	USB		JNA	J CG HQ Tokyo	J
6519,00	USB		UFL	Vladivostok R	RUS
6519,00	USB		OYR	Aasiaat R	GRL
6519,00	SSB			CS Ch 607	
6521,90	USB		CWF	Punta Carretas R	URG
6521,90	USB		PKC5	Pangkal Balam R	INS
6522,00	USB		OYR	Aasiat R	GRL
6522,00	USB			Tiksi MRSC	RUS
6522,00	USB		LSD836	Argentina R	ARG
6522,00	USB		UFH	Petrapavlovsk-Kamchatskii R	RUS
6522,00	USB		XVG	Hai Phong R	VTN
6522,00	USB		XSR	Haikou R	CHN
6522,00	SSB			CS Ch 608	
6522,00	USB		UGC	St. Petersburg R	RUS
6526,00	SSB			Bridgetown ACC	BRB
6526,00	SSB			Jamaica ACC , Kingston	JMC
6526,00	SSB			RDARA 12G	
6526,00	SSB			RDARA 14F	
6526,00	SSB			RDARA 2A	
6526,00	SSB			RDARA 2B	
6526,00	SSB			RDARA 3A	
6526,00	SSB			RDARA 4A	
6526,00	SSB			RDARA 6F	
6529,00	SSB			RDARA 3B	
6529,00	SSB			RDARA 6G	
6529,00	HFDL	B=1800Bd	H17**	Telde , Gran Canaria	E
6530,00	ALE USB	B=125Bd Ch=8 ChS=250Hz		AZE emergency net	AZE

Frequency	Mode	Mode Parameter	Callsign	User	Country
6532,00	USB			San Francisco ACC , CA	USA
6532,00	USB			Taipai ACC	TWN
6532,00	SSB			Hongkong ACC	HKG
6532,00	SSB			Manila ACC	PHL
6532,00	SSB			MWARA CWP	CWP
6532,00	SSB			Naha ACC	J
6532,00	SSB			Port Moresby ACC	PNG
6532,00	SSB			RDARA 12F	
6532,00	SSB			RDARA 2A	
6532,00	SSB			RDARA 2B	
6532,00	SSB			RDARA 3A	
6532,00	SSB			RDARA 4A	
6532,00	SSB			Seoul ACC	KOR
6532,00	SSB			Tokyo ACC	J
6532,00	HFDL	B=1800Bd	H07**	Shannon	IRL
6535,00	USB			Brasilia ACC	B
6535,00	USB			Canarias ACC	POR
6535,00	USB			Cayenne ACC	GUF
6535,00	USB			Dakar ACC	SEN
6535,00	USB			Recife ACC	B
6535,00	USB			Rio de Janeiro ACC	B
6535,00	USB			Manaus ACC	B
6535,00	USB			Paramaribo ACC	SUR
6535,00	USB			Casablanca ACC	MRC
6535,00	USB			Johannesburg ACC	AFS
6535,00	USB			Bouake ACC	CTI
6535,00	USB			Bangui ACC	CAF
6535,00	USB			Kano ACC	NIG
6535,00	USB			Niamey ACC	NGR
6535,00	USB			Nouadhibou ACC	MTN
6535,00	USB			Nouakchott ACC	MTN
6535,00	USB			Monrovia ACC	LBR
6535,00	USB			Ouagadougou ACC	BFA
6535,00	SSB			Abidjan ACC	CTI
6535,00	SSB			Bamako ACC	MLI
6535,00	SSB			Bissau ACC	GNB
6535,00	SSB			Canarias ACC , Las Palmas	CNR
6535,00	SSB			Dakar ACC	SEN
6535,00	SSB			Freetown ACC	SRL
6535,00	SSB			MWARA SAT	SAT
6535,00	SSB			RDARA 10D	
6535,00	SSB			RDARA 10A	
6535,00	SSB			RDARA 12C	
6535,00	SSB			RDARA 12J	
6535,00	SSB			RDARA 14B	
6535,00	SSB			RDARA 2C	
6535,00	SSB			RDARA 5D	
6535,00	SSB			RDARA 6G	
6535,00	SSB			RDARA 9C	
6535,00	SSB			Sal ACC	CPV
6535,00	HFDL	B=1800Bd	H05**	Auckland	NZL
6535,00	HFDL	B=1800Bd	H06**	Hat Yai	THA
6538,00	SSB			RDARA 11B	
6538,00	SSB			RDARA 3A	
6538,00	SSB			RDARA 3B	
6538,00	SSB			RDARA 9B	
6538,00	SSB			VOLMET AFI	AFI
6541,00	SSB			RDARA 10C	
6541,00	SSB			RDARA 13C	
6541,00	SSB			RDARA 14C	
6541,00	SSB			RDARA 2C	
6541,00	SSB			RDARA 6G	

Frequency	Mode	Mode Parameter	Callsign	User	Country
6544,00	SSB			RDARA 10D	
6544,00	SSB			RDARA 1C	
6544,00	SSB			RDARA 3A	
6544,00	SSB			RDARA 3B	
6544,00	SSB			RDARA 5A	
6544,00	SSB			RDARA 5C	
6544,00	SSB			RDARA 6C	
6544,80	STANAG 4539	B=2400Bd		UI MIL user	
6547,00	SSB			RDARA 10B	
6547,00	SSB			RDARA 10E	
6547,00	SSB			RDARA 12E	
6547,00	SSB			RDARA 12J	
6547,00	SSB			RDARA 13F	
6547,00	SSB			RDARA 13K	
6547,00	SSB			RDARA 14A	
6547,00	SSB			RDARA 2A	
6547,00	SSB			RDARA 2C	
6547,00	SSB			RDARA 5D	
6547,00	SSB			RDARA 6G	
6547,00	SSB			RDARA 9B	
6547,00	USB		EIP	Shanwik ACC	IRL
6550,00	ALE USB	B=125Bd Ch=8 ChS=250Hz			
6550,00	SSB			RDARA 11B	
6550,00	SSB			RDARA 13J	
6550,00	SSB			RDARA 1B	
6550,00	SSB			RDARA 3A	
6550,00	SSB			RDARA 3C	
6550,00	SSB			RDARA 5B	
6550,00	SSB			RDARA 6D	
6553,00	SSB			RDARA 10	
6553,00	SSB			RDARA 12E	
6553,00	SSB			RDARA 12F	
6553,00	SSB			RDARA 13E	
6553,00	SSB			RDARA 13F	
6553,00	SSB			RDARA 13K	
6553,00	SSB			RDARA 14A	
6553,00	SSB			RDARA 2A	
6553,00	SSB			RDARA 2C	
6553,00	SSB			RDARA 4B	
6553,00	SSB			RDARA 6G	
6553,00	SSB			RDARA 9	
6556,00	USB			Brisbane ACC	AUS
6556,00	SSB			Calcutta ACC	IND
6556,00	USB			Bali ACC	INS
6556,00	USB			Dhaka ACC	BGD
6556,00	USB			Guangzhou ACC	CHN
6556,00	SSB			Colombo ACC	CLN
6556,00	SSB			Jakarta ACC	INS
6556,00	SSB			Kathmandu ACC	NPL
6556,00	SSB			Kuala Lumpur ACC	MLA
6556,00	SSB			Kunming ACC	CHN
6556,00	SSB			Madras ACC	IND
6556,00	SSB			Male ACC	MLD
6556,00	SSB			MWARA SEA	SEA
6556,00	SSB			Perth ACC	AUS
6556,00	SSB			RDARA 10C	
6556,00	SSB			RDARA 13C	
6556,00	SSB			RDARA 3A	
6556,00	SSB			RDARA 3C	
6556,00	SSB			Singapore ACC	SNG
6556,00	SSB			Ujung Pandang ACC	INS

Frequency	Mode	Mode Parameter	Callsign	User	Country
6556,00	SSB			Yangon ACC	BRM
6559,00	SSB			MWARA AFI	AFI
6559,00	SSB			RDARA 11B	
6559,00	SSB			RDARA 13J	
6559,00	SSB			RDARA 14D	
6559,00	SSB			RDARA 2A	
6559,00	SSB			RDARA 6G	
6559,00	HFDL	B=1800Bd	H01**	San Francisco , CA	USA
6562,00	USB			Hong Kong ACC	HKG
6562,00	USB			Naha ACC	J
6562,00	USB			Port Moresby ACC	PNG
6562,00	USB			San Francisco ACC , CA	USA
6562,00	USB			Seoul ACC	KOR
6562,00	USB			Taipai ACC	TWN
6562,00	SSB			Manila ACC	PHL
6562,00	SSB			RDARA 10D	
6562,00	SSB			RDARA 13C	
6562,00	SSB			RDARA 2B	
6562,00	SSB			RDARA 2C	
6562,00	SSB			Tokyo ACC	J
6563,50	CW		RJD99	RUS N Severodvinsk	RUS
6565,00	SSB			Perth ACC	AUS
6565,00	SSB			RDARA 11B	
6565,00	SSB			RDARA 14E	
6565,00	SSB			RDARA 2A	
6565,00	SSB			RDARA 4	
6565,00	SSB			RDARA 6G	
6565,00	HARRIS AVS	Ch=24 ChS=112,5Hz			
6565,00	HFDL	B=1800Bd	H02**	Molokai , HI	HWA
6566,00	ALE USB	B=125Bd Ch=8 ChS=250Hz		MTN MOI net	MTN
6568,00	STANAG 4197	B=1800Bd Ch=16/39 ChS=112/56Hz		EGY N net	EGY
6568,00	SSB			RDARA 10C	
6568,00	SSB			RDARA 13C	
6568,00	SSB			RDARA 2B	
6568,00	SSB			RDARA 2C	
6568,00	SSB			RDARA 3B	
6568,00	SSB			RDARA 6D	
6568,00	SSB			RDARA 7C	
6568,00	USB			EGY MIL net	EGY
6571,00	USB			Beijing ACC	CHN
6571,00	USB			Guangzhou ACC	CHN
6571,00	USB			Hailar ACC	CHN
6571,00	USB			Jinan ACC	CHN
6571,00	USB			Lanzhou ACC	CHN
6571,00	USB			Kunming ACC	CHN
6571,00	SSB			Irkutsk ACC	RUS
6571,00	SSB			MWARA EA	EA
6571,00	SSB			Pyongyang ACC	KRE
6571,00	SSB			RDARA 12C	
6571,00	SSB			Shanghai ACC	CHN
6571,00	SSB			Ulaanbaatar ACC	MNG
6571,00	SSB			Wuhan ACC	CHN
6571,00	USB			Shenyang ACC	CHN
6571,00	USB			Taegu ACC	KOR
6571,00	USB			Urumqi ACC	CHN
6571,00	USB			Zhengzhou ACC	CHN
6574,00	SSB			MWARA AFI	AFI
6574,00	SSB			RDARA 13I	

Frequency	Mode	Mode Parameter	Callsign	User	Country
6574,00	SSB			RDARA 13M	
6574,00	SSB			RDARA 14D	
6574,00	SSB			RDARA 2A	
6574,00	SSB			RDARA 6G	
6577,00	USB			Merida ACC	MEX
6577,00	SSB			Barranquilla ACC	CLM
6577,00	SSB			Boyeros ACC	CUB
6577,00	SSB			MWARA CAR	CAR
6577,00	SSB		KEA5	New York ACC	USA
6577,00	SSB			Panama ACC	PNR
6577,00	SSB			Piarco ACC	TRD
6577,00	SSB			RDARA 13E	
6577,00	SSB			RDARA 2B	
6577,00	SSB			RDARA 2C	
6577,00	SSB			RDARA 3B	
6577,00	SSB			RDARA 4B	
6577,00	SSB			RDARA 6D	
6580,00	SSB			RDARA 10A	
6580,00	SSB			RDARA 13C	
6580,00	SSB			RDARA 13J	
6580,00	SSB			RDARA 13K	
6580,00	SSB			RDARA 14	
6580,00	SSB			RDARA 6G	
6580,00	SSB			RDARA 7E	
6580,00	SSB			RDARA 9C	
6580,00	SSB			VOLMET EUR	EUR
6580,00	*ALE USB*	*B=125Bd Ch=8 ChS=250Hz*		*JOR net*	*JOR*
6583,00	SSB			RDARA 2	
6583,00	SSB			RDARA 3	
6583,00	SSB			RDARA 6E	
6586,00	USB			Barranquilla ACC	CLM
6586,00	USB			Boyeros ACC	CTR
6586,00	USB			Cayenne ACC	GUF
6586,00	USB			Panama ACC	PNR
6586,00	USB			Paramaribo ACC	SUR
6586,00	USB			Piarco ACC	TRD
6586,00	USB			Georgtown ACC	GUY
6586,00	USB			Maiquetia ACC	VEN
6586,00	SSB			MWARA CAR	CAR
6586,00	SSB		KEA5	New York ACC , NY	USA
6586,00	SSB			RDARA 13G	
6586,00	SSB			RDARA 14C	
6586,00	SSB			RDARA 2C	
6586,00	SSB			RDARA 6G	
6586,00	SSB			RDARA 7	
6588,50	OFDM	B=2400Bd Ch=39 ChS=55Hz			
6589,00	HFDL	B=1800Bd	H11**	Albrook	PNR
6589,00	SSB			RDARA 3	
6592,00	USB			Barnaul ACC	RUS
6592,00	USB			Khanty-Mansiysk ACC	RUS
6592,00	USB			Kirensk ACC	RUS
6592,00	USB			Kolpashevo ACC	RUS
6592,00	SSB			Irkutsk ACC	RUS
6592,00	SSB			MWARA NCA	NCA
6592,00	SSB			Novosibirsk ACC	RUS
6592,00	SSB			RDARA 12C	
6592,00	USB			Krasnoyarsk ACC	RUS
6592,00	USB			Podkamennaya ACC	RUS
6592,00	USB			Surgut ACC	RUS
6592,00	USB			Yeniseysk ACC	RUS

Frequency	Mode	Mode Parameter	Callsign	User	Country
6595,00	SSB			Jakarta ACC	INS
6595,00	SSB			Medan ACC	INS
6595,00	SSB			Palembang ACC	INS
6595,00	SSB			RDARA 1B	
6595,00	SSB			RDARA 3B	
6595,00	SSB			RDARA 3C	
6595,00	SSB			RDARA 5B	
6595,00	SSB			RDARA 6D	
6595,00	USB			Shanwich ACC	IRL
6596,00	HFDL	B=1800Bd	H14**	Krasnoyarsk	RUS
6598,00	USB			Arkhangelsk ACC	RUS
6598,00	USB			Beirut ACC	LBN
6598,00	USB			Berlin ACC	D
6598,00	USB			Kiev ACC	UKR
6598,00	SSB			MWARA EUR	EUR
6598,00	SSB			RDARA 10B	
6598,00	SSB			RDARA 12E	
6598,00	SSB			RDARA 13H	
6598,00	SSB			RDARA 4B	
6598,00	SSB			RDARA 6G	
6598,00	SSB			RDARA 9B	
6598,00	USB			Lvov ACC	UKR
6598,00	USB			Minsk ACC	BLR
6598,00	USB			Moscow ACC	RUS
6598,00	USB			Murmansk ACC	RUS
6598,00	USB			Odessa ACC	UKR
6598,00	USB			Riga ACC	LVA
6598,00	USB			Simferopol ACC	UKR
6598,00	USB			Sofia ACC	BUL
6598,00	USB			St. Petersburg ACC	RUS
6598,00	USB			Syktyvkar ACC	RUS
6598,00	USB			Tunis ACC	TUN
6598,00	USB			Velikiye ACC	RUS
6598,00	USB			Vologda ACC	RUS
6598,00	USB			Vilnius ACC	LTU
6598,50	USB		DHM91	D AF Münster	D
6598,50	USB		DHO32	D AF Wunstorf	D
6598,50	USB		DHO78	D AF Kalkar	D
6598,50	USB		DHJ83	D AF Gatow	D
6600,50	ALE USB	B=125Bd Ch=8 ChS=250Hz		ROU MIL net	ROU
6601,00	SSB			RDARA 2	
6604,00	SSB		WSY70	New York VOLMET	USA
6604,00	SSB			RDARA 10A	
6604,00	SSB			RDARA 13N	
6604,00	SSB			RDARA 14B	
6604,00	SSB			RDARA 1D	
6604,00	SSB			RDARA 6G	
6604,00	SSB			RDARA 7C	
6604,00	SSB			VOLMET NAT	NAT
6604,00	USB		VFG	Gander VOLMET	CAN
6607,00	SSB			RDARA 3A	
6607,00	SSB			RDARA 6A	
6607,00	SSB			RDARA 6B	
6607,00	CW		4XZ	ISR N Haifa	ISR
6610,00	USB		VNJ	Casey Base , AUS	ANT
6610,00	USB			AUS Davis Base	ANT
6610,00	USB			Macquarie Island , AUS	ANT
6610,00	USB			Mawson Base , AUS	ANT
6610,00	SSB			RDARA 14F	
6610,00	SSB			RDARA 1D	
6610,00	SSB			RDARA 6G	

Frequency	Mode	Mode Parameter	Callsign	User	Country
6613,00	SSB			RDARA 13G	
6613,00	SSB			RDARA 3A	
6613,00	SSB			RDARA 6A	
6613,00	SSB			RDARA 6B	
6616,00	SSB			RDARA 12G	
6616,00	SSB			RDARA 14E	
6616,00	SSB			RDARA 4A	
6616,00	SSB			RDARA 6G	
6617,00	SSB		RLAP	Rostov VOLMET	RUS
6617,00	SSB		UHD	St. Petersburg VOLMET	RUS
6618,00	*ISR N HYBRID MODEM*	*B=2400Bd*	*4XZ*	*ISR N Haifa*	*ISR*
6619,00	SSB			RDARA 3A	
6619,00	SSB			RDARA 6B	
6619,00	HFDL	B=1800Bd	H10**	Muan	KOR
6622,00	SSB			MWARA NAT	NAT
6622,00	SSB			RDARA 12C	
6622,00	SSB			RDARA 13D	
6622,00	SSB			RDARA 6G	
6622,00	SSB			RDARA 7F	
6622,00	SSB			RDARA 9B	
6622,00	USB		EIP	Shanwik ACC	IRL
6622,00	SSB		VFG	Gander ACC	CAN
6625,00	SSB			MWARA MID	MID
6625,00	SSB			RDARA 3B	
6627,00	USB			Tasmanian Maritime R	AUS
6628,00	SSB			MWARA NAT	NAT
6628,00	SSB		KEA5	New York ACC	USA
6628,00	SSB			RDARA 12C	
6628,00	SSB			RDARA 13D	
6628,00	SSB			RDARA 13M	
6628,00	SSB			RDARA 14	
6628,00	SSB			RDARA 6G	
6628,00	SSB			RDARA 7E	
6628,00	USB		CSY	Santa Maria ACC	AZR
6628,00	HFDL	B=1800Bd	H13**	Santa Cruz	BOL
6630,00	USB			CHN air defense	CHN
6631,00	SSB			MWARA MID	MID
6631,00	SSB			RDARA 3B	
6631,00	SSB			RDARA 6C	
6631,00	SSB			Uralsk ACC	KAZ
6631,00	USB			Aden ACC	YEM
6631,00	USB			Amman ACC	JOR
6631,00	USB			Ankara ACC	TUR
6631,00	USB			Beirut ACC	LBN
6631,00	USB			Cairo ACC	EGY
6631,00	USB			Damascus ACC	SYR
6631,00	USB			Jeddah ACC	ARS
6631,00	USB			Kuwait ACC	KWT
6631,00	USB			Manama ACC	BHR
6631,00	USB			Odessa ACC	UKR
6631,00	USB			Simferopol ACC	UKR
6631,00	USB			Sanaa ACC	YEM
6631,00	USB			Tehran ACC	IRN
6631,00	USB			Tbilisi ACC	GEO
6631,00	USB			Yerevan ACC	ARM
6631,00	USB			Aktyubinsk ACC	KAZ
6631,00	USB			Almaty ACC	KAZ
6631,00	USB			Bishkek ACC	KGZ
6631,00	USB			Dushanbe ACC	TJK
6631,00	USB			Kuybyshev ACC	RUS
6631,00	USB			Kzyl-Orda ACC	KAZ

Frequency	Mode	Mode Parameter	Callsign	User	Country
6631,00	USB			Moscow ACC	RUS
6631,00	USB			Yerevan ACC	ARM
6634,00	SSB			RDARA 6G	
6637,00	SSB			W I	WW
6637,00	SSB			W II	WW
6637,00	SSB			W III	WW
6640,00	SSB			W II	WW
6640,00	SSB			W V	WW
6640,00	USB			San Francisco ACC , CA	USA
6642,00	CW			RUS Strategic AF net	RUS
6643,00	SSB			W I	WW
6643,00	SSB			W IV	WW
6644,00	ALE USB	B=125Bd Ch=8 ChS=250Hz		ISR AF net	ISR
6646,00	SSB			W II	WW
6646,00	SSB			W V	WW
6646,00	SSB			MAFF London	G
6646,00	HFDL	B=1800Bd	H09**	Utqiagvik , AK	ALS
6649,00	SSB			Antofagasta ACC	CHL
6649,00	SSB			Asuncion ACC	PRG
6649,00	SSB			Bogota ACC	CLM
6649,00	SSB			La Paz ACC	BOL
6649,00	SSB			Lima ACC	PRU
6649,00	SSB			Montevideo ACC	URG
6649,00	SSB			MWARA SAM	SAM
6649,00	SSB			Puerto Montt ACC	CHL
6649,00	SSB			Punta Arenas ACC	CHL
6649,00	SSB			Quito ACC	EQA
6649,00	SSB			RDARA 3A	
6649,00	USB			Barranquilla ACC	CLM
6649,00	USB			Maiquetia ACC	VEN
6649,00	USB			Buenos Aires ACC	ARG
6649,00	SSB			RDARA 6G	
6649,00	SSB			Santa Cruz ACC	BOL
6649,00	SSB			Santiago ACC	CHL
6649,00	USB			Cordoba ACC	ARG
6649,00	USB			Easter Island ACC	CHL
6649,00	USB			Talara ACC	PRU
6649,00	USB			Ushuaia ACC	ARG
6649,00	USB			Recife ACC	B
6652,00	SSB			RDARA 6G	
6652,00	SSB			RDARA 7B	
6652,00	HFDL	B=1800Bd	H16**	Agana	GUM
6655,00	USB			San Francisco ACC	USA
6655,00	SSB			MWARA NP	NP
6655,00	SSB			RDARA 2B	
6655,00	SSB			RDARA 6E	
6655,00	SSB			Tokyo ACC	J
6655,00	USB			San Francisco ACC , CA	USA
6658,00	SSB			RDARA 3C	
6658,00	SSB			RDARA 6A	
6661,00	SSB			MWARA NP	NP
6661,00	SSB			RDARA 2B	
6661,00	SSB			RDARA 6E	
6661,00	HFDL	B=1800Bd	H04**	Riverhead , NY	USA
6664,00	SSB			RDARA 3C	
6664,00	SSB			RDARA 5A	
6666,00	USB			EGY N net	EGY
6666,50	CHN 4+4	B=75Bd Ch=8 ChS=300/450Hz		CHN MIL	CHN
6667,00	SSB			RDARA 1E	
6667,00	SSB			RDARA 2B	

Frequency	Mode	Mode Parameter	Callsign	User	Country
6667,00	SSB			RDARA 6F	
6668,00	MIL 188-110A SER	B=2400Bd			
6670,00	SSB			RDARA 3C	
6670,00	MIL 188-110A SER	B=2400Bd			
6670,00	ALE USB	B=125Bd Ch=8 ChS=250Hz			
6673,00	USB			San Francisco ACC , CA	USA
6673,00	SSB			MWARA AFI	AFI
6673,00	SSB			MWARA CEP	CEP
6673,00	SSB			NOAA Hurricane Center Miami , FL	USA
6673,00	SSB			RDARA 10F	
6673,00	SSB			RDARA 12D	
6673,00	SSB			RDARA 13D	
6673,00	SSB			RDARA 14B	
6673,00	SSB			RDARA 2A	
6673,00	SSB			RDARA 6G	
6676,00	SSB		AXQ429	Australian Volmet	AUS
6676,00	SSB		HSD	Bangkok VOLMET	THA
6676,00	USB		AWB	Mumbai VOLMET	IND
6676,00	USB		AWC	Calcutta VOLMET	IND
6676,00	USB		ARA	Karachi VOLMET	PAK
6676,00	USB		9VA40	Singapore VOLMET	SNG
6676,00	USB			AUS VOLMET Brisbane	AUS
6676,00	SSB			VOLMET SEA	SEA
6678,00	USB			UN MINURSO net	AOE
6678,00	CCIR 493-4	B=100Bd S=170Hz		UN MINURSO net	AOE
6678,00	ALE USB	B=125Bd Ch=8 ChS=250Hz		MTN net	MTN
6679,00	USB		ZKAK	Auckland VOLMET	NZL
6679,00	SSB			Hongkong VOLMET	HKG
6679,00	SSB		JIA	Tokyo VOLMET	J
6679,00	SSB			VOLMET PAC	PAC
6679,00	SSB		KVM70	Honolulu VOLMET	HWA
6682,00	ALE USB	B=125Bd Ch=8 ChS=250Hz		CHL N net	CHL
6682,00	SSB			RDARA 6G	
6685,00	USB			RUS AF net	RUS
6685,00	ALE USB	B=125Bd Ch=8 ChS=250Hz		USA AF net	
6685,00	USB		PROSELO K	RUS AF Pskov	RUS
6685,00	USB		PRIRODNI J	RUS AF Butulirovka	RUS
6685,00	USB		KORSAR	RUS AF Moscow	RUS
6685,00	USB		KLARNETI ST	RUS AF Migalovo/Tver	RUS
6687,50	USB			IND CG net	IND
6688,00	USB			F AF TX Vernon	F
6688,00	USB		CAPITOLE	F AF Taverny	F
6688,00	USB			F AF net	F
6688,00	USB			EGY N net	EGY
6689,00	CW			RUS N AF net	RUS
6689,00	CW		59821	RUS N AF	RUS
6689,00	CW		59822	RUS N AF	RUS
6690,00	USB		DHN66	NATO Geilenkirchen	D
6690,00	MIL 188-110B SER	B=2400Bd			
6691,00	ALE USB	B=125Bd Ch=8 ChS=250Hz		G MIL net	G
6693,00	SSB			Jekaterinenburg VOLMET	RUS
6693,00	SSB		RQCI	Samara VOLMET	RUS
6693,00	SSB		UBB2	Syktyvkar VOLMET	RUS
6693,00	SSB		RVPE	Tyumen VOLMET	RUS

Frequency	Mode	Mode Parameter	Callsign	User	Country
6693,00	STANAG 4197	B=1800Bd Ch=16/39 ChS=112/56Hz			
6693,00	MIL 188-110B 39TONE	B=2400Bd Ch=39 ChS=56,25Hz			
6694,00	*STANAG 4285*	*B=2400Bd*	*CFH*	*CAN N Halifax*	*CAN*
6697,00	USB		MKL	G F AMCC Northwood	G
6697,50	ALE USB	B=125Bd Ch=8 ChS=250Hz		SNG N net	SNG
6699,00	STANAG 4481	B=75Bd S=850Hz			
6699,00	FSK	B=300Bd S=850Hz			
6700,00	USB			Mauritius MRCC	MAU
6700,00	USB		VEILLEUR	F AF Avord , E-3F net	F
6701,00	LINK 11 CLEW	B=2250Bd Ch=16 ChS=330/110/550Hz			
6702,00	STANAG 4481	B=75Bd S=850Hz			
6704,80	ALE 3G	B=2400Bd		UI MIL net	
6706,00	USB		CHR	CAN AF Trenton	CAN
6706,00	CW		RJE65	RUS N HQ Novorossiysk	RUS
6707,00	CW		4XZ	ISR N Haifa	ISR
6709,00	ALE USB	B=125Bd Ch=8 ChS=250Hz		G MIL net	G
6709,00	SYSTEME 3000 ALE	B=125Bd Ch=8 ChS=250Hz			F
6711,00	ALE USB	B=125Bd Ch=8 ChS=250Hz			
6712,00	ALE USB	B=125Bd Ch=8 ChS=250Hz		USA AF net	USA
6712,00	SSB			USA AF Strategic Command	USA
6712,00	HFDL	B=1800Bd	H03**	Reykjavik	ISL
6712,00	USB			F AF TX Vernon	F
6712,00	USB		CIRCUS VERT	F AF HQ Villacoublay	F
6714,30	ALE 3G	B=2400Bd		UI MIL net	
6715,20	*STANAG 4285*	*B=2400Bd*		*G MIL TX Inskip*	*G*
6718,00	*LINK 11 ISB*	*B=2250Bd Ch=16 ChS=330/110/550Hz*		*G AF*	*G*
6721,00	BAUDOT	B=75Bd S=850Hz		F AF	F
6722,00	ALE USB	B=125Bd Ch=8 ChS=250Hz		ROU F net	ROU
6722,00	MIL 188-110B SER	B=2400Bd		ROU F	ROU
6722,50	USB		IABH	I N ship Commandante Fascari	I
6722,50	USB		IAME	I N ship Maestrale	I
6722,50	*STANAG 4285*	*B=2400Bd*		*NOR N TX Gossen*	*NOR*
6727,00	USB		LBJ	NOR N Bodoe	NOR
6727,00	SSB		PBV	HOL N Valkenburg	HOL
6728,00	ALE USB	B=125Bd Ch=8 ChS=250Hz		AZE emergency net	AZE
6730,00	ALE USB	B=125Bd Ch=8 ChS=250Hz		AUT MIL net	AUT
6730,00	SSB			Aktyubinsk VOLMET	KAZ
6730,00	SSB			Baku VOLMET	AZE
6730,00	SSB			Karaganda VOLMET	KAZ
6730,00	SSB			Tbilisi VOLMET	GEO
6732,00	*STANAG 4481*	*B=75Bd S=850Hz*	*AJE*	*USA AF Crougthon*	*G*
6733,00	USB		IDR	I N Rome	I
6733,00	USB			G AF TASCOMM	G
6736,00	USB			Kinloss Rescue	G
6737,00	SSB			E AF Madrid	E
6737,00	SSB			E AF Sevilla	E
6737,00	*STANAG 4285*	*B=2400Bd*		*G MIL TX Inskip*	*G*

Frequency	Mode	Mode Parameter	Callsign	User	Country
6739,00	USB			USA AF Yokota AFB	J
6739,00	USB			USA AF Sigonella	I
6739,00	USB			USA AF Puerto Rico	PTR
6739,00	USB		AFS	USA AF Offutt AFB , Omaha , NE	USA
6739,00	ALE USB	B=125Bd Ch=8 ChS=250Hz		USA AF net	USA
6739,00	SSB			USA AF Ascension AFB	ASC
6739,00	SSB		AIE	USA AF Andrews AFB , Camp Springs , MD	USA
6739,00	SSB			USA AF Lajes	AZR
6739,00	USB			USA AF Diega Garcia	DGA
6739,00	USB			USA AF Guam	GUM
6739,00	USB			USA AF Hickam , HWA	USA
6741,50	SSB			RAF Buchan	G
6742,00	SSB			RAF Boulmer	G
6742,00	SSB			RAF Buchan	G
6742,00	SSB			RAF Neatishead	G
6742,00	CW		HMQ32		KOR
6742,00	CW		HMYL4	KRE cargo ship JI SONG7	KRE
6742,00	CW		HMYE9	KRE cargo ship SE JON BONG	KRE
6742,00	CW		HMYO8	KRE cargo ship KUM SUNG7	KRE
6742,00	CW		HMYL5	KRE cargo ship JISONG3	KRE
6742,00	CW		HMYP4	KRE cargo ship	KRE
6742,00	CW		HMJE	KRE cargo ship A BONG1	KRE
6745,00	USB		CHR	CAN AF Trenton	CAN
6745,00	USB			BGD N net	BGD
6745,00	ALE USB	B=125Bd Ch=8 ChS=250Hz		F AF net	F
6745,00	CODAN CHIRP	B=80Bd Ch=30 ChS=250Hz	1001		
6745,00	CODAN CHIRP	B=80Bd Ch=30 ChS=250Hz	9103		
6745,00	CODAN	B=2400Bd Ch=16 ChS=112,5Hz	9103		
6745,00	CODAN	B=2400Bd Ch=16 ChS=112,5Hz	1001		
6748,00	ALE USB	B=125Bd Ch=8 ChS=250Hz		I AF net	I
6750,00	USB			Shandong M	CHN
6751,00	USB			Cape Canaveral AFS , FL	USA
6751,00	MIL 188-110B 39TONE	B=2400Bd Ch=39 ChS=56,25Hz			
6752,00	CW		S	RUS AF marker	RUS
6752,00	ALE 3G	B=2400Bd		UI MIL net	
6753,00	CW		RFFN	RUS MIL	RUS
6753,00	SSB			Belawan ACC	INS
6753,00	SSB			CAN AF Vancouver	CAN
6754,00	USB		CHR	CAN AF Trenton	CAN
6754,00	ALE USB	B=125Bd Ch=8 ChS=250Hz		F N net	F
6755,00	CHN DATALINK	B=2400Bd Ch=30 ChS=75Hz		CHN N	CHN
6757,00	SSB			USA AF Strategic Command	USA
6759,00	STANAG 4481	B=75Bd S=850Hz		G N Inskip	G
6761,00	SSB			USA AF Strategic Command	USA
6763,20	STANAG 4285	B=2400Bd	DHJ58	D N Glücksburg TX Neuharlingersiel	D
6765,00	USB		HSW69	Bangkok M	THA
6765,00	MT63	B=20Bd Ch=64 ChS=15,625Hz		USA SHARES	USA
6765,00	MIL 188-110A SER	B=2400Bd		MFA Bucharest	ROU
6765,00	MIL 188-110A SER	B=2400Bd		ROU E	
6765,00	ALE USB	B=125Bd Ch=8 ChS=250Hz		USA SHARES net	USA
6766,70	STANAG 4285	B=2400Bd	DHJ58	D N TX Neuharlingersiel	D

Frequency	Mode	Mode Parameter	Callsign	User	Country
6766,70	*STANAG 4285*	*B=2400Bd*	*DHJ58*	*D N Glücksburg TX Marlow*	*D*
6767,00	CW			RUS N net	RUS
6767,00	CW		RCC	Petropavlovsk R	RUS
6768,00	PACTOR III	B=200Bd S=200Hz	WGM	Ft. Lauderdale , FL	USA
6774,00	CIS 4FSK	B=150Bd Ch=4 ChS=4000Hz			
6775,00	PSK	B=1200Bd		Yeakterinenburg	RUS
6776,20	*STANAG 4285*	*B=2400Bd*	*EBA*	*E N Madrid*	*E*
6777,50	ALE LSB	B=125Bd Ch=8 ChS=250Hz		GRC AF net	GRC
6778,00	ALE USB	B=125Bd Ch=8 ChS=250Hz		TUR Disaster and Emergency net	TUR
6778,20	STANAG 4285	B=2400Bd	DHJ58	D N Glücksburg TX	D
6779,00	RFSM8000	B=2400Bd		MFA Sofia	BUL
6780,00	CW				
6780,00	CW		NNA2AX	USA MARS	USA
6780,00	CW		XSG	Shanghai R	CHN
6781,00	ALE USB	B=125Bd Ch=8 ChS=250Hz		F N net	F
6781,20	*STANAG 4285*	*B=2400Bd*		*G MIL TX Inskip*	*G*
6781,20	*STANAG 4285*	*B=2400Bd*		*G MIL Crimond*	*G*
6781,20	*STANAG 4285*	*B=2400Bd*		*G MIL TX St. Eval*	*G*
6783,00	TADIRAN FSK	B=125Bd S=300Hz		ISR MIL	ISR
6783,45	CW		VA3KAH/B	ISM band beacon , Kah She Island , ON	CAN
6784,00	ALE USB	B=125Bd Ch=8 ChS=250Hz			
6784,00	MIL 188-110B 39TONE	B=2400Bd Ch=39 ChS=56,25Hz			
6786,00	*STANAG 4285*	*B=2400Bd*		*G MIL Crimond*	*G*
6786,00	*STANAG 4285*	*B=2400Bd*		*G MIL TX Inskip*	*G*
6786,00	BAUDOT	B=50Bd S=500Hz	NT9P**		RUS
6789,00	CW		VE3DJI/B	ISM band beacon Kingston , ON	CAN
6790,00	ALE USB	B=125Bd Ch=8 ChS=250Hz		ALG Sonatrach net	ALG
6790,00	USB			I Mil net	I
6790,00	CODAN	B=2400Bd Ch=16 ChS=112,5Hz		MFA Cairo	EGY
6790,00	MIL 188-110B SER	B=2400Bd		G MIL	G
6790,00	USB		VKE237	AUS HF Radio Club	AUS
6791,00	CIS MFSK 20	B=20Bd Ch=20 ChS=40Hz		RUS INTEL	RUS
6793,00	USB		VMD750	Austravel Safety Net	AUS
6793,00	CW			RUS Air Defense net	RUS
6795,00	ALE USB	B=125Bd Ch=8 ChS=250Hz		TUR Disaster and Emergency net	TUR
6795,00	*T600*	*B=50Bd S=500Hz*	*RMP*	*RUS N Kaliningrad*	*RUS*
6795,00	*ALE USB*	*B=125Bd Ch=8 ChS=250Hz*		*TUR Disaster and Emergency net*	*TUR*
6795,00	CIS ARQ	B=100Bd S=500Hz	RXZ32	RUS PTT	RUS
6796,00	USB		VKS737	AUS 4WD net	AUS
6797,00	MIL 188-110B SER	B=2400Bd		ROU STAR net	ROU
6800,00	CHN 4+4	B=75Bd Ch=8 ChS=300/450Hz		CHN MIL	
6800,00	*STANAG 4285*	*B=2400Bd*		*GRC N Athens*	*GRC*
6800,00	CIS 40,4-800	B=40,5Bd S=200Hz	UTN7	RUS MIL	RUS
6800,00	ALE USB	B=125Bd Ch=8 ChS=250Hz		USA SHARES net	USA
6805,00	USB			G N Sea Cadets	G
6805,00	*STANAG 4285*	*B=2400Bd*		*G MIL TX Akrotiri*	*CYP*
6806,00	ALE USB	B=125Bd Ch=8 ChS=250Hz		USA Civil Air Patrol net	USA
6807,00	ALE USB	B=125Bd Ch=8 ChS=250Hz		ROU diplo net	ROU

Frequency	Mode	Mode Parameter	Callsign	User	Country
6810,00	CW			RUS HFDF net	RUS
6812,00	CW			RUS HFDF net	RUS
6818,00					
6819,00	BAUDOT	B=200Bd S=850Hz			
6819,80	STANAG 4538	B=2400Bd			
6820,00	ALE USB	B=125Bd Ch=8 ChS=250Hz		USA AF net	
6820,00	ALE USB	B=125Bd Ch=8 ChS=250Hz		ALG MIL net	ALG
6820,00	MIL 188-110B 39TONE	B=2400Bd Ch=39 ChS=56,25Hz		ALG MIL net	ALG
6821,00	ALE USB	B=125Bd Ch=8 ChS=250Hz		LTU MIL net	LTU
6821,50	CW		PWZW	RUS Strategic Command Aerospace Defense	RUS
6823,00	USB			USA Army MARS	USA
6823,00	CIS MFSK 20	B=20Bd Ch=20 ChS=40Hz		RUS INTEL	RUS
6823,50	CW			RUS air defence net	RUS
6825,00	USB		VKJ	RFDS Meekathara	AUS
6825,00	CW		FAV22	F F Mont Valerien	F
6825,00	CW			RUS HFDF net	RUS
6826,00	SSB			USA AF Strategic Command	USA
6826,00	USB			LBY net	LBY
6828,00	ALE USB	B=125Bd Ch=8 ChS=250Hz		F N net	F
6828,00	MIL 188-110A SER	B=2400Bd		F N	F
6829,50	IRN N QPSK ADAPTIVE	B=468Bd		IRN N	
6830,00	MHF-50 MODEM	B=54,3Bd Ch=32 ChS=64,5Hz	ZSJ	AFS N Capetown	AFS
6830,00	CIS ARQ	B=100Bd S=500Hz		RUS PTT Moscow	RUS
6832,00	*MAHRS*	*B=2400Bd*	*DHJ58*	*D N Glücksburg TX Staberhuk*	*D*
6832,00	CW		RMP	RUS N HQ Kaliningrad	RUS
6832,00	CW		RFK99	RUS N	RUS
6832,60	STANAG 4481	B=50Bd S=850Hz	NAU	USA N San Juan	PTR
6834,00	PACTOR III	B=100Bd S=200Hz	OL3A		CZE
6834,00	PACTOR III	B=100Bd S=200Hz	OL1A	CZE MIL Prague	CZE
6834,00	STANAG 4481	B=75Bd S=850Hz	NAU	USA N San Juan	PTR
6836,00	CW		RMW46	RUS MIL	RUS
6837,00	CW		K4MT**		
6837,00	CW		NT9P**		RUS
6837,00	BAUDOT	B=50Bd S=500Hz			
6838,00	*ALE USB*	*B=125Bd Ch=8 ChS=250Hz*		*MLT N net*	*MLT*
6838,50	ALE USB	B=125Bd Ch=8 ChS=250Hz		HOL MIL net	HOL
6840,00	AM		EBC	San Fernando TS	E
6840,00	OQPSK	B=48000Bd		HF Trading New York	USA
6842,50	STANAG 4285	B=2400Bd	FUJ	F N Noumea	NCL
6842,70	*MIL 188-110B SER*	*B=2400Bd*		*S N Karlskrona*	*S*
6843,00	ALE USB	B=125Bd Ch=8 ChS=250Hz		CHL SENAPRED net	CHL
6845,00	USB		VJJ	RFDS Charleville	AUS
6845,00	CW		F**	RUS N Vladivostok	RUS
6845,00	ALE USB	B=125Bd Ch=8 ChS=250Hz		G MIL net	G
6848,00	ALE USB	B=125Bd Ch=8 ChS=250Hz		CHL N net	CHL
6850,00	USB		VNJ	Casey Base , AUS	ANT
6850,00	USB			AUS Davis Base	ANT
6850,00	USB			Macquarie Island , AUS	ANT
6850,00	USB			Mawson Base , AUS	ANT

Frequency	Mode	Mode Parameter	Callsign	User	Country
6851,00	CW		PRV	Preveza P	GRC
6852,00	CCIR 493-4	B=100Bd S=170Hz			
6854,00	ALE USB	B=125Bd Ch=8 ChS=250Hz			
6855,50	PACTOR III	B=100Bd S=200Hz	KDS	Ontario	CAN
6856,00	BAUDOT	B=200Bd S=850Hz		RUS INTEL	RUS
6856,00	*PSK*	*B=2400Bd*		*POL MIL*	*POL*
6857,00	SSB			MOD Moscow	RUS
6857,00	CW		RGT77	RUS MIL Moscow, RVSN	RUS
6858,00	PACTOR III	B=100Bd S=200Hz	KLT	Austin , TX	USA
6858,00	ALE USB	B=125Bd Ch=8 ChS=250Hz		USA Red Cross net	USA
6860,00	MIL 188-110B SER	B=2400Bd			
6860,00	CW		RMW56	RUS MIL	RUS
6861,00	ALE USB	B=125Bd Ch=8 ChS=250Hz		CHL DC net	CHL
6862,00	BAUDOT	B=75Bd S=850Hz	OSN	BEL N Oostende	BEL
6865,00	CIS ARQ	B=100Bd S=500Hz		RUS PTT	RUS
6866,00	CIS 50-500	B=50Bd S=500Hz		RUS	RUS
6868,00	ALE USB	B=125Bd Ch=8 ChS=250Hz		ISR AF net	ISR
6868,00	ALE USB	B=125Bd Ch=8 ChS=250Hz		UI net	
6869,00	CODAN	B=2400Bd Ch=16 ChS=112,5Hz			
6870,00	ALE USB	B=125Bd Ch=8 ChS=250Hz		VEN MIL net	VEN
6870,00	CIS 50-500	B=50Bd S=500Hz	RAZ2	RUS PTT	RUS
6870,00	BPSK	B=1200Bd			
6870,50	CHN 4+4	B=75Bd Ch=8 ChS=300/450Hz		CHN MIL	CHN
6873,00	CW		RMP	RUS N HQ Kaliningrad	RUS
6873,50	*ALE USB*	*B=125Bd Ch=8 ChS=250Hz*		*G MIL net*	*G*
6874,00	CW			RUS HFDF net	RUS
6875,00	USB			G N Sea Cadets	G
6877,00	ALE USB	B=125Bd Ch=8 ChS=250Hz		CHL SENAPRED net	CHL
6877,00	CW		RCB	RUS AF	RUS
6877,00	*CW*		*RIT*	*RUS N HQ Severomorsk*	*RUS*
6878,00	CIS 75-250	B=75Bd S=250Hz		RUS N Moscow	RUS
6879,00	ALE USB	B=125Bd Ch=8 ChS=250Hz		F N net	F
6880,00	USB		VKJ	RFDS Meekathara	AUS
6880,00	CODAN	B=2400Bd Ch=16 ChS=112,5Hz			
6880,00	USB			UN MONUSCO net	UGA
6881,00	USB			RUS F	RUS
6883,00	IRN QPSK	B=207Bd		IRN N	IRN
6885,00	CHN 4+4	B=75Bd Ch=8 ChS=300/450Hz		CHN MIL	CHN
6886,00	SSB			USA AF Strategic Command	USA
6887,50	*CIS 12*	*B=1440Bd Ch=12 ChS=200Hz*	*RMP*	*RUS N HQ Kaliningrad*	*RUS*
6890,00	RUS TACTICAL DATALINK	B=1200Bd S=800Hz Ch=1200		RUS MIL	RUS
6890,00	USB		VJT	RFDS Carnarvon	AUS
6890,00	USB		VNZ	RFDS Pt. Augusta	AUS
6890,00	OQPSK	B=48000Bd		HF Trading Chigago	USA
6893,00	STANAG 4529	B=1200Bd			
6893,00	ALE USB	B=125Bd Ch=8 ChS=250Hz		ALB MIL net	ALB
6893,20	*STANAG 4285*	*B=2400Bd*	*PBC*	*NLD N Goeree Island*	*HOL*

Frequency	Mode	Mode Parameter	Callsign	User	Country
6894,00	STANAG 4285	B=2400Bd		G MIL Crimond	G
6894,20	*STANAG 4285*	*B=2400Bd*	*PBC*	*NLD N Goeree Island*	*HOL*
6895,00	USB			CIS aero network day	CIS
6895,20	BAUDOT	B=75Bd S=850Hz	PBB	NLD N Den Helder	HOL
6897,00	ALE USB	B=125Bd Ch=8 ChS=250Hz		F N net	F
6897,00	*MAHRS*	*B=2400Bd*	*DHJ58*	*D N Glücksburg TX Staberhuk*	*D*
6897,50	USB			USA Army MARS	USA
6897,80	PACTOR II	B=100Bd S=200Hz		MRC meteo	MRC
6900,00					
6900,00	ALE USB	B=125Bd Ch=8 ChS=250Hz		MTN net	MTN
6900,00	*CIS 12*	*B=1440Bd Ch=12 ChS=200Hz*		*RUS N HQ Moscow*	*RUS*
6902,00	ALE USB	B=125Bd Ch=8 ChS=250Hz		MTN MOI net	MTN
6902,60	ALE USB	B=125Bd Ch=8 ChS=250Hz		USA diplo net	
6903,20	*STANAG 4285*	*B=2400Bd*		*G MIL St. Eval*	*G*
6903,50	*STANAG 4285*	*B=2400Bd*		*G MIL Crimond*	*G*
6904,00	ALE USB	B=125Bd Ch=8 ChS=250Hz		ISR AF net	ISR
6905,00	USB		VNJ	Casey Base , AUS	ANT
6905,00	USB			AUS Davis Base	ANT
6905,00	USB			Macquarie Island , AUS	ANT
6905,00	USB			Mawson Base , AUS	ANT
6905,00	ALE USB	B=125Bd Ch=8 ChS=250Hz		AUS Police NSW	AUS
6905,00	ALE USB	B=125Bd Ch=8 ChS=250Hz		AUS Police QLD	AUS
6905,00	CW		RQO61	RUS MIL	RUS
6908,00	SUI VFT	B=100Bd S=170Hz Ch=2 ChS=1000Hz		SUI MIL	SUI
6908,50	ALE USB	B=125Bd Ch=8 ChS=250Hz		USA Defence Logistic Agency net	USA
6908,50	CHN 4+4	B=75Bd Ch=8 ChS=300/450Hz		CHN MIL	CHN
6910,00	USB		AEM1WF	USA Army MARS	USA
6910,00	USB			HFoZ/Radtel net	AUS
6911,00	STANAG 4539 HDR	B=2400Bd		UI MIL net	
6912,00	CW		RBIZ	RUS N Repair Ship PM-138 Amur Class	RUS
6913,00	USB		AAR5GB	USA Army MARS region 5	USA
6913,00	USB			RUS MIL Rostov-na-Donu	RUS
6915,00	USB			G MIL Army Cadets	G
6915,00	STANAG 4285	B=2400Bd	TBB	TUR N Incirlik	TUR
6915,10	FAX 120/576		VCO	CAN CG Sydney	CAN
6917,00	*STANAG 4481*	*B=300Bd S=850Hz*		*I N*	*I*
6917,50	CW		L**	RUS N St. Petersburg	RUS
6917,50	CW		P**	RUS N Kaliningrad	RUS
6918,00	CIS 75-250	B=75Bd S=250Hz		RUS N Moscow	RUS
6922,00	USB			RUS MIL Rostov-na-Donu	RUS
6926,20	*STANAG 4285*	*B=2400Bd*	*EBA*	*E N Madrid*	*E*
6928,00	*CIS 12*	*B=1440Bd Ch=12 ChS=200Hz*	*RIT*	*RUS N HQ Severomorsk*	*RUS*
6930,00	ALE USB	B=125Bd Ch=8 ChS=250Hz		UKR net	UKR
6933,00	CIS OFDM 121	B=2541Bd Ch=121 ChS=25Hz		MFA Moscow	RUS
6933,00	ALE USB	B=125Bd Ch=8 ChS=250Hz		ISR AF net	ISR
6934,00	CW			RUS HFDF net	RUS

Frequency	Mode	Mode Parameter	Callsign	User	Country
6935,00	CW			RUS Operational Strategic Command Air Defense net	RUS
6935,00	*CIS 100-500*	*B=100Bd S=500Hz*		*Area west Kaliningrad*	
6935,00	*CIS 14*	*B=96Bd S=500Hz*		*Riga*	
6936,00	CW		P**	RUS N Kaliningrad	RUS
6936,00	CW			CHN MIL net	CHN
6939,00	CW			CHN MIL net	CHN
6940,00	ALE USB	B=125Bd Ch=8 ChS=250Hz		S MIL net	S
6942,00	*STANAG 4481*	*B=75Bd S=850Hz*	*NSY*	*USA N Niscemi*	*I*
6942,00	LINK 11 CLEW	B=2250Bd Ch=16 ChS=330/110/550Hz		USA N Jacksonville	USA
6942,50	MIL 188-110B SER	B=2400Bd			
6945,00	USB		AMBA	Samara ACC	RUS
6945,00	USB		ASSISTENT	Mosdok ACC	RUS
6945,00	USB		MELODIJA DWA	Kazan ACC	RUS
6945,00	USB			CIS aero network day	CIS
6945,00	USB		VJB	RFDS Derby	AUS
6948,00	ALE USB	B=125Bd Ch=8 ChS=250Hz		S MIL net	S
6950,00	ALE USB	B=125Bd Ch=8 ChS=250Hz		S FRO net	S
6950,00	USB		VJD	RFDS Alice Springs	AUS
6950,00	CIS 100-500	B=100Bd S=500Hz		RUS MIL Moscow	RUS
6951,50	CW			RUS HFDF net	RUS
6957,00	ALE LSB	B=125Bd Ch=8 ChS=250Hz		D Red Cross net	D
6957,00	PACTOR III	B=100Bd S=200Hz	DEK77	DRK Nienburg	D
6957,00	HF PAGER	B=5,86Bd		NVIS Club	RUS
6960,00	USB			BGD N net	BGD
6960,00	USB		VKL	RFDS Port Hedland	AUS
6960,50	R&S MODEM	B=2400Bd			
6962,00	T600	B=50Bd S=200Hz	RIT	RUS N HQ Severomorsk	RUS
6963,00	STANAG 4539	B=2400Bd		UI MIL user	
6964,00	USB			USA Army MARS	
6964,80	STANAG 4539	B=2400Bd		UI MIL user	
6965,00	USB		VJN	RFDS Cairns	AUS
6965,00	USB		VJJ	RFDS Charleville	AUS
6965,00	USB		VJI	RFDS Mount Isa	AUS
6969,50	RUS TACTICAL DATALINK	B=1200Bd S=800Hz Ch=1200		RUS MIL	RUS
6973,00	CW		RIT	RUS N HQ Severomorsk	RUS
6973,00	CW			RUS HFDF net	RUS
6974,00	*MIL 188-110B SER*	*B=2400Bd*		*POL MIL net*	*POL*
6974,00	CHN OFDM 30TONE	B=1800Bd Ch=30 ChS=75Hz		CHN MIL	CHN
6975,00	*CIS 12*	*B=1440Bd Ch=12 ChS=200Hz*		*RUS MIL St. Petersburg*	*RUS*
6977,00	USB		CIW681	CAN MIL	CAN
6977,00	USB			CAN CFARS net	CAN
6980,00	*4FSK*	*B=100Bd S=500Hz Ch=4 ChS=500Hz*		*Minsk*	
6980,00	CIS 75-250	B=75Bd S=250Hz		RUS N	RUS
6981,00	ALE USB	B=125Bd Ch=8 ChS=250Hz		USA deprtment of verteran affairs	USA
6981,00	CW			RUS MIL net	RUS
6983,50	ALE USB	B=125Bd Ch=8 ChS=250Hz		AUS Police QLD	AUS
6983,50	ALE USB	B=125Bd Ch=8 ChS=250Hz		AUS Police NT	AUS

Frequency	Mode	Mode Parameter	Callsign	User	Country
6983,50	ALE USB	B=125Bd Ch=8 ChS=250Hz		AUS Police WA	AUS
6985,00	ALE USB	B=125Bd Ch=8 ChS=250Hz		USA Army NG net Utah	USA
6985,00	ALE USB	B=125Bd Ch=8 ChS=250Hz		USA SAC net	USA
6986,50	USB			USA AF MARS	ALS
6989,00	CW		RAL2	RUS N HQ Astrakhan	RUS
6989,00	CW		RHQ2	RUS N AF	RUS
6989,00	CW		RGH2	RUS N AF	RUS
6989,00	CW		RMW2	RUS N AF	RUS
6990,00	VARA	Ch=1	SMZ31HA	S FRO Halmstad	S
6990,00	VARA	Ch=1	SMZ31VT	S FRO Bua	S
6990,00	VARA	Ch=1	SMZ31VK	S FRO net	S
6990,00	VARA	Ch=1	SMZ32AS	S FRO Alingsas	S
6990,00	VARA	Ch=1	SMZ31HL	S FRO net	S
6990,00	VARA	Ch=1	SMZ31KP	S FRO net	S
6990,00	VARA	Ch=1		S FRO net	S
6990,00	ALE USB	B=125Bd Ch=8 ChS=250Hz		Plath test frequency	D
6990,00	VARA	Ch=1	SMZ32GD	S FRO net	S
6990,00	VARA	Ch=1	SMZ12MB	S FRO Malmö	S
6990,00	ALE USB	B=125Bd Ch=8 ChS=250Hz		S FRO net	S
6990,00	VARA	Ch=1	SMZ31VA	S FRO net	S
6991,00	CW		FUT	Futility beacon service	
6991,50	USB		03T335	E REMER Alicante	E
6991,50	USB		04T434	E REMER Almeria	E
6991,50	USB		07T704	E REMER Ibiza	E
6991,50	USB		07T802	E REMER Mallorca	E
6991,50	USB		08T635	E REMER Barcelona	E
6991,50	USB		08T6672	E REMER Barcelona	E
6991,50	USB		10T201	E REMER Caceres	E
6991,50	USB		15T401	E REMER La Coruna	E
6991,50	USB		18T408	E REMER Granada	E
6991,50	USB		23O2	E REMER Jaen	E
6991,50	USB		25T1	E REMER Lleida	E
6991,50	USB		33T206	E REMER Asturias	MMR
6991,50	USB		33T201	E REMER Asturias	E
6991,50	USB		50T322	E REMER Zaragoza	E
6991,50	USB			E Civil Protection net REMER	E
6991,50	USB		16T4	E REMER Cuenca	E
6991,50	USB		36T4	E REMER Pontevedra	E
6991,50	USB		45O2	E REMER Toledo	E
6991,50	USB		46T4	E REMER Valencia	E
6991,50	USB		08T646	E REMER Barcelona	E
6991,50	USB		15T710	E REMER La Coruna	E
6991,50	USB		24T1	E REMER Leon	E
6992,00	THALES SKYMASTER ALE	B=125Bd Ch=8 ChS=250Hz			
6992,00	ALE USB	B=125Bd Ch=8 ChS=250Hz		G MIL net	G
6992,50	PACTOR	B=200Bd S=170Hz		ICRC Geneva	SUI
6992,50	PACTOR	B=200Bd S=170Hz		ICRC Zagreb	HRV
6992,50	USB		MRV	G N Sea Cadets	G
6993,00	USB		51T78	E REMER Ceuta	E
6993,00	USB		04T434	E REMER Almeria	E
6993,00	USB		25T1	E REMER Lleida	E
6993,00	USB		23O2	E REMER Jaen	E
6993,00	USB		10T201	E REMER Caceras	E
6993,00	CW		WH2XWF	Traveling Ionospheric Disturbance Detector	USA
6993,00	USB		45O2	E REMER Toledo	E
6993,00	USB			E Civil Protection net REMER	E

Frequency	Mode	Mode Parameter	Callsign	User	Country
6993,00	USB		40T112	E REMER Segovia	E
6993,00	USB		25T1	E REMER Lleida	E
6993,00	USB		28T542	E REMER Madrid	E
6993,00	USB		23T2	E REMER Jaen	E
6993,00	USB		38T2	E REMER Santa Cruz de Tenerife	E
6993,00	USB		28T858	E REMER Madrid	E
6993,00	USB		07T8	E REMER Menorca	E
6993,00	USB		25T04	E REMER Lleida	E
6993,00	USB		43T09	E REMER Tarragona	E
6993,00	USB		44O1	E REMER Teruel	E
6993,50	CIS 75-200	B=75Bd S=200Hz		RUS N Moscow	RUS
6995,00	CIS 50-500	B=50Bd S=500Hz	RVQ	RUS GOV	RUS
6995,00	CW		RVQ	RUS	RUS
6995,00	CW		RUH	RUS	RUS
6995,20	*MAHRS*	*B=2400Bd*	*DHJ58*	*D N Glücksburg TX Staberhuk*	*D*
6996,00	USB			USA Army MARS net	
6996,00	CIS 50-500 BAUDOT	B=50Bd S=500Hz		RUS	RUS
6996,00	USB		AAN4MEM	USA Army MARS	USA
6996,00	USB		AAN4TYS	USA Army MARS	USA
6997,80	BAUDOT	B=45,45Bd S=170Hz		UI AMS	
6998,00	*T600*	*B=50Bd S=250Hz*	*RMP*	*RUS N Kaliningrad*	*RUS*
7000,00	CLOVER 2000	B=500Bd Ch=8 ChS=250Hz	2711**		
7000,00	*STANAG 4285*	*B=2400Bd*	*FUV*	*F N Djibouti*	*DJI*
7000,00	CLOVER 2000	B=500Bd Ch=8 ChS=250Hz	TI1ME		
7000,00	CLOVER 2000	B=500Bd Ch=8 ChS=250Hz	SP1FT		
7000,00	CW		D**	RUS N Sevastopol	UKR
7008,00	CIS 75-250	B=75Bd S=250Hz		RUS N Moscow	RUS
7011,00	*CIS 12*	*B=1440Bd Ch=12 ChS=200Hz*		*RUS N HQ Moscow*	*RUS*
7018,00	*CIS 50-1000*	*B=50Bd S=1000Hz*	*REA4*	*RUS AF HQ Moscow*	*RUS*
7019,00	CW		RAL2	RUS N HQ Astrakhan	RUS
7030,00	*CIS 12*	*B=1440Bd Ch=12 ChS=200Hz*		*Minsk*	*BLR*
7038,50	CW		O**	RUS N	
7038,50	CW		OK0EU	CZE QRP beacon net	CZE
7038,60	WSPR	B=1,46Bd Ch=4 ChS=1,46Hz		WSPR net	
7038,80	CW		P**	RUS N Kaliningrad	RUS
7038,90	CW		S**	RUS N Severomorsk	RUS
7039,00	CW		D**	RUS N Sevastopol	UKR
7039,00	CW		C**	RUS N Moscow	RUS
7039,10	CW		A**	RUS N Astrakhan	
7039,20	CW		F**	RUS N Vladivostok	RUS
7039,30	CW		K**	RUS N Petropavlovsk Kamchatskiy	RUS
7039,40	CW		M**	RUS N Magadan	RUS
7040,00	PSK31	B=31Bd		PSK31 channel	
7042,50	MFSK			MFSK modes user	
7043,00	BAUDOT	B=45,45Bd S=170Hz		RTTY user	
7045,00	LONGCHAT	B=40Bd Ch=1		LongChat net	
7046,00	ROS USB	B=1Bd Ch=144 ChS=15,625Hz		ROS frequency	
7047,30	ROBUST PACKET RADIO	B=200Bd Ch=8 ChS=60Hz		Robust Packet Radio net, HF-APRS	
7047,50	CW		W1AW	ARRL morse training	USA
7047,50	FT4	B=5,86Bd Ch=8 ChS=5,86Hz		FT4 channel	

Frequency	Mode	Mode Parameter	Callsign	User	Country
7049,50	ALE USB	B=125Bd Ch=8 ChS=250Hz		HFLINK network	
7051,00	T600	B=50Bd S=200Hz		RUS N	RUS
7055,00	LONGCHAT	B=40Bd Ch=1		LongChat net	
7056,00	USB			RUS MIL Rostov-na-Donu	RUS
7064,50	CW		H**		RUS
7071,00	ALE USB	B=125Bd Ch=8 ChS=250Hz			
7074,00	FT8	B=5,86Bd Ch=8 ChS=5,86Hz		FT8 channel	
7075,00	USB			Wireless Institute Civil Emergency Service WICEN	AUS
7076,00	JT65	B=2,69Bd Ch=65 ChS=2,69Hz		JT65 channel	
7078,00	JT9	B=1,736Bd Ch=9 ChS=1,736Hz		JT9 channel	
7078,00	JS8	B=6,25Bd Ch=8 ChS=6,25Hz		JS8 user	
7080,00	CW		4XZ	ISR N Haifa	ISR
7086,00	LSB			Weather & Cocktail net caribbean	
7088,00	LINK 11 SLEW	B=2400Bd		Tripoli area	LBY
7095,00	USB			PARA National Traffic System net	PHL
7095,00	USB			PARA DUCWNET	PHL
7102,00	ALE USB	B=125Bd Ch=8 ChS=250Hz		HFLINK network	
7103,00	CCIR 493-4	B=100Bd S=170Hz			
7105,00	VARA	Ch=1		VARA chat freqeuency	
7106,00	CCIR 493-4	B=100Bd S=170Hz			
7109,00	CCIR 493-4	B=100Bd S=170Hz			
7110,00	LSB			Ham radio emergency frequency	R1
7112,00	CCIR 493-4	B=100Bd S=170Hz			
7117,00	CIS 100-1000	B=100Bd S=1000Hz	REA4	RUS AF HQ Moscow	RUS
7121,00	CCIR 493-4	B=100Bd S=170Hz			
7124,00	CCIR 493-4	B=100Bd S=170Hz			
7127,00	CCIR 493-4	B=100Bd S=170Hz			
7130,00	ARD9800 MODEM	B=2250Bd Ch=36 ChS=62,5Hz		AOR Digital Voice net	USA
7130,00	CCIR 493-4	B=100Bd S=170Hz			
7133,00	CCIR 493-4	B=100Bd S=170Hz			
7139,00	CCIR 493-4	B=100Bd S=170Hz			
7142,00	CCIR 493-4	B=100Bd S=170Hz			
7145,00	CCIR 493-4	B=100Bd S=170Hz			
7148,00	CCIR 493-4	B=100Bd S=170Hz			
7157,00	CCIR 493-4	B=100Bd S=170Hz			
7160,00	CCIR 493-4	B=100Bd S=170Hz			
7166,00	ALE USB	B=125Bd Ch=8 ChS=250Hz			
7173,00	ALE USB	B=125Bd Ch=8 ChS=250Hz			
7175,00	HFPAGER	B=5,86Bd Ch=18		RUS HFPager net	RUS
7177,00	ARD9800 MODEM	B=2250Bd Ch=36 ChS=62,5Hz		AOR Digital Voice net	USA
7177,00	CIS 12	B=1440Bd Ch=12 ChS=200Hz	RCV	RUS N HQ Sevastopol	UKR
7180,00	LSB			Arctic Öcean RIng	NOR
7185,00	ALE USB	B=125Bd Ch=8 ChS=250Hz		HFLINK network	
7196,00	CW			RUS MIL net	RUS
7232,00	LSB			Tar Heel emergency net , NC	USA
7262,50	*STANAG 4285*	*B=2400Bd*	*TBB*	*TUR N Izmir*	*TUR*
7268,00	LSB		WX4NHC	Huricane Watch net Miami , FL	USA
7280,00	CW			RUS MIL net	RUS

Frequency	Mode	Mode Parameter	Callsign	User	Country
7286,00	ARD9800 MODEM	B=2250Bd Ch=36 ChS=62,5Hz		AOR Digital Voice net	USA
7291,00	ARD9800 MODEM	B=2250Bd Ch=36 ChS=62,5Hz		AOR Digital Voice net	USA
7296,00	ALE USB	B=125Bd Ch=8 ChS=250Hz		HFLINK network	
7302,00	USB			USA AF MARS	
7304,60	ALE USB	B=125Bd Ch=8 ChS=250Hz		ARINC Urgent Link net	USA
7307,00	*STANAG 4285*	*B=2400Bd*	*JWT*	*NOR N Stavanger*	*NOR*
7313,00	SSB			US Army MARS	
7313,50	SSB			USA AF MARS	
7314,00	CW		REA4	RUS AF HQ Moscow	RUS
7315,00	SSB			USA AF MARS	HWA
7315,60	ALE USB	B=125Bd Ch=8 ChS=250Hz		ARINC Urgent Link net	USA
7317,00	ALE USB	B=125Bd Ch=8 ChS=250Hz		USA AF net	USA
7318,00	QPSK	B=100Bd		POL INTEL	POL
7318,60	ALE USB	B=125Bd Ch=8 ChS=250Hz		ARINC Urgent Link net	USA
7324,00	SSB			USA AF MARS	USA
7325,00	ALE USB	B=125Bd Ch=8 ChS=250Hz		USA FAA net	USA
7325,00	BAUDOT	B=70Bd S=850Hz		MARSCOMM	USA
7327,50	PACTOR III	B=200Bd S=200Hz	WGM	Hollywood , FL	USA
7329,00	USB			USA AF MARS	
7331,00	SSB			USA AF MARS	ALS
7335,00	CW , SSB		CHU	Ottawa TS	CAN
7336,20	*STANAG 4285*	*B=2400Bd*		*G MIL TX St. Eval*	*G*
7348,00	ALE USB	B=125Bd Ch=8 ChS=250Hz		USA FEMA net	USA
7349,00	CCIR 493-4	B=100Bd S=170Hz			
7357,00	SSB			USA AF MARS	
7358,00	SSB			US Army MARS	
7360,00	SSB			USA AF MARS	GUM
7360,00	*STANAG 4285*	*B=2400Bd*		*G F Inskip*	*G*
7361,00	CW		RFFCH	RUS MIL	RUS
7361,00	CW		RFFR	RUS MIL	RUS
7361,00	CW		RFFO	RUS MIL	RUS
7361,00	CW		RFFQ	RUS MIL	RUS
7362,00	ALE USB	B=125Bd Ch=8 ChS=250Hz			D
7366,20	*STANAG 4285*	*B=2400Bd*		*G MIL TX St. Eval*	*G*
7372,50	USB			USA MARS	USA
7373,00	T600	B=50Bd S=200Hz	RCV	RUS N Sevastopol	UKR
7378,00	CW		FDE2	F AF Tours	F
7380,00	PACTOR III	B=100Bd S=200Hz	WQAB964	Sailmail San Diego , CA	USA
7391,00	ALE USB	B=125Bd Ch=8 ChS=250Hz		USA Shares net	USA
7393,00	ALE USB	B=125Bd Ch=8 ChS=250Hz		USA Department of Agriculture net	USA
7395,00	FAX 120/576		HSW64	Bangkok M	THA
7400,00	HAARP waveform			HAARP	USA
7400,50	ALE USB	B=125Bd Ch=8 ChS=250Hz		G MIL net	G
7402,00	CW		REA4	RUS AF Moscow	RUS
7405,00	PACTOR III	B=100Bd S=200Hz	9Z4DH	Sailmail Chaguaramas	TRD
7405,20	*STANAG 4285*	*B=2400Bd*		*TUR N TX Izmir*	*TUR*
7407,00	SSB			USA AF MARS	J
7410,00	SYSTEME 3000 ALE	B=125Bd Ch=8 ChS=250Hz			F

Frequency	Mode	Mode Parameter	Callsign	User	Country
7412,50	ALE LSB	B=125Bd Ch=8 ChS=250Hz		GRC AF net	GRC
7415,20	*STANAG 4285*	*B=2400Bd*	*DHJ58*	*D N Glücksburg TX Marlow*	*D*
7416,00	CW		RDL	RUS N Moscow	RUS
7416,00	CW		4XZ	ISR N Haifa	ISR
7421,50	ALE USB	B=125Bd Ch=8 ChS=250Hz		GRC AF net	GRC
7425,00	BAUDOT	B=50Bd S=500Hz			
7428,00	ALE USB	B=125Bd Ch=8 ChS=250Hz		USA FEMA net	USA
7430,00	*CIS 100-500*	*B=100Bd S=500Hz*		*RUS MIL West Riga*	*LVA*
7433,00	T600	B=50Bd S=100Hz	RCV	RUS N HQ Sevastopol	UKR
7433,50	FAX 120/576		HLL2	Seoul M	KOR
7436,00	*CIS 12*	*B=1440Bd Ch=12 ChS=200Hz*	*RCV*	*RUS N HQ Sevastopol*	*UKR*
7438,00	ALE USB	B=125Bd Ch=8 ChS=250Hz		S MIL net	S
7445,30	ALE USB	B=125Bd Ch=8 ChS=250Hz		F	F
7450,00	*CIS 100-1000*	*B=100Bd S=1000Hz*		*Moscow*	*RUS*
7452,70	MIL 188-110B APP C	B=2400Bd		S MIL	S
7454,20	*STANAG 4285*	*B=2400Bd*	*DHJ58*	*D N Glücksburg TX Marlow*	*D*
7454,20	*STANAG 4285*	*B=2400Bd*	*DHJ58*	*D N Glücksburg TX Neuharlingersiel*	*D*
7455,00	STANAG 4481	B=50Bd S=850Hz	NAU	USA N San Juan	PTR
7457,00	SSB			USA AF MARS	
7457,00	USB		AFA2AJ	USA AF MARS	USA
7461,20	*STANAG 4285*	*B=2400Bd*	*DHJ58*	*D N Glücksburg TX Marlow*	*D*
7465,00	USB		VJJ	RFDS Charleville	AUS
7465,00	CIS 100-500 BAUDOT	B=100Bd S=500Hz		RIS diplo	RUS
7465,00	USB		VJN	RFDS Cairns	AUS
7467,00	*CW*		*RIT*	*RUS N HQ Severomorsk*	*RUS*
7473,50	USB			USA AF MARS	
7475,00	SSB			USA AF Strategic Command	USA
7477,00	ALE USB	B=125Bd Ch=8 ChS=250Hz		USA NG net	USA
7477,50	*STANAG 4481*	*B=75Bd S=850Hz*		*USA AF rougthon*	*G*
7479,50	STANAG 4481	B=75Bd S=850Hz	NSY	USA N Niscemi , Sicily	I
7480,00	ALE USB	B=125Bd Ch=8 ChS=250Hz		USA Red Cross net	USA
7482,50	ALE USB	B=125Bd Ch=8 ChS=250Hz			D
7490,00	*STANAG 4285*	*B=2400Bd*		*G MIL TX Inskip*	*G*
7490,00	ALE USB	B=125Bd Ch=8 ChS=250Hz			D
7500,00	ALE USB	B=125Bd Ch=8 ChS=250Hz		EGY Mil net	EGY
7505,00	ALE USB	B=125Bd Ch=8 ChS=250Hz		CLM net	CLM
7505,80	ALE USB	B=125Bd Ch=8 ChS=250Hz		F	
7506,00	USB		PEGASO	E Combat Air Command Torrejon de Ardoz	E
7508,00	BAUDOT	B=75Bd S=170Hz	ZSJ	AFS N Capetown	AFS
7508,70	CW		D**	RUS N Sevastopol	UKR
7508,80	CW		P**	RUS N Kaliningrad	RUS
7508,90	CW		S**	RUS N Severomorsk	RUS
7509,00	CW		C**	RUS N Moscow	RUS
7509,10	CW		A**	RUS N Astrakhan	RUS
7510,00	*CIS 100-500*	*B=100Bd S=500Hz*		*RUS MIL Moscow*	*RUS*
7511,20	*STANAG 4285*	*B=2400Bd*	*DHJ58*	*D N Glücksburg TX Neuharlingersiel*	*D*

Frequency	Mode	Mode Parameter	Callsign	User	Country
7512,00	ALE USB	B=125Bd Ch=8 ChS=250Hz		I AF net	I
7512,00	ALE USB	B=125Bd Ch=8 ChS=250Hz		AZE emergency net	AZE
7513,50	ALE LSB	B=125Bd Ch=8 ChS=250Hz		GRC AF net	GRC
7515,00	CW		REA4	RUS AF Moscow	RUS
7515,20	*STANAG 4285*	*B=2400Bd*	*DHJ58*	*D N Glücksburg TX Marlow*	*D*
7518,00	MIL 188-110B SER	B=2400Bd		CZR E Budapest	UNG
7519,00	ALE USB	B=125Bd Ch=8 ChS=250Hz		ALG MIL net	ALG
7519,00	MIL 188-110A	B=2400Bd		ALG MIL net	ALG
7521,00	CW		RGT77	RUS MIL Moscow RVSN	RUS
7523,00	CW		RBL88	RUS N	RUS
7525,00	SSB			NASA launch tracking net	USA
7525,00	*ALE USB*	*B=125Bd Ch=8 ChS=250Hz*		*MTN net*	*MTN*
7525,00	ALE USB	B=125Bd Ch=8 ChS=250Hz		USA Army NG net Utah	USA
7525,50	ALE USB	B=125Bd Ch=8 ChS=250Hz			
7527,00	ALE USB	B=125Bd Ch=8 ChS=250Hz		USA customs net	USA
7527,00	SSB			USA AF MARS	
7527,00	ALE USB	B=125Bd Ch=8 ChS=250Hz		USA COTHEN net	USA
7531,00	PANTHER-H	B=2400Bd			
7531,00	KOR 2FSK	B=1200Bd S=850Hz Ch=1		KOR MIL Busan	KOR
7535,00	FAX 120/576		VMW	Wiluna M	AUS
7535,00	USB			USA N Norfolk , VA	USA
7535,00	*ALE USB*	*B=125Bd Ch=8 ChS=250Hz*		*G MIL net*	*G*
7536,80	STANAG 4538	B=2400Bd			
7540,00	SSB			USA AF MARS	
7540,00	MIL 188-110B SER	B=2400Bd			
7540,00	ALE USB	B=125Bd Ch=8 ChS=250Hz			
7542,00	USB			CIS aero network day	CIS
7542,80	CW			Morse training	F
7543,00	USB			E Civil Protection net REMER	E
7545,00	SSB			US Army MARS	
7545,00	SSB			USA AF MARS	
7545,00	STANAG 4481	B=75Bd S=850Hz	NSY	USA N Niscemi , Sicily	I
7547,00	ALE USB	B=125Bd Ch=8 ChS=250Hz			
7551,20	*STANAG 4285*	*B=2400Bd*	*DHJ58*	*D N TX Neuharlingersiel*	*D*
7554,00	CHN 4+4	B=75Bd Ch=8 ChS=300/450Hz	XSV70	CHN Great Wall Base	ANT
7554,70	*STANAG 4285*	*B=2400Bd*	*FUG*	*F N Saissac*	*FF*
7556,00	USB			PHL CG ship BRP TERESA MAGBANUA	PHL
7556,00	USB			PHL CG net	PHL
7556,00	USB			PHL CG ship BRD MALABRIGO	PHL
7556,20	STANAG 4285	B=2400Bd		F N Toulon	F
7560,00	ALE USB	B=125Bd Ch=8 ChS=250Hz		CLM net	CLM
7564,00	CW		RBL88	RUS MIL	RUS
7565,60	ALE 3G	B=2400Bd		UI MIL net	
7566,00	CW		RCV	RUS N HQ Sevastopol	UKR
7567,00	CW		RIT	RUS N HQ Severomorsk	RUS
7567,50	ALE USB	B=125Bd Ch=8 ChS=250Hz		G MIL net	G
7570,00	CW		CGA984	Essex County, ON	CAN

Frequency	Mode	Mode Parameter	Callsign	User	Country
7570,00	T600	B=50Bd S=100Hz	RCV	RUS N HQ Sevastopol	UKR
7571,00	T600	B=50Bd S=200Hz	RCV	RUS N HQ Sevastopol	UKR
7572,00	USB		AZOR	E AF	E
7572,00	USB		CRATER	E AF	E
7572,00	USB		MATADOR	E AF EVA 2 Villatobas	E
7572,00	USB		DRAGON	E AF EVA 13 Sierra Espuna	E
7572,00	USB		PRIMUS	E AF EADA Zaragoza	E
7572,00	USB		ORION	E AF Motril , Granada.	E
7572,00	USB		PEGASO	E Combat Air Command Torrejon de Ardoz	E
7573,00	CW		RBL88	RUS N	RUS
7574,00	CW		RGT77	RUS MIL Moscow, RVSN	RUS
7575,50	CIS OFDM 160	B=4800Bd Ch=160 ChS=37,5Hz			RUS
7579,20	*STANAG 4285*	*B=2400Bd*		*G MIL TX Akrotiri*	*CYP*
7583,00	*MIL 188-110B SER*	*B=2400Bd*		*S MIL Karlskrona*	*S*
7585,00	USB			EGY N net	EGY
7585,50	PACTOR II	B=100Bd S=200Hz	OEH91	AUT Red Cross Voralberg	AUT
7585,50	PACTOR II	B=100Bd S=200Hz	OEH20	AUT Red Cross Vienna	AUT
7588,50	USB			USA Army MARS	
7591,00	BAUDOT	B=50Bd S=500Hz		RUS MIL Moscow	RUS
7594,00	CW		RCV	RUS N HQ Sevastopol	UKR
7595,00	SUI VFT	B=100Bd S=170Hz Ch=2 ChS=1000Hz		SUI AF	SUI
7596,00	ALE USB	B=125Bd Ch=8 ChS=250Hz		GRC AF net	GRC
7597,00	*STANAG 4481*	*B=75Bd S=850Hz*	*NPG*	*USA N Dixon , CA*	*USA*
7597,00	ALE USB	B=125Bd Ch=8 ChS=250Hz			
7597,00	ALE LSB	B=125Bd Ch=8 ChS=250Hz		D Red Cross net	D
7599,00	CHN 4+4	B=75Bd Ch=8 ChS=300/450Hz		CHN MIL	
7600,00	AM		HD2IOA	Guayaquil TS	EQA
7602,00	ALE USB	B=125Bd Ch=8 ChS=250Hz		USA Civil Air Patrol net	USA
7602,00	CW		RIT	RUS N HQ Severomorsk	RUS
7604,00	ALE USB	B=125Bd Ch=8 ChS=250Hz		ALG MIL net	ALG
7605,00	CIS 100-500	B=100Bd S=500Hz		RUS MIL Moscow	RUS
7606,00	LINK 22	B=2250Bd			
7608,00	ALE USB	B=125Bd Ch=8 ChS=250Hz		ALG MIL net	ALG
7608,00	ALE USB	B=125Bd Ch=8 ChS=250Hz		POL MIL net	POL
7608,00					
7610,00	CIS ARQ	B=100Bd S=500Hz		RUS PTT Archangelsk	RUS
7611,00	ALE USB	B=125Bd Ch=8 ChS=250Hz		ALG MIL net	ALG
7611,00	CW		V**	RUS N Khiva	RUS
7617,00	MIL 188-110B 39TONE	B=2400Bd Ch=39 ChS=56,25Hz			
7617,00	*ALE USB*	*B=125Bd Ch=8 ChS=250Hz*		*TUR Disaster and Emergency net*	*TUR*
7622,80	STANAG 4538	B=2400Bd			
7625,00	*CIS 100-500*	*B=100Bd S=500Hz*	*RMP*	*RUS N Kaliningrad*	*RUS*
7625,00	CIS ARQ	B=100Bd S=500Hz		RUS PTT	RUS
7626,00	ALE USB	B=125Bd Ch=8 ChS=250Hz		AZE emergency net	AZE
7627,50	ALE USB	B=125Bd Ch=8 ChS=250Hz			
7628,00	USB			F MARPAT net	F
7628,00	CW			RUS MIL net	RUS

Frequency	Mode	Mode Parameter	Callsign	User	Country
7629,00	ALE USB	B=125Bd Ch=8 ChS=250Hz			
7630,00	ALE USB	B=125Bd Ch=8 ChS=250Hz		ALG MIL net	ALG
7630,00	USB			USA AF MARS	
7630,00	*CIS ARQ*	*B=100Bd S=500Hz*	*RDP5*	*RUS PTT Kaliningrad*	*RUS*
7630,00	USB			F	
7630,00	STANAG 4197	B=1800Bd Ch=16/39 ChS=112/56Hz		USA N	
7632,00	ALE USB	B=125Bd Ch=8 ChS=250Hz		USA AF net	
7632,00	CW		RJS	RUS N HQ Vladivostok	RUS
7633,00	USB			BRM N net	BRM
7633,50	SSB			USA AF MARS	
7635,00	CIS ARQ	B=100Bd S=500Hz		RUS PTT	RUS
7637,50	CW			RUS MIL net	RUS
7640,00	SYSTEME 3000	B=2400Bd		F MIL	F
7640,00	CLOVER2000	Ch=4		UI net	
7641,00	ALE USB	B=125Bd Ch=8 ChS=250Hz		POL MIL net	POL
7642,00	ALE USB	B=125Bd Ch=8 ChS=250Hz		USA SHARES net	USA
7646,00	*BAUDOT*	*B=50Bd S=425Hz*	*DDH7*	*DWD TX Pinneberg*	*D*
7647,50	ALE USB	B=125Bd Ch=8 ChS=250Hz			
7648,00	STANAG 4538	B=2400Bd			
7650,00	CHN 4+4	B=75Bd Ch=8 ChS=300/450Hz		CHN MIL	CHN
7650,00	ALE USB	B=125Bd Ch=8 ChS=250Hz		USA NG	USA
7650,00	*ALE USB*	*B=125Bd Ch=8 ChS=250Hz*		*F Airbus net*	*F*
7650,00	ALE USB	B=125Bd Ch=8 ChS=250Hz		UNAMA net	AFG
7650,00	CIS 100-500	B=100Bd S=500Hz		RUS MIL	RUS
7651,00	ALE USB	B=125Bd Ch=8 ChS=250Hz		ISR AF net	ISR
7652,00	USB		VMD750	Austravel Safety Net	AUS
7655,00	ALE USB	B=125Bd Ch=8 ChS=250Hz		ISR AF net	ISR
7656,30	ALE USB	B=125Bd Ch=8 ChS=250Hz		F	
7657,00	*T600*	*B=50Bd S=250Hz*		*RUS N Murmansk*	*RUS*
7657,00	ALE USB	B=125Bd Ch=8 ChS=250Hz		AUS Police NSW	AUS
7657,00	ALE USB	B=125Bd Ch=8 ChS=250Hz		AUS Police NT	AUS
7657,00	ALE USB	B=125Bd Ch=8 ChS=250Hz		AUS Police QLD	AUS
7657,00	ALE USB	B=125Bd Ch=8 ChS=250Hz		AUS Police WA	AUS
7657,00	ALE USB	B=125Bd Ch=8 ChS=250Hz		SAPOL Communication Infrastructure	AUS
7657,00	CW		RDL	RUS N Moscow	RUS
7659,00	RUS MFSK32 OFDM60	B=40Bd Ch=16		MFA Moscow	RUS
7659,50	BAUDOT	B=200Bd S=850Hz			
7660,00	ALE USB	B=125Bd Ch=8 ChS=250Hz		SAPOL Communication Infrastructure	AUS
7660,00	ALE USB	B=125Bd Ch=8 ChS=250Hz		AUS Police NSW	AUS
7660,00	ALE USB	B=125Bd Ch=8 ChS=250Hz		AUS Police WA	AUS

Frequency	Mode	Mode Parameter	Callsign	User	Country
7660,00	ALE USB	B=125Bd Ch=8 ChS=250Hz		AUS Police NT	AUS
7660,00	STANAG 4539	B=2400Bd		USA MIL	
7660,00	MIL 188-110A SER	B=2400Bd		USA MIL	
7661,00	MIL 188-110B SER	B=2400Bd		SUI F Suva Reka	SRB
7661,00	ALE USB	B=125Bd Ch=8 ChS=250Hz		SUI F net	
7665,00	ALE LSB	B=125Bd Ch=8 ChS=250Hz			
7665,00	ALE USB	B=125Bd Ch=8 ChS=250Hz		USA Civil Air Patrol net	USA
7665,00	CW		RIW	RUS N HQ Moscow	RUS
7667,00	RUS TACTICAL DATALINK	B=1200Bd S=800Hz		RUS MIL	RUS
7674,00	CW		RGT77	RUS MIL Moscow, RVSN	RUS
7676,00	SSB			NASA launch support aircraft	USA
7676,00	CIS 12	B=1440Bd Ch=12 ChS=200Hz		RUS N Vladivostok	RUS
7677,50	ALE USB	B=125Bd Ch=8 ChS=250Hz		ALG MIL net	ALG
7681,00	*STANAG 4285*	*B=2400Bd*		*G MIL TX Crimond*	*G*
7685,00	CIS 12	B=1440Bd Ch=12 ChS=200Hz		RUS MIL Kosovo	SRB
7690,00	ALE USB	B=125Bd Ch=8 ChS=250Hz		TUR Red Crescent net	TUR
7692,00	MIL 188-110B SER	B=2400Bd			
7695,00	CIS ARQ	B=100Bd S=500Hz		RUS PTT	RUS
7696,00	CW		RJF94	RUS F	RUS
7697,00	CW			RUS HFDF net	RUS
7697,00	ALE USB	B=125Bd Ch=8 ChS=250Hz		USA Red Cross net	USA
7697,10	ALE USB	B=125Bd Ch=8 ChS=250Hz			USA
7698,00	SYSTEME 3000	B=2400Bd			
7700,00	CHN OFDM 30TONE	B=1500Bd Ch=30 ChS=60Hz		CHN N	CHN
7701,00	MIL 188-110B SER	B=2400Bd			
7704,80	CW		WI2XER	Skycast Services	USA
7710,00	USB			G MIL Army Cadets	G
7711,00	STANAG 4481	B=50Bd S=850Hz		USA MIL	
7712,00	USB		TBRS	TUR CG boat TCSG27	TUR
7712,00	USB		TBN	TUR CG	TUR
7713,00	CW		EHY66	TJK MIL	TJK
7713,00	*STANAG 4285*	*B=2400Bd*		*S MIL*	*S*
7713,20	*STANAG 4285*	*B=2400Bd*		*TUR MIL Ankara*	*TUR*
7718,50	USB			USA Army MARS	
7720,00	USB			BGD Mil Coxs Bazar	BGD
7720,00	BARRETT SELCALL	B=100Bd S=190Hz		BGD Mil Coxs Bazar	BGD
7725,20	*STANAG 4285*	*B=2400Bd*	*DHJ58*	*D N Glücksburg TX Marlow*	*D*
7730,00	CODAN	B=2400Bd Ch=16 ChS=112,5Hz		MFA Cairo	EGY
7732,00	MIL 188-110B SER	B=2400Bd	CS004A**		
7732,00	MIL 188-110B SER	B=2400Bd	BS008C1**		
7733,00	CHN OFDM 30TONE	B=1800Bd Ch=30 ChS=60Hz		CHN N	CHN
7739,00	CHN OFDM 30TONE	B=1800Bd Ch=30 ChS=75Hz		CHN MIL	CHN
7739,00	ALE USB	B=125Bd Ch=8 ChS=250Hz		ALG oilfield net	ALG
7745,00	ALE USB	B=125Bd Ch=8 ChS=250Hz		ARS AF net	ARS

Frequency	Mode	Mode Parameter	Callsign	User	Country
7745,00	*STANAG 4285*	*B=2400Bd*	*TBB*	*TUR N TX Incirlik*	*TUR*
7745,00	*ALE USB*	*B=125Bd Ch=8 ChS=250Hz*		*JOR net*	*JOR*
7746,00	CW		REA4	RUS AF Moscow	RUS
7746,50	STANAG 4285	B=2400Bd	TBB	TUR N Incirlik	TUR
7749,00	ALE USB	B=125Bd Ch=8 ChS=250Hz		ALG MIL net	ALG
7753,00	USB			G MIL Army Cadets	G
7754,00	ALE USB	B=125Bd Ch=8 ChS=250Hz		CHN MIL net	CHN
7755,00	ALE USB	B=125Bd Ch=8 ChS=250Hz			
7756,00	LINK 22	B=2250Bd			
7757,00	CW			RUS MIL net	RUS
7759,00	ALE USB	B=125Bd Ch=8 ChS=250Hz		NIG oil/gas net	NIG
7763,00	QPSK	B=1200Bd			
7765,00	SSB			NASA rocket booster recovery	USA
7765,00	ALE USB	B=125Bd Ch=8 ChS=250Hz		S FRO net	S
7766,70	SITOR	B=100Bd S=170Hz		EGY E Paris	F
7767,00	CHN N MIL 188-110 MOD	B=2400Bd		CHN N	CHN
7770,00	THALES SKYMASTER ALE	B=125Bd Ch=8 ChS=250Hz			
7772,50	USB			USA Army MARS	
7775,00	SSB		ZHF33	BAS Signy Base , G	SOK
7775,00	ALE USB	B=125Bd Ch=8 ChS=250Hz		POL MIL net	POL
7775,00	MIL 188-110A SER	B=2400Bd		POL MIL net	POL
7778,00	LINK 11 CLEW	B=2250Bd Ch=16 ChS=330/110/550Hz			
7778,50	ALE USB	B=125Bd Ch=8 ChS=250Hz		G MIL net	G
7778,50	ALE USB	B=125Bd Ch=8 ChS=250Hz		USA FBI net	USA
7779,00	CIS MFSK 13	B=31,25Bd Ch=13 ChS=125Hz		MFA Moscow	RUS
7780,00	MFSK16	B=31Bd S=31,25Hz Ch=16		RUS	RUS
7780,30	*MIL 188-110B SER*	*B=2400Bd*		*S N Karlskrona*	*S*
7784,00	STANAG 4285	B=2400Bd	NSS	USA N TX Davidsonville , MD	USA
7785,00	ALE USB	B=125Bd Ch=8 ChS=250Hz		PHL MOI net	PHL
7785,00	CW		REA4	RUS AF Moscow	RUS
7788,00	CHN 4+4	B=75Bd Ch=8 ChS=300/450Hz		CHN MIL	CHN
7788,00	USB			EGY N net	EGY
7790,00	BAUDOT	B=50Bd S=500Hz	NT9P**		RUS
7793,00	MIL 188-110A SER	B=2400Bd		ROU E Kiev	UKR
7793,00	MIL 188-110A SER	B=2400Bd		MFA Bucharest	ROU
7793,00	ALE USB	B=125Bd Ch=8 ChS=250Hz		ALG gas & oil net Sonatrach	ALG
7795,00	FAX 120/576		JMH2	Tokyo M	J
7796,70	SITOR A	B=100Bd S=170Hz		MFA Cairo	EGY
7800,00	ALE USB	B=125Bd Ch=8 ChS=250Hz		SAPOL Communication Infrastructure	AUS
7801,00	CW		RCV	RUS N HQ Sevastopol	UKR
7804,00	CHN N MIL 188-110 MOD	B=2400Bd		CHN N	CHN
7805,00	ALE USB	B=125Bd Ch=8 ChS=250Hz		USA Army NG Wyoming	USA

Frequency	Mode	Mode Parameter	Callsign	User	Country
7806,00	ALE USB	B=125Bd Ch=8 ChS=250Hz		USA FEMA net	USA
7807,20	*STANAG 4539*	*B=2400Bd*		*Croughton*	*G*
7808,00	CW			RUS Operational Strategic Command Air Defense net	RUS
7809,00	MIL 188-110B SER	B=2400Bd		IRQ Border Guard net	IRQ
7809,00	ALE USB	B=125Bd Ch=8 ChS=250Hz		IRQ Border Guard net	IRQ
7810,00	*ALE USB*	*B=125Bd Ch=8 ChS=250Hz*		*TUR Disaster and Emergency net*	*TUR*
7812,00	USB			BGD N net	BGD
7812,00	CODAN CHIRP	B=80Bd Ch=30 ChS=250Hz		BGD N net	BGD
7812,00	CODAN	B=2400Bd Ch=16 ChS=112,5Hz		BGD N net	BGD
7812,70	*MIL 188-110B SER*	*B=2400Bd*		*S MIL*	*S*
7813,20	*STANAG 4539*	*B=2400Bd*		*Croughton*	*G*
7813,30	ALE USB	B=125Bd Ch=8 ChS=250Hz		G MIL net	G
7814,00	ALE USB	B=125Bd Ch=8 ChS=250Hz		MRC MIL net	MRC
7815,00	CW		RBL88	RUS N	RUS
7815,00	CW		RFN77	RUS N	RUS
7815,00	CW		RGR99	RUS N	RUS
7815,00	CLOVER2500	Ch=4	40H1		
7815,00	CLOVER2500	Ch=4	40C1		
7816,00	ALE USB	B=125Bd Ch=8 ChS=250Hz		ALG MIL net	ALG
7816,70	SITOR A	B=100Bd S=170Hz		EGY E London	G
7817,00	THALES SKYMASTER ALE	Ch=8			
7819,00	ALE USB	B=125Bd Ch=8 ChS=250Hz		USA Army NG net Utah	USA
7819,00	ALE USB	B=125Bd Ch=8 ChS=250Hz		USA SAC net	USA
7819,60	USB			USA CG net	USA
7822,00	PACTOR II	B=200Bd S=200Hz	XJN714	Sailmail Lunenburg , NS	CAN
7822,00	CW		RGT77	RUS MIL Moscow, RVSN	RUS
7822,20	*STANAG 4539*	*B=2400Bd*		*Croughton*	*G*
7823,00	CIS 4FSK	B=150Bd Ch=4 ChS=4000Hz			
7827,00	CW		RGT77	RUS MIL Moscow, RVSN	RUS
7830,00	CHN 4+4	B=75Bd Ch=8 ChS=300/450Hz		CHN MIL	
7830,40	CW		U		
7831,00	RTTY			USA AF MARS	
7832,00	MIL 188-110B 39TONE	B=2400Bd Ch=39 ChS=56,25Hz			
7833,00	ALE USB	B=125Bd Ch=8 ChS=250Hz		F AF net	F
7835,00	ALE USB	B=125Bd Ch=8 ChS=250Hz		S MIL net	S
7835,70	*STANAG 4285*	*B=2400Bd*	*IDR*	*I N TX Rome*	*I*
7836,00	T600	B=50Bd S=200Hz		RUS N Bishkek	KGZ
7836,00	CCIR 493-4	B=100Bd S=170Hz			
7840,00	BAUDOT	B=200Bd S=850Hz			
7843,20	*STANAG 4285*	*B=2400Bd*	*EBA*	*E N Madrid*	*E*
7846,00	*LINK 11 ISB*	*B=2250Bd Ch=16 ChS=330/110/550Hz*	*PBC*	*NLD N Goeree Island*	*HOL*
7848,00	USB			USA Army MARS	
7850,00	USB			Grenada MRSC	GND
7850,00	USB			Saint Vincent and the Grenadines CG Base	VCT

Frequency	Mode	Mode Parameter	Callsign	User	Country
7850,00	CW , SSB		CHU	Ottawa TS	CAN
7852,00	BAUDOT	B=50Bd S=500Hz			
7853,00	CW			RUS MIL net	RUS
7853,00	CW		RGT77	RUS MIL Moscow, RVSN	RUS
7860,00	ALE USB	B=125Bd Ch=8 ChS=250Hz		G MIL net	G
7865,00	ALE USB	B=125Bd Ch=8 ChS=250Hz		S FRO net	S
7868,50	STANAG 4285	B=2400Bd	TBB	TUR N Izmir	TUR
7870,00	USB			CIS aero network day	CIS
7871,70	STANAG 4285	B=2400Bd	ICV	I N TX Tavolara	I
7872,00	MIL 188-110B SER	B=2400Bd		POL F	POL
7872,00	MIL 188-110B SER	B=2400Bd		POL F	POL
7872,00	USB			POL F	POL
7874,00	CIS 20	B=75Bd Ch=20 ChS=120Hz		RUS MIL net	RUS
7874,70	STANAG 4285	B=2400Bd	IDR	I N Rome	I
7880,00	FAX 120/288/576		DDK3	DWD TX Pinneberg	D
7882,50	USB			USA Army MARS	USA
7885,00	ALE USB	B=125Bd Ch=8 ChS=250Hz		CHL SENAPRED net	CHL
7885,00	ALE USB	B=125Bd Ch=8 ChS=250Hz		CHN diplo net	CHN
7887,30	CIS 75-1000	B=75Bd S=1000Hz		RUS MIL St. Petersburg	RUS
7889,00	LINK 22	B=2250Bd			
7890,00	RUS TACTICAL DATALINK	B=1200Bd S=800Hz Ch=1200		RUS MIL	RUS
7892,00	ALE USB	B=125Bd Ch=8 ChS=250Hz		TUR Disaster and Emergency net	TUR
7894,00	ALE USB	B=125Bd Ch=8 ChS=250Hz		UN MINUSCA net	
7895,00	ALE USB	B=125Bd Ch=8 ChS=250Hz		F N net	F
7896,00	ALE USB	B=125Bd Ch=8 ChS=250Hz		F N net	F
7896,00	CHN 4+4	B=75Bd Ch=8 ChS=300/450Hz		CHN MIL	
7898,00	PACTOR IV	B=1800Bd	DEK77	DRK Nienburg	D
7898,00	PACTOR IV	B=1800Bd	DEK30	DRK KIel	D
7898,00	ALE LSB	B=125Bd Ch=8 ChS=250Hz		German Red Cross net	D
7899,00	USB		VMS469	Reids Radiodata net	AUS
7899,00	ALE USB	B=125Bd Ch=8 ChS=250Hz		ALG MIL net	ALG
7899,00	USB		VKE237	AUS HF Radio Club	AUS
7900,00	CIS 50-500	B=50Bd S=500Hz	RNI2	RUS GOV	RUS
7900,00	CW		RNI2	RUS N	RUS
7900,00	CIS 50-500	B=50Bd S=500Hz	RAZ2	RUS PTT	RUS
7901,00	ALE USB	B=125Bd Ch=8 ChS=250Hz		ALG MIL net	ALG
7903,00	USB		XVB	Ben Thuy R	VTN
7903,00	USB		XVA	Ca Mau R	VTN
7903,00	USB		XVU	Can Tho R	VTN
7903,00	USB		XVC	Cua Ong R	VTN
7903,00	USB		XVT	Da Nang R	VTN
7903,00	USB		XVG	Hai Phong R	VTN
7903,00	USB		XVS	Ho Chi Minh R	VTN
7903,00	USB		XVQ	Hon Gai R	VTN
7903,00	USB		XVD	Hue R	VTN
7903,00	USB		XVM	Mong Cai R	VTN
7903,00	USB		XVK	Kien Giang R	VTN
7903,00	USB		XVN	Nha Trang R	VTN
7903,00	USB			Phan Rang R	VTN

Frequency	Mode	Mode Parameter	Callsign	User	Country
7903,00	USB		XVP	Phan Thiet R	VTN
7903,00	USB		XVY	Phu Yen R	VTN
7903,00	USB		XVI	Quy Nhon R	VTN
7903,00	USB			Thanh Hoa R	VTN
7903,00	USB		XVR	Vung Tau R	VTN
7903,50	ALE USB	B=125Bd Ch=8 ChS=250Hz		G MIL net	G
7903,50	ALE USB	B=125Bd Ch=8 ChS=250Hz		USA FBI net	USA
7905,00	MIL 188-110B SER	B=2400Bd		ROU F	ROU
7906,00	USB		XVM	Mong Cai R	VTN
7906,00	USB		XVP	Phan Thiet R	VTN
7906,00	USB		XVU	Can Tho R	VTN
7906,00	USB		XVK	Kien Giang R	VTN
7906,00	USB		XVI	Quy Nhon R	VTN
7906,00	USB		XVD	Hue R	VTN
7906,00	USB		XVQ	Hon Gai R	VTN
7906,00	USB		XVY	Phu Yen R	VTN
7906,00	USB		XVA	Ca Mau R	VTN
7906,00	USB		XVN	Nha Trang R	VTN
7906,00	USB		XVG	Hai Phong R	VTN
7906,00	USB		XVR	Vung Tau R	VTN
7906,00	USB		XVT	Da Nang R	VTN
7906,00	USB		XVS	Ho Chi Minh Ville R	VTN
7906,00	USB			Phan Rang R	VTN
7906,00	USB		cua ong	Cua Ong R	VTN
7906,40	PACTOR II	B=100Bd S=200Hz		UNHCR net	
7906,80	FAX 90/576		RBW41	Murmansk M	RUS
7907,00	*STANAG 4285*	*B=2400Bd*		*G MIL St. Eval*	*G*
7912,00	USB			Phan Rang R	VTN
7914,00	CW			RUS Operational Strategic Command Air Defense net	RUS
7915,00	USB		XVB	Ben Thuy R	VTN
7915,00	ALE USB	B=125Bd Ch=8 ChS=250Hz		CHL MOI net	CHL
7915,00	USB			USA AF MARS	USA
7915,00	USB			BGD N net	BGD
7918,00	CCIR 493-4	B=100Bd S=170Hz			
7919,80	STANAG 4538	B=2400Bd			
7920,00	*STANAG 4285*	*B=2400Bd*		*G MIL St. Eval*	*G*
7920,00	*STANAG 4285*	*B=2400Bd*		*G MIL Akrotiri*	*CYP*
7921,00	CCIR 493-4	B=100Bd S=170Hz			
7922,50	USB		VNJ	Casey Base , AUS	ANT
7922,50	USB			AUS Davis Base	ANT
7922,50	USB			Macquarie Island , AUS	ANT
7922,50	USB			Mawson Base , AUS	ANT
7926,00	CIS OFDM 45	B=2400Bd Ch=45 ChS=62,5Hz		RUS	RUS
7927,00	USB		XVS	Ho Chi Minh R	VTN
7927,00	USB		XVI	Quy Nhon R	VTN
7927,00	CIS SELCALL	B=150Bd S=250Hz Ch=4 ChS=1000Hz		CIS	RUS
7932,00	CW			RUS MIL net	RUS
7932,00	CW			RUS MIL net	RUS
7932,00	ALE USB	B=125Bd Ch=8 ChS=250Hz		USA Red Cross net	USA
7933,00	CCIR 493-4	B=100Bd S=170Hz			
7933,00	CCIR 493-4	B=100Bd S=170Hz			
7935,00	ALE USB	B=125Bd Ch=8 ChS=250Hz		USA Red Cross net	USA
7936,00	CCIR 493-4	B=100Bd S=170Hz			
7936,00	CCIR 493-4	B=100Bd S=170Hz			
7937,00	*STANAG 4285*	*B=2400Bd*		*G MIL Crimond*	*G*

Frequency	Mode	Mode Parameter	Callsign	User	Country
7938,00	CCIR 493-4	B=100Bd S=170Hz			
7938,00	USB		AFA5KK	USA AF MARS	USA
7938,00	MIL 188-110A	B=2400Bd	AFA5KK	USA AF MARS	USA
7939,00	CCIR 493-4	B=100Bd S=170Hz			
7940,00	ALE USB	B=125Bd Ch=8 ChS=250Hz		MTN net	MTN
7940,00	CHN DATALINK	B=2400Bd Ch=30 ChS=75Hz		CHN N	CHN
7941,40	PACTOR II	B=200Bd S=200Hz	WPTG385	Sailmail Corpus Christi , TX	USA
7942,00	CW		REA4	RUS AF Moscow	RUS
7942,00	USB		XVP	Phan Thiet R	VTN
7952,00	CIS OFDM 121	B=2541Bd Ch=121 ChS=25Hz		RUS E Lisbon	POR
7953,00	ALE USB	B=125Bd Ch=8 ChS=250Hz		USA Army MARS	USA
7954,00	USB		XVR	Vung Tau R	VTN
7957,00	CW			RUS	
7957,40	PACTOR II	B=200Bd S=200Hz	KUZ533	Sailmail Honululu	HWA
7957,40	PACTOR III	B=100Bd S=200Hz	RC01	Sailmail Maputo	MOZ
7960,00	USB			Thanh Hoa R	TUN
7961,40	PACTOR II	B=200Bd S=200Hz	KZN508	Sailmail Rockhill , SC	USA
7962,00	PACTOR III	B=100Bd S=200Hz	V8V2222	Sailmail Brunai Bay	BRU
7962,00	CW		RGT77	RUS MIL Moscow, RVSN	RUS
7963,00	CCIR 493-4	B=100Bd S=170Hz			
7963,00	USB		XVI	Quy Nhon R	VTN
7965,00	ALE USB	B=125Bd Ch=8 ChS=250Hz		ISR AF net	ISR
7966,00	USB		XVY	Phu Yen R	VTN
7969,00	USB		XVA	Ca Mau R	VTN
7969,00	ALE USB	B=125Bd Ch=8 ChS=250Hz		ALG Oilfield net	ALG
7969,00	CW			RUS air defence	RUS
7969,00	T600	B=50Bd S=200Hz	RDL	RUS N Moscow	RUS
7969,00	CW			RUS MIL net	RUS
7971,40	PACTOR II	B=200Bd S=200Hz	WRD719	Sailmail Palo Alto	USA
7971,40	PACTOR III	B=100Bd S=200Hz	WQLI952	Sailmail Watsonville , CA	USA
7972,00	USB		XVT	Da Nang R	VTN
7972,50	BAUDOT	B=50Bd S=500Hz			
7973,00	MFSK11	B=125Bd Ch=11 ChS=250Hz			
7975,00	USB		XVN	Nha Trang R	VTN
7978,00	USB		XVK	Kien Giang R	VTN
7980,00	MIL 188-110A SER	B=2400Bd		F N	F
7980,00	RUS TACTICAL DATALINK	B=1200Bd S=800Hz Ch=1200		RUS MIL	RUS
7980,00	ALE USB	B=125Bd Ch=8 ChS=250Hz		F N net	F
7981,00	USB		XVU	Can Tho R	VTN
7981,00	8PSK	B=400Bd			
7981,00	ALE USB	B=125Bd Ch=8 ChS=250Hz		S MIL net	S
7981,40	PACTOR II	B=200Bd S=200Hz	KZN508	Sailmail Rockhill , SC	USA
7984,00	IRN QPSK	B=468Bd		IRN N	IR
7989,00	ALE USB	B=125Bd Ch=8 ChS=250Hz		CLM net	CLM
7990,00	USB		XVP	Phan Thiet R	VTN
7991,00	ALE USB	B=125Bd Ch=8 ChS=250Hz		USA SHARES net	USA
7993,00	ALE USB	B=125Bd Ch=8 ChS=250Hz		TWN N net	TWN
7993,50	ALE USB	B=125Bd Ch=8 ChS=250Hz			

Frequency	Mode	Mode Parameter	Callsign	User	Country
7994,00	CW			RUS Operational Strategic Command Air Defense net	RUS
7994,00	CW			RUS air defence net	RUS
7994,00	ALE USB	B=125Bd Ch=8 ChS=250Hz			
7995,00	USB			McMurdo Ops	
7995,00	PACTOR III	B=100Bd S=200Hz	WHV382	Sailmail Friday Habor , WA	USA
7996,00	CCIR 493-4	B=100Bd S=170Hz			
7996,00	USB		XVN	Nha Trang R	VTN
7997,00	8PSK	B=2400Bd			
7998,00	ALE USB	B=125Bd Ch=8 ChS=250Hz			
7998,90	*CIS 50-1000*	*B=50Bd S=1000Hz*		*RUS MIL Moscow*	*RUS*
7999,00	USB		XVB	Ben Thuy R	VTN
7999,00	USB		XVA	Ca Mau R	VTN
7999,00	CCIR 493-4	B=100Bd S=170Hz			
7999,00	DPRK ARQ	B=1200Bd S=600Hz		DPRK E Warsaw	POL
8000,00	HAARP WAVEFORM			HAARP	ALS
8000,00	ALE USB	B=125Bd Ch=8 ChS=250Hz		F N net	
8000,00	8PSK	B=1800Bd			
8000,00	CCIR 493-4	B=100Bd S=170Hz	2155		
8000,00	CCIR 493-4	B=100Bd S=170Hz	2156		
8001,00	CCIR 493-4	B=100Bd S=170Hz			
8002,00	CW		4XZ	ISR N Haifa	ISR
8002,00	*CIS 12*	*B=1440Bd Ch=12 ChS=200Hz*		*RUS N HQ Moscow*	*RUS*
8003,00	CCIR 493-4	B=100Bd S=170Hz			
8005,00	CLOVER2000	Ch=4	EMGA11		
8005,00	CLOVER 2000	B=500Bd Ch=8			
8005,00	ALE USB	B=125Bd Ch=8 ChS=250Hz		ROU net	
8005,00	*STANAG 4285*	*B=2400Bd*		*G MIL TX Crimond*	*G*
8006,00	*T600*	*B=50Bd S=200Hz*	*RMP*	*RUS N Kaliningrad*	*RUS*
8006,00	CW		JG2XA	Experimental station Tokyo	J
8006,00	CCIR 493-4	B=100Bd S=170Hz			
8008,00	*CIS 12*	*B=1440Bd Ch=12 ChS=200Hz*		*RUS N HQ Moscow*	*RUS*
8009,00	CCIR 493-4	B=100Bd S=170Hz			
8009,00	ALE USB	B=125Bd Ch=8 ChS=250Hz		CHN net	
8009,00	THALES SKYMASTER ALE	B=125Bd Ch=8 ChS=250Hz			
8009,00	SYSTEME 3000	B=2000Bd			
8009,40	PACTOR II	B=200Bd S=200Hz	WPUC469	Sailmail South Daytona , FL	USA
8010,00	CHN 4FSK	B=100Bd Ch=4 ChS=500Hz			
8010,00	MIL 188-110B SER	B=2400Bd	BUCURESTI2**	ROU MOI Bucharest	ROU
8010,00	MIL 188-110B SER	B=2400Bd	SIBIU**	ROU MOI Sibiu	ROU
8010,00	MIL 188-110B SER	B=2400Bd	MARAMURES**	ROU MOI Maramures	ROU
8010,00	MIL 188-110B SER	B=2400Bd	PRAHOVA**	ROU MOI Prahova	ROU
8010,00	CHN 4FSK	B=100Bd Ch=4 ChS=500Hz			
8011,50	ALE USB	B=125Bd Ch=8 ChS=250Hz		POL MIL net	POL
8012,00	ALE USB	B=125Bd Ch=8 ChS=250Hz		USA Civil Air Patrol net	USA
8012,00	CCIR 493-4	B=100Bd S=170Hz			

Frequency	Mode	Mode Parameter	Callsign	User	Country
8013,20	*STANAG 4285*	*B=2400Bd*		*G MIL TX INskip*	*G*
8013,50	STANAG 4285	B=2400Bd		G MIL St. Eval	G
8014,00	CW		RFK76	RUS N	RUS
8014,00	CW		RHQ33	RUS N ship PRIAZOVYE	RUS
8014,00	ALE USB	B=125Bd Ch=8 ChS=250Hz		TUN diplo net	TUN
8015,00	CW			RUS Operational Strategic Command Air Defense net	RUS
8015,00	CLOVER200	Ch=4	EMGA11		
8015,00	CLOVER 2000	B=500Bd Ch=8			
8015,00	CCIR 493-4	B=100Bd S=170Hz			
8016,00	ALE USB	B=125Bd Ch=8 ChS=250Hz		HRV NPRD net	HRV
8017,00	CIS 50-500	B=50Bd S=500Hz			
8018,00	CCIR 493-4	B=100Bd S=170Hz			
8018,00	*CIS 75-200*	*B=75Bd S=200Hz*		*RUS N Kaliningrad*	*RUS*
8018,00	SYSTEME 3000	B=2400Bd			
8020,00	ALE USB	B=125Bd Ch=8 ChS=250Hz		I AF net	I
8020,00	CW		RLT96	RUS	
8020,40	PACTOR II	B=200Bd S=200Hz	WHV861	Sailmail San Luis Obispo , CA	USA
8021,00	CCIR 493-4	B=100Bd S=170Hz			
8021,00	MIL 188-110B 39TONE	B=2400Bd Ch=39 ChS=56,25Hz		MOD Vienna	AUT
8021,00	ALE USB	B=125Bd Ch=8 ChS=250Hz		ROU diplo net	ROU
8021,00	MIL 188-110A SER	B=2400Bd		MFA Bucharest	ROU
8021,70	SITOR A	B=100Bd S=170Hz		MFA Cairo	EGY
8021,70	SITOR B	B=100Bd S=170Hz		MFA Cairo	EGY
8022,00	USB		VKS737	AUS 4WD net	AUS
8023,70	SITOR A	B=100Bd S=170Hz		MFA Cairo	EGY
8024,00	CCIR 493-4	B=100Bd S=170Hz			
8026,00	ALE LSB	B=125Bd Ch=8 ChS=250Hz		Public health net , TX	USA
8027,00	USB		VMS469	Reids Radiodata net	AUS
8027,00	CCIR 493-4	B=100Bd S=170Hz			
8029,00	CW		W**	RUS AF net	RUS
8029,00	CW		TRL5**	RUS AF	RUS
8029,00	CW			RUS Strategic AF net	RUS
8030,00	CCIR 493-4	B=100Bd S=170Hz			
8030,00	LINK 11 CLEW	B=2250Bd Ch=16 ChS=330/110/550Hz			
8031,00	PACTOR	B=200Bd S=170Hz		UNHCR	
8031,00	PACTOR	B=200Bd S=170Hz		UNHCR Geneva	SUI
8033,00	ALE USB	B=125Bd Ch=8 ChS=250Hz		ROU diplo net	ROU
8033,00	MIL 188-110A SER	B=2400Bd		ROU E Athens	GRC
8033,00	MIL 188-110A SER	B=2400Bd		MFA Bucharest	ROU
8033,00	USB			RUS Strategic AF	RUS
8033,00	CCIR 493-4	B=100Bd S=170Hz			
8033,70	ALE USB	B=125Bd Ch=8 ChS=250Hz			
8034,00					
8034,00	USB			RUS	
8036,50	*STANAG 4285*	*B=2400Bd*	*TBB*	*TUR N Ankara*	*TUR*
8037,00	ALE USB	B=125Bd Ch=8 ChS=250Hz		USA Army MARS net	USA
8038,00	USB			G AF VOLMET St. Eval	G
8038,00	CIS 50-500	B=50Bd S=500Hz		RUS	RUS
8040,00	*FAX 120/576*		*GYA*	*G N London*	*G*
8041,00	CCIR 493-4	B=100Bd S=170Hz			
8043,00	USB			HFoZ/Radtel net	AUS

Frequency	Mode	Mode Parameter	Callsign	User	Country
8043,00	*IRN QPSK*	*B=207Bd*		*IRN N*	*IRN*
8047,00	ALE USB	B=125Bd Ch=8 ChS=250Hz		USA National Guard net	USA
8048,00	CW		RCV	RUS N HQ Sevastopol	UKR
8048,00	CW		RQN3	RUS N CHMS Sevastopol	UKR
8048,00	*T600*	*B=50Bd S=200Hz*	*RMP*	*RUS N Kaliningrad*	*RUS*
8050,00	ALE USB	B=125Bd Ch=8 ChS=250Hz		USA FEMA net	USA
8050,00	*STANAG 4285*	*B=2400Bd*		*TUR F Ankara*	*TUR*
8052,20	STANAG 4285	B=2400Bd		TUR F Ankara	TUR
8054,00	ALE USB	B=125Bd Ch=8 ChS=250Hz			
8054,50	MFSK	B=40Bd Ch=60 ChS=40Hz			
8055,00	ALE USB	B=125Bd Ch=8 ChS=250Hz		MTN MOI net	MTN
8055,00	PACTOR III	B=100Bd S=200Hz	KLT	Austin , TX	USA
8055,00	CIS MFSK 13	B=31,25Bd Ch=13 ChS=125Hz		MFA Moscow	RUS
8056,00	ALE USB	B=125Bd Ch=8 ChS=250Hz		USA Defence Logistc Agency	USA
8056,00	SYSTEME 3000	B=2400Bd		F F	F
8056,60	ALE USB	B=125Bd Ch=8 ChS=250Hz		USA diplo net	USA
8056,70	SITOR A	B=100Bd S=170Hz		MFA Cairo	EGY
8056,70	STANAG 4481	B=75Bd S=850Hz		G MIL Crimond	G
8056,90	PACTOR III	B=200Bd S=200Hz	WGM	Ft. Lauderdale , FL	USA
8057,00	ALE USB	B=125Bd Ch=8 ChS=250Hz		PAK N net	
8058,00	MIL 188-110B SER	B=2400Bd		MFA Bucharest	ROU
8058,00	CCIR 493-4	B=100Bd S=170Hz			
8058,00	BAUDOT	B=50Bd S=500Hz	NT9P**		RUS
8058,60	ALE USB	B=125Bd Ch=8 ChS=250Hz		USA diplo net	USA
8061,00	MFSK12	B=10Bd Ch=12 ChS=40Hz			RUS
8061,00	ALE USB	B=125Bd Ch=8 ChS=250Hz		CLM net	CLM
8061,00	ALE USB	B=125Bd Ch=8 ChS=250Hz		ALG MIL net	ALG
8062,00	BPSK	B=500Bd			
8065,00	ALE USB	B=125Bd Ch=8 ChS=250Hz		USA Army NG net Utah	USA
8066,00	ALE USB	B=125Bd Ch=8 ChS=250Hz		ROU diplo net	ROU
8066,00	MIL 188-110A SER	B=2400Bd		ROU E Vienna	AUT
8066,00	MIL 188-110A SER	B=2400Bd		MFA Bucharest	ROU
8066,00	CCIR 493-4	B=100Bd S=170Hz			
8066,10	FSK	B=75Bd S=125Hz			
8066,70	SITOR A	B=100Bd S=170Hz		MFA Cairo	EGY
8067,20	CIS OFDM 60	B=2400Bd Ch=60 ChS=44,5Hz		RUS diplo	RUS
8067,80	CCIR 493-4	B=100Bd S=170Hz		CHN MIL Spratly Island	CHN
8070,00	USB		CBP22	La Paz VOLMET	BOL
8070,00	CLOVER2000	Ch=4	EFS1		
8070,00	CLOVER2000	Ch=4	2BC!		
8072,00	ALIS	B=228,66Bd S=170Hz	2191		
8072,00	CCIR 493-4	B=100Bd S=170Hz			
8073,00	ALE USB	B=125Bd Ch=8 ChS=250Hz		ALG AF net	ALG
8073,00	CHN 4+4	B=75Bd Ch=8 ChS=300/450Hz	XSV85	CHN N	CHN

Frequency	Mode	Mode Parameter	Callsign	User	Country
8074,00	*STANAG 4285*	*B=2400Bd*		*G MIL TX St. Eval*	*G*
8075,00	ALE USB	B=125Bd Ch=8 ChS=250Hz		PHL MOI net	PHL
8075,00	CCIR 493-4	B=100Bd S=170Hz			
8075,50	ALE USB	B=125Bd Ch=8 ChS=250Hz		USA Urgentlink net	USA
8077,00	ALE USB	B=125Bd Ch=8 ChS=250Hz		ALG AF net	ALG
8078,00	CW		REA4	RUS AF Moscow	RUS
8078,00	CCIR 493-4	B=100Bd S=170Hz			
8078,70	SITOR A	B=100Bd S=170Hz		EGY E Washington	USA
8079,00	*CIS 100-500*	*B=100Bd S=500Hz*		*RUS N Archangelsk*	*RUS*
8079,00	CIS ARQ	B=100Bd S=500Hz		RUS PTT	RUS
8080,00	*LINK 11 CLEW*	*B=2250Bd Ch=16 ChS=330/110/550Hz*		*USA MIL Norffolk , VA*	*USA*
8080,00	USB			PHL N net	PHL
8080,00	ALE USB	B=125Bd Ch=8 ChS=250Hz		F N net	F
8081,00	CCIR 493-4	B=100Bd S=170Hz			
8081,00	LINK 11 CLEW	B=2250Bd Ch=16 ChS=330/110/550Hz			
8081,00	*CIS 12*	*B=1440Bd Ch=12 ChS=200Hz*		*RUS N HQ Moscow*	*RUS*
8082,00	*STANAG 4285*	*B=2400Bd*		*G F Penkale Sands*	*G*
8082,00	ALE USB	B=125Bd Ch=8 ChS=250Hz		Queensland Deptartment of Community Safety	AUS
8084,00	CCIR 493-4	B=100Bd S=170Hz			
8085,00	FAX 120/576		GYA	G N London TX Inskip	G
8085,50	ALE USB	B=125Bd Ch=8 ChS=250Hz		ARINC Urgent Link net	USA
8086,00	ALE USB	B=125Bd Ch=8 ChS=250Hz		ALG MIL net	ALG
8087,20	2FSK	B=100Bd S=170Hz		Naples	I
8088,00	ALE USB	B=125Bd Ch=8 ChS=250Hz		TUN diplo net	TUN
8088,70	*STANAG 4285*	*B=2400Bd*	*IDR*	*I N Rome*	*I*
8089,00	*STANAG 4285*	*B=2400Bd*		*G MIL Akrotiri*	*CYP*
8090,00	CCIR 493-4	B=100Bd S=170Hz			
8090,00	USB		NABOR	RUS Strategic AF	RUS
8090,00	USB			RUS strategic AF net	RUS
8090,00	USB		SHAPORA	RUS Strategic AF	RUS
8092,00	*ALE USB*	*B=125Bd Ch=8 ChS=250Hz*		*TUR Disaster and Emergency net*	*TUR*
8093,00	STANAG 4481	B=50Bd S=850Hz	NKW	USA N Diego Garcia	DGA
8095,00	USB			CIS aero network day	CIS
8100,00	USB			Mauritius MRCC	MAU
8100,00	*MAHRS*	*B=2400Bd*	*DHJ58*	*D N Glücksburg TX Staberhuk*	*D*
8100,00	ALE USB	B=125Bd Ch=8 ChS=250Hz		GANOB net	EGY
8101,00	USB		XVG	Hai Phong R	VTN
8101,00	SSB			CS , SS Ch 1	
8101,00	USB			Yachting net	
8101,90	CIS 100-1000	B=100Bd S=1000Hz		RUS	RUS
8104,00	USB			RUS N Pacific Fleet	RUS
8104,00	USB		XVS	Ho Chi Minh R	VTN
8104,00	SSB			CS , SS Ch 2	
8104,00	*MAHRS*	*B=2400Bd*	*DHJ58*	*D N Glücksburg TX Staberhuk*	*D*
8104,00	USB			Caribbean calling frequency and safety net	
8104,00	USB			Trans Atlantic Cruiser's net	

Frequency	Mode	Mode Parameter	Callsign	User	Country
8104,00	CIS 68TONE	B=50Bd Ch=68 ChS=47Hz		RUS diplo	RUS
8104,00	USB		KPK	SSCA Glenn Tuttle R Punta Gorda, FL	USA
8104,00	USB		KJM	SSCA Jim West R Ellijay, GAL	USA
8104,00	USB		KPK	SSCA Trans-Atlantic Cruisers Net	USA
8105,00	FAX 120/576		SVO	Olympia R	GRC
8105,00	CIS 100-1000	B=100Bd S=1000Hz			RUS
8105,00	CIS 68TONE	B=50Bd Ch=68 ChS=47Hz		RUS diplo	RUS
8107,00	*ALE USB*	*B=125Bd Ch=8 ChS=250Hz*		*G MIL net*	*G*
8107,00	SSB			CS , SS Ch 3	
8107,00	CW			RUS HFDF net	RUS
8107,00	MIL 188-110B APP C	B=2400Bd			
8108,00	*CIS 75-250*	*B=75Bd S=250Hz*		*RUS N Moscow*	*RUS*
8108,00	CW		RGT77	RUS MIL Moscow RVSN	RUS
8109,00	ALE USB	B=125Bd Ch=8 ChS=250Hz		AZE emergency net	AZE
8109,00	ALE USB	B=125Bd Ch=8 ChS=250Hz		LBY MIL net	LBY
8110,00	USB		VNJ	Casey Base , AUS	ANT
8110,00	USB			AUS Davis Base	ANT
8110,00	USB			Macquarie Island , AUS	ANT
8110,00	USB			Mawson Base , AUS	ANT
8110,00	SSB			CS , SS Ch 4	
8111,00	CIS 20	B=75Bd Ch=20 ChS=120Hz		RUS MIL net	RUS
8112,00	CW		W	RUS AF marker	RUS
8112,00	CW			RUS Strategic AF net	RUS
8113,00	USB		VMW	Wiluna M	AUS
8113,00	SSB			CS , SS Ch 5	
8113,50	CW		RMP	RUS N HQ Kaliningrad	RUS
8115,00	ALE USB	B=125Bd Ch=8 ChS=250Hz		SUI MIL net	SUI
8115,00	CIS 50-500	B=50Bd S=500Hz	RYF2	RUS PTT	RUS
8116,00	SSB			CS , SS Ch 6	
8116,00	USB			Gulf Harbor R	NZL
8117,00	USB			Taipai WX	TWN
8118,00	RUS HYBRID MFSK OFDM	B=40Bd Ch=16		Moscow	RUS
8118,00	RUS HYBRID MFSK OFDM	B=40Bd Ch=16		Praha	CZE
8118,00	BAUDOT	B=200Bd S=500Hz		RUS INTEL	RUS
8119,00	SSB			CS , SS Ch 7	
8120,00	CLOVER 2500	B=625Bd Ch=8 ChS=312,5Hz	40C1	MIL net	
8120,00	ALE USB	B=125Bd Ch=8 ChS=250Hz			
8120,00	USB			G AF Air Cadets	G
8120,00	CW		RMX	RUS AF	RUS
8120,00	CW		RCH84	RUS N	RUS
8120,00	CW		RJF94	RUS N Moscow	RUS
8120,00	CLOVER2500	Ch=4	40P1	MIL net	
8120,00	CLOVER2500	Ch=4	40Q1	MIL net	
8120,00	CW		RAA	RUS N HQ Moscow	RUS
8120,00	CW		RJH77	RUS N Archangelsk	RUS
8121,00	USB			CHN air defense	CHN
8122,00	STANAG 4285	B=2400Bd		AUS MIL Darwin	AUS
8122,00	SSB			CS , SS Ch 8	
8122,00	USB			Yachtsmen Mediternean net	
8123,20	*STANAG 4285*	*B=2400Bd*	*IDN*	*I N Naples*	*I*

Frequency	Mode	Mode Parameter	Callsign	User	Country
8123,20	*STANAG 4285*	*B=2400Bd*	*NSS*	*NATO HQ Naples*	*I*
8125,00	USB		XVT	Da Nang R	VTN
8125,00	ALE USB	B=125Bd Ch=8 ChS=250Hz		PEMEX net	MEX
8125,00	SSB			CS , SS Ch 9	
8125,00	ALE USB	B=125Bd Ch=8 ChS=250Hz		G MIL net	G
8127,00	STANAG 4481	B=75Bd S=850Hz		G MIL Crimond	G
8127,00	*STANAG 4285*	*B=2400Bd*		*G Mil St. Eval*	*G*
8128,00	SSB			CS , SS Ch 10	
8128,00	STANAG 4285	B=2400Bd		G MIL Crimond	G
8129,50	CIS 75-125	B=75Bd S=125Hz			RUS
8129,50	CIS OFDM 45	B=2400Bd Ch=45 ChS=62,5Hz		RUS	RUS
8131,00	USB		XVT	Da Nang R	VTN
8131,00	USB		BALANS	RUS Strategic AF	RUS
8131,00	USB		KATOLIK	RUS Strategic AF	RUS
8131,00	SSB			CS , SS Ch 11	
8133,00	USB		BENEDETTO**	I F	I
8133,00	USB		CESARE**	I F	I
8133,00	ALE ISB	B=125Bd Ch=8 ChS=125Hz		ARS AF net	ARS
8134,00	USB		ZAV3	Vlore R	ALB
8134,00	SSB			CS , SS Ch 12	
8135,60	PACTOR III	B=200Bd S=850Hz			
8136,00	CW		RDL	RUS N Moscow	RUS
8136,20	PACKET RADIO	B=300Bd S=200Hz	CE4TST		
8136,20	PACKET RADIO	B=300Bd S=200Hz	HQ2CEN		EGY
8137,00	USB		XVR	Vung Tau R	VTN
8137,00	SSB			CS , SS Ch 13	
8138,00	CIS 1200	B=1200Bd		RUS	RUS
8138,70	*STANAG 4285*	*B=2400Bd*	*IDR*	*I N TX Rome*	*I*
8139,00	ALE USB	B=125Bd Ch=8 ChS=250Hz		USA SAC net	USA
8140,00	USB		XVT	Da Nang R	VTN
8140,00	SSB			CS , SS Ch 14	
8140,20	STANAG 4285	B=2400Bd	DHJ58	D N Glücksburg TX Marlow	D
8143,00	USB		XVC	Cua Ong R	VTN
8143,00	T600	B=50Bd S=250Hz		RUS N Vladivostok	RUS
8143,00	SSB			CS , SS Ch 15	
8143,00	*CIS 4FSK*	*B=100Bd Ch=4 ChS=500Hz*		*RUS MIL Archangelsk*	*RUS*
8143,00	USB		HLF	Seoul R	KOR
8143,00	ALE USB	B=125Bd Ch=8 ChS=250Hz		PAK N net	
8143,00	USB			Panama Pacific net	
8143,00	R&S MODEM	B=2400Bd		PAK N	PAK
8144,00	MIL 188-110A SER	B=2400Bd		AUS MHFCS North West Cape	AUS
8145,00	*STANAG 4481*	*B=75Bd S=850Hz*	*AJE*	*USA AF Croughton*	*G*
8145,00	STANAG 4481	B=75Bd S=850Hz	NSY	USA N Niscemi , Sicily	I
8146,00	USB		XVN	Nha Trang R	VTN
8146,00	SSB			CS , SS Ch 16	IRL
8148,00	ALE USB	B=125Bd Ch=8 ChS=250Hz		SUI MIL net	SUI
8148,50	USB		AFA4JV	USA AF MARS	USA
8148,50	USB		AFA4YU	USA AF MARS	USA
8149,00	USB		XVD	Hue R	VTN
8149,00	SSB			CS , SS Ch 17	
8149,20	*STANAG 4285*	*B=2400Bd*	*IDR*	*I N Rome*	*I*
8150,00	STANAG 4285	B=2400Bd		AUS MHFCS Humpty Doo	AUS
8150,00	*ALE USB*	*B=125Bd Ch=8 ChS=250Hz*		*TUR Disaster and Emergency net*	*TUR*

Frequency	Mode	Mode Parameter	Callsign	User	Country
8151,00	ALE USB	B=125Bd Ch=8 ChS=250Hz		G MIL net	G
8151,70	STANAG 4285	B=2400Bd	OUA	DNK N Frederickshavn	DNK
8152,00	ALE USB	B=125Bd Ch=8 ChS=250Hz		S MIL net	S
8152,00	SSB			CS , SS Ch 18	
8152,00	USB			Cruiseheimer's net (East coast and Bahamas)	
8152,00	USB			Cruiseheimer's net	
8152,00	USB			Trans Atlantic Cruisers net	
8152,00	CIS OFDM 45	Ch=45 ChS=40Hz		MFA Moscow	RUS
8152,00	CW		RMP	RUS N HQ Kaliningrad	RUS
8152,00	USB		KPK	SSCA Trans-Atlantic Cruisers Net	USA
8155,00	USB		XVM	Mong Cai R	VTN
8155,00	SSB			CS , SS Ch 19	
8156,00	USB			BAH MIL net	BAH
8156,20	*STANAG 4285*	*B=2400Bd*	*IDR*	*I N Rome*	*I*
8157,00	QPSK	B=2400Bd		?CHN	CHN
8157,00	CW		RIS9	RUS N	RUS
8158,00	USB		XVK	Kien Giang R	VTN
8158,00					
8158,00	SSB			CS , SS Ch 20	
8159,00	CW		4XZ	ISR N Haifa	ISR
8160,00	ALE USB	B=125Bd Ch=8 ChS=250Hz		HFoZ/Radtel net	AUS
8160,00	USB			HFoZ/Radtel net	AUS
8161,00	ALE USB	B=125Bd Ch=8 ChS=250Hz		G MIL net	G
8161,00	USB		XVY	Phu Yen R	VTN
8161,00	SSB			CS , SS Ch 21	
8161,00	USB			Namba/Sheila net	
8162,00	CW		W**	RUS AF net	RUS
8162,00	CW			RUS Strategic AF net	RUS
8164,00	SSB			CS , SS Ch 22	
8164,50	ALE USB	B=125Bd Ch=8 ChS=250Hz			S
8165,00	ALE USB	B=125Bd Ch=8 ChS=250Hz		R&S test net	
8165,00	USB		VNZ	RFDS Pt. Augusta	AUS
8165,00	ALE USB	B=125Bd Ch=8 ChS=250Hz			
8167,00	USB		XVN	Nha Trang R	VTN
8167,00	ALE USB	B=125Bd Ch=8 ChS=250Hz		G Mil net	G
8167,00	SSB			CS , SS Ch 23	
8170,00	USB		XVU	Can Tho R	VTN
8170,00	SSB			CS , SS Ch 24	
8170,00	*STANAG 4285*	*B=2400Bd*		*G MIL Akrotiri*	*CYP*
8170,00	CW			RUS Strategic AF net	RUS
8170,20	*STANAG 4285*	*B=2400Bd*		*NOR N Bergen*	*NOR*
8170,20	*STANAG 4285*	*B=2400Bd*	*JWT*	*NOR N Stavanger*	*NOR*
8170,70	*STANAG 4285*	*B=2400Bd*		*NOR N Oslo*	*NOR*
8171,00	CIS 4FSK	B=150Bd Ch=4 ChS=4000Hz			
8171,50	ALE USB	B=125Bd Ch=8 ChS=250Hz		USA MIL net	USA
8172,00	ALE LSB	B=125Bd Ch=8 ChS=250Hz		ARS	ARS
8173,00	USB		XVQ	Hon Gai R	VTN
8173,00	SSB			CS , SS Ch 25	
8173,00	USB			Pacific Magellan net	
8174,00	ALE USB	B=125Bd Ch=8 ChS=250Hz		ALG MIL net	ALG

Frequency	Mode	Mode Parameter	Callsign	User	Country
8175,00	PACTOR III	B=100Bd S=200Hz	OL1A	CZE MIL Prague	CZE
8176,00	USB		VMC	Charleville M	AUS
8176,00	USB		XVT	Da Nang R	VTN
8176,00	SSB			CS , SS Ch 26	
8176,00	ALE USB	B=125Bd Ch=8 ChS=250Hz		AUS Police WA	AUS
8178,00	SSB			USA AF MARS	J
8179,00	SSB			CS , SS Ch 27	
8180,00	ALE USB	B=125Bd Ch=8 ChS=250Hz		TUN diplo net	TUN
8180,00	CHN 16TONE PSK	B=75Bd Ch=16		CHN F	CHN
8180,00	STANAG 4285	B=2400Bd		F N net	F
8181,50	ALE USB	B=125Bd Ch=8 ChS=250Hz		USA National Guard net	USA
8182,00	SSB			CS , SS Ch 28	
8182,00	*ALE USB*	*B=125Bd Ch=8 ChS=250Hz*		*G MIL net*	*G*
8185,00	SSB			CS , SS Ch 29	
8185,00	PACTOR III	B=100Bd S=200Hz	HPPM2	Sailmail Chiriqui	PNR
8186,00	ARCOTEL FARCOS	B=1800Bd		D F	
8187,00	OLIVIA 16/500	B=31,25Bd Ch=16	MFJ04	G N Sea Cadets network control station Epsom Downs	G
8187,00	OLIVIA 16/500	B=31,25Bd Ch=16	MRZ51	G N Sea Cadets	G
8187,00	OLIVIA 16/500	B=31,25Bd Ch=16 ChS=31,25Hz	MRZ53	G N Sea Cadets	G
8187,00	OLIVIA 16/500	B=31,25Bd Ch=16 ChS=31,25Hz	MFJ99	G N Sea Cadets	G
8188,00	SSB			CS , SS Ch 30	
8188,00	USB			Northwest Caribbean	
8188,50	PACTOR III	B=100Bd S=200Hz	HPPM3	Sailmail Panama	PNR
8189,00	ALE USB	B=125Bd Ch=8 ChS=250Hz		ALG MIL net	ALG
8190,00	ALE USB	B=125Bd Ch=8 ChS=250Hz		Guardia di Financa net	I
8190,00	CIS ARQ	B=100Bd S=500Hz		RUS PTT	RUS
8190,00	ALE ISB	B=125Bd Ch=8 ChS=250Hz		ARS AF net	ARS
8190,00	ALE LSB	B=125Bd Ch=8 ChS=250Hz		ARS AF net	ARS
8191,00	SSB			CS , SS Ch 31	
8191,00	*CIS 50-1000*	*B=50Bd S=1000Hz*	*REA4*	*RUS AF HQ Moscow*	*RUS*
8191,00	CW		RMP	RUS N HQ Kaliningrad	RUS
8191,50	CW		9MB	MLA N Johour Baharu	MLA
8191,50	CW		REA4	RUS AF HQ Moscow	RUS
8192,00	BAUDOT	B=50Bd S=850Hz	9MR	MLA N Johour Baharu	MLA
8194,00	USB		PEGASO	E Combat Air Command Torrejon de Ardoz	E
8195,00	ALE USB	B=125Bd Ch=8 ChS=250Hz		ROU MOI net	ROU
8195,00	MIL 188-110B SER	B=2400Bd		ROU MOI Bucharest	ROU
8195,00	SSB			SS Ch 801	
8198,00	SSB			SS Ch 802	
8201,00	SSB			SS Ch 803	
8203,70	*STANAG 4285*	*B=2400Bd*		*GRC N Crete*	*GRC*
8204,00	SSB			SS Ch 804	
8204,00	CW		Y**	RUS AF strategic net , BEAR net	RUS
8204,00	STANAG 4481	B=75Bd S=850Hz	NSY	USA N Niscemi , Sicily	I
8204,50	STANAG 4481	B=75Bd S=850Hz	AJE	USA AF AWS Croughton	G
8207,00	SSB			SS Ch 805	
8207,00	*ALE USB*	*B=125Bd Ch=8 ChS=250Hz*		*MLT N net*	*MLT*
8207,00	*STANAG 4285*	*B=2400Bd*	*TBB*	*TUR N Ankara*	*TUR*
8210,00	SSB			SS Ch 806	

Frequency	Mode	Mode Parameter	Callsign	User	Country
8211,00	USB		LBJ	NOR N Bodoe	NOR
8213,00	SSB			SS Ch 807	
8214,00	*MAHRS*	*B=2400Bd*	*DHJ58*	*D N Glücksburg TX Staberhuk*	*D*
8216,00	SSB			SS Ch 808	
8219,00	SSB			SS Ch 809	
8222,00	SSB			SS Ch 810	
8223,50	SYSTEME 3000	B=2400Bd			
8225,00	SSB			SS Ch 811	
8228,00	SSB			SS Ch 812	
8228,00	8PSK	B=2400Bd			
8231,00	SSB			SS Ch 813	
8234,00	SSB			SS Ch 814	
8234,00	*ALE USB*	*B=125Bd Ch=8 ChS=250Hz*		*ISR AF net*	*ISR*
8235,00	QPSK	B=750Bd		KOR MIL	KOR
8235,00	USB			KOR MIL	KOR
8237,00	SSB			SS Ch 815	
8237,00	USB		BUSSOL	RUS Black Sea fleet	RUS
8240,00	SSB			SS Ch 816	
8243,00	SSB			SS Ch 817	
8246,00	SSB			SS Ch 818	
8249,00	SSB			SS Ch 819	
8250,00	8PSK	B=2400Bd			
8252,00	SSB			SS Ch 820	
8255,00	SSB			SS Ch 821	
8258,00	MIL 188-110B 39TONE	B=2400Bd Ch=39 ChS=56,25Hz		MRC F	MRC
8258,00	LSB			MRC F	MRC
8258,00	SSB			SS Ch 822	
8259,20	*STANAG 4285*	*B=2400Bd*	*EBA*	*E N Madrid*	*E*
8260,00	USB			RUS MIL	
8260,00	CIS 12	B=1440Bd Ch=12 ChS=200Hz		RUS MIL	
8261,00	SSB			SS Ch 823	
8264,00	SSB			SS Ch 824	
8265,10	PACTOR III	B=100Bd S=200Hz	KDS	Ontario	CAN
8267,00	SSB			SS Ch 825	
8270,00	STANAG 4197	B=1800Bd Ch=16/39 ChS=112/56Hz			
8270,00	SSB			SS Ch 826	
8273,00	SSB			SS Ch 827	
8274,00	8PSK	B=2400Bd			
8276,00	SSB			SS Ch 828	
8279,00	SSB			SS Ch 829	
8279,20	*STANAG 4285*	*B=2400Bd*		*GRC Mil Athens*	*GRC*
8281,00	ALE USB	B=125Bd Ch=8 ChS=250Hz			
8282,00	SSB			SS Ch 830	
8284,70	*STANAG 4285*	*B=2400Bd*	*IDR*	*I N Rome*	*I*
8285,00	SSB			SS Ch 831	
8286,00	USB			USA MARS net	USA
8286,00	MIL 188-110A SER	B=2400Bd		USA MARS net	USA
8288,00	SSB			SS Ch 832	
8289,40	PACTOR III	B=100Bd S=200Hz	FOHXM	Sailmail Manihi Atoll	OCE
8291,00	USB		PKD5	Benoa R , Bali	INS
8291,00	USB		FFD	Saint Denis R	REU
8291,00	USB			Cape Town R	AFS
8291,00	USB		PKX	Jakarta R , Jawa	INS
8291,00	USB		PKE5	Ternate R , Maluku	INS
8291,00	USB		CWS	URG N Montevideo	URG
8291,00	USB		HCY	Ayora R , Santa Cruz	EQA

Frequency	Mode	Mode Parameter	Callsign	User	Country
8291,00	USB		ZLM	Taupo Maritime R	NZL
8291,00	USB		NMF	USA CG Boston , MA	USA
8291,00	USB		CWF	Punta Carretas R	URG
8291,00	USB		JNA	J CG HQ Tokyo	J
8291,00	USB			AUS coast net	AUS
8291,00	USB		VFF	Iqaluit R	CAN
8291,00	USB		VRC	Hong Kong R	CHN
8291,00	USB		PKY4	Sorong R , Irian Jaya	INS
8291,00	USB		NMA	USA CG Miami , FL	USA
8291,00	USB		PKG	Bandjarmasin R , Kalimantan	INS
8291,00	USB			USA CG Honolulu	HWA
8291,00	USB		PKO	Tarakan R , Kalimantan	INS
8291,00	USB			USA CG Point Reyes	USA
8291,00	USB		PKR	Semarang R , Djawa	INS
8291,00	USB		PNK	Jayapura R , Irian Jaya	INS
8291,00	USB		P2M	Port Moresbay R	PNG
8291,00	USB		PKY3	Manokwari R , Irian Jaya	INS
8291,00	USB		PKM	Bitung R , Sulawesi	INS
8291,00	SSB			Distress/calling frequency	
8291,00	SSB			CS Ch 833	
8291,00	SSB			SS Ch 833	
8291,00	USB		PKF3	Kendari R	INS
8291,00	USB		PKM44	Pantolan R , Sulawesi	INS
8291,00	USB		PKP38	Batu Ampar R	INS
8291,00	USB		PKC4	Panjang R , Sumatera	INS
8291,00	USB		PKA	Sabang R , We	INS
8291,00	USB			Benete R , Sumbawa	INS
8291,00	USB		PKK	Kupang R , Timor	INS
8291,00	USB		FJA	Mahina R	OCE
8291,00	USB		PKB	Belawan R , Sumatera	INS
8291,00	SITOR B	B=100Bd S=170Hz	SVO	Olympia R	GRC
8292,00	ALE USB	B=125Bd Ch=8 ChS=250Hz		PEMEX net	MEX
8293,90	ALE USB	B=125Bd Ch=8 ChS=250Hz		UI net	
8294,00	USB		ZOE	Tristan da Cunha R	TRC
8294,00	USB		VIA	Adelaide R	AUS
8294,00	USB		L2A	ARG N Buenos Aires	ARG
8294,00	USB		XVT	Da Nang R	VTN
8294,00	USB		XVG	Hai Phong R	VTN
8294,00	USB		XVS	Ho Chi Minh R	VTN
8294,00	SSB			simplex frequency ship - ship	
8294,00	PACTOR III	B=100Bd S=200Hz	PYB45	Sao Paulo	B
8294,00	NECODE SELCALL	B=100Bd S=170Hz			USA
8294,00	USB		WQCN860	Ocean-Pro R Naples , FL	USA
8294,00	USB		VAX498	South Bound II R Burlington , ONT	CAN
8294,00	USB		XVM	Mong Cai R	VTN
8294,00	USB		XVP	Phan Thiet R	VTN
8294,00	USB		XVU	Can Tho R	VTN
8294,00	USB		XVK	Kien Giang R	VTN
8294,00	USB		XVI	Quy Nhon R	VTN
8294,00	USB		XVD	Hue R	VTN
8294,00	USB		XVQ	Hon Gai R	VTN
8294,00	USB		XVY	Phu Yen R	VTN
8294,00	USB		XVA	Ca Mau R	VTN
8294,00	USB		XVN	Nha Trang R	VTN
8294,00	USB		XVR	Vung Tau R	VTN
8294,00	USB		XVC	Cua Ong R	VTN
8294,00	USB			Phan Rang R	VTN
8297,00	USB			D yacht net	
8297,00	SSB			simplex frequency ship - ship	

Frequency	Mode	Mode Parameter	Callsign	User	Country
8297,00	USB		ZLM	Taupo Maritime R	NZL
8300,00	*STANAG 4285*	*B=2400Bd*	*6WW*	*F N Dakar*	*SEN*
8300,20	*STANAG 4285*	*B=2400Bd*		*G MIL TX St. Eval*	*G*
8302,00	*STANAG 4285*	*B=2400Bd*		*G MIL Crimond*	*G*
8302,00	FAX 120/576		XSG	Shanghai R	CHN
8304,00	SSB		IGJ	I N Augusta	I
8308,50	PACTOR III	B=100Bd S=200Hz	ZKN2SM	Sailmail Niue	NIU
8309,20	STANAG 4285	B=2400Bd Ch=1		G MIL TX Falkland	FLK
8310,00	SITOR A	B=100Bd S=170Hz		J CG Tokyo	J
8310,50	FAX 120/576		XSQ	Guangzhou R	CHN
8313,00	ALE USB	B=125Bd Ch=8 ChS=250Hz		SVK AF net	SVK
8313,00	J PSK	B=1500Bd		J MIL	J
8315,00	*LINK 11 ISB*	*B=2250Bd Ch=16 ChS=330/110/550Hz*		*I F Leece*	*I*
8319,20	STANAG 4285	B=2400Bd	PBB	NLD N Den Helder	HOL
8322,00	LINK 11 ISB	B=2250Bd Ch=16 ChS=330/110/550Hz			
8324,00	USB		PBB	NLD N Den Helder	HOL
8324,00	USB		PBC	NLD N Goeree Island	HOL
8326,00	STANAG 4285	B=2400Bd	FUJ	F N Noumea	NCL
8330,00	CW		RCRE	RUS N ship FEODOR GOLOVIN	RUS
8331,20	*STANAG 4285*	*B=2400Bd*		*NOR N Bodo*	*NOR*
8333,00	PACTOR III	B=200Bd S=850Hz			
8333,20	*STANAG 4481*	*B=75Bd S=850Hz*		*G F St. Eval*	*G*
8334,50	*STANAG 4285*	*B=2400Bd*		*G MIL TX St. Eval*	*G*
8334,50	STANAG 4285	B=2400Bd		G MIL TX Inskip	G
8336,00	ALE USB	B=125Bd Ch=8 ChS=250Hz		G MIL net	G
8337,00	CW		RMP	RUS N HQ Kaliningrad	RUS
8337,50	USB		PBB	NLD N Den Helder	NLD
8337,50	USB		PBC	NLD N Goeree Island	NLD
8337,50	BAUDOT	B=75Bd S=850Hz	PBH	NLD N Den Helder	NLD
8338,00	*CIS 12*	*B=1440Bd Ch=12 ChS=200Hz*		*RUS N ship*	
8340,00	MIL 188-110B SER	B=2400Bd		MEX MIL	MEX
8340,00	ALE LSB	B=125Bd Ch=8 ChS=250Hz			
8342,00	BAUDOT	B=150Bd S=850Hz	FUJ	F N Noumea	NCL
8344,00	CW		RMTO	RUS N SAR ship ALEKSANDR FROLOV	RUS
8344,00	USB			EGY CG net	EGY
8344,50	CW		UDYY	RUS cargo ship IGARKA	RUS
8344,50	CW		UHVA	RUS cargo ship AVALON	RUS
8344,50	CW		UHZK	RUS ship VOSTOK3	UKR
8345,00	CW		RFF73	RUS N ship	RUS
8345,00	CW		RGR77	RUS N ship BURYA	RUS
8345,00	CW		RCIZ	RUS N ship	RUS
8345,00	CW		RBEG	RUS N ship	RUS
8345,00	CW		RFH77	RUS N ship LADNYY	RUS
8345,00	CW		RMRV	RUS N ship SMOLNYY	RUS
8345,00	CW		RME	RUS N	RUS
8345,00	CW		RFME	RUS N ship	USA
8345,00	CW		RHI99	RUS N ship ALEXANDER SHABATIN	RUS
8345,00	CW		RBDF	RUS N ship	RUS
8345,00	CW		RFH61	RUS N ship PEREKOP	RUS
8345,00	CW		UCTA5	RUS N ship YELYNA	RUS
8345,00	CW		RCRQ	RUS N cargo ship YAUZA	RUS
8345,00	CW		RCRE	RUS N ship ADMIRAL FEDOR GOLOVIN	RUS
8345,00	CW		RAL65	RUS N ship KAMA	RUS
8345,00	CW		RIR96	RUS N ship AZOV	RUS
8345,00	CW		RHL80	RUS N ship KALININGRAD	RUS

Frequency	Mode	Mode Parameter	Callsign	User	Country
8345,00	CW		RKO81	RUS N ship LENA	RUS
8345,00	CW		RHC84	RUS N ship EKVATOR	RUS
8345,00	CW		RJS81	RUS N ship	RUS
8345,00	CW		RMHW	RUS N ship	RUS
8345,00	CW		RMMA	RUS N ship VIKTOR LEONOV	RUS
8345,00	CW		RMC99	RUS N ship YEVGENIY KHVOROV	RUS
8345,00	CW		RMGB	RUS N ship IMAN	RUS
8345,00	CW		RLD69	RUS N ship GEORGY POBEDONOSETS BDK 45	RUS
8345,00	CW		RJQ84	RUS N ship VIKHR	RUS
8345,00	CW		RHO62	RUS N ship ADMIRAL VLADIMIRSKIY	RUS
8345,00	CW		RAL48	RUS N ship NIKOLEI CHIKER	RUS
8345,00	CW		RCIV	RUS N landing ship SARATOV	RUS
8345,00	CW		RJT22	RUS CG Moscow	RUS
8345,00	CW		RGR35	RUS N destroyer VICE ADMIRAL KULAKOV	RUS
8345,00	CW		RJP98	RUS N ship	RUS
8348,00	CW		RHO62	RUS N ship ADMIRAL VLADIMIRSKIY	RUS
8350,00	ALE USB	B=125Bd Ch=8 ChS=250Hz		IRQ MIL net	IRQ
8353,00	ALE USB	B=125Bd Ch=8 ChS=250Hz		PAK N net	
8356,00	ISR N HYBRID MODEM	B=2400Bd	4XZ	ISR N Haifa	ISR
8361,00	ALE USB	B=125Bd Ch=8 ChS=250Hz		TUN N net	TUN
8362,20	STANAG 4285	B=2400Bd		G N Crimond	G
8364,00	USB			NATO F	
8366,00	CW			SS calling frequency Ch 1	
8366,50	CW			SS calling frequency Ch 2	
8367,00	CW			SS calling frequency Ch 3	
8367,50	CW			SS calling frequency Ch 4	
8368,00	CW			SS calling frequency Ch 5	
8368,50	CW			SS calling frequency Ch 6	
8369,00	CW			SS calling frequency Ch 7	
8369,50	CW			SS calling frequency Ch 8	
8370,00	CW			SS calling frequency Ch 9	
8370,30	ALE USB	B=125Bd Ch=8 ChS=250Hz		TUN N net	TUN
8370,50	CW			SS calling frequency Ch 10	
8374,00	FMCW			Fort Caswell, NC	USA
8375,00	USB			BGD N net	BGD
8375,00	FMCW			Georgetown, SC	USA
8376,50	SITOR B	B=100Bd S=170Hz	ZLM	Taupo Maritime R	NZL
8376,50	SITOR , CW	B=100Bd S=170Hz	JNA	J CG HQ Tokyo	J
8376,50	SITOR , CW	B=100Bd S=170Hz	VRC	Hong Kong R	CHN
8376,50	SITOR , CW	B=100Bd S=170Hz	PKR3	Cilacap R , Jawa	INS
8376,50	SITOR , CW	B=100Bd S=170Hz	PKR	Semarang R , Djawa	INS
8376,50	SITOR , CW	B=100Bd S=170Hz	PKD	Surabaya R , Djawa	INS
8376,50	SITOR , CW	B=100Bd S=170Hz	PKN	Balikpapan R , Kalimantan	INS
8376,50	SITOR , CW	B=100Bd S=170Hz	PNK	Jayapura R , Irian Jaya	INS
8376,50	SITOR , CW	B=100Bd S=170Hz	PKY4	Sorong R , Irian Jaya	INS
8376,50	SITOR , CW	B=100Bd S=170Hz	PKM	Bitung R , Sulawesi	INS
8376,50	.			Distress/safety frequency	WW
8376,50	SITOR , CW	B=100Bd S=170Hz	PKK	Kupang R , Timor	INS
8377,00	.			SS Ch 2	
8377,50	.			SS Ch 3	
8378,00	.			SS Ch 4	
8378,50	.			SS Ch 5	
8379,00	SITOR A	B=100Bd S=170Hz	ZLM	Taupo Maritime R	NZL
8379,00	.			SS Ch 6	
8379,50	.			SS Ch 7	
8380,00	.			SS Ch 8	

Frequency	Mode	Mode Parameter	Callsign	User	Country
8380,50	.			SS Ch 9	
8381,00	.			SS Ch 10	
8381,20	*STANAG 4285*	*B=2400Bd*	*EBA*	*E N Madrid*	*E*
8381,50	.			SS Ch 11	
8382,00	.			SS Ch 12	
8382,50	.			SS Ch 13	
8383,00	.			SS Ch 14	
8383,50	.			SS Ch 15	
8384,00	.			SS Ch 16	
8384,50	.			SS Ch 17	
8385,00	.			SS Ch 18	
8385,00	USB			TUR MIL net	TUR
8385,00	STANAG 4285	B=2400Bd		TUR MIL net	TUR
8385,50	.			SS Ch 19	
8386,00	.			SS Ch 20	
8386,00	ALE USB	B=125Bd Ch=8 ChS=250Hz		SNG N net	SNG
8386,50	.			SS Ch 21	
8387,00	.			SS Ch 22	
8387,50	.			SS Ch 23	
8388,00	.			SS Ch 24	
8388,50	.			SS Ch 25	
8389,00	.			SS Ch 26	
8389,50	.			SS Ch 27	
8390,00	.			SS Ch 28	
8390,50	.			SS Ch 29	
8391,00	.			SS Ch 30	
8391,50	.			SS Ch 31	
8392,00	.			SS Ch 32	
8392,50	.			SS Ch 33	
8393,00	.			SS Ch 34	
8393,50	.			SS Ch 35	
8394,00	.			SS Ch 36	
8394,50	.			SS Ch 37	
8395,00	.			SS Ch 38	
8395,50	.			SS Ch 39	
8396,00	.			SS Ch 40	
8396,50	.			SS Ch 1	
8397,00	.			SS Ch 2	
8397,50	.			SS Ch 3	
8397,70	MIL 188-110B SER	B=2400Bd	HWK**	S N Karlskrona	S
8398,00	.			SS Ch 4	
8398,50	.			SS Ch 5	
8399,00	PACTOR II	B=200Bd S=200Hz	V8V2222	Sailmail Brunai Bay	BRU
8399,00	.			SS Ch 6	
8399,50	.			SS Ch 7	
8400,00	ALE USB	B=125Bd Ch=8 ChS=250Hz		EGY Mil net	EGY
8400,00	CHN N MIL 188-110 MOD	B=2400Bd		CHN N	CHN
8400,00	.			SS Ch 8	
8400,00	ALE USB	B=125Bd Ch=8 ChS=250Hz			
8400,50	.			SS Ch 9	
8401,00	.			SS Ch 10	
8401,50	.			SS Ch 11	
8402,00	.			SS Ch 12	
8402,00	USB			PHL CG net	PHL
8402,00	USB			PHL CG ship BRP MALAPASCUA	PHL
8402,50	.			SS Ch 13	
8403,00	USB			OMA N net	OMA
8403,00	.			SS Ch 14	

Frequency	Mode	Mode Parameter	Callsign	User	Country
8403,50	.			SS Ch 15	
8404,00	.			SS Ch 16	
8404,00	HC-265				
8404,00	USB			EGY N net	EGY
8404,50	.			SS Ch 17	
8405,00	.			SS Ch 18	
8405,20	STANAG 4529	B=1200Bd			
8405,50	.			SS Ch 19	
8406,00	.			SS Ch 20	
8406,50	.			SS Ch 21	
8407,00	.			SS Ch 22	
8407,50	.			SS Ch 23	
8408,00	.			SS Ch 24	
8408,00	CIS 75-200	B=75Bd S=200Hz		RUS N ship	
8408,00	USB		IALS	I N ship ALISEO F574	E
8408,20	*STANAG 4285*	*B=2400Bd*	*PBC*	*NLD N Goeree Island*	*HOL*
8408,20	STANAG 4285	B=2400Bd	FUO	F N Toulon	F
8408,30	STANAG 4529	B=1200Bd			
8408,50	.			SS Ch 25	
8409,00	.			SS Ch 26	
8409,50	.			SS Ch 27	
8410,00	.			SS Ch 28	
8410,00	CW			RUS N calling frequency	RUS
8410,50	.			SS Ch 29	
8411,00	.			SS Ch 30	
8411,50	.			SS Ch 31	
8412,00	.			SS Ch 32	
8412,50	.			SS Ch 33	
8412,50	FAX 120/576		XSQ	Guangzhou R	CHN
8413,00	.			SS Ch 34	
8413,50	.			SS Ch 35	
8414,00	.			SS Ch 36	
8414,50	DSC	B=100Bd S=170Hz	JNA	J CG HQ Tokyo	J
8414,50	USB		VFF	Iqaluit R	CAN
8414,50	.			DSC distress/safety frequency	WW
8415,00	.			SS DSC frequency	WW
8415,50	.			SS DSC frequency	WW
8416,00					
8416,50	SITOR B	B=100Bd S=170Hz	LGV	Vardoe R	NOR
8416,50	SITOR-B	B=100Bd S=170Hz	LGS	Svalbard R	NOR
8416,50	SITOR B	B=100Bd S=170Hz	UFL	Vladivostok R	RUS
8416,50	SITOR B	B=100Bd S=170Hz	VFF	CAN CG Iqaluit	CAN
8416,50	.			MSI frequency	WW
8416,50	SITOR-B	B=100Bd S=170Hz	LGI	Hammerfest R	NOR
8416,50	SITOR B	B=100Bd S=170Hz	UAT	Moscow R	RUS
8416,50	SITOR B	B=100Bd S=170Hz	SUZ	Serapeum R	EGY
8416,50	SITOR B	B=100Bd S=170Hz	NMO	USA CG Honolulu	HWA
8416,50	SITOR B	B=100Bd S=170Hz	NMC	USA CG Pt. Reyes , CA	USA
8416,50	SITOR B	B=100Bd S=170Hz	NMF	USA CG Boston , MA	USA
8416,50	USB		L2C	ARG N Buenos Aires	ARG
8417,00	SITOR A	B=100Bd S=170Hz	PKN2	Balikpapan R , Kalimantan	INS
8417,00	CW		RMP	RUS N HQ Kaliningrad	RUS
8417,00	.			CS Ch 2	
8417,00	SITOR , CW	B=100Bd S=170Hz	PKP	Dumai R , Sumatera	INS
8417,00	SITOR , CW	B=100Bd S=170Hz	PKK	Kupang R , Timor	INS
8417,00	CW		RMP	RUS N HQ Kaliningrad	RUS
8417,50	SITOR , CW	B=100Bd S=170Hz	SUK	Kosseir R	EGY
8417,50	.			CS Ch 3	
8417,50	SITOR , CW	B=100Bd S=170Hz	XSV	Tianjing R	CHN
8418,00	.			CS Ch 4	
8418,00	SITOR , CW	B=100Bd S=170Hz	PKD	Surabaya R , Djawa	INS
8418,50	.			CS Ch 5	

Frequency	Mode	Mode Parameter	Callsign	User	Country
8419,00	.			CS Ch 6	
8419,50	SITOR , CW	B=100Bd S=170Hz	JNA	J CG HQ Tokyo	J
8419,50	SITOR , CW	B=100Bd S=170Hz	PKR	Semarang R , Djawa	INS
8419,50	.			CS Ch 7	
8419,50	SITOR , CW	B=100Bd S=170Hz	4PB	Colombo R	CLN
8420,00	SITOR , CW	B=100Bd S=170Hz	PKR3	Cilacap R , Jawa	INS
8420,00	.			CS Ch 8	
8420,00	SITOR , CW	B=100Bd S=170Hz	PKJ28	Sei Kolak Kijang R	INS
8420,50	.			CS Ch 9	
8421,00	.			CS Ch 10	
8421,00	SITOR , CW	B=100Bd S=170Hz	XSQ	Guangzhou R	CHN
8421,00	CIS 8181	B=81Bd S=500Hz		RUS MIL	RUS
8421,00	SITOR , CW	B=100Bd S=170Hz	XSV	Tianjin R	CHN
8421,50	.			CS Ch 11	
8421,50	SITOR , CW	B=100Bd S=170Hz	JCS	Choshi R	J
8421,50	SITOR , CW	B=100Bd S=170Hz	SVO49	Olympia R	GRC
8422,00	PACTOR III	B=200Bd S=200Hz	OSY	Sailmail Brugge	BEL
8422,00	SITOR , CW	B=100Bd S=170Hz	EQI	Abbas R	IRN
8422,00	SITOR , CW	B=100Bd S=170Hz	PKE37	Sanana R , Kep Sula	INS
8422,00	SITOR , CW	B=100Bd S=170Hz	PKY23	Fak Fak R	INS
8422,00	SITOR , CW	B=100Bd S=170Hz	PKY5	Merauke R , Irian Jaya	INS
8422,00	SITOR , CW	B=100Bd S=170Hz	PKF3	Kendari R	INS
8422,00	.			CS Ch 12	
8422,00	SITOR , CW	B=100Bd S=170Hz	PKP38	Batu Ampar R	INS
8422,50	.			CS Ch 13	
8423,00	SITOR , CW	B=100Bd S=170Hz	UFL	Vladivostok R	RUS
8423,00	SITOR , CW	B=100Bd S=170Hz	PKM	Bitung R , Sulawesi	INS
8423,00	.			CS Ch 14	
8423,00	SITOR , CW	B=100Bd S=170Hz	SVO491	Olympia R	GRC
8423,50	.			CS Ch 15	
8424,00	SITOR , CW	B=100Bd S=170Hz	UDK2	Murmansk R	RUS
8424,00	SITOR , CW	B=100Bd S=170Hz	SVO4	Olympia R	GRC
8424,50	.			CS Ch 16	
8424,50	.			CS Ch 17	
8425,00	SITOR , CW	B=100Bd S=170Hz	PKY3	Manokwari R , Irian Jaya	INS
8425,00	.			CS Ch 18	
8425,00	SITOR , CW	B=100Bd S=170Hz	PKM25	Tahuna R	INS
8425,00	SITOR , CW	B=100Bd S=170Hz	PKC4	Panjang R , Sumatera	INS
8425,50	.			CS Ch 19	
8425,50	SITOR , CW	B=100Bd S=170Hz	XSG	Shanghai R	CHN
8426,00	SITOR , CW	B=100Bd S=170Hz	PKY2	Biak R , Irian Jaya	INS
8426,00	.			CS Ch 20	
8426,00	SITOR , CW	B=100Bd S=170Hz	PKB3	Sibolga R , Sumatera	INS
8426,00	SITOR A , CW	B=100Bd S=170Hz	UIW	Kaliningrad R	RUS
8426,50	.			CS Ch 21	
8427,00	SITOR , CW	B=100Bd S=170Hz	PKN	Balikpapan R , Kalimantan	INS
8427,00	.			CS Ch 22	
8427,00	BAUDOT	B=45,45Bd S=170Hz	KPH	San Francisco R , CA (Bolinas/Port Reyes)	USA
8427,00	SITOR A	B=100Bd S=170Hz	KPH	San Francisco R , CA (Bolinas/Port Reyes)	USA
8427,50	SITOR , CW	B=100Bd S=170Hz	A9M	Bahrain R	BHR
8427,50	.			CS Ch 23	
8428,00	USB		XSX	Keelung R	TWN
8428,00				CS Ch 24	
8428,50	SITOR , CW	B=100Bd S=170Hz	UFM3	Nevelsk R	RUS
8428,50	.			CS Ch 25	
8429,00	SITOR , CW	B=100Bd S=170Hz	XSQ	Guangzhou R	CHN
8429,00	DSC	B=100Bd S=170Hz	PKZ2	Cirebon R , Jawa	INS
8429,00	SITOR , CW	B=100Bd S=170Hz	PKS	Pontianak R , Kalimantan	INS
8429,00	.			CS Ch 26	
8429,00	SITOR , CW	B=100Bd S=170Hz	PKM44	Pantolan R , Sulawesi	INS
8429,50	.			CS Ch 27	

Frequency	Mode	Mode Parameter	Callsign	User	Country
8430,00	.			CS Ch 28	
8430,00	SITOR , CW	B=100Bd S=170Hz	XSG	Shanghai R	CHN
8430,50	.			CS Ch 29	
8431,00	DSC	B=100Bd S=170Hz	PKZ34	Cigading R , Jawa	INS
8431,00	SITOR , CW	B=100Bd S=170Hz	PKD3	Lembar R , Lombok	INS
8431,00	.			CS Ch 30	
8431,00	SITOR , CW	B=100Bd S=170Hz	4PB	Colombo R	CLN
8431,00	SITOR , CW	B=100Bd S=170Hz	TAH	Istanbul R	TUR
8431,00	SITOR , CW	B=100Bd S=170Hz	XSQ	Guangzhou R	CHN
8431,50	.			CS Ch 31	
8431,50	SITOR , CW	B=100Bd S=170Hz	UAT	Moscow R	RUS
8432,00	.			CS Ch 32	
8432,00	CW		4XZ	ISR N Haifa	ISR
8432,50	.			CS Ch 33	
8433,00	CW		XSG	Shanghai R	CHN
8433,00	SITOR , CW	B=100Bd S=170Hz	PKD5	Benoa R , Bali	INS
8433,00	.			CS Ch 34	
8433,00	SITOR , CW	B=100Bd S=170Hz	WOO	Ocean Gate R , NJ	USA
8433,50	.			CS Ch 35	
8434,00	.			CS Ch 36	
8434,00	SITOR , CW	B=100Bd S=170Hz	TAH	Istanbul R	TUR
8434,50	.			CS Ch 37	
8435,00	.			CS Ch 38	
8435,00	SITOR , CW	B=100Bd S=170Hz	4PB	Colombo R	CLN
8435,00	SITOR , CW	B=100Bd S=170Hz	XSQ	Guangzhou R	CHN
8435,50	.			CS Ch 39	
8436,00	T600	B=50Bd S=250Hz		RUS N	RUS
8436,00	.			CS Ch 40	
8436,00	SITOR , CW	B=100Bd S=170Hz	XSG	Shanghai R	CHN
8436,50	DSC	B=100Bd S=170Hz	JNA	J CG HQ Tokyo	J
8436,50	DSC	B=100Bd S=170Hz	PKX	Jakarta R , Jawa	INS
8436,50	DSC	B=100Bd S=170Hz	PKR	Semarang R , Djawa	INS
8436,50	DSC	B=100Bd S=170Hz	PKN	Balikpapan R , Kalimantan	INS
8436,50	DSC	B=100Bd S=170Hz	PNK	Jayapura R , Irian Jaya	INS
8436,50	DSC	B=100Bd S=170Hz	PKY4	Sorong R , Irian Jaya	INS
8436,50	DSC	B=100Bd S=170Hz	PKE	Amboina R , Ceram	INS
8436,50	DSC	B=100Bd S=170Hz	PKM	Bitung R , Sulawesi	INS
8436,50	DSC	B=100Bd S=170Hz	PKF	Makassar R , Sulawesi	INS
8436,50	DSC	B=100Bd S=170Hz	PKB	Belawan R , Sumatera	INS
8436,50	DSC	B=100Bd S=170Hz	PKP	Dumai R , Sumatera	INS
8436,50	DSC	B=100Bd S=170Hz	PKK	Kupang R , Timor	INS
8437,00	SITOR , CW	B=100Bd S=170Hz	UFL	Vladivostok R	RUS
8437,00	CW		PKC	Palembang R , Sumatera	INS
8438,60	PACTOR III	B=100Bd S=200Hz	WNU	Austin , TX	USA
8439,00	*BAUDOT*	*B=75Bd S=850Hz*	*PBC*	*HOL N Goeree Island*	*HOL*
8439,00	*BAUDOT*	*B=75Bd S=850Hz*	*PBB*	*NLD N Den Helder*	*HOL*
8441,00	CW		C6N	Nassau R	BAH
8441,00	CW		XYR7	Rangoon R	BRM
8442,00	MFSK17	B=125Bd Ch=17 ChS=125Hz			
8442,00	PACTOR II	B=200Bd S=200Hz	VZX	Sailmail Darawank	AUS
8442,50	CW		TCR	Tophane Kulesi R	TUR
8443,20	FAX 120/576	S=1000Hz	RBW41	Murmansk M	RUS
8445,00	CW		A4M	Muscat R	OMA
8445,00	CW		PKK	Kupang R , Timor	INS
8445,00	CW		PKN6	Samarinda R , Kalimantan	INS
8445,00	CW		PKR3	Cilacap R , Jawa	INS
8445,90	FAX 60/120/288/576		RBW	Murmansk M	RUS
8449,00	CW		TJC6	Douala R	CME
8450,00	USB		6YX	JMC CG Jamaica	JMC
8451,00	STANAG 4285	B=2400Bd	FUO	F N Toulon	F

Frequency	Mode	Mode Parameter	Callsign	User	Country
8453,00	*STANAG 4285*	*B=2400Bd*	*FUO*	*F N Toulon*	*F*
8453,00	*STANAG 4285*	*B=2400Bd*	*FUG8*	*F N Saissac*	*F*
8453,00	STANAG 4285	B=2400Bd	RFFME*	F N Saissac	F
8454,00	SITOR-B	B=100Bd S=170Hz	UIW23	Kaliningrad R	RUS
8457,00	CW		PKG	Bandjarmasin R , Kalimantan	INS
8457,00	CW		PKP	Dumai R , Sumatera	INS
8457,00	CW		PKY5	Merauke R , Irian Jaya	INS
8457,00	CW			RUS air defence net	RUS
8458,00	BAUDOT	B=75Bd S=850Hz	OSN	BEL N Oostende	BEL
8458,20	*STANAG 4285*	*B=2400Bd*	*OSN*	*BEL N Brügge*	*BEL*
8458,50	8PSK	B=2400Bd			
8459,00	FAX 120/576		NOJ	USA CG Kodiak , ALS	USA
8461,00	CW		PKD	Surabaya R , Djawa	INS
8461,00	CW		PKR	Semarang R , Djawa	INS
8461,00	CW		PKY4	Sorong R , Irian Jaya	INS
8462,00	*BAUDOT*	*B=50Bd S=850Hz*	*9MR*	*MLA N Johor Baharu*	*MLA*
8463,20	*STANAG 4285*	*B=2400Bd*	*EBA*	*E N Madrid*	*E*
8463,20	*STANAG 4285*	*B=2400Bd*	*FUG*	*F N Saissac*	*F*
8467,00	GTOR	B=200Bd S=200Hz	ERMBEL	B N Belem Brasil	B
8467,00	GTOR	B=200Bd S=200Hz	NPABOC	B N Patrol Ship P-62 BOCAINA	B
8467,50	FAX 120/576		JJC	Tokyo M	J
8473,00	CW		PKE	Amboina R , Ceram	INS
8473,00	CW		PKS	Pontanak R , Kalimantan	INS
8473,50	STANAG 4285	B=2400Bd	FUX	F N Le Port	REU
8475,50	STANAG 4285	B=2400Bd	FUX	F N Le Port	REU
8476,20	*STANAG 4285*	*B=2400Bd*	*CTA*	*POR N Lisbon*	*POR*
8478,50	STANAG 4285	B=2400Bd	FUF	F N Fort de France	MRT
8478,50	ALE USB	B=125Bd Ch=8 ChS=250Hz		G MIL net	G
8481,00	ALE USB	B=125Bd Ch=8 ChS=250Hz		CHN MIL net	CHN
8484,00	CW		HLG	Seoul R	KOR
8484,00	CHN OFDM 30TONE	B=1800Bd Ch=30 ChS=75Hz		CHN MIL	CHN
8484,40	STANAG 4285	B=2400Bd	CTA	POR N TX Ponta Delgada	AZR
8488,00	STANAG 4285	B=2400Bd	CTA	POR N TX Ponta Delgada	AZR
8488,00	STANAG 4285	B=2400Bd Ch=2	CTA	POR N Lisbon	POR
8488,20	STANAG 4285	B=2400Bd	CKN	CAN MIL Aldergrove	CAN
8490,00	CW		AQP5	PAK N Karachi	PAK
8492,00	*MAHRS*	*B=2400Bd*	*DHJ58*	*D N Glücksburg TX Staberhuk*	*D*
8492,00	*STANAG 4285*	*B=2400Bd*	*EBA*	*E N Madrid*	*E*
8492,20	*STANAG 4285*	*B=2400Bd*	*FUG*	*F N Saissac*	*F*
8494,60	CW		F**	RUS N Vladivostok	RUS
8494,70	CW		D**	RUS N Sevastopol	UKR
8494,80	CW		P**	RUS N Kaliningrad	RUS
8494,80	CW		C**	RUS N Moscow	RUS
8495,00	CW		S**	RUS N Severomorsk	RUS
8495,00	CW		Q**	RUS N	RUS
8495,10	CW		A**	RUS N Astrakhan	
8495,30	CW		K**	RUS N Petropavlovsk	RUS
8495,40	CW		M**	RUS N Magadan	RUS
8497,80	CW		L**	RUS N St. Petersburg	RUS
8498,00	CIS 1200	B=1200Bd		RUS	RUS
8500,00	ALE USB	B=125Bd Ch=8 ChS=250Hz		VEN N net	
8500,00	MIL 188-110B SER	B=2400Bd		VEN N	
8500,00	USB			EGY N net	EGY
8502,00	CW		XSG	Shanghai R	CHN
8502,00	USB		NMG	USA CG New Orleans , LA	USA
8503,90	FAX 120/576		NMG	USA CG New Orleans , LA	USA
8505,00	*MAHRS*	*B=2400Bd*	*DHJ58*	*D N Glücksburg TX Staberhuk*	*D*
8505,70	*STANAG 4285*	*B=2400Bd*	*PBC*	*NLD N Den Helder*	*HOL*

Frequency	Mode	Mode Parameter	Callsign	User	Country
8508,00	T600	B=50Bd S=250Hz	RDL	RUS N Moscow	RUS
8508,00	CW		JFM	Muroto R	J
8512,00	CW			RUS Mil net	RUS
8512,20	STANAG 4285	B=2400Bd	PBC	NLD N Goeree Island	HOL
8517,00	STANAG 4285	B=2400Bd	FUG	F N Saissac	F
8518,80	F N FSK	B=50Bd S=850Hz Ch=1	FUG	F N Saissac	F
8520,00	USB			BGD N net	BGD
8520,00	CODAN CHIRP	B=80Bd Ch=30 ChS=250Hz	4102		
8520,00	CODAN CHIRP	B=80Bd Ch=30 ChS=250Hz	2031	BRM Mil net	E
8520,00	CODAN	B=2400Bd Ch=16 ChS=112,5Hz			
8523,40	CW		JOR	Nagasaki R	J
8524,20	STANAG 4285	B=2400Bd	PBC	NLD N Goeree Island	HOL
8528,00	CHN OFDM 30TONE	B=1500Bd Ch=30 ChS=60Hz		CHN	CHN
8530,00	CW		UJE	Nizhnij Novgorod	RUS
8532,00	ALE USB	B=125Bd Ch=8 ChS=250Hz		G MIL net	G
8535,00	T600	B=50Bd S=100Hz	RIW	RUS N HQ Moscow	RUS
8536,00	CW		RCB	Mineralnye Vody Air	RUS
8536,00	CW		RJC38	RUS F	RUS
8536,00	CW		RJC48	RUS F	RUS
8536,00	CW		RJF94	RUS F	RUS
8536,00	CW		RCH84		
8538,00	STANAG 4481	B=50Bd S=850Hz	NPN	USA N Guam	GUM
8538,20	STANAG 4285	B=2400Bd	PJK	HOL N Willemstad	ATN
8538,50	STANAG 4285	B=2400Bd		AUS MHFCS Humpty Doo	AUS
8540,20	STANAG 4285	B=2400Bd	CFH	CAN N Halifax	CAN
8542,00	CW		PKX	Jakarta R , Jawa	INS
8542,20	STANAG 4285	B=2400Bd	FUO	F N Toulon	F
8548,00	STANAG 4285	B=2400Bd		G MIL Ascension	SHN
8550,00	SITOR , CW	B=100Bd S=170Hz	UFH	Petrapavlovsk-Kamchatskii R	RUS
8550,00	BAUDOT	B=75Bd S=850Hz	PBB	NLD N Den Helder	HOL
8551,00	J OFDM 30+2	B=1500Bd Ch=32 ChS=25Hz		J N broadcast	J
8551,00	J OFDM 30+2	B=1500Bd Ch=32 ChS=25Hz		J MIL	J
8552,60	STANAG 4285	B=2400Bd	FUF	F N Fort de France	MRT
8553,20	STANAG 4285	B=2400Bd	EBA	E N Madrid	E
8553,40	STANAG 4285	B=2400Bd		AUS MHFCS Humpty Doo	AUS
8557,00	STANAG 4285	B=2400Bd		AUS MHFCS Humpty Doo	AUS
8561,00	MFSK17	B=125Bd Ch=17 ChS=125Hz			
8562,00	CW		UYT2	Kiev R	UKR
8563,00	STANAG 4285	B=2400Bd	FUG	F N Saissac	FF
8564,20	STANAG 4285	B=2400Bd	CFH	CAN N Halifax	CAN
8566,00	CW		ZSJ4	AFS N Capetown	AFS
8566,80	F N FSK	B=50Bd S=850Hz Ch=1	FUG	F N Saissac	F
8568,00	STANAG 4285	B=2400Bd	FUV	F N Djibouti	DJI
8570,00	PACTOR I	B=100Bd S=200Hz		B N net	B
8571,00	CW		UWS	Kiev R	UKR
8573,60	STANAG 4285	B=2400Bd	FUX	F N Le Port	REU
8580,00	MHF-50 MODEM	B=54,3Bd Ch=32 ChS=64,5Hz	ZSJ	AFS N Capetown	AFS
8581,70	PACTOR I	B=100Bd S=170Hz	VIGNA13**		
8581,70	PACTOR I	B=100Bd S=170Hz	BIDARA**		
8581,70	PACTOR I	B=100Bd S=170Hz	VIGNA5**		
8582,20	SITOR B	B=100Bd S=170Hz	PWZ33	B N Rio de Janeiro	B
8585,00	CW		RDL	RUS N Moscow	RUS

Frequency	Mode	Mode Parameter	Callsign	User	Country
8585,00	SITOR , CW	B=100Bd S=170Hz	UFL	Vladivostok R	RUS
8585,00	T600	B=50Bd S=1000Hz	RDL	RUS N Moscow	RUS
8586,80	STANAG 4538	B=2400Bd			
8587,00	IRN BPSK	B=207Bd		IRN	
8587,00	CIS OFDM 60	B=2400Bd Ch=60 ChS=44,5Hz		RUS diplo	RUS
8588,00	J PSK	B=1500Bd		J MIL	J
8590,00	SITOR , CW	B=100Bd S=170Hz	UFL	Vladivostok R	RUS
8590,00	USB			BGD N net	BGD
8596,00	CCIR 493-4	B=100Bd S=170Hz			
8600,00	CW		ELH	Harbel R	LBR
8600,00	CW		XSV	Tianjin R	CHN
8600,00	USB			EGY N net	EGY
8603,00	MHF-50 MODEM	B=54,3Bd Ch=32 ChS=64,5Hz	ZSJ	AFS N Capetown	AFS
8610,00	STANAG 4285	B=2400Bd	TBA	TUR N Ankara	TUR
8612,00	*CIS 12*	*B=1440Bd Ch=12 ChS=200Hz*	*RMP*	*RUS N HQ Kaliningrad*	*RUS*
8612,20	*STANAG 4285*	*B=2400Bd*		*G MIL Crimond*	*G*
8613,00	8PSK	B=2400Bd			
8617,00	SSB			Bangalore ACC	IND
8617,00	SSB			Mumbai ACC	IND
8617,00	SSB			Car ACC	ADM
8617,00	SSB			Hyderabad ACC	IND
8617,00	SSB			Madras ACC	IND
8617,00	SSB			Port Blair ACC	ADM
8618,20	*STANAG 4285*	*B=2400Bd*	*JXU*	*NOR N Bodo*	*NOR*
8622,50	ALE USB	B=125Bd Ch=8 ChS=250Hz		USA FEMA net	USA
8622,50	CW		RMP	RUS N HQ Kaliningrad	RUS
8623,00	*CIS 12*	*B=1440Bd Ch=12 ChS=200Hz*		*RUS MIL Moscow*	*RUS*
8623,00	*CIS 12 ARQ*	*B=1440Bd Ch=12 ChS=200Hz*		*RUS MIL Moscow*	*RUS*
8623,50	*LINK 11 CLEW*	*B=2250Bd Ch=16 ChS=330/110/550Hz*		*S N Karlskrona*	*S*
8625,00	STANAG 4285	B=2400Bd	FUM	F N Papeete , Tahiti	OCE
8626,00	CW		PKI2	Jakarta R	INS
8626,00	CW		PKC5	Pangkal Balam R	INS
8626,00	CW		PKC2	Plaju R , Sumatera	INS
8627,00	CW		RJQ86	RUS N ship	RUS
8627,00	CW		RMP	RUS N HQ Kaliningrad	RUS
8629,00	CW		XSD	Yingkou R	CHN
8630,00	CW		URA2	Nikolayev R	UKR
8632,00	CW		RJF	RUS MIL	RUS
8632,50	*STANAG 4285*	*B=2400Bd*	*FUO*	*F N Toulon*	*F*
8634,00	CW		VTG6	IND N Mumbai	IND
8636,00	CW		HLW	Seoul R	KOR
8638,00	AUS MIL FSK MODEM	B=300Bd S=345Hz		AUS MIL	AUS
8638,50	PACTOR II	B=100Bd S=200Hz			
8641,00	ALE USB	B=125Bd Ch=8 ChS=250Hz			
8642,00	CW		CTV	Monsanto R	POR
8642,00	CW		KPH	San Francisco R , CA (Bolinas/Port Reyes)	USA
8646,00	CW		VTP6	IND N Vishakhapatnam	IND
8646,00	*T600*	*B=50Bd S=250Hz*		*RUS N Moscow*	*RUS*
8646,00	*ALE USB*	*B=125Bd Ch=8 ChS=250Hz*		*TUR Disaster and Emergency net*	*TUR*
8646,00	STANAG 4285	B=2400Bd	FUJ	F N Noumea	NCL
8651,00	CIS OFDM 60	B=2400Bd Ch=60 ChS=44,5Hz		Khabarvorsk	RUS

Frequency	Mode	Mode Parameter	Callsign	User	Country
8652,50	CW		PZN3	Paramaribo R	SUR
8653,60	CW		JCS	Choshi R	J
8654,00	STANAG 4481	B=50Bd S=850Hz	NPN	USA N Guam	GUM
8656,00	T600	B=50Bd S=200Hz		RUS N Smolensk	RUS
8656,00	*ALE USB*	*B=125Bd Ch=8 ChS=250Hz*		*TUR Disaster and Emergency net*	
8657,00	CW		UAI	RUS	
8658,00	FAX 120/576		JFX	Kagoshima R	J
8658,00	FAX 120/576		JFW	Fukushima R	J
8658,00	8PSK	B=2800Bd			
8658,00	FAX 120/576		JFC	Misaki R	J
8665,00	CW		XSG	Shanghai R	CHN
8672,20	FAX 120/576		GYA	G N Northwood	G
8672,20	*STANAG 4285*	*B=2400Bd*		*G F Inskip*	G
8673,00	*T600*	*B=50Bd S=500Hz*	*RCV*	*RUS N Sevastopol*	*RUS*
8676,00	CLOVER 2000	B=500Bd Ch=8 ChS=250Hz			
8676,00	CIS 500 FSK BURST	B=500Bd S=1000Hz		RUS N	RUS
8676,00	STANAG 4481	B=50Bd S=850Hz	NPM	USA N Lualualei	HWA
8676,50	MIL 188-110A SER	B=2400Bd		AUS MHFCS Humpty Doo	AUS
8676,50	MIL 188.110A SER	B=2400Bd		AUS MHFCS North West Cape	AUS
8677,00	FAX 120/576		CBV	Valpariso R	CHL
8677,20	*STANAG 4529*	*B=1200Bd*	*EBA*	*E N Madrid*	*E*
8680,00	CW		XUK2	Kompong Som Ville R	CBG
8680,00	PSK 63	B=63Bd Ch=2 ChS=400Hz		POL Intel	POL
8682,00	FAX 120/576		NMC	USA CG Pt. Reyes , CA	USA
8684,00	CW		RCV	RUS N HQ Sevastopol	UKR
8684,00	CW		LZL4	Bourgas R	BUL
8684,50	PACTOR III	B=100Bd S=200Hz	VZX	Sailmail Darawank R	AUS
8685,00	*MAHRS*	*B=2400Bd*	*DHJ58*	*D N Glücksburg TX Staberhuk*	*D*
8686,00	CW		PKM44	Pantolan R , Sulawesi	INS
8686,00	CW		JCT	Choshi R	J
8686,00	CW		PKA	Sabang R , We	INS
8686,00	CW		PKB	Belawan R , Sumatera	INS
8686,00	CW		PKF	Makassar R , Sulawesi	INS
8686,00	CW		PKM9	Donggala R , Sulawesi	INS
8689,00	USB			EGY N net	EGY
8692,00	CIS 500 FSK BURST	B=500Bd S=1000Hz		RUS N	RUS
8692,00	*MAHRS*	*B=2400Bd*	*DHJ58*	*D N Glücksburg TX Staberhuk*	*D*
8694,00	CW		PKP2	Telukbayur R , Sumatera	INS
8694,00	CW		PKM	Bitung R , Sulawesi	INS
8694,00	CW		PNK	Jayapura R , Irian Jaya	INS
8694,00	SITOR , CW	B=100Bd S=170Hz	XSZ	Dalian R	CHN
8694,00	STANAG 4481	B=50Bd S=850Hz	NPG	USA N Dixon , CA	USA
8696,00	FAX 120/576		CBM	Magallanes Zonal R	CHL
8697,00	*STANAG 4285*	*B=2400Bd*	*CTA*	*POR N Lisbon*	*POR*
8698,20	*STANAG 4285*	*B=2400Bd*	*CTA*	*POR N Lisbon*	*POR*
8698,20	*STANAG 4285*	*B=2400Bd*	*JWT*	*NOR N Stavanger*	*NOR*
8701,00	*STANAG 4285 LSB*	*B=2400Bd*	*CTA08*	*POR N Lisbon*	*F*
8702,00	CW		ODR3	Beyrouth R	LBN
8702,00	SSB		3XO28	Kamsar R	GUF
8703,50	J PSK	B=1500Bd		J MIL	J
8705,40	CW		PKY2	Biak R , Irian Jaya	INS
8706,00	CW		3CA7	Banapa R	GUI
8706,40	USB		USI	Kherson R	UKR
8707,00	USB		JNA	J CG HQ Tokyo	J
8707,00	USB		CBV	Valpariso R	CHL
8707,00	SSB			CS Ch 834	

Frequency	Mode	Mode Parameter	Callsign	User	Country
8707,00	USB		UUT	Odessa R	UKR
8707,00	MIL 188-110A SER	B=2400Bd		S MIL	S
8707,00	STANAG 4285	B=2400Bd		F MIL	F
8707,80	CCIR 493-4	B=100Bd S=170Hz		CHN MIL Spratly Island	CHN
8710,00	USB		SXE	Aspropyrgos Attikis R	GRC
8710,00	USB			Peiraias CG R	GRC
8710,00	USB		HSA	Bangkok R	THA
8710,00	USB		JNA	J CG HQ Tokyo	J
8710,00	USB		L3A	ARG N Comodoro Rivadavia	ARG
8710,00	SSB			CS Ch 835	
8713,00	USB		L2A	ARG N Mar del Plata	ARG
8713,00	USB		CBV	Valpariso R	CHL
8713,00	SSB			CS Ch 836	
8714,00	CW			RUS Mil net	RUS
8715,20	STANAG 4285	B=2400Bd		I N	
8715,50	MCPSK	Ch=7 ChS=300Hz			
8716,00	USB		SVO41	Olympic R	GRC
8716,00	USB		XSZ	Dalian R	CHN
8716,00	SSB			CS Ch 837	
8716,00	PACTOR III	B=100Bd S=200Hz	DZO38	Manila R	PHL
8718,00	CW		TJC7	Douala R	CME
8719,00	USB		A9M	Bahrain R	BHR
8719,00	USB		V5W	Walvis Bay R	NMB
8719,00	USB		SDJ	Stockholm R	S
8719,00	USB		9MG	Penang R	MLA
8719,00	USB		CBV	Valpariso R	CHL
8719,00	SSB			Sevastopol R	
8719,00	SSB			CS Ch 801	
8722,00	USB			Lisbon R	POR
8722,00	USB		WDI	Mobile R	USA
8722,00	CW		TRA8	Libreville R	GAB
8722,00	SSB			CS Ch 802	
8722,00	SSB		4PB	Colombo R	CLN
8722,00	SSB		D4A	S. Vincente de Cabo R	CPV
8722,00	SSB		SVO42	Olympia R	GRC
8722,00	ALE USB	B=125Bd Ch=8 ChS=250Hz		MLT N net	MLT
8724,00	STANAG 4481	B=50Bd S=850Hz		MFA New Delhi	IND
8725,00	USB			Cape Town R	AFS
8725,00	USB		SDJ	Stockholm R	S
8725,00	USB		HLS	Seoul R	KOR
8725,00	SSB			CS Ch 803	
8725,00	USB		P2M	Port Moresbay R	PNG
8728,00	USB			Bilbao R via Madrid	E
8728,00	USB		LSD836	Argentina R	ARG
8728,00	USB		KLB	Seattle R , WA	USA
8728,00	SSB			CS Ch 804	
8728,00	USB		3AC	Monaco R	MCO
8731,00	USB		EQJ	Chahbahar R	IRN
8731,00	USB			Cape Town R	AFS
8731,00	USB			San Lorenzo R	EQA
8731,00	USB		HCY	Ayora R , Santa Cruz	EQA
8731,00	USB			Cristobal R	EQA
8731,00	USB		UFL	Vladivostok R	RUS
8731,00	USB			Floreana R , Santa Maria	EQA
8731,00	USB			Seymour R , Baltra	EQA
8731,00	USB			Villamil R , Isabela	EQA
8731,00	USB		KLB	Seattle R , WA	USA
8731,00	SSB			CS Ch 805	
8731,00	USB		UAT	Moscow R	RUS
8731,00	USB		HCG	Guayaquil R	EQA
8731,00	USB		P2M	Port Moresbay R	PNG

Frequency	Mode	Mode Parameter	Callsign	User	Country
8731,30	SSB		ETC	Assab R	ETH
8734,00	USB		A9M	Bahrain R	BHR
8734,00	CW		USI	Kherson R	UKR
8734,00	USB		HSA	Bangkok R	THA
8734,00	USB		LSD836	Argentina R	ARG
8734,00	SSB			CS Ch 806	
8734,00	SSB		SVO43	Olympia R	GRC
8735,70	STANAG 4285	B=2400Bd	VWGZ	IND N Vishakapatnam	IND
8737,00	USB			Saint Helena R	SHN
8737,00	USB		CBV	Valpariso R	CHL
8737,00	SSB			CS Ch 807	
8737,00	USB		5BA	Cyprus R	CYP
8737,00	CW		RMP	RUS N HQ Kaliningrad	RUS
8740,00	USB			OMA N net	OMA
8740,00	USB		YQI	Constanta R	ROU
8740,00	USB			Cape Town R	AFS
8740,00	USB		YKM7	Lattakia R	SYR
8740,00	USB		HCY	Ayora R , Santa Cruz	EQA
8740,00	USB		LSD836	Argentina R	ARG
8740,00	SSB			CS Ch 808	
8740,00	SSB		4PB	Colombo R	CLN
8740,00	SSB		SVO44	Olympia R	GRC
8740,00	USB		ZSC	Capetown R	AFS
8743,00	USB		CBV	Valpariso R	CHL
8743,00	USB		HSW	Bangkok M	THA
8743,00	SSB			CS Ch 809	
8743,00	SSB		SVO45	Olympia R	GRC
8746,00	USB		A9M	Bahrain R	BHR
8746,00	USB		LZW	Varna R	BUL
8746,00	USB		EQO	Now Shar R	IRN
8746,00	USB		9WH	Kota Kinabalu R	MLA
8746,00	USB		PKM44	Pantolan R , Sulawesi	INS
8746,00	SSB			CS Ch 810	
8746,00	USB		3DP	Suva R	FJI
8746,00	SSB		EQM	Bushehr R	IRN
8746,00	USB		PKB	Belawan R , Sumatera	INS
8746,00	USB		PKF3	Kendari R	INS
8746,00	SSB		TAH	Istanbul R	TUR
8746,80	USB		PKN	Balikpapan R , Kalimantan	INS
8747,00	CW		4XZ	ISR N Haifa	ISR
8749,00	USB		LSD836	Argentina R	ARG
8749,00	USB		OYR	Aasiaat R	GRL
8749,00	SSB			CS Ch 811	
8749,00	SSB		TAH	Istanbul R	TUR
8750,00	CLOVER 2000	B=500Bd Ch=8 ChS=250Hz			EGY
8752,00	USB		VFF	Iqaluit R	CAN
8752,00	SSB			CS Ch 812	
8753,00	USB		PKX	Jakarta R , Jawa	INS
8755,00	USB		9WH	Kota Kinabalu R	MLA
8755,00	USB		XVT	Da Nang R	VTN
8755,00	USB		CBV	Valpariso R	CHL
8755,00	USB		XSL	Fuzhou R	CHN
8755,00	USB		XSV	Tianjin R	CHN
8755,00	SSB			CS Ch 813	
8755,00	SSB		D4A	S. Vincente de Cabo R	CPV
8758,00	SSB			Liepaja R	LVA
8758,00	USB		LSD836	Argentina R	ARG
8758,00	USB		HSA	Bangkok R	THA
8758,00	SSB		SVO46	Olympic R	GRC
8758,00	USB			Benete R , Sumbawa	INS
8758,00	SSB			CS Ch 814	

Frequency	Mode	Mode Parameter	Callsign	User	Country
8759,00	IRN N QPSK ADAPTIVE	B=468Bd		IRN N	
8759,20	USB		PKI2	Jakarta R	INS
8759,20	USB		PKC2	Plaju R , Sumatera	INS
8761,00	USB		ESA	Tallin R	EST
8761,00	USB		OSU41	Oostende R	BEL
8761,00	USB		CBV	Valpariso R	CHL
8761,00	USB		XVN	Nha Trang R	VTN
8761,00	SSB			CS Ch 815	
8763,00	*LINK 11 ISB*	*B=2250Bd Ch=16 ChS=330/110/550Hz*		*UK MIL E London*	*G*
8764,00	USB		UIW	Kaliningrad R	RUS
8764,00	USB		XSU	Yantai R	CHN
8764,00	USB		NMN	USA CG Portsmouth , VA	USA
8764,00	USB		NMA	USA CG Miami , FL	USA
8764,00	USB		NMF	USA CG Boston , MA	USA
8764,00	USB		NMC	USA CG Pt. Reynes , CA	USA
8764,00	USB		PKP38	Batu Ampar R	INS
8764,00	USB		XSA2	Nanjing R	CHN
8764,00	USB		XSR	Haikou R	CHN
8764,00	SSB			CS Ch 816	
8764,00	USB		PKE5	Ternate R , Maluku	INS
8764,00	USB		PKG	Bandjarmasin R , Kalimantan	INS
8764,00	USB		PKP	Dumai R , Sumatera	INS
8764,00	USB			USA CG Honolulu	HWA
8764,00	USB		NMO	USA CG Honolulu	HWA
8765,00	MIL 188-110B SER	B=2400Bd	MORTON25**	POL MIL	POL
8765,00	MIL 188-110B SER	B=2400Bd	IGIELIT37**	POL MIL	POL
8766,00	FAX 120/576		NMN	USA CG Chesapeake , VA	USA
8767,00	USB		EQO	Now Shar R	IRN
8767,00	USB		STP	Port Sudan R	SDN
8767,00	USB		SUH	Alexandria R	EGY
8767,00	USB		ZBR	Bermuda Harbour R	BER
8767,00	USB		CBV	Valpariso R	CHL
8767,00	SSB			CS Ch 817	
8767,20	USB		9WW	Kuching R	MLA
8770,00	USB		LZW	Varna R	BUL
8770,00	USB		STP	Port Sudan R	SDN
8770,00	USB		9MG	Penang R	MLA
8770,00	SSB			CS Ch 818	
8770,00	SSB		S7Q	Seychelles R	SEY
8770,00	USB		XSG	Shanghai R	CHN
8770,00	USB		YJM	Port Vila R	VUT
8770,00	SSB		TJC	Douala R	CME
8773,00	ALE USB	B=125Bd Ch=8 ChS=250Hz		G MIL net	G
8773,00	USB		UDK2	Murmansk R	RUS
8773,00	USB		CBT25	Lebu R	CHL
8773,00	SSB			CS Ch 819	
8773,00	USB		PPR	Rio R	B
8773,00	SSB		SVN47	Olympia R	GRC
8773,00	PACTOR III	B=100Bd S=200Hz	PYB45	Sao Paulo	B
8776,00	USB		EQC	Amirabad R	IRN
8776,00	USB		LSD836	Argentina R	ARG
8776,00	SSB		EQN	Khomeini R	IRN
8776,00	SSB			CS Ch 820	
8776,00	*USB*		*SVO*	*Olympia R*	*GRC*
8779,00	USB		OBC3	Callao MRCC	PRU
8779,00	USB		PKD5	Benoa R , Bali	INS

Frequency	Mode	Mode Parameter	Callsign	User	Country
8779,00	USB			Cape Town R	AFS
8779,00	USB		OBF4	Mollendo MRSC	PRU
8779,00	USB		9WH	Kota Kinabalu R	MLA
8779,00	USB		PKE5	Ternate R , Maluku	INS
8779,00	USB		OBY2	Paita MRSC	PRU
8779,00	USB		9MG	Penang R	MLA
8779,00	USB		LSD836	Argentina R	ARG
8779,00	USB		UFH	Petrapavlovsk-Kamchatskii R	RUS
8779,00	USB		CBV	Valpariso R	CHL
8779,00	USB		PKR	Semarang R , Djawa	INS
8779,00	USB		PKO	Tarakan R , Kalimantan	INS
8779,00	USB		PNK	Jayapura R , Irian Jaya	INS
8779,00	USB		PKY3	Manokwari R , Irian Jaya	INS
8779,00	USB		PKM	Bitung R , Sulawesi	INS
8779,00	USB		PKF3	Kendari R	INS
8779,00	USB		PKM44	Pantolan R , Sulawesi	INS
8779,00	USB		PKP38	Batu Ampar R	INS
8779,00	USB		PKC4	Panjang R , Sumatera	INS
8779,00	USB		PKA	Sabang R , We	INS
8779,00	USB			Benete R , Sumbawa	INS
8779,00	USB		PKK	Kupang R , Timor	INS
8779,00	SSB			CS Ch 821	
8779,00	SSB		4PB	Colombo R	CLN
8779,00	USB		9WW	Kuching R	MLA
8779,00	SSB		A4M	Muscat R	OMA
8779,00	SSB		A7D	Doha R	QAT
8779,00	USB		CBM	Magallanes Zonal R	CHL
8779,00	SSB		D3E	Luanda R	AGL
8779,00	USB		PPR	Rio R	B
8779,00	USB		E5R	Rarotonga R	CKH
8780,00	ISR N HYBRID MODEM	B=2400Bd	4XZ	ISR N Haifa	ISR
8780,90	USB		XFL	Mazatlan P	MEX
8782,00	USB		EQJ	Chahbahar R	IRN
8782,00	USB		KFS	Palo Alto R	USA
8782,00	SSB			CS Ch 822	
8782,00	USB		PPR	Rio R	B
8782,00	USB		XSQ	Guangzhou R	CHN
8784,00	CW		RJS24	RUS N	RUS
8784,00	CW		RMJ99	RUS N ship	RUS
8784,00	CW		RCLH	RUS N Ship SIBIRYAKOV	RUS
8784,00	CW		RBDF	RUS N ship	RUS
8784,00	CW		RMKW	RUS N tug SERGEI BALK	RUS
8784,00	CW		UCBC5	RUS N ship YAKOV GREBELSKIY	RUS
8785,00	USB		YQI	Constanta R	ROU
8785,00	USB		XVG	Hai Phong R	VTN
8785,00	SSB			CS Ch 823	
8787,00	T600	B=50Bd S=200Hz		RUS N	RUS
8788,00	USB		EQL	Anzali R	IRN
8788,00	SSB			CS Ch 824	
8788,00	SSB		A4M	Muscat R	OMA
8788,00	USB		WLO	Mobile R , AL	USA
8788,00	STANAG 4481	B=50Bd S=850Hz			
8791,00	FAX 120/576		XSG	Shanghai R	CHN
8791,00	USB		EQC	Amirabad R	IRN
8791,00	USB		SDJ	Stockholm R	S
8791,00	USB		XSG	Shanghai R	CHN
8791,00	SSB			CS Ch 825	
8791,00	USB		E5R	Rarotonga R	CKH
8791,00	USB		UUT	Odessa R	UKR
8792,00	STANAG 4481	B=50Bd S=850Hz			
8794,00	USB		XST	Qingdao R	CHN

Frequency	Mode	Mode Parameter	Callsign	User	Country
8794,00	SSB			CS Ch 826	
8794,00	USB		PKA	Sabang R , We	INS
8794,00	USB		PKD5	Benoa R , Bali	INS
8794,00	USB		PKE	Amboina R , Ceram	INS
8794,00	USB		PKK	Kupang R , Timor	INS
8794,00	SSB		VFA	Inuvik R	CAN
8795,00	ALE USB	B=125Bd Ch=8 ChS=250Hz		ROU E net	ROU
8795,00	MIL 188-110A SER	B=2400Bd		ROU E Prague	CZE
8795,00	MIL 188-110A SER	B=2400Bd		MFA Bucharest	ROU
8796,40	USB		XFL	Mazatlan P	MEX
8796,40	USB		PKD	Surabaya R , Djawa	INS
8797,00	USB		J2A	Djibouti R	DJI
8797,00	USB		HLS	Seoul R	KOR
8797,00	CW		EQN	Khomeini R	IRN
8797,00	SSB			CS Ch 827	
8797,00	*ALE USB*	*B=125Bd Ch=8 ChS=250Hz*		*ISR AF net*	*ISR*
8798,00	USB		SDJ	Stockholm R	S
8800,00	USB		CND	Agadir R	MRC
8800,00	USB		3BM	Mauritius R	MAU
8800,00	USB		PKO	Tarakan R , Kalimantan	INS
8800,00	USB		PKY3	Manokwari R , Irian Jaya	INS
8800,00	SSB			CS Ch 828	
8800,00	SSB		CNP	Casablanca R	MRC
8800,00	USB		PKR	Semarang R , Djawa	INS
8800,00	USB		PKY4	Sorong R , Irian Jaya	INS
8800,00	USB		PNK	Jayapura R , Irian Jaya	INS
8800,00	USB		PPR	Rio R	B
8800,00	SSB		TJC	Douala R	CME
8801,00	ALE USB	B=125Bd Ch=8 ChS=250Hz		CHN net	CHN
8801,50	2FSK	B=400Bd S=800Hz			
8802,00	*MAHRS*	*B=2400Bd*	*DHJ58*	*D N Glücksburg TX Staberhuk*	*D*
8802,60	SSB		ODR7	Beyrouth R	LBN
8803,00	USB		EQI	Abbas R	IRN
8803,00	CW		EQM	Bushehr R	IRN
8803,00	SSB			CS Ch 829	
8803,00	SSB		ZPB	Asuncion R	PRG
8804,00	USB		VESTNIK	RUS N Kaliningrad	RUS
8805,00	ALE USB	B=125Bd Ch=8 ChS=250Hz		POL MIL net	
8806,00	USB		EQL	Anzali R	IRN
8806,00	USB		UFM3	Nevelsk R	RUS
8806,00	USB		XVR	Vung Tau R	VTN
8806,00	USB			Benete R , Sumbawa	INS
8806,00	SSB			CS Ch 830	
8806,00	USB		PKC4	Panjang R , Sumatera	INS
8806,00	USB		PKM	Bitung R , Sulawesi	INS
8806,00	USB		PPR	Rio R	B
8806,00	USB		XSG	Shanghai R	CHN
8808,00	*CIS 50-1000*	*B=50Bd S=1000Hz*		*UKR F*	*UKR*
8808,80	USB		PKM	Bitung R , Sulawesi	INS
8808,80	USB		PKC	Palembang R , Sumatera	INS
8809,00	USB		EQI	Abbas R	IRN
8809,00	USB		9MG	Penang R	MLA
8809,00	SSB			CS Ch 831	
8809,00	SSB		GNI	Niton R	G
8809,00	SSB		TAH	Istanbul R	TUR
8809,70	PACTOR I	B=100Bd S=170Hz	ACIDUS**		
8809,70	PACTOR I	B=100Bd S=170Hz	HEVEA5**		
8810,50	USB		LBJ	NOR N Bodoe	NOR

Frequency	Mode	Mode Parameter	Callsign	User	Country
8810,50	STANAG 4285	B=2400Bd	LBJ	NOR N Bodoe	NOR
8812,00	USB		LZW	Varna R	BUL
8812,00	DSC	B=100Bd S=170Hz	HLS	Seoul R	KOR
8812,00	USB		CWF	Punta Carretas R	URG
8812,00	USB		UFL	Vladivostok R	RUS
8812,00	USB		XVS	Ho Chi Minh R	VTN
8812,00	SSB			CS Ch 832	
8812,00	USB		TAH	Istanbul R	TUR
8812,00	USB		IAR	Rome R	I
8815,00	ALE USB	B=125Bd Ch=8 ChS=250Hz			
8816,00	CW		RJC38	RUS N AF	RUS
8816,00	CW		52411**	RUS N AF plane AN-26	RUS
8816,00	CW		52416**	RUS N AF plane AN-26	RUS
8816,00	CW		RJF94	RUS N AF	RUS
8816,00	SSB			RDARA 12C	
8816,00	SSB			RDARA 13J	
8816,00	SSB			RDARA 14A	
8816,00	SSB			RDARA 4A	
8816,00	SSB			RDARA 6G	
8816,00	CW			RUS N AF net	RUS
8816,00	CW		26910**	RUS N AF plane	RUS
8816,00	CW		10710**	RUS N AF plane	RUS
8816,00	CW		42322**	RUS N AF plane	RUS
8816,00	CW		10492**	RUS N AF plane AN-12	RUS
8816,00	CW		52860**	RUS N AF plane	RUS
8816,00	CW		54485**	RUS N AF plane	RUS
8816,00	CW		54483**	RUS N AF plane	RUS
8816,00	CW		RCB	RUS AF	RUS
8816,00	CW		10712**	RUS N AF plane AN-12	RUS
8816,00	CW		75225**	RUS AF plane	RUS
8816,00	CW		RMX	RUSN AF	RUS
8818,00	SSB			Aktyubinsk VOLMET	KAZ
8818,00	SSB			Baku VOLMET	AZE
8818,00	SSB			Karaganda VOLMET	KAZ
8818,00	USB		RDFG	Tashkent VOLMET	UZB
8818,00	SSB			Tbilisi VOLMET	GEO
8819,00	SSB			RDARA 10	
8819,00	SSB			RDARA 13C	
8819,00	SSB			RDARA 2B	
8819,00	SSB			RDARA 2C	
8819,00	SSB			RDARA 9B	
8819,00	USB		RDFG	Tashkent VOLMET	UZB
8820,00	ALE USB	B=125Bd Ch=8 ChS=250Hz			EGY
8821,00	CW		S	RUS AF marker	RUS
8822,00	SSB			RDARA 11B	
8822,00	SSB			RDARA 13G	
8822,00	SSB			RDARA 14	
8822,00	SSB			RDARA 2A	
8822,00	SSB			RDARA 3B	
8822,00	SSB			RDARA 5A	
8822,00	SSB			RDARA 5C	
8822,00	USB			Norilsk Control	RUS
8822,00	USB		BOING SEATTLE	Boing Seattle	USA
8822,00	USB		BOING EVERETT	Boing Seattle	USA
8825,00	ALE USB	B=125Bd Ch=8 ChS=250Hz		VEN N net	VEN
8825,00	SSB			MWARA NAT	NAT
8825,00	SSB		KEA5	New York ACC	USA

Frequency	Mode	Mode Parameter	Callsign	User	Country
8825,00	SSB			RDARA 13H	
8825,00	SSB			RDARA 14F	
8825,00	SSB			RDARA 6G	
8825,00	SSB		CSY	Santa Maria ACC	AZR
8825,00	HFDL	B=1800Bd	H06**	Hat Yai	THA
8828,00	USB		ZKAK	Auckland VOLMET	NZL
8828,00	SSB			Hongkong VOLMET	HKG
8828,00	SSB			RDARA 13N	
8828,00	SSB			RDARA 1D	
8828,00	SSB		JIA	Tokyo VOLMET	J
8828,00	SSB			VOLMET PAC	PAC
8828,00	SSB		KVM70	Honululu VOLMET	HWA
8830,00	CW			RUS AF net	RUS
8831,00	SSB			MWARA NAT	NAT
8831,00	SSB			RDARA 13F	
8831,00	SSB			RDARA 14F	
8831,00	SSB			RDARA 6G	
8831,00	SSB		VFG	Gander ACC	CAN
8831,00	USB		EIP	Shanwik ACC	IRL
8834,00	SSB			RDARA 10	
8834,00	SSB			RDARA 13C	
8834,00	SSB			RDARA 2B	
8834,00	SSB			RDARA 2C	
8834,00	SSB			RDARA 6C	
8834,00	SSB			RDARA 7C	
8834,00	HFDL	B=1800Bd	H08**	Johannesburg	AFS
8837,00	SSB			RDARA 10B	
8837,00	SSB			RDARA 13M	
8837,00	SSB			RDARA 3A	
8837,00	SSB			RDARA 3C	
8837,00	SSB			RDARA 4A	
8837,00	SSB			RDARA 9B	
8839,00	STANAG 4197	B=1800Bd Ch=16/39 ChS=112/56Hz		EGY N net	EGY
8839,00	USB			EGY N net	EGY
8840,00	SSB			RDARA 1C	
8840,00	SSB			RDARA 6	
8843,00	SSB			MWARA CEP	CEP
8843,00	SSB			RDARA 10E	
8843,00	SSB			RDARA 13C	
8843,00	SSB			RDARA 13J	
8843,00	SSB			RDARA 13K	
8843,00	SSB			RDARA 14D	
8843,00	SSB			RDARA 5D	
8843,00	SSB			RDARA 6G	
8843,00	SSB			San Francisco ACC , CA	USA
8843,00	HFDL	B=1800Bd	H07**	Shannon	IRL
8846,00	USB			Barranquilla ACC	CLM
8846,00	USB			Cayenne ACC	GUF
8846,00	USB			Georgetown ACC	GUY
8846,00	USB			Maiquetia ACC	VEN
8846,00	USB			Panama ACC	PNR
8846,00	USB			Paramaribo ACC	SUR
8846,00	USB			Piarco ACC	TRD
8846,00	SSB			Boyeros ACC	CUB
8846,00	SSB			MWARA CAR	CAR
8846,00	SSB		KEA5	New York ACC	USA
8846,00	SSB			RDARA 2	
8846,00	SSB			RDARA 3	
8846,00	SSB			RDARA 7F	
8846,00	SSB			RDARA 9	

Frequency	Mode	Mode Parameter	Callsign	User	Country
8847,00	USB			RUS AF net	RUS
8847,00	ALE USB	B=125Bd Ch=8 ChS=250Hz		ISR AF net	ISR
8847,00	USB		KORSAR	RUS AF Moscow	RUS
8847,00	USB		KLARNETI ST	RUS AF Migalovo/Tver	RUS
8847,00	USB		KASTA	RUS AF	RUS
8847,00	USB		MAGNETR ON	RUS AF	RUS
8847,00	USB		LADIA	RUS AF	RUS
8847,00	USB		DAVLENIE	RUS AF Taganrog	RUS
8849,00	SSB			RDARA 13K	
8849,00	SSB			VOLMET SEA	SEA
8849,00	LSB		BSQ	Bejing VOLMET	CHN
8849,00	USB			Guangzhou VOLMET	CHN
8849,00	USB		VRK	Honkong VOLMET	CHN
8850,00	ALE USB	B=125Bd Ch=8 ChS=250Hz			
8852,00	SSB			RDARA 12E	
8852,00	SSB			RDARA 3B	
8852,00	SSB			RDARA 3C	
8852,00	SSB			RDARA 9	
8852,00	SSB			VOLMET AFI	AFI
8852,00	ALE USB	B=125Bd Ch=8 ChS=250Hz		ISR AF net	
8852,00	MIL 188-110B SER	B=2400Bd		ISR AF net	
8853,00	USB			EGY N net	EGY
8855,00	USB			Buenos Aires ACC	ARG
8855,00	USB			Campo Grande ACC	B
8855,00	USB			Recife ACC	B
8855,00	USB			Salvador ACC	B
8855,00	SSB			Asuncion ACC	PRG
8855,00	SSB			Belem ACC	B
8855,00	SSB			Bogota ACC	CLM
8855,00	SSB			Brasilia ACC	B
8855,00	SSB			Cayenne ACC	GUF
8855,00	SSB			Georgtown ACC	GUY
8855,00	SSB			Iquitos ACC	PRU
8855,00	SSB			Johannesburg ACC	AFS
8855,00	SSB			La Paz ACC	BOL
8855,00	SSB			Leticia ACC	CLM
8855,00	SSB			Lima ACC	PRU
8855,00	SSB			Maiquetia ACC , Caracas	VEN
8855,00	SSB			Manaus ACC	B
8855,00	SSB			Montevideo ACC	URG
8855,00	SSB			MWARA SAM	SAM
8855,00	SSB			Paramaribo ACC	SUR
8855,00	SSB			Piarco ACC	TRD
8855,00	SSB			Porte Alegre ACC	B
8855,00	SSB			Porto Velho ACC	B
8855,00	SSB			RDARA 10A	
8855,00	SSB			RDARA 14	
8855,00	SSB			RDARA 2	
8855,00	SSB			Recife ACC	B
8855,00	SSB			Rio de Janeiro ACC	B
8855,00	SSB			Santa Cruz ACC	BOL
8858,00	SSB			RDARA 10D	
8858,00	SSB			RDARA 13E	
8858,00	SSB			RDARA 13F	
8858,00	SSB			RDARA 14D	
8858,00	SSB			RDARA 4A	
8858,00	SSB			RDARA 6G	

Frequency	Mode	Mode Parameter	Callsign	User	Country
8861,00	USB			Rio de Janeiro ACC	B
8861,00	USB			Brasilia ACC	B
8861,00	USB			Cayenne ACC	GUF
8861,00	USB			Bouake ACC	CTI
8861,00	USB			Kano ACC	NIG
8861,00	SSB			Abidjan ACC	CTI
8861,00	SSB			Bamako ACC	MLI
8861,00	SSB			Banjul ACC	GMB
8861,00	SSB			Bissau ACC	GNB
8861,00	SSB			Canarias ACC , Las Palmas	AZR
8861,00	SSB			Casablanca ACC	MRC
8861,00	SSB			Conakry ACC	GUI
8861,00	SSB			Dakar ACC	SEN
8861,00	SSB			Freetown ACC	SRL
8861,00	SSB			Irkutsk VOLMET	RUS
8861,00	SSB			Jakutsk VOLMET	RUS
8861,00	SSB			Johannesburg ACC	AFS
8861,00	SSB		UGEF	Khabarovsk VOLMET	RUS
8861,00	SSB			Magadan VOLMET	RUS
8861,00	SSB			Manaus ACC	B
8861,00	SSB			MWARA SAT	SAT
8861,00	SSB			Nouakchott ACC	MTN
8861,00	SSB			Paramaribo ACC	SUR
8861,00	SSB			RDARA 3A	
8861,00	SSB			RDARA 3B	
8861,00	SSB			RDARA 6E	
8861,00	SSB			RDARA 9B	
8861,00	SSB			Recife ACC	B
8861,00	SSB			Sal ACC	CPV
8861,00	USB			Niamey ACC	NGR
8861,00	USB			Nouadhibou ACC	MTN
8861,00	USB			Ouagadougou ACC	BFA
8861,00	USB			Roberts ACC	LBR
8864,00	SSB			Iceland ACC , Reykjavik	ISL
8864,00	SSB			MWARA NAT	NAT
8864,00	SSB		KEA5	New York ACC	USA
8864,00	SSB			RDARA 13F	
8864,00	SSB			RDARA 2B	
8864,00	SSB			RDARA 6B	
8864,00	SSB			RDARA 6F	
8864,00	SSB			RDARA 7E	
8864,00	SSB		CSY	Santa Maria ACC	AZR
8864,00	USB		EIP	Shanwik ACC	IRL
8864,00	SSB		VFG	Gander ACC	CAN
8867,00	SSB			MWARA SP	SP
8867,00	SSB			RDARA 10C	
8867,00	SSB			RDARA 13D	
8867,00	SSB			RDARA 13M	
8867,00	SSB			RDARA 6G	
8867,00	USB			San Francisco ACC , CA	USA
8870,00	USB			San Francisco ACC , CA	USA
8870,00	SSB			RDARA 14	
8870,00	SSB			RDARA 5	
8870,00	SSB			RDARA 6G	
8870,00	SSB			VOLMET NAT	NAT
8870,00	STANAG 4197	B=1800Bd Ch=16/39 ChS=112/56Hz			
8873,00	USB			Niamey ACC	NGR
8873,00	SSB			RDARA 12E	
8873,00	SSB			RDARA 12F	
8873,00	SSB			RDARA 13I	

Frequency	Mode	Mode Parameter	Callsign	User	Country
8873,00	SSB			RDARA 4	
8873,00	SSB			RDARA 6G	
8873,00	SSB			RDARA 9C	
8873,00	SSB			RDARA 9D	
8873,00	USB		CXO	Carrasco VOLMET	URG
8875,00	CW			MNG MIL net	MNG
8875,00	RUS TACTICAL DATALINK	B=1200Bd S=800Hz Ch=1200		RUS MIL	RUS
8876,00	SSB			NOAA Hurricane Center Kansas City , KS	USA
8876,00	SSB			NOAA Hurricane Center Miami , FL	USA
8876,00	SSB			RDARA 10A	
8876,00	SSB			RDARA 12D	
8876,00	SSB			RDARA 14G	
8876,00	SSB			RDARA 2A	
8879,00	USB			Antananarivo ACC	MDG
8879,00	SSB			Brisbane ACC	AUS
8879,00	SSB			Beira ACC	MOZ
8879,00	SSB			Mumbai ACC	IND
8879,00	SSB			Cocos Island ACC	ICO
8879,00	USB			Jeddah ACC	ARS
8879,00	USB			Chennai ACC	IND
8879,00	USB			Mahajanga ACC	MDG
8879,00	USB			Male ACC	MLD
8879,00	USB			St.Denis ACC	REU
8879,00	USB			Toamasina ACC	MDG
8879,00	SSB			Colombo ACC	CLN
8879,00	SSB			Dar es Salaam ACC	TZA
8879,00	SSB			Harare ACC	ZWE
8879,00	SSB			Iceland ACC , Reykjavik	ISL
8879,00	SSB			Kigali ACC	RRW
8879,00	SSB			Lilongwe ACC	MWI
8879,00	SSB			Lusaka ACC	ZMB
8879,00	SSB			Mauritius ACC , Plaisance	MAU
8879,00	SSB			Moroni ACC	COM
8879,00	SSB			MWARA INO	INO
8879,00	SSB			MWARA NAT	NAT
8879,00	SSB			Perth ACC	AUS
8879,00	SSB			RDARA 3B	
8879,00	SSB			Seychelles ACC , Mahe	SEY
8879,00	USB		EIP	Shanwik ACC	IRL
8879,00	SSB		5YD	Nairobi ACC	KEN
8879,00	SSB		VFG	Gander ACC	CAN
8879,00	USB			Mumbai ACC	IND
8880,00	ALE USB	B=125Bd Ch=8 ChS=250Hz		GANOB net	EGY
8880,00	CCIR 493-4	B=100Bd S=170Hz			
8880,00	CCIR 493-4	B=100Bd S=170Hz			
8882,00	SSB			RDARA 2C	
8882,00	SSB			RDARA 6D	
8885,00	SSB			RDARA 11B	
8885,00	SSB			RDARA 13G	
8885,00	SSB			RDARA 14C	
8885,00	SSB			RDARA 5	
8885,00	SSB			RDARA 6B	
8885,00	HFDL	B=1800Bd	H15**	Al Muharraq	BHR
8886,00	HFDL	B=1800Bd	H14**	Krasnoyarsk	RUS
8888,00	SSB			Jekaterinenburg VOLMET	RUS
8888,00	SSB		UNNN	Novosibirsk VOLMET	RUS
8888,00	SSB			RDARA 2	
8888,00	SSB			RDARA 6G	
8888,00	SSB			RDARA 7	
8888,00	SSB		RQCI	Samara VOLMET	RUS

Frequency	Mode	Mode Parameter	Callsign	User	Country
8888,00	SSB		UBB2	Syktyvkar VOLMET	RUS
8888,00	SSB		RVPE	Tyumen VOLMET	RUS
8891,00	SSB			Bodo ACC	NOR
8891,00	SSB			Iceland ACC , Reykjavik	ISL
8891,00	SSB			MWARA NAT	NAT
8891,00	SSB			RDARA 14E	
8891,00	SSB			RDARA 6A	
8891,00	USB		EIP	Shanwik ACC	IRL
8891,00	SSB		VFG	Gander ACC	CAN
8894,00	USB			N djamena ACC	TCD
8894,00	USB			Gao ACC	MLI
8894,00	SSB			Alger ACC	ALG
8894,00	SSB			Kano ACC	NIG
8894,00	SSB			MWARA AFI	AFI
8894,00	SSB			Niamey ACC	NGR
8894,00	SSB			RDARA 12F	
8894,00	SSB			RDARA 14A	
8894,00	SSB			RDARA 3C	
8894,00	SSB			Tamanrasset ACC	ALG
8894,00	SSB			Timimoun ACC	ALG
8894,00	SSB			Tripoli ACC	LBY
8894,00	SSB			Tunis ACC	ALG
8894,00	STANAG 4481	B=75Bd S=850Hz		USA AF Barford St. John	G
8894,00	HFDL	B=1800Bd	H11**	Albrook	PNR
8895,00	CW			RUS Strategic AF net	RUS
8895,00	ALE LSB	B=125Bd Ch=8 ChS=250Hz		TWN N net	TWN
8895,00	CW		W**	RUS AF net	RUS
8897,00	USB			Guangzhou ACC	CHN
8897,00	USB			Hailar ACC	CHN
8897,00	USB			Jinan ACC	CHN
8897,00	USB			Lanzhou ACC	CHN
8897,00	USB			Shenyang ACC	CHN
8897,00	SSB			Bejing ACC	CHN
8897,00	SSB			Irkutsk ACC	RUS
8897,00	SSB			Kunming ACC	CHN
8897,00	SSB			MWARA EA	EA
8897,00	SSB			Pyongyang ACC	KRE
8897,00	SSB			Shanghai ACC	CHN
8897,00	SSB			Ulaanbaatar ACC	MNG
8897,00	SSB			Urumchi ACC	CHN
8897,00	USB			Taegu ACC	CHN
8897,00	USB			Wuhan ACC	CHN
8897,00	USB			Zhengzhou ACC	CHN
8897,00	USB			Auckland ACC	AUS
8897,00	USB			Brisbane ACC	AUS
8897,00	USB			Nadi ACC	FJI
8897,00	USB			Pascua (Easter Island) ACC	CHL
8897,00	USB			Port Vila ACC	VUT
8897,00	USB			Rarotonga ACC	CKH
8897,00	USB			San Francisco ACC , CA	USA
8897,00	USB			Tahiti ACC	OCE
8897,00	USB			Wallis ACC	WAL
8900,00	SSB			RDARA 10D	
8900,00	SSB			RDARA 13G	
8900,00	SSB			RDARA 14B	
8900,00	SSB			RDARA 3A	
8903,00	SSB			Accra ACC	GHA
8903,00	SSB			Bangui ACC	CAF
8903,00	SSB			Douala ACC	CME
8903,00	SSB			Entebbe ACC	UGD
8903,00	SSB			Goma ACC	

Frequency	Mode	Mode Parameter	Callsign	User	Country
8903,00	USB			Franceville ACC	GAB
8903,00	SSB			Harare ACC	ZWE
8903,00	SSB			Hongkong ACC	HKG
8903,00	SSB			Johannesburg ACC	AFS
8903,00	SSB			Kano ACC	
8903,00	SSB			Kinshasa ACC	ZAI
8903,00	SSB			Kisangani ACC	
8903,00	SSB			Lagos ACC	NIG
8903,00	SSB			Libreville ACC	GAB
8903,00	SSB			Luanda ACC	AGL
8903,00	SSB			Lubumbashi ACC	ZAI
8903,00	SSB			Lusaka ACC	ZMB
8903,00	SSB			Malabo ACC	GNE
8903,00	SSB			Manila ACC	PHL
8903,00	SSB			MWARA AFI	AFI
8903,00	SSB			MWARA CWP	CWP
8903,00	SSB			Naha ACC	J
8903,00	SSB			Niamey ACC	NIG
8903,00	SSB			N`djamena ACC	TCD
8903,00	SSB			Port Moresby ACC	PNG
8903,00	SSB			RDARA 10B	
8903,00	SSB			RDARA 13M	
8903,00	SSB			Sao Tome ACC	STP
8903,00	SSB			Tokyo ACC	J
8903,00	USB			Garoua ACC	CME
8903,00	USB			San Francisco ACC , CA	USA
8903,00	USB			Seoul ACC	KOR
8903,00	USB			Taipei ACC	TWN
8903,00	USB			Maiduguri ACC	NIG
8903,00	USB			Maroua ACC	CME
8903,00	USB			Niamtougou ACC	TGO
8903,00	USB			Pointe Noire ACC	COG
8903,00	USB			Port Gentil ACC	GAB
8903,00	USB			Roberts ACC	LBR
8903,00	USB			Windhoek ACC	NMB
8903,00	USB			Yaounde ACC	CME
8906,00	USB			Paramaribo ACC	SUR
8906,00	SSB			MWARA NAT	NAT
8906,00	SSB		KEA5	New York ACC	USA
8906,00	SSB			RDARA 13H	
8906,00	SSB			RDARA 6A	
8906,00	SSB			RDARA 6E	
8906,00	SSB			RDARA 7B	
8906,00	SSB			RDARA 9B	
8906,00	SSB		CSY	Santa Maria ACC	AZR
8906,00	USB		EIP	Shanwik ACC	IRL
8906,00	USB		LVE	Cordoba VOLMET	ARG
8906,00	SSB			Canarias ACC , Las Palmas	CNR
8906,00	SSB		VFG	Gander ACC	CAN
8906,00	USB			Piarco ACC	TRD
8909,00	USB		PROCELKA	RUS Strategic AF	RUS
8909,00	USB		OCHISTKA	RUS Strategic AF	RUS
8909,00	CW			RUS AF	RUS
8909,00	SSB			RDARA 2A	
8909,00	SSB			RDARA 6E	
8911,00	8PSK	B=1800Bd			
8912,00	SSB			RDARA 11B	
8912,00	SSB			RDARA 13D	
8912,00	SSB			RDARA 14C	
8912,00	SSB			RDARA 5B	

Frequency	Mode	Mode Parameter	Callsign	User	Country
8912,00	SSB			RDARA 6G	
8912,00	HFDL	B=1800Bd	H02**	Molokai , HI	HWA
8912,00	HFDL	B=1800Bd	H04**	Riverhead , NY	USA
8912,00	ALE USB	B=125Bd Ch=8 ChS=250Hz		USA COTHEN net	USA
8915,00	SSB			San Francisco ACC	USA
8915,00	USB			Tokyo ACC	J
8915,00	SSB			RDARA 3C	
8915,00	SSB			RDARA 5A	
8915,00	USB			San Francisco ACC , CA	USA
8917,00	*FSK*	*B=150Bd S=550Hz*		*RUS AF North Minsk*	*BLR*
8917,00	*FSK*	*B=150Bd S=550Hz*		*RUS AF*	
8918,00	USB			Barranquilla ACC	CLM
8918,00	USB			Boyeros ACC	CUB
8918,00	USB			Merida ACC	MEX
8918,00	SSB			MWARA CAR	CAR
8918,00	SSB			MWARA MID	MID
8918,00	USB		KEA5	New York ACC	USA
8918,00	USB			Panama ACC	PNR
8918,00	USB			Piarco ACC	TRD
8918,00	SSB			RDARA 6C	
8919,00	USB		BOING EVERETT	Boing Seattle	USA
8919,00	USB		BOING SEATTLE	Boing Seattle	USA
8921,00	USB			PanAM R	USA
8921,00	ALE USB	B=125Bd Ch=8 ChS=250Hz			
8921,00	SSB			W I	WW
8921,00	SSB			W III	WW
8921,00	HFDL	B=1800Bd	H05**	Auckland	NZL
8924,00	SSB			W I	WW
8924,00	SSB			W IV	WW
8927,00	HFDL	B=1800Bd	H01**	San Francisco , CA	USA
8927,00	SSB			W II	WW
8927,00	SSB			W V	WW
8927,00	HFDL	B=1800Bd	H09**	Utqiagvik , AK	ALS
8927,00	HFDL	B=1800Bd	H16**	Agana	GUM
8930,00	*USB*		*SDJ*	*Stockholm R*	*S*
8930,00	SSB			W I	WW
8930,00	SSB			W III	WW
8933,00	SSB			W II	WW
8933,00	SSB			W V	WW
8933,00	ALE USB	B=125Bd Ch=8 ChS=250Hz		ROU diplo net	ROU
8933,00	MIL 188-110A SER	B=2400Bd		ROU E Moscow	RUS
8933,00	MIL 188-110A SER	B=2400Bd		MFA Bucharest	ROU
8933,00	USB			San Francisco ACC , CA	USA
8935,00	ALE USB	B=125Bd Ch=8 ChS=250Hz		USA MIL net	USA
8936,00	USB			Airbus Toulouse	F
8936,00	SSB			W I	WW
8936,00	SSB			W II	WW
8936,00	HFDL	B=1800Bd	H02**	Molokai , HI	HWA
8936,00	USB			IBERIA Madrid ACC	E
8936,00	HFDL	B=1800Bd	H09**	Utqiagvik , AK	ALS
8938,00	USB			AUS Davis Base	ANT
8938,00	USB			Mawson Base , AUS	ANT
8938,00	USB		LWL	Comodoro VOLMET	ARG
8939,00	SSB			RDARA 10B	
8939,00	SSB			RDARA 13C	
8939,00	SSB			RDARA 2A	

Frequency	Mode	Mode Parameter	Callsign	User	Country
8939,00	SSB			RDARA 2C	
8939,00	SSB			RDARA 6F	
8939,00	SSB		RLAP	Rostov VOLMET	RUS
8939,00	SSB		UHD	St. Petersburg VOLMET	RUS
8939,00	HFDL	B=1800Bd	H10**	Muan	KOR
8939,00	ALE USB	B=125Bd Ch=8 ChS=250Hz		MTN net	MTN
8942,00	USB			Guangzhou ACC	CHN
8942,00	USB			Irkutsk ACC	RUS
8942,00	USB			Pyongyang ACC	KRE
8942,00	USB			Ulaanbaatar ACC	MNG
8942,00	USB			Seoul ACC	KOR
8942,00	USB			Jakarta ACC	INS
8942,00	USB			Kuala Lumpur ACC	MLA
8942,00	USB			Bali ACC	INS
8942,00	USB			Kota Kinabalu ACC	
8942,00	SSB			Bangkok ACC	THA
8942,00	SSB			Hanoi ACC	VTN
8942,00	SSB			Ho Chi Min Ville ACC	VTN
8942,00	SSB			Hongkong ACC	HKG
8942,00	SSB			Manila ACC	PHL
8942,00	SSB			MWARA SEA	SEA
8942,00	SSB			RDARA 3A	
8942,00	SSB			Singapore ACC	SNG
8942,00	SSB			Tokyo ACC	J
8942,00	SSB			Vientiane ACC	LAO
8942,00	HFDL	B=1800Bd	H07**	Shannon	IRL
8945,00	SSB			RDARA 10F	
8945,00	SSB			RDARA 13K	
8945,00	SSB			RDARA 14E	
8945,00	SSB			VOLMET MID	MID
8948,00	SSB			RDARA 12C	
8948,00	SSB			RDARA 6A	
8948,00	HFDL	B=1800Bd	H17**	Telde , Gran Canaria	E
8950,00	*ALE USB*	*B=125Bd Ch=8 ChS=250Hz*		*TUR Disaster and Emergency net*	*TUR*
8951,00	USB			Aden ACC	YEM
8951,00	USB			Amman ACC	JOR
8951,00	USB			Ankara ACC	TUR
8951,00	SSB			Aktyubinsk ACC	KAZ
8951,00	SSB			Alma Ata ACC	KAZ
8951,00	SSB			Bishkek ACC	KGZ
8951,00	SSB			Dushanbe ACC	TJK
8951,00	SSB			Kyzl-Orda ACC	RUS
8951,00	SSB			MWARA MID	MID
8951,00	SSB			Uralsk ACC	KAZ
8951,00	USB			Beirut ACC	LBN
8951,00	USB			Cairo ACC	EGY
8951,00	USB			Damascus ACC	SYR
8951,00	USB			Jeddah ACC	ARS
8951,00	USB			Kuwait ACC	KWT
8951,00	USB			Manama ACC	BHR
8951,00	USB			Odess ACC	UKR
8951,00	USB			Sanaa ACC	YEM
8951,00	USB			Simferopol ACC	UKR
8951,00	USB			Tehran ACC	IRN
8951,00	USB			Tbilisi ACC	GEO
8951,00	USB			Yerevan ACC	ARM
8951,00	USB			Dushanbe ACC	TJK
8951,00	USB			Kuybyshev ACC	RUS
8951,00	USB			Moscow ACC	RUS
8951,00	USB			Samarkhand ACC	UZB

Frequency	Mode	Mode Parameter	Callsign	User	Country
8951,00	USB			San Francisco ACC , CA	USA
8954,00	SSB			RDARA 10E	
8954,00	SSB			RDARA 12J	
8954,00	SSB			RDARA 14B	
8954,00	SSB			RDARA 3	
8957,00	SSB			Bolivia ACC	BOL
8957,00	SSB			RDARA 12C	
8957,00	SSB			RDARA 13D	
8957,00	SSB			RDARA 14G	
8957,00	SSB			RDARA 3B	
8957,00	SSB			RDARA 6D	
8957,00	SSB			VOLMET EUR	EUR
8957,00	USB		EIP	Shannon VOLMET	IRL
8957,00	HFDL	B=1800Bd	H13**	Santa Cruz	BOL
8960,00	SSB			RDARA 6G	
8960,00	SSB			RDARA 7F	
8962,00	STANAG 4481	B=75Bd S=850Hz		POR N Lisbon TX Azores	POR
8964,00	*CODAN*	*B=2400Bd Ch=16 ChS=112,5Hz*		*IRN net*	*IRN*
8964,00	*ALE USB*	*B=125Bd Ch=8 ChS=250Hz*		*IRN net*	*IRN*
8964,00	*CODAN CHIRP*	*B=80Bd Ch=30 ChS=81Hz*		*IRN net*	*IRN*
8965,00	USB		DHJ83	D AF Gatow	D
8965,00	ALE USB	B=125Bd Ch=8 ChS=250Hz		USA AF net	
8965,00	USB		BALIZA	ROU AF net	ROU
8967,00	SSB			ARS AF Khamis Mushayt	ARS
8967,00	SSB			Seychelles ACC , Mahe	SEY
8967,00	SSB		AHF4	USAF Albrook	PNZ
8968,00	MIL 188-110A SER	B=2400Bd		USA MIL net	USA
8968,00	NOKIA MSG TERMINAL	B=301,1Bd S=760Hz		MFA Helsinki	
8968,00	MIL 188-110A SER	B=2400Bd Ch=1	NSY	USA N Niscemi , Sicily	I
8971,00	ALE USB	B=125Bd Ch=8 ChS=250Hz		I AF net	I
8971,00	*LINK 11 CLEW*	*B=2250Bd Ch=16 ChS=330/110/550Hz*		*NOR N North of Trondheim*	*NOR*
8972,00	USB		CIRCUS VERT	F AF HQ Villacoublay	F
8973,00	SSB			POR AF Lajes AFB	AZR
8974,00	SSB			NZL AF Ohakea	NZL
8974,00	USB			UKR AF net	UKR
8974,00	USB		0A**	IRL N AF Dublin	IRL
8977,00	HFDL	B=1800Bd	H03**	Reykjavik	ISL
8980,00	ALE USB	B=125Bd Ch=8 ChS=250Hz		USA CG net	USA
8980,00	SSB			NASA Cape R	USA
8980,00	ALE USB	B=125Bd Ch=8 ChS=250Hz		ALG net	ALG
8983,00	USB		DHJ78	D N Nordholz	D
8983,00	USB			D N AF net	D
8984,00	SSB			Nouadhibou ACC	MTN
8984,00	SSB			Nouakchott ACC	MTN
8984,00	SSB			USCG Boston , MA	USA
8984,00	SSB			USCG Charleston ,SC	USA
8984,00	SSB			USCG Guatanamo Bay	GTB
8984,00	SSB			USCG Kodiak	ALS
8984,00	SSB			USCG Miami , FL	USA
8984,00	SSB			USCG New Orleans ,	USA
8984,00	SSB			USCG San Francisco , CA	USA

Frequency	Mode	Mode Parameter	Callsign	User	Country
8986,00	ALE USB	B=125Bd Ch=8 ChS=250Hz		F DGA net	F
8988,00	CW		RJC38	RUS N AF MUrmansk	RUS
8989,00	USB		CHR	CAN AF Trenton	CAN
8989,00	SSB			BEL AF Melsbroek	BEL
8989,00	SSB			BEL AF Rouveroy	BEL
8989,00	SSB		ONY77	BEL AF Koksijde	BEL
8989,00	USB			Jeddah ACC	ARS
8990,00	ALE USB	B=125Bd Ch=8 ChS=250Hz		ALG police net	ALG
8990,00	CW			RUS Strategic AF net	RUS
8992,00	USB			USA AF Sigonella	I
8992,00	USB			USA AF Lajes AFB	AZR
8992,00	USB			USA AF Puerto Rico	PTR
8992,00	USB			USA AF Ascension AFB	ASC
8992,00	USB			USA AF Diego Garcia	DGA
8992,00	SSB			USA AF Croughton	G
8992,00	ALE USB	B=125Bd Ch=8 ChS=250Hz		USA AF net	USA
8992,00	SSB		AFA3	USA AF Andrews AFB , Camp Springs , MD	USA
8992,00	SSB		AFS	USA AF Offutt AFB , Omaha , NE	USA
8992,00	SSB		AIE	USA AF Andersen AFB	GUM
8992,00	SSB		AIF80	USA AF Yokota AB	J
8992,00	SSB			USA AF Guam	GUM
8992,00	SSB			USA AF McDill AFB , FL	USA
8992,00	USB			F AF TX Vernon	F
8992,00	USB			USA AF Hickam , HWA	USA
8992,00	USB		AKA	USAF Elmendorf AFB , Anchorage	ALS
8992,00	USB		CIRCUS VERT	F AF HQ Villacoublay	F
8993,00	CW		FAV22	F F Mont-Valerien	F
8994,00	STANAG 4481	B=50Bd S=850Hz	NSS	USA N TX Davidsonville , MD	USA
8999,50	ALE 3G	B=2400Bd		UI MIL net	
9007,00	USB		CHR	CAN AF Trenton	CAN
9007,00	SSB		CJX	CAN AF St. John`s	CAN
9007,00	STANAG 4481	B=75Bd S=850Hz	NAU	USA N San Juan	PTR
9007,00	LINK 11 CLEW	B=2250Bd Ch=16 ChS=330/110/550Hz		J Mil Yokosuka	J
9010,00	USB			B AF net	B
9011,00	SSB			NOAA Hurricane Center Miami , FL	USA
9017,00	SSB			USA AF Strategic Command	USA
9018,30	MIL 188-110B	B=2400Bd		IRQ N net	IRQ
9018,30	*ALE USB*	*B=125Bd Ch=8 ChS=250Hz*		*IRQ N net*	*IRQ*
9018,50	ALE USB	B=125Bd Ch=8 ChS=250Hz			
9019,00	*ALE USB*	*B=125Bd Ch=8 ChS=250Hz*		*G MIL net*	*G*
9022,00	SSB			NASA launch support aircraft	USA
9024,00	CIS OFDM 60	B=2400Bd Ch=60 ChS=44,5Hz		RUS diplo Moscow	RUS
9024,00	SSB		UHD	St. Petersburg ACC	RUS
9024,00	SSB			Murmansk ACC	RUS
9024,00	SSB			Arkchangelsk ACC	RUS
9024,00	SSB			Petrozavodsk ACC	RUS
9024,00	SSB			Vologda ACC	RUS
9024,00	SSB			Velikiye Luki ACC	RUS
9025,00	USB			TUR F net	TUR
9025,00	ALE USB	B=125Bd Ch=8 ChS=250Hz		USA AF net	
9026,00	CCIR 493-4	B=100Bd S=170Hz	2111		
9026,00	CCIR 493-4	B=100Bd S=170Hz	2112		

Frequency	Mode	Mode Parameter	Callsign	User	Country
9026,00	CCIR 493-4	B=100Bd S=170Hz	2113		
9026,00	CCIR 493-4	B=100Bd S=170Hz	2228	BRM Mil net	E
9026,00	ALE USB	B=125Bd Ch=8 ChS=250Hz			
9027,00	CW			RUS Strategic AF net	RUS
9027,00	SSB			USA AF Strategic Command	USA
9027,00	CW			RUS Strategic AF net	RUS
9031,00	USB			G AF TASCOMM	G
9032,00	USB			McMurdo ACC	
9032,00	USB		NPX	USARP Amundson ScottSouth Pole Station	ANT
9032,00	USB		NPX	USARP Amundson-ScottSouth Pole Station	ANT
9035,00	USB			UN MONUSCO net	UGA
9037,00	ALE USB	B=125Bd Ch=8 ChS=250Hz		GRC AF net	GRC
9037,00	USB			Kinloss Rescue	G
9038,00	ALE USB	B=125Bd Ch=8 ChS=250Hz		USA MIL net	
9040,00	USB			BGD N net	BGD
9040,00	CODAN CHIRP	B=80Bd Ch=30 ChS=81Hz			
9040,00	*ALE USB*	*B=125Bd Ch=8 ChS=250Hz*		*CHN N net Horn of Africa*	*CHN*
9041,00	RUS DBPSK 16FSK	B=1866Bd Ch=16 ChS=175Hz		RUS INTEL	RUS
9043,00	USB			Cape Canaveral AFS , FL	USA
9043,00	SSB			NASA launch support aircraft	USA
9044,00	*T600*	*B=50Bd S=200Hz*		*RUS N Smolensk*	*RUS*
9044,00	CW		RMW32	RUS N	RUS
9044,00	CW		RMW56	RUS N	RUS
9045,00	MIL 188-110B 39TONE	B=2400Bd Ch=39 ChS=56,25Hz		?CHN	
9046,00	USB			Arabic net	
9046,00	*ALE USB*	*B=125Bd Ch=8 ChS=250Hz*		*IRN net*	*IRN*
9047,00	ALE USB	B=125Bd Ch=8 ChS=250Hz		USA Civil Air Patrol net	USA
9047,00	SSB			USA AF MARS	ALS
9047,00	OFDM	B=1560Bd Ch=26 ChS=75Hz			
9047,00	CCIR 493-4	B=100Bd S=170Hz			
9049,50	ALE USB	B=125Bd Ch=8 ChS=250Hz		USA SHARES net	USA
9050,00	MIL 188-110B SER	B=2400Bd		?CHN	
9050,00	*ALE USB*	*B=125Bd Ch=8 ChS=250Hz*		*F Airbus net*	*F*
9050,00	ALE USB	B=125Bd Ch=8 ChS=250Hz		ALG AF net	ALG
9050,00	RFSM8000	B=2400Bd		MFA Sofia	BUL
9051,00	*MAHRS*	*B=2400Bd*	*DHJ58*	*D N Glücksburg TX Staberhuk*	*D*
9051,50	MIL 188-110A SER	B=2400Bd		AUS MHFCS North West Cape	AUS
9053,00	ALE USB	B=125Bd Ch=8 ChS=250Hz		AUS Police WA	AUS
9054,00	CCIR 493-4	B=100Bd S=170Hz			
9054,00	CHN 4+4	B=75Bd Ch=8 ChS=300/450Hz		CHN N	CHN
9055,00	CW			RUS HFDF net	RUS
9055,00	RFSM8000	B=2400Bd		BUL E Algers	ALG
9055,00	RFSM8000	B=2400Bd		BUL E	
9056,00	STANAG 4285	B=2400Bd		AUS MIL Riverina	AUS
9056,00	STANAG 4285	B=2400Bd		AUS MHFCS Townsville	AUS
9057,00	SSB			USA AF Strategic Command	USA
9057,00	ALE USB	B=125Bd Ch=8 ChS=250Hz		USA SHARES net	USA

Frequency	Mode	Mode Parameter	Callsign	User	Country
9060,00	ALE USB	B=125Bd Ch=8 ChS=250Hz		ALG MIL net	ALG
9060,00	*CIS 12*	*B=1440Bd Ch=12 ChS=200Hz*		*RUS N ship*	
9064,00	ALE USB	B=125Bd Ch=8 ChS=250Hz		USA SHARES net	USA
9065,00	USB				
9065,00	WINDRM	B=1912,5Bd Ch=51 ChS=46,875Hz		CUB Intelligence Directorate	CUB
9065,00	*CIS 100-500*	*B=100Bd S=500Hz*		*RUS MIL Novosibirsk*	*RUS*
9065,00	CIS ARQ	B=100Bd S=500Hz		RUS PTT	RUS
9065,00	MIL 188-110B SER	B=2400Bd			
9067,00	BAUDOT	B=100Bd S=500Hz			
9068,00	CW		RDL	RUS N Moscow	RUS
9069,00	ALE USB	B=125Bd Ch=8 ChS=250Hz		POL MIL net	POL
9069,00	STANAG 4285	B=2400Bd		AUS MHFCS Humpty Doo	AUS
9069,00	STANAG 4285	B=2400Bd		AUS MHFCS Riverina	AUS
9070,00	PACTOR IV	B=100Bd S=200Hz		USA SHARES Net	USA
9070,00	CIS OFDM 60	B=2400Bd Ch=60 ChS=44,5Hz		RUS diplo	
9071,00	OFDM	B=2541Bd Ch=121 ChS=25Hz		MFA Moscow	RUS
9071,00	QPSK	B=2400Bd		MFA Moscow	RUS
9072,00	THALES ROBUST WAVEFORM	B=1000Bd Ch=8 ChS=250Hz		MRC MIL net	MRC
9072,00	CHN DATALINK 2400 BD	B=2400Bd			CHN
9074,20	*CIS 100-1000*	*B=100Bd S=1000Hz*		*CIS F Kaliningrad*	*RUS*
9075,00	USB			USA AF MARS	
9075,00	*CIS 4FSK*	*B=100Bd Ch=4 ChS=500Hz*		*RUS MIL Kaliningrad*	*RUS*
9075,00	CIS 3000	B=3000Bd			RUS
9075,00	STANAG 4481	B=75Bd S=850Hz	NPM	USA N Lualualei	HWA
9077,00	ALE USB	B=125Bd Ch=8 ChS=250Hz		S FRO net	S
9077,50	CIS OFDM 45	B=2400Bd Ch=45 ChS=62,5Hz		RUS	RUS
9077,50	*CW*		*RIT*	*RUS N HQ Severomorsk*	*RUS*
9078,70	MIL 188-110B SER	B=2400Bd	HWK**	S N Karlskrona	S
9080,00	ALE USB	B=125Bd Ch=8 ChS=250Hz		ALG AF net	ALG
9080,00	STANAG 4197	B=1800Bd Ch=16/39 ChS=112/56Hz		EGY N net	EGY
9080,00	USB			EGY N net	EGY
9080,00	STANAG 4197	B=1800Bd Ch=16/39 ChS=112/56Hz		EGY N net	EGY
9081,00	USB			HFoZ/Radtel net	AUS
9084,00	ALE USB	B=125Bd Ch=8 ChS=250Hz		CHL SENAPRED net	CHL
9084,00	SYSTEME 3000	B=2400Bd		F F	
9088,00	CW			RUS HFDF net	RUS
9090,00	USB			BGD Mil Coxs Bazar	BGD
9090,00	CCIR 493-4	B=100Bd S=170Hz			
9090,70	MIL 188-110B SER	B=2400Bd	HWK**	S N Karlskrona	S
9091,00	CW			RUS Operational Strategic Command Air Defense net	RUS
9091,00	CIS MFSK 13	B=31,25Bd Ch=13 ChS=125Hz		MFA Moscow	RUS

Frequency	Mode	Mode Parameter	Callsign	User	Country
9091,50	ALE USB	B=125Bd Ch=8 ChS=250Hz			
9093,50	STANAG 4285	B=2400Bd	FUO	F N Toulon	F
9095,00	STANAG 4285	B=2400Bd	6WW	F N Dakar	SEN
9095,70	STANAG 4285	B=2400Bd	FUO	F N Toulon	F
9097,00	ALE USB	B=125Bd Ch=8 ChS=250Hz		G MIL net	G
9098,70	STANAG 4285	B=2400Bd	FUG	F N Saissac	F
9100,00	HAARP WAVEFORM			HAARP	USA
9100,00	ALE USB	B=125Bd Ch=8 ChS=250Hz		USA Army NG net Utah	USA
9100,60	STANAG 4481	B=150Bd S=850Hz	FUG	F N Saissac	F
9105,00	MIL 188-110A SER	B=2400Bd			
9106,00	ALE USB	B=125Bd Ch=8 ChS=250Hz		ROU diplo net	ROU
9106,00	ALE USB	B=125Bd Ch=8 ChS=250Hz		USA National Telecommunication Coordination net	USA
9106,00	ALE USB	B=125Bd Ch=8 ChS=250Hz		USA customs	USA
9106,00	CIS 500-1000	B=500Bd S=1000Hz		RUS N	RUS
9106,00	ALE USB	B=125Bd Ch=8 ChS=250Hz		USA SHARES net	USA
9106,00	ALE USB	B=125Bd Ch=8 ChS=250Hz		USA COTHEN net	USA
9110,00	FAX 120/576		NMF	USA CG Boston MA	USA
9110,00	IRN N PSK	B=207Bd		IRN N	IRN
9112,00	MIL 188-110B SER	B=2400Bd		MFA Bucharest	ROU
9112,00	MIL 188-110B SER	B=2400Bd		ROU E Tel Aviv	ISR
9112,00	STANAG 4481	B=50Bd S=850Hz	NPM	USA N Lualualei	HWA
9114,00	CHN 4+4	B=75Bd Ch=8 ChS=300/450Hz		CHN MIL net	CHN
9114,00	STANAG 4481	B=50Bd S=850Hz	NPM	USA N Lualualei	HWA
9114,00	ALE USB	B=125Bd Ch=8 ChS=250Hz		USA FAA	USA
9115,00	CIS 12	B=2880Bd Ch=12 ChS=200Hz		RUS Mil	RUS
9115,00	RFSM8000	B=2400Bd		ROU Diplo	
9115,00	ALE USB	B=125Bd Ch=8 ChS=250Hz		ROU diplo net	ROU
9115,00	MIL 188-110A SER	B=2400Bd		ROU E Ankara	TUR
9115,00	MIL 188-110A SER	B=2400Bd		MFA Bucharest	ROU
9119,00	CIS 4FSK	B=150Bd Ch=4 ChS=4000Hz			
9119,50	ALE USB	B=125Bd Ch=8 ChS=250Hz		USA NG	USA
9120,00	ALE USB	B=125Bd Ch=8 ChS=250Hz		MTN MOI net	MTN
9120,00	ALE USB	B=125Bd Ch=8 ChS=250Hz		MTN net	MTN
9120,00	MIL 188-110A SER	B=2400Bd		ROU E Beirut	LBN
9120,00	MIL 188-110A SER	B=2400Bd		MFA Bucharest	ROU
9121,00	ALE USB	B=125Bd Ch=8 ChS=250Hz		USA Defence Logistic Agency	USA
9121,00	ALE USB	B=125Bd Ch=8 ChS=250Hz		USA Army NG net Utah	USA
9122,00	CIS MFSK 66	B=40Bd Ch=66 ChS=40Hz		RUS diplo	RUS
9127,00	8PSK	B=1800Bd			
9128,00	CW			RUS Strategic AF net	RUS
9130,00	CIS 50-500	B=50Bd S=500Hz	RWI	RUS GOV Moscow	RUS
9130,70	STANAG 4285	B=2400Bd		DNK N Frederickshaven	DNK

Frequency	Mode	Mode Parameter	Callsign	User	Country
9132,00	SSB			NASA launch support aircraft	USA
9132,00	*SYSTEME 3000*	*B=2400Bd*		*F MIL Vannes*	*F*
9135,00	*ALE USB*	*B=125Bd Ch=8 ChS=250Hz*		*TUR Disaster and Emergency net*	*TUR*
9135,00	CW		RIW	RUS N HQ Moscow	RUS
9140,00	ALE USB	B=125Bd Ch=8 ChS=250Hz		CHL SENAPRED net	CHL
9140,00	*CIS 4FSK*	*B=100Bd Ch=4 ChS=500Hz*		*RUS MIL Moscow*	*RUS*
9140,00	CIS ARQ	B=100Bd S=500Hz		RUS PTT	RUS
9140,00	*CIS 100-500*	*B=100Bd S=500Hz*		*RUS MIL St. Petersburg*	*RUS*
9141,70	MIL 188-110B	B=2400Bd			
9142,00	CW			RUS HFDF net	RUS
9145,00	CW		RIW	RUS N HQ Moscow	RUS
9145,00	CW		RKV95	RUS N	RUS
9145,20	*STANAG 4285*	*B=2400Bd*		*G MIL TX St. Eval*	*G*
9146,60	MIL 188-110B SER	B=2400Bd		E N	E
9146,60	USB			E N	E
9150,00	ALE USB	B=125Bd Ch=8 ChS=250Hz		GANOB net	EGY
9151,00	CIS OFDM 35	B=2800Bd Ch=35 ChS=50Hz		RUS diplo	RUS
9151,00	CHN 4+4	B=75Bd Ch=8 ChS=300/450Hz			CHN
9152,70	SITOR A	B=100Bd S=170Hz		EGY E Belgrade	YUG
9153,00	CHN4+4	B=75Bd Ch=8 ChS=300/450Hz	XSV70	CHN Great Wall Base	ANT
9155,00	WINDRM	B=1912,5Bd Ch=51 ChS=46,875Hz		CUB Intelligence Directorate	CUB
9155,40	STANAG 4285	B=2400Bd		AUS MHFCS Humpty Doo	AUS
9159,00	STANAG 4285	B=2400Bd		AUS MHFCS Humpty Doo	AUS
9159,00	MIL 188-110B SER	B=2400Bd		AUS MIL	AUS
9159,00	ALE USB	B=125Bd Ch=8 ChS=250Hz		CHN net	CHN
9160,00	*CIS 100-500*	*B=100Bd S=500Hz*		*RUS MIL E Moscow*	*RUS*
9160,00	CW			RUS HFDF net	RUS
9160,00	CIS ARQ	B=100Bd S=500Hz		RUS PTT	RUS
9160,00	ALE USB	B=125Bd Ch=8 ChS=250Hz		ARS MIL net	ARS
9162,00	CW		P5O	INS N net	INS
9164,00	BAUDOT	B=200Bd S=1000Hz		RUS INTEL	RUS
9165,00	FAX 120/576		HLL2	Seoul M	KOR
9167,00	*RUS HYBRID MODEM MFSK-PSK*	*B=40Bd Ch=16*		*Moscow*	*RUS*
9167,00	*RUS HYBRID MODEM PSK-PSK*	*B=125Bd Ch=2 ChS=250Hz*		*Moscow*	*RUS*
9169,00	T600	B=50Bd S=200Hz		RUS N HQ Vladivostok	RUS
9169,00	ALE LSB	B=125Bd Ch=8 ChS=250Hz		TWN N net	TWN
9175,00	STANAG 4538	B=2400Bd			
9175,00	CIS 50-250	B=50Bd S=250Hz		RUS MIL	RUS
9180,00	ALE USB	B=125Bd Ch=8 ChS=250Hz		TUR Disaster and Emergency net	TUR
9180,00	PACTOR III	B=100Bd S=200Hz	WHX	Droop Mountain , WV	USA
9182,00	ALE USB	B=125Bd Ch=8 ChS=250Hz		CHN net	
9183,50	ALE USB	B=125Bd Ch=8 ChS=250Hz		USA FBI net	USA
9187,00	ALE USB	B=125Bd Ch=8 ChS=250Hz		SUI MIL net	
9187,00	MIL 188-110B SER	B=2400Bd		SUI MIL	

Frequency	Mode	Mode Parameter	Callsign	User	Country
9190,00	ALE USB	B=125Bd Ch=8 ChS=250Hz		S FRO net	S
9190,00	*STANAG 4285*	*B=2400Bd*		*G MIL TX Inskip*	*G*
9193,00	CIS 100-1000	B=100Bd S=1000Hz		RUS	
9198,00	ALE USB	B=125Bd Ch=8 ChS=250Hz		CHL N	CHL
9201,00	ALE USB	B=125Bd Ch=8 ChS=250Hz		CHN MIL net	CHN
9201,00	BAUDOT	B=50Bd S=500Hz	NT9P**		RUS
9202,00	CIS AKULA	B=500Bd S=1000Hz		RUS N	RUS
9204,00	CIS OFDM 80	B=2400Bd Ch=80 ChS=37,5Hz			
9205,20	PACKET RADIO	B=300Bd S=200Hz	CE4ANK	Ankara	TUR
9205,20	PACKET RADIO	B=300Bd S=200Hz	HQ2CEN		
9205,20	PACKET RADIO	B=300Bd S=200Hz	CE4ALG		
9205,50	PACKET RADIO	B=300Bd S=200Hz	HQ2CEN		
9205,50	PACKET RADIO	B=300Bd S=200Hz	CE4ANK		
9205,50	PACKET RADIO	B=300Bd S=200Hz	CE4ALG		
9206,00	*CIS 12*	*B=1440Bd Ch=12 ChS=200Hz*		*RUS MIL Moscow*	*RUS*
9206,00	BAUDOT	B=50Bd S=500Hz	NT9P**		RUS
9207,00	MIL 188-110C APP D	B=7200Bd			
9210,00	MIL 188-110C APP D	B=4800Bd			
9210,00	ALE USB	B=125Bd Ch=8 ChS=250Hz		ARS AF net	ARS
9211,50	PACTOR I	B=100Bd S=170Hz Ch=1		ALG MOI	ALG
9213,00	CIS OFDM 60	B=2400Bd Ch=60 ChS=44,5Hz			RUS
9219,00	*ALE USB*	*B=125Bd Ch=8 ChS=250Hz*		*ISR AF net*	*ISR*
9219,00	CHN 4+4	B=75Bd Ch=8 ChS=300/450Hz		CHN MIL	
9220,00	SSB			USA AF Strategic Command	USA
9222,00	CW			RUS Operational Strategic Command Air Defense net	RUS
9224,00	*T600*	*B=50Bd S=200Hz*		*RUS N Murmansk*	*RUS*
9228,00	CW			RUS HFDF net	RUS
9229,00	MIL 188-110B SER	B=2400Bd		S MIL	S
9230,00	*CIS 100-500*	*B=100Bd S=500Hz*		*RUS MIL Krasnodar*	*RUS*
9232,00	CIS 12	B=1440Bd Ch=12 ChS=200Hz	RIT	RUS N HQ Severomorsk	RUS
9235,00	ALE USB	B=125Bd Ch=8 ChS=250Hz		S FRO net	S
9235,50	ALE USB	B=125Bd Ch=8 ChS=250Hz		USA Army NG net	USA
9240,00	*ALE USB*	*B=125Bd Ch=8 ChS=250Hz*		*TUR Disaster and Emergency net*	*TUR*
9240,00	WINDRM	B=1912,5Bd Ch=51 ChS=46,875Hz		CUB Intelligence Directorate	CUB
9240,30	*CIS 100-1000*	*B=100Bd S=1000Hz*		*Moscow*	*RUS*
9241,00	CW		RJS	RUS NHQ Vladivostok	RUS
9244,00	USB			EGY N net	EGY
9248,50	*CIS 12*	*B=1440Bd Ch=12 ChS=200Hz*	*RCV*	*RUS N HQ Sevastopol*	*UKR*
9250,00	ALE USB	B=125Bd Ch=8 ChS=250Hz		LBY GMRA net	LBY
9251,00	CHN 4+4	B=75Bd Ch=8 ChS=300/450Hz		CHN N Red Sea	

Frequency	Mode	Mode Parameter	Callsign	User	Country
9259,00	CIS OFDM 121	B=2541Bd Ch=121 ChS=25Hz		MFA Moscow	RUS
9270,00	ALE USB	B=125Bd Ch=8 ChS=250Hz		USA Department of Agriculture net	USA
9273,00	CHN OFDM 30TONE	B=1800Bd Ch=30 ChS=60Hz		CHM MIL	CHN
9276,00	CIS MFSK 64	B=40Bd Ch=64 ChS=46Hz		RUS E	
9286,00	ALE USB	B=125Bd Ch=8 ChS=250Hz		G MIL net	G
9290,00	ALE USB	B=125Bd Ch=8 ChS=250Hz		USA AF MARS net	USA
9293,00	ALE USB	B=125Bd Ch=8 ChS=250Hz		ISR AF net	ISR
9295,00	ALE USB	B=125Bd Ch=8 ChS=250Hz		USA National Guard net	USA
9295,00	ALE USB	B=125Bd Ch=8 ChS=250Hz		USA Army NG net Utah	USA
9296,00	BAUDOT	B=50Bd S=500Hz	NT9P**		RUS
9299,00	STANAG 4197	B=1800Bd Ch=16/39 ChS=112/56Hz			EGY
9300,00	ALE USB	B=125Bd Ch=8 ChS=250Hz		POL F net	
9300,00	CIS 4FSK	B=150Bd Ch=4 ChS=4000Hz			RUS
9300,00	CHN 4+4	B=75Bd Ch=8 ChS=300/450Hz		CHN	
9300,00	ALE USB	B=125Bd Ch=8 ChS=250Hz		Queensland Deptartment of Community Safety	AUS
9300,00	CIS 200-500 BAUDOT	B=200Bd S=500Hz		RUS diplo	RUS
9300,00	CIS 50-500 BAUDOT	B=50Bd S=500Hz		RUS diplo	RUS
9303,50	USB			USA AF MARS	
9304,90	*CIS 50-500*	*B=50Bd S=500Hz*		*RUS F Moscow*	*RUS*
9305,00	ALE USB	B=125Bd Ch=8 ChS=250Hz		USA SAC net	USA
9309,00	ALE USB	B=125Bd Ch=8 ChS=250Hz		BUL MOI net	BUL
9310,00	MIL 188-110B SER	B=2400Bd		CHN diplo	
9311,00	CW		RBL88	RUS N	RUS
9311,10	ALE USB	B=125Bd Ch=8 ChS=250Hz		USA FBI net	USA
9315,00	ALE USB	B=125Bd Ch=8 ChS=250Hz		ALG oil/gas fields	ALG
9316,00	CW			RUS HFDF net	RUS
9323,00	USB		VMD750	Austravel Safety Net	AUS
9330,00	WINDRM	B=1912,5Bd Ch=51 ChS=46,875Hz		CUB Intelligence Directorate	CUB
9335,00	*STANAG 4285*	*B=2400Bd*		*G MIL TX Crimond*	*G*
9336,00	CW		RAA	RUS N HQ Moscow	RUS
9338,00	STANAG 4481	B=50Bd S=350Hz	NAU	USA N San Juan	PTR
9338,00	STANAG 4481	B=50Bd S=850Hz	NPM	USA N Lualualei	HWA
9338,50	ALE USB	B=125Bd Ch=8 ChS=250Hz		USA CG net	USA
9345,00	ALE USB	B=125Bd Ch=8 ChS=250Hz		TUN MIL net	TUN
9346,00	T600	B=50Bd S=250Hz	RDL	RUS N Moscow	RUS
9346,00	CW		RDL	RUS N Moscow	RUS
9350,00	CW			RUS HFDF net	RUS
9353,70	MIL 188-110B APP C	B=2400Bd		S MIL	S

Frequency	Mode	Mode Parameter	Callsign	User	Country
9354,00	ALE USB	B=125Bd Ch=8 ChS=250Hz			
9355,00	*STANAG 4285*	*B=2400Bd*		*G MIL TX Akrotiri*	*CYP*
9358,00	CW			RUS HFDF net	RUS
9360,00	CIS 50-500	B=50Bd S=500Hz	REA	RUS GOV	RUS
9360,00	USB		VMS469	Reids Radiodata net	AUS
9360,00	ISR N HYBRID MODEM	B=2400Bd	4XZ	ISR N Haifa	ISR
9373,00	CW		RMP	RUS N HQ Kaliningrad	RUS
9375,00	ALE USB	B=125Bd Ch=8 ChS=250Hz		F N net	OCE
9380,00	STANAG 4481	B=75Bd S=850Hz	NPN	USA N Guam	GUM
9384,00	LSB		OL1B	CZR diplo	
9384,00	PACTOR III	B=100Bd S=200Hz	OL1B	CZR diplo	
9389,00	ALE USB	B=125Bd Ch=8 ChS=250Hz		POL MIL net	
9393,00	ARCOTEL FARCOS	B=1800Bd			
9395,00	STANAG 4285	B=2400Bd		AUS MHFCS Humpty Doo	AUS
9395,00	CIS 75-200	B=75Bd S=200Hz		CIS ship frequency	
9396,00	CW		RHQ33	RUS N	RUS
9402,00	ALE USB	B=125Bd Ch=8 ChS=250Hz		CHN MIL net	CHN
9413,00	ALE USB	B=125Bd Ch=8 ChS=250Hz		ROU diplo net	ROU
9414,50	ALE USB	B=125Bd Ch=8 ChS=250Hz		Public health net , TX	USA
9428,00	MIL 188-110A SER	B=2400Bd		ROU E Tbilisi	GEO
9428,00	MIL 188-110A SER	B=2400Bd		MFA Bucharest	ROU
9428,00	ALE USB	B=125Bd Ch=8 ChS=250Hz		ROU diplo net	ROU
9441,00	ALE USB	B=125Bd Ch=8 ChS=250Hz		AUT MIL net	AUT
9443,00	T600	B=50Bd S=250Hz	RIT	RUS N HQ Severomorsk	RUS
9445,00	ALE USB	B=125Bd Ch=8 ChS=250Hz		USA FEMA net	USA
9448,00	*T600*	*B=50Bd S=200Hz*		*RUS N Moscow*	*RUS*
9450,00	PSK 125	B=125Bd	KBC	KBC Trucker Radio	USA
9452,00	*ALE USB*	*B=125Bd Ch=8 ChS=250Hz*		*F N net*	*F*
9454,00	CW			CHN air defense net	CHN
9462,00	ALE USB	B=125Bd Ch=8 ChS=250Hz		USA FEMA net	USA
9471,00	CIS OFDM 45	B=2400Bd Ch=45 ChS=62,5Hz		RUS diplo	RUS
9493,00	ALE USB	B=125Bd Ch=8 ChS=250Hz		F N net	F
9500,00	*CIS 12*	*B=1440Bd Ch=12 ChS=200Hz*		*RUS MIL Moscow*	*RUS*
9555,00	*DPRK ARQ*	*B=600Bd S=600Hz*		*DPRK E Moscow*	*RUS*
9574,00	CIS 50-200	B=50Bd S=200Hz	RMP	RUS N HQ Kaliningrad	RUS
9595,00	CW		REA4	RUS AF Moscow	RUS
9600,00	HAARP waveform			HAARP	USA
9670,00	AM			Radio DARC	D
9684,00	*CIS 12*	*B=1440Bd Ch=12 ChS=200Hz*	*RMP*	*RUS N HQ Kaliningrad*	*RUS*
9689,00	*CIS 12*	*B=1440Bd Ch=12 ChS=200Hz*		*RUS MIL Moscow*	*RUS*
9700,00	CW		RMX	RUS N AF	RUS
9719,00	USB		SDJ	Stockholm R	S
9733,00	*CIS 12*	*B=1440Bd Ch=12 ChS=200Hz*		*RUS N HQ Moscow*	*RUS*

Frequency	Mode	Mode Parameter	Callsign	User	Country
9742,10	ALE USB	B=125Bd Ch=8 ChS=250Hz		ROU diplo net	ROU
9755,00	DPRK ARQ	B=1200Bd S=600Hz		DPRK E Warsaw	POL
9770,00	VS: PANTHER-H	B=2400Bd			
9830,00	STANAG 4481	B=50Bd S=850Hz	NAU	USA N San Juan	PTR
9830,00	STANAG 4481	B=50Bd S=850Hz	NSS	USA N TX Davidsonville , MD	USA
9836,00	T600	B=50Bd S=200Hz	RDL	RUS N Moscow TX Kaliningrad	RUS
9870,00	ALE USB	B=125Bd Ch=8 ChS=250Hz			
9870,00	STANAG 4529	B=1200Bd			
9870,00	ALE USB	B=125Bd Ch=8 ChS=250Hz		JOR net	JOR
9877,00	ALE USB	B=125Bd Ch=8 ChS=250Hz		USA FEMA net	USA
9890,00	*STANAG 4285*	*B=2400Bd*	*CFH*	*CAN N Halifax*	*CAN*
9900,00	CIS 8181	B=81Bd S=500Hz		RUS MIL	RUS
9901,00	CHN 4+4	B=75Bd Ch=8 ChS=300/450Hz	VSF92		
9905,00	ALE USB	B=125Bd Ch=8 ChS=250Hz		G MIL net	G
9906,00	*STANAG 4285*	*B=2400Bd*	*FUG*	*F N Saissac*	*F*
9909,00	ALE USB	B=125Bd Ch=8 ChS=250Hz		POL MIL net	POL
9910,60	STANAG 4285	B=2400Bd	FUJ	F N Noumea	NCL
9920,70	*STANAG 4285*	*B=2400Bd*	*IDR*	*I N Rome*	*I*
9928,00	ALE USB	B=125Bd Ch=8 ChS=250Hz		USA FEMA net	USA
9929,00	*CIS 75-250*	*B=75Bd S=250Hz*		*RUS N Moscow*	*RUS*
9931,00	ALE USB	B=125Bd Ch=8 ChS=250Hz		ALG AF net	ALG
9933,00	CW		RAL2	RUS N HQ Astrakhan	RUS
9935,00	*CIS 50-500*	*B=50Bd S=500Hz*		*Moscow*	*RUS*
9940,00	USB		VNJ	Casey Base , AUS	ANT
9940,00	USB			AUS Davis Base	ANT
9940,00	USB			Macquarie Island , AUS	ANT
9940,00	USB			Mawson Base , AUS	ANT
9950,00	CIS 4FSK	B=150Bd Ch=4 ChS=4000Hz		Moscow	RUS
9959,00	ALE USB	B=125Bd Ch=8 ChS=250Hz			
9963,00	CW		4XZ	ISR N Haifa	ISR
9970,00	ALE USB	B=125Bd Ch=8 ChS=250Hz		CHL MOI net	CHL
9970,00	STANAG 4285	B=2400Bd		G MIL Crimond	G
9982,50	FAX 120/576		KVM70	USA CG Honululu	HWA
9985,00	USB		VKE237	AUS HF Radio Club	AUS
9986,50	*CIS 75-200*	*B=75Bd S=200Hz*		*RUS N Murmansk*	*RUS*
9996,00	*CW*		*RWM*	*Moscow TS*	*RUS*
10000,00	AM		LOL	Buenos Aires TS	ARG
10000,00	USB		PPE	Rio De Janeiro TS	B
10000,00	AM		WWV	Fort Collins TS , CO	USA
10000,00	AM		WWVH	Kekaha Kauai TS	HWA
10000,00	CW , AM		BPM	Xi`an TS	CHN
10000,00	8PSK	B=1800Bd			
10000,00	CIS 12	B=1440Bd Ch=12 ChS=200Hz		RUS MIL	RUS
10000,00	STANAG 4285	B=2400Bd	FUV	F N Djibouti	DJI
10000,00	*F N FSK*	*B=50Bd S=850Hz Ch=1*	*FUG*	*F N Saissac*	*F*

10000 – 15000 kHz

Frequency	Mode	Mode Parameter	Callsign	User	Country
10000,00	AM		LOL	Buenos Aires TS	ARG
10000,00	USB		PPE	Rio De Janeiro TS	B
10000,00	AM		WWV	Fort Collins TS , CO	USA
10000,00	AM		WWVH	Kekaha Kauai TS	HWA
10000,00	CW , AM		BPM	Xi`an TS	CHN
10000,00	8PSK	B=1800Bd			
10000,00	CIS 12	B=1440Bd Ch=12 ChS=200Hz		RUS MIL	RUS
10000,00	STANAG 4285	B=2400Bd	FUV	F N Djibouti	DJI
10000,00	*F N FSK*	*B=50Bd S=850Hz Ch=1*	*FUG*	*F N Saissac*	*F*
10006,00	SSB			RDARA 10	
10006,00	SSB			RDARA 13G	
10006,00	SSB			RDARA 6A	
10009,00	SSB			RDARA 13K	
10009,00	SSB			RDARA 2B	
10009,00	SSB			RDARA 2C	
10009,00	SSB			RDARA 7B	
10009,00	SSB			RDARA 9B	
10012,00	SSB			RDARA 10	
10012,00	SSB			RDARA 13J	
10012,00	SSB			RDARA 5	
10015,00	SSB			RDARA 12D	
10015,00	SSB			RDARA 2	
10015,00	SSB			RDARA 6C	
10015,00	ALE USB	B=125Bd Ch=8 ChS=250Hz		MTN net	MTN
10018,00	USB			Almaty ACC	KAZ
10018,00	SSB			Ashkabad ACC	TKM
10018,00	SSB			Mumbai ACC	IND
10018,00	SSB			Dushanbe ACC	TJK
10018,00	SSB			Kabul ACC	AFG
10018,00	SSB			Karachi ACC	PAK
10018,00	SSB			Kathmandu ACC	NPL
10018,00	SSB			Kuwait ACC	KWT
10018,00	SSB			Lahore ACC	PAK
10018,00	SSB			Muscat ACC	OMA
10018,00	SSB			MWARA MID	MID
10018,00	SSB			RDARA 13J	
10018,00	SSB			RDARA 13K	
10018,00	SSB			RDARA 6G	
10018,00	SSB			RDARA 9	
10018,00	SSB			Samarkand ACC	UZB
10018,00	SSB			Seychelles ACC , Mahe	SEY
10018,00	SSB			Tashkent ACC	UZB
10018,00	SSB			Teheran ACC	IRN
10018,00	SSB			Urumchi ACC	CHN
10018,00	USB			Addis Ababa ACC	ETH
10018,00	USB			Aden ACC	YEM
10018,00	USB			Asmara ACC	ERT
10018,00	USB			Bahrain ACC	BHR
10018,00	USB			Benghazi ACC	LBY
10018,00	USB			Mumbai ACC	IND
10018,00	USB			Bujumbura ACC	BDI
10018,00	USB			Cairo ACC	EGY
10018,00	USB			Comoros ACC	COM
10018,00	USB			Dar es Salaam ACC	TZA
10018,00	USB			Entebbe ACC	UGA

Frequency	Mode	Mode Parameter	Callsign	User	Country
10018,00	USB			Hargeisa ACC	SOM
10018,00	USB			Djibouti ACC	DJI
10018,00	USB			Jeddah ACC	ARS
10018,00	USB			Khartoum ACC	SDN
10018,00	USB			Kigali ACC	RRW
10018,00	USB			Kisimayu ACC	SOM
10018,00	USB			Male ACC	MLD
10018,00	USB			Mogadishu ACC	SOM
10018,00	USB			Nairobi ACC	KEN
10018,00	USB			Port Sudan ACC	SDN
10018,00	USB			Sanaa ACC	YEM
10018,00	USB			Seychelles ACC	SEY
10018,00	USB			Tripoli ACC	LBY
10018,00	USB			Bishkek ACC	KGZ
10018,00	USB			New Delhi ACC	IND
10018,00	USB			Odessa ACC	UKR
10018,00	USB			Tbilisi ACC	GEO
10018,00	USB			Yerevan ACC	ARM
10018,00	USB			Abadan ACC	IRN
10021,00	CW		S	RUS AF marker	RUS
10021,00	SSB			RDARA 12C	
10021,00	SSB			RDARA 13G	
10021,00	SSB			RDARA 6B	
10024,00	USB			Barranquilla ACC	CLM
10024,00	USB			Maiquetia ACC	VEN
10024,00	USB			Buenos Aires ACC	ARG
10024,00	USB			Easter Island ACC	CHL
10024,00	SSB			Antofagasta ACC	CHL
10024,00	SSB			Asuncion ACC	PRG
10024,00	SSB			Bogota ACC	CLM
10024,00	SSB			Cordoba ACC	ARG
10024,00	SSB			La Paz ACC	BOL
10024,00	SSB			Lima ACC	PRU
10024,00	SSB			MWARA SAM	SAM
10024,00	SSB			Punta Arenas ACC	CHL
10024,00	SSB			Purerto Montt ACC	CHL
10024,00	SSB			Quito ACC	EQA
10024,00	SSB			RDARA 2B	
10024,00	SSB			RDARA 2C	
10024,00	SSB			RDARA 3B	
10024,00	SSB			RDARA 9B	
10024,00	SSB			Santa Cruz ACC	BOL
10024,00	SSB			Santiago ACC	CHL
10024,00	USB			Talara ACC	PRU
10024,00	USB			CENAMER Tegucigalpa ACC	HND
10024,00	USB			Ushuaia ACC	ARG
10027,00	SSB			W I	WW
10027,00	SSB			W II	WW
10027,00	HFDL	B=1800Bd	H02**	Molokai , HI	HWA
10027,00	USB			IBERIA Madrid ACC	E
10027,00	HFDL	B=1800Bd	H09**	Utqiagvik , AK	ALS
10027,00	USB			LDOC Iberia	E
10030,00	SSB			W I	WW
10030,00	SSB			W IV	WW
10033,00	USB			PanAM R	USA
10033,00	SSB			ARINC Honolulu	HWA
10033,00	SSB			W II	WW
10033,00	SSB			W V	WW
10036,00	SSB			RDARA 13G	
10036,00	SSB			RDARA 13H	
10036,00	SSB			RDARA 1E	
10036,00	SSB			RDARA 6E	

Frequency	Mode	Mode Parameter	Callsign	User	Country
10039,00	SSB			RDARA 12C	
10039,00	SSB			RDARA 3B	
10039,00	SSB			RDARA 3C	
10039,00	SSB			RDARA 4A	
10039,00	SSB			RDARA 9B	
10039,00	USB			Chita ACC	RUS
10039,00	USB			Chulman ACC	RUS
10039,00	USB			Ekimchan ACC	RUS
10039,00	USB			Irkutsk ACC	RUS
10039,00	USB			Kirensk ACC	RUS
10039,00	USB			Khabarovsk ACC	RUS
10039,00	USB			Pyongyang ACC	KRE
10039,00	USB			Ulaanbaatar ACC	MNG
10039,00	USB			Ulan Ude ACC	RUS
10042,00	SSB			MWARA EA	EA
10042,00	SSB			RDARA 10F	
10042,00	SSB			RDARA 13C	
10042,00	SSB			RDARA 13J	
10042,00	SSB			RDARA 13K	
10042,00	SSB			RDARA 9C	
10042,00	USB			Beijing ACC	CHN
10042,00	USB			Guangzhou ACC	CHN
10042,00	USB			Hailar ACC	CHN
10042,00	USB			Irkutsk ACC	RUS
10042,00	USB			Jinan ACC	CHN
10042,00	USB			Kunming ACC	CHN
10042,00	USB			Lanzhou ACC	CHN
10042,00	USB			Shanghai ACC	CHN
10042,00	USB			Shenyang ACC	CHN
10042,00	USB			Taegu ACC	KOR
10042,00	USB			Pyongyang ACC	KRE
10042,00	USB			Ulaanbaatar ACC	MNG
10042,00	USB			Urumqi ACC	CHN
10042,00	USB			Wuhan ACC	CHN
10042,00	USB			Zhengzhou ACC	CHN
10045,00	SSB			RDARA 11B	
10045,00	SSB			RDARA 13H	
10045,00	SSB			RDARA 14	
10045,00	SSB			RDARA 2	
10045,00	SSB			RDARA 3A	
10045,00	USB		BOING SEATTLE	Boing Seattle	USA
10045,00	USB		BOING EVERETT	Boing Seattle	USA
10048,00	SSB			MWARA NP	NP
10048,00	SSB			RDARA 13A	
10048,00	SSB			RDARA 13B	
10048,00	SSB			RDARA 2A	
10048,00	SSB			RDARA 5D	
10048,00	SSB			San Francisco ACC , CA	USA
10048,00	SSB			Tokyo ACC	J
10050,00	CHN 16TONE PSK	B=75Bd Ch=16		CHN MIL	CHN
10051,00	USB		WSY70	New York VOLMET	USA
10051,00	SSB			RDARA 13I	
10051,00	SSB			RDARA 6A	
10051,00	SSB			RDARA 6E	
10051,00	SSB			VOLMET NAT	NAT
10051,00	USB		VFG	Gander VOLMET	CAN
10051,70	SITOR A	B=100Bd S=170Hz		MFA Cairo	EGY
10054,00	SSB			RDARA 12	
10054,00	SSB			RDARA 2A	
10054,00	SSB			RDARA 2C	

Frequency	Mode	Mode Parameter	Callsign	User	Country
10054,00	SSB			RDARA 6G	
10057,00	USB			San Francisco ACC , CA	USA
10057,00	SSB		TNL	Brazzaville VOLMET	COG
10057,00	SSB			MWARA CEP	CEP
10057,00	SSB			RDARA 3A	
10057,00	SSB			VOLMET AFI	AFI
10057,00	USB		5ST	Antananarivo VOLMET	MDG
10060,00	SSB			RDARA 13K	
10060,00	SSB			RDARA 1D	
10060,00	SSB			RDARA 6F	
10060,00	HFDL	B=1800Bd	H10**	Muan	KOR
10063,00	HFDL	B=1800Bd	H11**	Albrook	PNR
10063,00	SSB			RDARA 12E	
10063,00	SSB			RDARA 4B	
10063,00	SSB			RDARA 6G	
10066,00	SSB			Bangkok ACC	THA
10066,00	SSB			Calcutta ACC	IND
10066,00	SSB			Colombo ACC	CLN
10066,00	SSB			Guangzhou ACC	CHN
10066,00	SSB			Jakarta ACC	INS
10066,00	SSB			Kathmandu ACC	NPL
10066,00	SSB			Kuala Lumpur ACC	MLA
10066,00	SSB			Kunmig ACC	CHN
10066,00	SSB			Madras ACC	IND
10066,00	SSB			Male ACC	MLD
10066,00	SSB			MWARA SEA	SEA
10066,00	SSB			RDARA 10A	
10066,00	SSB			RDARA 13M	
10066,00	SSB			RDARA 1B	
10066,00	SSB			Singapore ACC	SNG
10066,00	SSB			Yangon ACC	BRM
10066,00	HFDL	B=1800Bd	H06**	Hat Yai	THA
10066,00	USB			Bali ACC	INS
10066,00	USB			Dhaka ACC	BGD
10066,00	USB			Brisbane ACC	AUS
10066,00	USB			Jakarta ACC	INS
10066,00	USB			Ujung Pandang ACC	INS
10069,00	SSB			W I	WW
10069,00	SSB			W IV	WW
10072,00	SSB			W I	WW
10072,00	SSB			W III	WW
10075,00	SSB			W II	WW
10075,00	SSB			W V	WW
10075,00	HFDL	B=1800Bd	H15**	Al Muharraq	BHR
10078,00	SSB			W I	WW
10078,00	SSB			W III	WW
10081,00	HFDL	B=1800Bd	H01**	San Francisco , CA	USA
10081,00	SSB			MWARA CWP	CWP
10081,00	SSB			RDARA 13F	
10081,00	SSB			RDARA 4A	
10081,00	SSB			RDARA 6A	
10081,00	SSB			RDARA 7C	
10081,00	HFDL	B=1800Bd	H07**	Shannon	IRL
10084,00	SSB			MWARA EUR	EUR
10084,00	SSB			MWARA SP	SP
10084,00	SSB			RDARA 13D	
10084,00	SSB			RDARA 6E	
10084,00	USB			Arkhangelsk ACC	RUS
10084,00	USB			Beirut ACC	LBN
10084,00	HFDL	B=1800Bd	H05**	Auckland	NZL
10084,00	USB			Berlin ACC	D
10084,00	USB			Kiev ACC	UKR

Frequency	Mode	Mode Parameter	Callsign	User	Country
10084,00	USB			Lvov ACC	UKR
10084,00	USB			Minsk ACC	BLR
10084,00	USB			Moscow ACC	RUS
10084,00	USB			Murmansk ACC	RUS
10084,00	USB			Odessa ACC	UKR
10084,00	USB			Simferopol ACC	UKR
10084,00	USB			Riga ACC	LVA
10084,00	USB			Sofia ACC	BUL
10084,00	USB			St. Petersburg ACC	RUS
10084,00	USB			Syktyvkar ACC	RUS
10084,00	USB			Velikiye ACC	RUS
10084,00	USB			Vologda ACC	RUS
10084,00	USB			Tunis ACC	TUN
10084,00	USB			Vilnius ACC	LTU
10087,00	SSB			RDARA 14	
10087,00	SSB			RDARA 3	
10087,00	SSB			VOLMET SAM	SAM
10087,00	HFDL	B=1800Bd	H14**	Krasnoyarsk	RUS
10090,00	SSB			RDARA 12E	
10090,00	SSB			RDARA 12F	
10090,00	SSB			VOLMET NCA	NCA
10093,00	SSB			RDARA 11B	
10093,00	SSB			RDARA 13N	
10093,00	SSB			RDARA 5B	
10093,00	SSB			RDARA 6B	
10093,00	HFDL	B=1800Bd	H09**	Utqiagvik , AK	ALS
10096,00	SSB			Asuncion ACC	PRG
10096,00	SSB			Belem ACC	B
10096,00	SSB			Bogota ACC	CLM
10096,00	SSB			Brasilia ACC	B
10096,00	SSB			Campo Grande ACC	B
10096,00	SSB			Georgtown ACC	GUY
10096,00	SSB			La Paz ACC	BOL
10096,00	SSB			Maiquetia ACC , Caracas	VEN
10096,00	SSB			Manaus ACC	B
10096,00	SSB			Montevideo ACC	URG
10096,00	SSB			MWARA NCA	NCA
10096,00	SSB			MWARA SAM	SAM
10096,00	SSB			Paramaribo ACC	SUR
10096,00	SSB			Piarco ACC	ARG
10096,00	SSB			Porto Alegre ACC	B
10096,00	SSB			Porto Velho ACC	B
10096,00	SSB			RDARA 7D	
10096,00	SSB			Recife ACC	B
10096,00	SSB			Salvador ACC	B
10096,00	SSB			Santa Cruz ACC	BOL
10096,00	USB			Cayenne ACC	GUF
10096,00	USB			Iquitos ACC	PRU
10096,00	USB			Leticia ACC	CLM
10096,00	USB			Rio de Janeiro ACC	B
10096,00	USB			Buenos Aires ACC	ARG
10096,00	USB			Lima ACC	PRU
10096,00	USB			Barnaul ACC	RUS
10096,00	USB			Irkutsk ACC	RUS
10096,00	USB			Khanty-Mansiysk ACC	RUS
10096,00	USB			Kirensk ACC	RUS
10096,00	USB			Krasnoyarsk ACC	RUS
10096,00	USB			Novosibirsk ACC	RUS
10096,00	USB			Podkamennaya ACC	RUS
10096,00	USB			Surgut ACC	RUS
10096,00	USB			Yeniseysk ACC	RUS

Frequency	Mode	Mode Parameter	Callsign	User	Country
10100,00	ALE USB	B=125Bd Ch=8 ChS=250Hz		ALG MIL net	ALG
10100,80	*BAUDOT*	*B=50Bd S=450Hz*	*DDK9*	*DWD TX Pinneberg*	*D*
10102,00	CW		RBL88	RUS N	RUS
10103,00	*CIS 12*	*B=1440Bd Ch=12 ChS=200Hz*		*RUS MIL Moscow*	*RUS*
10107,00	ALE USB	B=125Bd Ch=8 ChS=250Hz		USA AF net	
10110,00	ALE LSB	B=125Bd Ch=8 ChS=250Hz			AFG
10110,20	*STANAG 4285*	*B=2400Bd*		*G MIL Akrotiri*	*CYP*
10111,90	CIS 100-1000	B=100Bd S=1000Hz		RUS	
10114,00	ALE USB	B=125Bd Ch=8 ChS=250Hz		ALG net	ALG
10114,70	CIS 14	B=96Bd S=1000Hz		PTT Moscow	RUS
10114,70	PACTOR II	B=100Bd S=200Hz	STAT151**	MFA Tunis	TUN
10114,80	*CIS 100-1000*	*B=100Bd S=1000Hz*		*RUS MIL Moscow*	*RUS*
10115,00	ALE USB	B=125Bd Ch=8 ChS=250Hz		VEN F nez	VEN
10115,00	*4FSK*	*B=100Bd Ch=4 ChS=500Hz*		*RUS MIL Moscow*	*RUS*
10115,00	USB			Wireless Institute Civil Emergency Service WICEN	AUS
10117,00	FAX 120/576		BAF4	Bejing M	CHN
10121,00	*CIS 12*	*B=1440Bd Ch=12 ChS=200Hz*		*RUS MIL Moscow*	*RUS*
10126,00	CW			RUS HFDF net	RUS
10130,00	THALES SKYMASTER ALE	B=125Bd Ch=8 ChS=250Hz			
10130,00	STANAG 4285	B=2400Bd	NSS	USA N TX Davidsonville , MD	USA
10130,00	FAX 90/576		RBW48	Murmansk M	RUS
10130,00	ROS USB	B=1Bd Ch=144 ChS=15,625Hz		ROS frequency	
10130,00	JS8	B=6,25Bd Ch=8 ChS=6,25Hz		JS8 user	
10131,00	ALE USB	B=125Bd Ch=8 ChS=250Hz		HFLINK network	
10133,00	VARA	Ch=1		VARA chat freqeuency	
10135,00	ALE USB	B=125Bd Ch=8 ChS=250Hz		CHL SENAPRED net	CHL
10136,00	FT8	B=5,86Bd Ch=8 ChS=5,86Hz		FT8 channel	
10138,00	USB			Ham radio emergency frequency	R1
10138,00	JT65	B=2,69Bd Ch=65 ChS=2,69Hz		JT65 channel	
10138,70	WSPR	B=1,46Bd Ch=4 ChS=1,46Hz		WSPR net	
10140,00	SSB			USA AF MARS	ALS
10140,00	JT9	B=1,736Bd Ch=9 ChS=1,736Hz		JT9 channel	
10140,00	FT4	B=5,86Bd Ch=8 ChS=5,86Hz		FT4 channel	
10141,00	PSK31	B=31Bd		PSK31 channel	
10141,00	MFSK			MFSK modes user	
10143,00	BAUDOT	B=45,45Bd S=170Hz		RTTY user	
10144,00	CW		DK0WCY	Aurora Beacon	D
10145,00	HFPAGER	B=5,86Bd Ch=18		RUS HFPager net	RUS
10145,00	ALE USB	B=125Bd Ch=8 ChS=250Hz		HFLINK network	
10146,00	CW		HB9TC	Ticino B	SUI

Frequency	Mode	Mode Parameter	Callsign	User	Country
10147,30	ROBUST PACKET RADIO	B=200Bd Ch=8 ChS=60Hz		Robust Packet Radio net, HF-APRS	
10150,00	PACTOR III	B=100Bd S=200Hz	9Z4DH	Sailmail Chaguaramas	TRD
10150,00	ISR N HYBRID MODEM	B=2400Bd	4XZ	ISR N Haifa	ISR
10151,00	ALE LSB	B=125Bd Ch=8 ChS=250Hz		CHL N net	CHL
10151,00	CIS VFT: M/S 1.0	B=50Bd S=150Hz Ch=6 ChS=500Hz		RUS	RUS
10153,00	STANAG 4481	B=50Bd S=850Hz	NAU	USA N San Juan	PTR
10154,00	ALE USB	B=125Bd Ch=8 ChS=250Hz		USA Army NG net Utah	USA
10155,00	STANAG 4285	B=2400Bd		I N Rome	I
10155,00	STANAG 4481	B=50Bd S=850Hz	NAU	USA N San Juan	PTR
10155,00	T600	B=50Bd S=100Hz	RIW	RUS N HQ Moscow	RUS
10156,00	ISR N HYBRID MODEM	B=2400Bd	4XZ	ISR N Haifa	ISR
10160,00	ALE USB	B=125Bd Ch=8 ChS=250Hz		CHL SENAPRED net	CHL
10160,00	*ALE USB*	*B=125Bd Ch=8 ChS=250Hz*		*TUR Disaster and Emergency net*	*TUR*
10160,00	QPSK	B=4800Bd		Luxembourg area	LUX
10161,00	CIS OFDM 45	B=2400Bd Ch=45 ChS=62,5Hz		RUS diplo	RUS
10161,60	ALE USB	B=125Bd Ch=8 ChS=250Hz		ARINC Urgent Link net	USA
10162,00	ALE USB	B=125Bd Ch=8 ChS=250Hz		USA Civil Air Patrol net	USA
10163,00	CHN DATALINK	B=2400Bd Ch=30 ChS=75Hz		CHN N	CHN
10164,00	*T600*	*B=50Bd S=200Hz*	*RDL*	*RUS N HQ Moscow*	*RUS*
10165,00	RFSM8000	B=2400Bd		MFA Sofia	BUL
10165,00	RFSM8000	B=2400Bd		MFA Sofia	BUL
10167,00	LINK 11 CLEW	B=2250Bd Ch=16 ChS=330/110/550Hz			
10167,00	CW		K4MT**	RUS MIL	RUS
10167,00	BAUDOT	B=50Bd S=500Hz	K4MT**	RUS MIL	RUS
10170,00	*VFT: CIS 100-1440*	*B=100Bd S=1440Hz Ch=3 ChS=480Hz*		*Moscow*	*RUS*
10170,00	RFSM8000	B=2400Bd		BUL E Rome	I
10170,60	ALE USB	B=125Bd Ch=8 ChS=250Hz		ARINC Urgent Link net	USA
10175,00	ALE USB	B=125Bd Ch=8 ChS=250Hz		TUR Disaster and Emergency net	TUR
10175,00	*ALE USB*	*B=125Bd Ch=8 ChS=250Hz*		*TUR Disaster and Emergency net*	*TUR*
10175,00	RFSM8000	B=2400Bd		BUL E Kiev	UKR
10176,00	ALE USB	B=125Bd Ch=8 ChS=250Hz		TUN diplo net	TUN
10178,50	USB			USA AF MARS	
10180,00	USB		VKS737	AUS 4WD net	AUS
10180,00	ALE LSB	B=125Bd Ch=8 ChS=250Hz		TWN N net	TWN
10181,00	ALE USB	B=125Bd Ch=8 ChS=250Hz		USA AF net	
10184,20	*STANAG 4285*	*B=2400Bd*	*IDR*	*I N Rome*	*I*
10186,20	*STANAG 4285*	*B=2400Bd*	*FUG*	*F N Saissac*	*F*
10190,50	ALE 3G	B=2400Bd		I MIL	I
10191,00	CIS OFDM 160	B=4800Bd Ch=160 ChS=37,5Hz			
10193,50	CIS OFDM 44	B=3520Bd Ch=44 ChS=100Hz			

Frequency	Mode	Mode Parameter	Callsign	User	Country
10194,00	ALE USB	B=125Bd Ch=8 ChS=250Hz		USA FEMA net	USA
10194,00	T600	B=50Bd S=200Hz	RDL	RUS N Moscow	RUS
10195,20	STANAG 4285	B=2400Bd	PBB	NLD N TX Den Helder	HOL
10200,00	ALE USB	B=125Bd Ch=8 ChS=250Hz		F Airbus net	F
10202,00	ALE USB	B=125Bd Ch=8 ChS=250Hz		Public health net , TX	USA
10202,00	ALE USB	B=125Bd Ch=8 ChS=250Hz		USA FEMA net	USA
10202,00	CW			RUS HFDF net	RUS
10203,00	USB		VMD750	Austravel Safety Net	AUS
10203,00	CW		RJS	RUS N HQ Vladivostok	RUS
10203,70	STANAG 4285	B=2400Bd	PBB	NOR N TX Oslo	NOR
10203,70	STANAG 4285	B=2400Bd	JXU	NOR N TX Bodo	NOR
10204,00	USB			EGY N net	EGY
10204,00	STANAG 4197	B=1800Bd Ch=16/39 ChS=112/56Hz		EGY N net	EGY
10204,00	USB			EGY CG net	EGY
10206,00	MIL 188-110B 39TONE	B=2400Bd Ch=39 ChS=56,25Hz			
10206,00	PACTOR III	B=100Bd S=200Hz	WQAB964	Sailmail San Diego , CA	USA
10208,20	STANAG 4285	B=2400Bd		G MIL TX St. Eval	G
10212,00	MIL 188-110B SER	B=2400Bd		AUS MIL	AUS
10217,50	ALE USB	B=125Bd Ch=8 ChS=250Hz			
10218,00	OFDM	B=2400Bd Ch=121 ChS=25Hz		MFA Moscow	RUS
10218,00	QPSK	B=2400Bd		MFA Moscow	RUS
10223,70	SITOR A	B=100Bd S=170Hz		MFA Cairo	EGY
10227,50	ALE USB	B=125Bd Ch=8 ChS=250Hz		POL MIL net	POL
10231,00	CW		4XZ	ISR N Haifa	ISR
10232,00	ALE USB	B=125Bd Ch=8 ChS=250Hz			
10232,10	FSK2	B=200Bd S=400Hz			
10234,00	ALE USB	B=125Bd Ch=8 ChS=250Hz		CHL SENAPRED net	CHL
10235,00	LINK 11 CLEW	B=2250Bd Ch=16 ChS=330/110/550Hz		I MIL Verona	I
10235,00					
10236,00	T600	B=50Bd S=200Hz	RCV	RUS N Sevastopol	UKR
10242,00	ALE USB	B=125Bd Ch=8 ChS=250Hz		USA COTHEN net	USA
10243,00	RUS HYBRID MODEM MFSK-PSK	B=40Bd Ch=16		Moscow	RUS
10243,00	RUS HYBRID MODEM PSK-PSK	B=125Bd Ch=2 ChS=250Hz		Moscow	RUS
10244,00	ALE USB	B=125Bd Ch=8 ChS=250Hz		ALG oil/gas net	
10247,00	CIS 12	B=1440Bd Ch=12 ChS=200Hz		RUS MIL	RUS
10250,00	ALE LSB	B=125Bd Ch=8 ChS=250Hz		MOZ gas project	MOZ
10250,00	HFT MODEM		WI2XNX	10Band LLC	USA
10250,00	ALE USB	B=125Bd Ch=8 ChS=250Hz			EGY
10250,00	CLOVER 2000	B=500Bd Ch=8 ChS=250Hz			
10251,00	T600	B=50Bd S=200Hz		RUS N Murmansk	RUS

Frequency	Mode	Mode Parameter	Callsign	User	Country
10251,00	CIS MFSK 11	B=125Bd Ch=11 ChS=250Hz		Moscow	RUS
10254,00	SYSTEME 3000 ALE	B=125Bd Ch=8 ChS=250Hz			F
10260,00	LINK 11 CLEW	B=2250Bd Ch=16 ChS=330/110/550Hz		G MIL Plymouth	G
10263,00	CW		RAL2	RUS N HQ Astrakhan	RUS
10263,00	CW		RBI2	RUS N	RUS
10263,00	CW		RGH2	RUS N	RUS
10263,00	CW		RHQ2	RUS N	RUS
10263,00	CW		RBL66	RUS N	RUS
10263,00	CW		RBL673	RUS N	RUS
10263,00	RUS DBPSK 16FSK	B=1866Bd Ch=16 ChS=175Hz		RUS INTEL	RUS
10264,20	*STANAG 4285*	*B=2400Bd*	*FUE*	*F N Brest*	*F*
10264,30	CIS 50-500	B=50Bd S=500Hz			
10265,00	STANAG 4481	B=75Bd S=850Hz	NDT	USN Yokosuka	J
10269,00	*STANAG 4285*	*B=2400Bd*	*FUE*	*F N Brest*	*F*
10270,00	SSB			USA AF MARS	
10270,00	ALE USB	B=125Bd Ch=8 ChS=250Hz			
10272,00	STANAG 4481	B=75Bd S=850Hz		G MIL Crimond	G
10272,50	ALE USB	B=125Bd Ch=8 ChS=250Hz		D red Cross net	D
10273,00	ALE USB	B=125Bd Ch=8 ChS=250Hz		AUT MIL net	AUT
10273,00	SSB			USA AF MARS	
10275,00	ALE LSB	B=125Bd Ch=8 ChS=250Hz		ALG oil/gas field	ALG
10280,00	CIS 1200	B=1200Bd		RUS	RUS
10280,00	*T600*	*B=50Bd S=250Hz*	*RMP*	*RUS N HQ Kaliningrad*	*RUS*
10282,00	STANAG 4285	B=2400Bd		G MIL St. Eval	G
10283,00	PACTOR	B=200Bd S=170Hz		ICRC net	
10284,00	*CIS 75-200*	*B=75Bd S=200Hz*		*RUS F Moscow*	*RUS*
10284,00	PACTOR II	B=100Bd S=200Hz		ICRC Pristina	SER
10286,00	STANAG 4285	B=2400Bd	NSS	USA N TX Davidsonville , MD	USA
10286,00	*STANAG 4481*	*B=75Bd S=850Hz*		*USA AF Crougthon*	*G*
10291,50	ALE USB	B=125Bd Ch=8 ChS=250Hz		CAN CFARS net	CAN
10292,00	CW		RIR98	RUS N repair ship PM-56 Armur Class	RUS
10295,00	ALE USB	B=125Bd Ch=8 ChS=250Hz		SAPOL Communication Infrastructure	AUS
10295,00	ALE USB	B=125Bd Ch=8 ChS=250Hz		AUS Police QLD	AUS
10296,00	*STANAG 4285*	*B=2400Bd*	*JWT*	*NOR N Stavanger*	*NOR*
10299,00	ALE LSB	B=125Bd Ch=8 ChS=250Hz			
10300,00	MIL 188-110B SER	B=2400Bd	CENTR5**	MFA Bucharest	ROU
10300,00	MIL 188-110B SER	B=2400Bd		ROU E Damascus	SYR
10300,00	*MIL 188-110B SER*	*B=2400Bd*	*CENTR6***	*MFA Bucharest*	*ROU*
10302,00	MIL 188-110B SER	B=2400Bd			
10305,00	SSB			NASA space missile tactical net	USA
10306,50	CW		CGA984	Essex County, ON	CAN
10308,00	CIS 12	B=1440Bd Ch=12 ChS=200Hz		RUS MIL	RUS
10309,00	CW		RHQ33	RUS N	RUS
10309,00	CW		RJH41	RUS N	RUS
10310,00	*CIS 50-500*	*B=50Bd S=500Hz*		*RUS MIL Kazan area*	*RUS*
10310,00	PACTOR III	B=100Bd S=200Hz	OL1A	CZE MIL Prague	CZE
10310,60	*ALE USB*	*B=125Bd Ch=8 ChS=250Hz*		*F N net*	*F*

Frequency	Mode	Mode Parameter	Callsign	User	Country
10311,00	MIL 188-110B 39TONE	B=2400Bd Ch=39 ChS=56,25Hz		TUN MIL	TUN
10315,00	*BAUDOT*	*B=50Bd S=500Hz*		*Moscow*	*RUS*
10315,00	PACTOR III	B=100Bd S=200Hz	WHV382	Sailmail Friday Habor , WA	USA
10317,00	*STANAG 4481*	*B=75Bd S=850Hz*		*G MIL Crimond*	*G*
10318,00	STANAG 4481	B=75Bd S=850Hz			
10318,00	RUS HYBRID MODEM MFSK-PSK	B=40Bd Ch=16 ChS=40Hz		Moscow	RUS
10320,00	PACTOR II	B=200Bd S=200Hz	WHV861	Sailmail San Luis Obispo , CA	USA
10323,00	PACTOR II	B=200Bd S=200Hz	V8V2222	Sailmail Brunai Bay	BRU
10323,00	ALE USB	B=125Bd Ch=8 ChS=250Hz		USA 3. Grey Wolve Brigade	POL
10323,00	MIL 188-110B SER	B=2400Bd		CHN	CHN
10324,00	ALE USB	B=125Bd Ch=8 ChS=250Hz			
10325,00	PACTOR II	B=200Bd S=200Hz	KUZ533	Sailmail Honululu	HWA
10326,00	*ALE USB*	*B=125Bd Ch=8 ChS=250Hz*		*IRN net*	*IRN*
10326,00	*CODAN CHIRP*	*B=80Bd Ch=30 ChS=250Hz*		*IRN net*	*IRN*
10326,00	*CODAN*	*B=2400Bd Ch=16 ChS=112,5Hz*		*IRN net*	*IRN*
10329,00	PACTOR II	B=200Bd S=200Hz	HPPM2	Sailmail Chiriqui	PNR
10330,00	CLOVER 2000	B=500Bd Ch=8 ChS=250Hz	5GSI11		MTN
10330,00	CLOVER2000	Ch=4	8GSI11		MTN
10330,00	USB				LBY
10330,00	CLOVER2000	Ch=4		MTN MIL net	MTN
10330,00	ALE USB	B=125Bd Ch=8 ChS=250Hz			
10330,00	MIL 188-110C	B=2400Bd		S MIL	S
10331,00	PACTOR II	B=200Bd S=200Hz	KZN508	Sailmail Rockhill , SC	USA
10332,50	USB		AAR1HF	USA Army MARS	USA
10332,50	USB		AAR4GC	USA Army MARS	USA
10332,50	ALE USB	B=125Bd Ch=8 ChS=250Hz		USA MIL net	
10333,00	USB			EGY CG net	EGY
10335,00	PACTOR III	B=200Bd S=200Hz	RC01	Sailmail Maputo	MOZ
10335,00	CHN DBPSK	B=2400Bd Ch=1		CHN AF net	CHN
10336,00	CW			RUS Operational Strategic Command Air Defense net	RUS
10338,00	CW		RIW	RUS N HQ Moscow	RUS
10340,00	ALE USB	B=125Bd Ch=8 ChS=250Hz		ALG MOI net	ALG
10340,00	ALE USB	B=125Bd Ch=8 ChS=250Hz			
10341,00	ALE USB	B=125Bd Ch=8 ChS=250Hz		USA FEMA net	USA
10341,50	ALE USB	B=125Bd Ch=8 ChS=250Hz			
10342,00	CIS OFDM 160	B=4800Bd Ch=160 ChS=37,5Hz			RUS
10343,00	PACTOR II	B=200Bd S=200Hz	WRD719	Sailmail Palo Alto	USA
10343,00	PACTOR III	B=100Bd S=200Hz	WQLI952	Sailmail Watsonville , CA	USA
10344,50	MIL 188-110B SER	B=2400Bd		G MIL	G
10344,50	*ALE USB*	*B=125Bd Ch=8 ChS=250Hz*		*G MIL net*	*G*
10345,00	WINDRM	B=1912,5Bd Ch=51 ChS=46,875Hz		CUB Intelligence Directorate	CUB
10350,00	MIL 188-110B SER	B=2400Bd		MFA Bucharest	ROU
10352,00	MIL 188-110B SER	B=2400Bd		MFA Bucharest	ROU
10352,00	MIL 188-110B SER	B=2400Bd		ROU E Vienna	AUT

Frequency	Mode	Mode Parameter	Callsign	User	Country
10358,00	ALE USB	B=125Bd Ch=8 ChS=250Hz		FEMA net	USA
10358,20	*STANAG 4285*	*B=2400Bd*		*G MIL TX Inskip*	*G*
10361,40	PACTOR II	B=200Bd S=200Hz	WPTG385	Sailmail Corpus Christi , TX	USA
10366,40	PACTOR II	B=200Bd S=200Hz	WPUC469	Sailmail South Daytona , FL	USA
10368,00	AUS ISB MODEM	B=600Bd S=345Hz		AUS MIL	AUS
10368,20	STANAG 4285	B=2400Bd	PBC	NLD N Goeree Island	HOL
10370,00	ALE USB	B=125Bd Ch=8 ChS=250Hz		POL MIL net	
10373,00	*STANAG 4285*	*B=2400Bd*		*G MIL Ascension Island*	*SHN*
10377,00	CW		RGT77	RUS MIL Moscow, RVSN	RUS
10380,00	CLOVER 2000	B=500Bd Ch=8 ChS=250Hz			
10380,00	*CIS 100-500*	*B=100Bd S=500Hz*		*RUS N Murmansk*	*RUS*
10380,00	ALE USB	B=125Bd Ch=8 ChS=250Hz		GANOB net	EGY
10388,00	CW		RIW	RUS N HQ Moscow	RUS
10388,00	*CIS 75-200*	*B=75Bd S=200Hz*	*RIW*	*RUS N HQ Moscow*	*RUS*
10390,00	CIS 300-500	B=300Bd S=500Hz		RUS AF	
10390,00	ALE USB	B=125Bd Ch=8 ChS=250Hz		SVK AF net	SVK
10390,00	CODAN CHIRP	B=80Bd Ch=30 ChS=81Hz	20111**		
10390,00	CODAN CHIRP	B=80Bd Ch=30 ChS=81Hz	487173**		
10391,50	STANAG 4481	B=75Bd S=850Hz	NPG	USA N Dixon , CA	USA
10394,00	CIS 100-1000	B=100Bd S=1000Hz		RUS S Moscow	RUS
10401,00	*CIS 12*	*B=1440Bd Ch=12 ChS=200Hz*	*RMP*	*RUS N HQ Kaliningrad*	*RUS*
10405,00	FSK	B=600Bd S=850Hz		AUS MIL	AUS
10405,70	PACTOR II FEC	B=100Bd S=200Hz	AQP	PAK N Karachi	PAK
10407,00	STANAG 4285	B=2400Bd		AUS MHFCS Townsville	AUS
10410,00	ALE USB	B=125Bd Ch=8 ChS=250Hz		G MIL net	G
10410,00	*STANAG 4285*	*B=2400Bd*		*G MIL St. Eval*	*G*
10411,00	STANAG 4197	B=1800Bd Ch=16/39 ChS=112/56Hz			
10411,00	*ALE USB*	*B=125Bd Ch=8 ChS=250Hz*		*CHN net*	*CHN*
10412,60	SYSTEME 3000	B=2400Bd			
10415,00	*STANAG 4285*	*B=2400Bd*		*G MIL TX Akrotiri*	*CYP*
10417,50	ALE USB	B=125Bd Ch=8 ChS=250Hz		UKR net	UKR
10420,00	ALE USB	B=125Bd Ch=8 ChS=250Hz		G MIL net	G
10424,00	STANAG 4285	B=2400Bd		AUS MHFCS Humpty Doo	AUS
10424,00	MIL 188-110A SER	B=2400Bd		AUS MHFCS North West Cape	AUS
10424,00	ALE USB	B=125Bd Ch=8 ChS=250Hz			
10425,00	ALE USB	B=125Bd Ch=8 ChS=250Hz			
10425,00	CW		RAL2	RUS N HQ Astrakhan	RUS
10426,50	BAUDOT	B=50Bd S=400Hz	FDI	F AF Aix	F
10428,00	ALE USB	B=125Bd Ch=8 ChS=250Hz		CHN net	CHN
10428,00	STANAG 4481	B=75Bd S=850Hz	NPG	USA N Dixon , CA	USA
10430,00	STANAG 4481	B=75Bd S=850Hz	NPG	USA N Dixon , CA	USA
10432,00	CIS OFDM 93	B=2400Bd Ch=93 ChS=31,25Hz			RUS
10435,80	CW			RUS HFDF net	RUS
10436,00	CW			RUS HFDF net	RUS

Frequency	Mode	Mode Parameter	Callsign	User	Country
10443,50	*CIS 50-1000*	*B=50Bd S=1000Hz*		*RUS MIL Moscow*	*RUS*
10447,00	CW		PWZW	RUS Strategic Command Aerospace Defense	RUS
10447,00	ALE USB	B=125Bd Ch=8 ChS=250Hz		ROU diplo net	ROU
10448,00	ALE USB	B=125Bd Ch=8 ChS=250Hz		USA FEMA net	USA
10449,00	*ALE USB*	*B=125Bd Ch=8 ChS=250Hz*		*CHN net*	*CHN*
10449,00	*CHN DATALINK*	*B=2400Bd Ch=30 ChS=75Hz*		*Kunming*	*CHN*
10450,00	ALE USB	B=125Bd Ch=8 ChS=250Hz		USA CAP net	USA
10450,00	PACTOR II	B=200Bd S=200Hz	HPPM1	Sailmail Chiriqui	PNR
10450,00	CW		RIW	RUS N HQ Moscow	RUS
10450,00	PACTOR III	B=100Bd S=200Hz	HPPM3	Sailmail Panama	PNR
10450,00	USB			UN MONUSCO net	UGA
10450,00	USB			UN MONUSCO net	UGA
10452,00	ALE USB	B=125Bd Ch=8 ChS=250Hz		ROU diplo net	ROU
10452,00	MIL 188-110B SER	B=2400Bd		ROU E Tripoli	LBY
10452,00	MIL 188-110B SER	B=2400Bd		MFA Bucharest	ROU
10452,00	SSB			USA AF Strategic Command	USA
10452,00	*T600*	*B=50Bd S=200Hz*		*RUS N Sveromorsk*	*UKR*
10456,00	CW			RUS HFDF net	RUS
10457,00	STANAG 4197	B=1800Bd Ch=16/39 ChS=112/56Hz		EGY N net	EGY
10460,00	USB			RUS AF net	RUS
10462,00	STANAG 4285	B=2400Bd		AUS MHFCS Humpty Doo	AUS
10463,00	ALE USB	B=125Bd Ch=8 ChS=250Hz			
10465,00	STANAG 4285	B=2400Bd		AUS MHFCS Humpty Doo	AUS
10465,00	USB			RUS pirates net	RUS
10467,50	*PSK BURSTS*	*B=2400Bd*		*Paris*	*F*
10467,70	*STANAG 4285*	*B=2400Bd*	*FUE*	*F N Brest*	*F*
10471,00	CIS MFSK 68	B=50Bd Ch=68 ChS=47Hz			RUS
10472,00	ALE USB	B=125Bd Ch=8 ChS=250Hz		POL F net	
10475,00	*CIS 100-500*	*B=100Bd S=500Hz*		*RUS MIL E Samara*	*RUS*
10475,00	CIS ARQ	B=100Bd S=500Hz		RUS PTT	RUS
10476,20	PACTOR II	B=200Bd S=200Hz	VZX	Sailmail Darawank	AUS
10477,00	ALE USB	B=125Bd Ch=8 ChS=250Hz		AZE emergency net	AZE
10477,00	ALE USB	B=125Bd Ch=8 ChS=250Hz		CHN MIL	
10478,00	*CIS 100-500*	*B=100Bd S=500Hz*		*GEO MIL Tbilisi*	*GEO*
10480,00	ALE USB	B=125Bd Ch=8 ChS=250Hz		ROU diplo net	ROU
10480,00	MIL 188-110A SER	B=2400Bd		ROU E Copenhagen	DNK
10480,00	MIL 188-110A SER	B=2400Bd		MFA Bucharest	ROU
10484,00	STANAG 4285	B=2400Bd		I E Alger	ALG
10492,00	CW		RFH61	RUS N	RUS
10493,00	ALE USB	B=125Bd Ch=8 ChS=250Hz		USA FEMA net	USA
10498,00	SYSTEME 3000	B=2400Bd		F F	F
10498,00	ALE USB	B=125Bd Ch=8 ChS=250Hz		F N net	F
10498,50	ALE USB	B=125Bd Ch=8 ChS=250Hz		USA FBI net	USA
10500,00	CIS 50-500	B=50Bd S=500Hz	RAZ2	RUS GOV	RUS
10500,00	USB			EGY CG net	EGY

Frequency	Mode	Mode Parameter	Callsign	User	Country
10500,00	2FSK	B=300Bd S=1200Hz			TUR
10502,00	ALE USB	B=125Bd Ch=8 ChS=250Hz		AUS Police NT	AUS
10504,00	ALE USB	B=125Bd Ch=8 ChS=250Hz		USA CAP net	USA
10505,00	ALE USB	B=125Bd Ch=8 ChS=250Hz		AUS Police NT	AUS
10505,00	ALE USB	B=125Bd Ch=8 ChS=250Hz		AUS Police NSW	AUS
10505,00	ALE USB	B=125Bd Ch=8 ChS=250Hz		AUS Police QLD	AUS
10505,00	ALE USB	B=125Bd Ch=8 ChS=250Hz		SAPOL Communication Infrastructure	AUS
10505,00	ALE USB	B=125Bd Ch=8 ChS=250Hz		CHL MOI net	CHL
10509,60	ALE USB	B=125Bd Ch=8 ChS=250Hz		F N net	F
10510,00	ALE USB	B=125Bd Ch=8 ChS=250Hz		USA CAP net	USA
10512,00	PACTOR II FEC	B=100Bd S=200Hz	ARL	PAK N Karachi	PAK
10512,00	ALE USB	B=125Bd Ch=8 ChS=250Hz		PAK N net	
10513,00	*CIS 12*	*B=1440Bd Ch=12 ChS=200Hz*		*RUS MIL E Samara*	*RUS*
10518,00	ALE USB	B=125Bd Ch=8 ChS=250Hz		USA CAP net	USA
10519,00	ALE USB	B=125Bd Ch=8 ChS=250Hz		ALG AF net	ALG
10519,00	USB			EGY N net	EGY
10522,00	PACTOR III	B=100Bd S=200Hz	RC01	Sailmail Maputo	MOZ
10523,00	PACTOR II	B=200Bd S=200Hz	XJN714	Sailmail Lunenburg , NS	CAN
10524,60	ALE USB	B=125Bd Ch=8 ChS=250Hz		CHN net	CHN
10525,00	STANAG 4285	B=2400Bd		G MIL Tx St. Eval	G
10526,00	*CW*		*RIT*	*RUS N HQ Severomorsk*	*RUS*
10527,00	*ALE USB*	*B=125Bd Ch=8 ChS=250Hz*		*Wuhan area*	*CHN*
10527,00	*MIL 188-110B SER*	*B=2400Bd Ch=39 ChS=56,25Hz*		*Wuhan area*	*CHN*
10528,00	ALE USB	B=125Bd Ch=8 ChS=250Hz		ALG AF net	ALG
10535,00	*T600*	*B=50Bd S=250Hz*		*RUS N Moscow*	*RUS*
10535,00	CW			RUS HFDF net	RUS
10535,00	*T600*	*B=50Bd S=200Hz*		*RUS N Moscow*	*RUS*
10536,00	BAUDOT	B=75Bd S=850Hz	CFH	CAN N Halifax	CAN
10540,00	*T600*	*B=75Bd S=200Hz*		*RUS N HQ Moscow*	*RUS*
10540,00	CW		RJD52	RUS N HQ Astrakhan	RUS
10540,00	CW		RAA	RUS N HQ Moscow	RUS
10540,00	*CW*		*RIT*	*RUS N HQ Severomorsk*	*RUS*
10540,00	CW		RIW	RUS N HQ Moscow	RUS
10542,00	ALE USB	B=125Bd Ch=8 ChS=250Hz		USA Civil Air Patrol net	USA
10542,00	CW		RDL	RUS N Moscow tx Vileyka	BLR
10543,00	*CW*		*RCV*	*RUS N HQ Sevastopol*	*UKR*
10550,00	ALE USB	B=125Bd Ch=8 ChS=250Hz		USA Civil Air Patrol net	USA
10551,00	CHN 4+4	B=75Bd Ch=8 ChS=300/450Hz		CHN MIL	
10552,00	ALE USB	B=125Bd Ch=8 ChS=250Hz		ALG AF net	ALG
10555,00	FAX 120/576		VMW	Wiluna M	AUS

Frequency	Mode	Mode Parameter	Callsign	User	Country
10557,00	ALE USB	B=125Bd Ch=8 ChS=250Hz		USA Civil Air Patrol net	USA
10559,20	*STANAG 4285*	*B=2400Bd*	*DHJ58*	*D N Glücksburg TX Neuharlingersiel*	*D*
10559,50	STANAG 4285	B=2400Bd	DHJ58	D N Glücksburg TX Neuharlingersiel	D
10560,00	CW		RAA	RUS N HQ Moscow	RUS
10561,20	*STANAG 4285*	*B=2400Bd*	*DHJ58*	*D N Glücksburg TX Neuharlingersiel*	*D*
10561,20	*STANAG 4285*	*B=2400Bd*	*DHJ58*	*D N Glücksburg TX Marlow*	*D*
10568,20	*STANAG 4285*	*B=2400Bd*	*IDR*	*I N Rome*	*I*
10576,00	SSB			USA AF MARS	
10583,00	ALE USB	B=125Bd Ch=8 ChS=250Hz		ALG MIL net	AG
10584,00	*CIS 12*	*B=1440Bd Ch=12 ChS=200Hz*		*RUS MIL Moscow*	*RUS*
10584,00	*CIS 12*	*B=1440Bd Ch=12 ChS=200Hz*		*RUS MIL Moscow*	*RUS*
10588,00	ALE USB	B=125Bd Ch=8 ChS=250Hz		USA FEMA net	USA
10595,00	STANAG 4285	B=2400Bd		AUS MHFCS Humpty Doo	AUS
10595,00	*CIS 100-500*	*B=100Bd S=500Hz*		*RUS MIL Moscow*	*RUS*
10595,00	CIS ARQ	B=100Bd S=500Hz	RKD48	RUS PTT	RUS
10600,00	*ALE USB*	*B=125Bd Ch=8 ChS=250Hz*		*F Airbus net*	
10600,00	ALE USB	B=125Bd Ch=8 ChS=250Hz		LBY GMRA net	LBY
10600,00	CLOVER 2000	B=500Bd Ch=8 ChS=250Hz	BASE21**		
10600,00	CLOVER 2000	B=500Bd Ch=8 ChS=250Hz	ELSAL001 **		
10603,00	CHN OFDM 30TONE	B=1800Bd Ch=30 ChS=60Hz		CHN MIL	CHN
10604,20	*STANAG 4285*	*B=2400Bd*		*G MIL TX St. Eval*	*G*
10614,00	*CIS 12*	*B=1440Bd Ch=12 ChS=200Hz*		*RUS MIL Krasnoyarsk*	*RUS*
10620,00	PACTOR III	B=100Bd S=200Hz	CEV773	Sailmail Los Lagos	CHL
10621,00	USB			EGY CG net	EGY
10622,00	CIS BPSK	B=1200Bd Ch=1			
10623,00	PACTOR III	B=100Bd S=200Hz	CEV773	Sailmail Los Lagos	CHL
10625,00	CIS MFSK 13	B=31,25Bd Ch=13 ChS=125Hz		MFA Moscow	RUS
10626,00	ALE USB	B=125Bd Ch=8 ChS=250Hz		F N net	F
10627,00	ALE USB	B=125Bd Ch=8 ChS=250Hz		Queensland Deptartment of Community Safety	AUS
10636,50	ALE USB	B=125Bd Ch=8 ChS=250Hz			
10638,00	ALE USB	B=125Bd Ch=8 ChS=250Hz		GRC MIL net	GRC
10640,00	OQPSK	B=48000Bd		HF Trading	USA
10648,00	ALE USB	B=125Bd Ch=8 ChS=250Hz		TUR Ministry of Health net	TUR
10652,00	CHN N MIL 188-110 MOD	B=2400Bd		CHN N	CHN
10653,00	CW		REA4	RUS AF Moscow	RUS
10661,00	CIS OFDM 60	B=2400Bd Ch=60 ChS=44,5Hz		MFA Moscow	RUS
10662,00	*CIS 12*	*B=1440Bd Ch=12 ChS=200Hz*		*RUS MIL Moscow*	*RUS*
10672,00	T600	B=50Bd S=200Hz	RJD56	RUS N HQ Murmansk	RUS
10672,00	CIS 100-500	B=100Bd S=500Hz		RUS MIL	RUS
10673,00	*CIS OFDM 60*	*B=2400Bd Ch=60 ChS=44,5Hz*		*MFA Moscow*	*RUS*
10676,00	CW		RJD52	Astrakhan R	RUS
10677,00	MIL 188-110A SER	B=2400Bd		ARS AF	ARS

Frequency	Mode	Mode Parameter	Callsign	User	Country
10683,00	CW		RGT77	RUS MIL Moscow, RVSN	RUS
10687,00	CHN 4+4	B=75Bd Ch=8 ChS=300/450Hz			
10695,00	ALE USB	B=125Bd Ch=8 ChS=250Hz		F AF net	F
10700,00	*T600*	*B=50Bd S=200Hz*		*RUS MIL Moscow*	*RUS*
10704,80	*CIS 100-1000*	*B=100Bd S=1000Hz*		*RUS MIL Novosibirsk*	*RUS*
10706,00	ALE USB	B=125Bd Ch=8 ChS=250Hz		ARINC Urgent Link net	USA
10711,00	USB			USA N Mayport , FL	USA
10712,00	*T600*	*B=50Bd S=200Hz*	*RDL*	*RUS N Moscow*	*RUS*
10713,00	ALE USB	B=125Bd Ch=8 ChS=250Hz		POL MIL net	POL
10713,00	ALE USB	B=125Bd Ch=8 ChS=250Hz		CHN MIL net	CHN
10715,00	WINDRM	B=1912,5Bd Ch=51 ChS=46,875Hz		CUB Intelligence Directorate	CUB
10717,00	*CIS 12*	*B=1440Bd Ch=12 ChS=200Hz*		*RUS MIL Moscow*	*RUS*
10720,20	*STANAG 4285*	*B=2400Bd*	*DHJ58*	*D N Glücksburg TX Marlow*	*D*
10722,00	*IRN N PSK*	*B=207Bd*		*IRN N*	*IRN*
10722,00	CW			CHN MIL net	CHN
10723,00	IRN N QPSK ADAPTIVE	B=468Bd		IRN N	IRN
10726,00	*CIS 12*	*B=1440Bd Ch=12 ChS=200Hz*		*RUS MIL Moscow*	*RUS*
10726,00	*T600*	*B=50Bd S=250Hz*		*RUS N Moscow*	*RUS*
10733,20	*STANAG 4285*	*B=2400Bd*	*PBB*	*HOL N Den Helder TX Zeewolde*	*HOL*
10733,20	STANAG 4285	B=2400Bd	PBB	NLD N Den Helder TX Flevo	HOL
10733,60	ALE USB	B=125Bd Ch=8 ChS=250Hz		USA diplo net	
10735,00	CW		REA4	RUS AF Moscow	RUS
10740,00	*STANAG 4285*	*B=2400Bd*		*G MIL St. Eval*	*G*
10746,40	ALE USB	B=125Bd Ch=8 ChS=250Hz		FIN diplo net	FIN
10750,00	ALE USB	B=125Bd Ch=8 ChS=250Hz		GANOB net	EGY
10755,00	ALE LSB	B=125Bd Ch=8 ChS=250Hz		TWN N net	TWN
10760,00	*STANAG 4285*	*B=2400Bd*	*TBB*	*TUR N Izmir*	*TUR*
10762,00	*T600*	*B=50Bd S=200Hz*	*RMP*	*RUS N Kaliningrad*	*RUS*
10767,00	CIS OFDM 60	B=2400Bd Ch=60 ChS=44,5Hz			RUS
10772,50	ALE USB	B=125Bd Ch=8 ChS=250Hz			
10775,00	ALE USB	B=125Bd Ch=8 ChS=250Hz		S FRO net	S
10780,00	USB			USA AF Cape Radio	USA
10787,00	CIS MFSK 20	B=10Bd Ch=20 ChS=40Hz		RUS INTEL	RUS
10790,00	CIS OFDM 79	B=2400Bd Ch=79 ChS=37,5Hz		RUS	RUS
10795,00	ALE USB	B=125Bd Ch=8 ChS=250Hz		USA MIL net	USA
10796,00	CW		RJD52	RUS N HQ Astrakhan	RUS
10796,00	CW		RAA	RUS N HQ Moscow	RUS
10796,00	*CW*		*RIT*	*RUS N HQ Severomorsk*	*RUS*
10797,50	ALE USB	B=125Bd Ch=8 ChS=250Hz		USA National Guard net	USA
10800,00	PSK 63	B=63Bd Ch=2 ChS=400Hz		POL Intel	POL
10814,00	CW			RUS HFDF net	RUS

Frequency	Mode	Mode Parameter	Callsign	User	Country
10816,00	ALE USB	B=125Bd Ch=8 ChS=250Hz			
10819,50	ALE USB	B=125Bd Ch=8 ChS=250Hz		USA NG net	USA
10820,00	MFSK17	B=125Bd Ch=17 ChS=125Hz			
10821,40	STANAG 4285	B=2400Bd		AUS MHFCS Humpty Doo	AUS
10822,00	CW		C**	RUS N Moscow	RUS
10822,50	QPSK	B=1200Bd		UI	
10825,00	STANAG 4285	B=2400Bd		AUS MHFCS Humpty Doo	AUS
10828,00	ALE USB	B=125Bd Ch=8 ChS=250Hz		USA FEMA net	USA
10830,00	STANAG 4481	B=50Bd S=850Hz	NDT	USA N Totsuka	J
10830,00	ALE USB	B=125Bd Ch=8 ChS=250Hz		S FRO net	S
10831,00	USB			EGY N	EGY
10831,00	STANAG 4197	B=1800Bd Ch=16/39 ChS=112/56Hz		EGY N net	EGY
10832,00	CCIR 493-4	B=100Bd S=170Hz	1247		
10832,00	CCIR 493-4	B=100Bd S=170Hz	1246		
10832,00	CCIR 493-4	B=100Bd S=170Hz	1244		
10832,00	CLOVER 2000	B=500Bd Ch=8 ChS=250Hz			
10833,00	CHN 4+4	B=75Bd Ch=8 ChS=300/450Hz		CHN MIL	
10836,00	ALE USB	B=125Bd Ch=8 ChS=250Hz		G MIL net	G
10843,50	USB			G AF Air Cadets	G
10848,50	ALE 3G	B=2400Bd		AUS N	
10850,00	*ALE USB*	*B=125Bd Ch=8 ChS=250Hz*		*F Airbus net*	*F*
10850,00	CIS 3000	B=3000Bd		MFA Moscow	RUS
10851,70	STANAG 4285	B=2400Bd	IHMW	I N ship VIAREGGIO	I
10851,70	USB		IHMW	I N ship VIAREGGIO	I
10851,70	*STANAG 4285*	*B=2400Bd*	*IDR*	*I N Rome*	*I*
10854,00	LINK 11 ISB	B=2250Bd Ch=16 ChS=330/110/550Hz			
10863,00	CHN 4+4	B=75Bd Ch=8 ChS=300/450Hz		CHN MIL	CHN
10866,00	ALE USB	B=125Bd Ch=8 ChS=250Hz		ROU diplo net	ROU
10871,50	CW		U**		RUS
10871,70	CW		D**	RUS N Sevastopol	UKR
10871,80	CW		P**	RUS N Kaliningrad	RUS
10871,90	CW		S**	RUS N Severomorsk	RUS
10872,00	CW		C**	RUS N Moscow	RUS
10872,10	CW		A**	RUS N Astrakhan	RUS
10872,20	CW		F**	RUS N Vladivostok	RUS
10872,30	CW		K**	RUS N	RUS
10872,40	CW		M**	RUS N Magadan	RUS
10873,00	*CIS 12*	*B=1440Bd Ch=12 ChS=200Hz*	*RIT*	*RUS N HQ Severomorsk*	*RUS*
10884,00	STANAG 4197	B=1800Bd Ch=16/39 ChS=112/56Hz		EGY N net	EGY
10884,00	USB			EGY N	EGY
10885,00	*CIS 12*	*B=1440Bd Ch=12 ChS=200Hz*	*RIT*	*RUS N HQ Severomorsk*	*RUS*
10893,50	*ALE USB*	*B=125Bd Ch=8 ChS=250Hz*		*G MIL net TX Inskip*	*G*
10895,00	STANAG 4481	B=75Bd S=850Hz		US N Isabella	PTR

Frequency	Mode	Mode Parameter	Callsign	User	Country
10898,00	CIS 68TONE	B=50Bd Ch=68 ChS=47Hz		RUS INTEL Moscow	RUS
10899,00	ALE USB	B=125Bd Ch=8 ChS=250Hz		USA FEMA net	USA
10900,00	ALE USB	B=125Bd Ch=8 ChS=250Hz		B Army net	B
10900,00	USB			EGY N net	EGY
10911,00	CIS 500 FSK BURST	B=500Bd S=1000Hz		RUS N	RUS
10913,50	ALE USB	B=125Bd Ch=8 ChS=250Hz		USA FBI net	USA
10913,50	USB			USA FBI net	USA
10914,00	CHN N MIL 188-110 MOD	B=2400Bd		CHN N	CHN
10914,50	ALE USB	B=125Bd Ch=8 ChS=250Hz		B N net	B
10915,00	PACTOR III	B=100Bd S=200Hz	OL1A	CZE MIL Prague	CZE
10916,00	CIS 12	B=1440Bd Ch=12 ChS=200Hz	RCV	RUS N HQ Sevastopol	UKR
10917,40	STANAG 4285	B=2400Bd		AUS MHFCS Humpty Doo	
10919,50	CHN 4+4	B=75Bd Ch=8 ChS=300/450Hz		CHN MIL	CHN
10921,00	AUS MIL MODEM			AUS MIL	AUS
10922,00	*CIS 12*	*B=1440Bd Ch=12 ChS=200Hz*	*RMP*	*RUS N HQ Kaliningrad*	*RUS*
10928,00	ALE USB	B=125Bd Ch=8 ChS=250Hz		USA FEMA net	USA
10933,00	CW		RGT77	RUS MIL Moscow RVSN	RUS
10934,00	*MIL 188-110C SER*	*B=2400Bd*		*SW London*	*G*
10935,70	MIL 188-110B SER	B=2400Bd		S MIL	S
10939,00	MFSK17	B=125Bd Ch=17 ChS=125Hz			
10942,20	STANAG 4285	B=2400Bd	DHJ58	D N Glücksburg TX Neuhalringersiel	D
10942,20	STANAG 4285	B=2400Bd	DHJ58	D N Glücksburg TX Marlow	D
10943,00	CIS MFSK 20	B=20Bd Ch=20 ChS=40Hz		RUS INTEL	RUS
10943,20	*STANAG 4285*	*B=2400Bd*	*CFH*	*CAN N Halifax*	*CAN*
10945,00	*BAUDOT*	*B=75Bd S=850Hz*	*CFH*	*CAN F Halifax*	*CAN*
10946,00	USB		CAPITOLE	F AF Taverny	F
10948,00	CW			RUS HFDF net	RUS
10953,00	CHN 4+4	B=75Bd Ch=8 ChS=300/450Hz		CHN MIL	
10962,00	CW			RUS	
10963,20	*STANAG 4285*	*B=2400Bd Ch=1*		*DNK N TX Frederickshavn*	*DNK*
10965,00	ALE USB	B=125Bd Ch=8 ChS=250Hz		F AF net	F
10974,00	STANAG 4481	B=75Bd S=850Hz	NSY	USA N Niscemi , Sicily	I
10977,00	ALE USB	B=125Bd Ch=8 ChS=250Hz		SUI MIL net	SUI
10981,00	CW		REA4	RUS AF Moscow	RUS
10982,00	PACTOR II	B=200Bd S=200Hz	WHV861	Sailmail San Luis Obispo , CA	USA
10983,50	ALE USB	B=125Bd Ch=8 ChS=250Hz		G MIL net	G
10984,00	*CIS 12*	*B=1440Bd Ch=12 ChS=200Hz*	*RCV*	*RUS N HQ Sevastopol*	*UKR*
10985,00	*CW*			*RUS N Kaliningrad*	*RUS*
10986,00	ALE USB	B=125Bd Ch=8 ChS=250Hz		USA CG net	USA
10987,00	CW			RUS	
10990,00	J 16 TONE QPSK	B=1200Bd Ch=16 ChS=100Hz			
11000,00	CW		RIW	RUS N HQ Moscow	RUS

Frequency	Mode	Mode Parameter	Callsign	User	Country
11000,00	*STANAG 4285*	*B=2400Bd*	EBA	*E N Madrid*	*E*
11000,50	ALE USB	B=125Bd Ch=8 ChS=250Hz			
11000,50	ALE USB	B=125Bd Ch=8 ChS=250Hz			
11004,00	*CIS 12*	*B=1440Bd Ch=12 ChS=200Hz*		*RUS N HQ Moscow*	*RUS*
11005,00	CIS 12	B=1440Bd Ch=12 ChS=200Hz		RUS N HQ Moscow	RUS
11005,50	*STANAG 4285*	*B=2400Bd*		*NOR N Gossen*	*NOR*
11005,70	*STANAG 4285*	*B=2400Bd*		*NOR N Andoya*	*N*
11006,00	*STANAG 4285*	*B=2400Bd*		*NOR N Bergen*	*NOR*
11010,00	BR 6028	B=75Bd S=170Hz	ZLO	NZL N Irirangi	NZL
11011,00	RFSM8000	B=2400Bd		MFA Sofia	BUL
11012,00	*CIS 12*	*B=1440Bd Ch=12 ChS=200Hz*		*RUS N HQ Moscow*	*RUS*
11012,00	RUS DBPSK 16FSK	B=1866Bd Ch=16 ChS=175Hz		RUS INTEL	RUS
11013,50	USB			HFoZ/Radtel net	AUS
11015,00	*STANAG 4285*	*B=2400Bd*		*G MIL TX Crimond*	*G*
11016,50	USB			HFoZ/Radtel net	AUS
11017,00	ALE USB	B=125Bd Ch=8 ChS=250Hz		ALG AF net	ALG
11017,00	MIL 188-110A SER	B=2400Bd	CM1	ALG AF Blida	ALG
11017,00	MIL 188-110A SER	B=2400Bd	COF	ALG AF Cheraga	ALG
11018,00	CIS 74.5-250	B=74,5Bd S=250Hz			
11020,00	STANAG 4285	B=2400Bd		TUR F	TUR
11020,00	RFSM8000	B=2400Bd		MFA Sofia	ROU
11020,00	RFSM8000	B=2400Bd		MFA Sofia	BUL
11020,00	RFSM8000	B=2400Bd		MFA Sofia	BUL
11020,00	*CIS 12*	*B=1440Bd Ch=12 ChS=200Hz*		*RUS N St. Petersburg*	*RUS*
11021,00	MIL 188-110B SER	B=2400Bd		USA CG	USA
11025,00	BAUDOT	B=100Bd S=500Hz	RKV23		RUS
11025,00	FSK ARQ	B=100Bd S=500Hz			
11025,00	CIS ARQ	B=100Bd S=500Hz	RKV23	RUS PTT Tyumen	RUS
11026,00	HC-265				
11026,00	RFSM8000	B=2400Bd		BUL E	
11026,00	RFSM8000	B=2400Bd		BUL E Berlin	D
11030,00	FAX 120/576		VMC	Charleville M	AUS
11030,25	ALE USB	B=125Bd Ch=8 ChS=250Hz		CHN net	CHN
11030,50	*ALE USB*	*B=125Bd Ch=8 ChS=250Hz*		*AUT MIL net*	*AUT*
11032,00	CIS OFDM 45	B=2400Bd Ch=45 ChS=62,5Hz		Moscow	RUS
11032,00	RFSM8000	B=2400Bd		MFA Sofia	
11032,00	RFSM8000	B=2400Bd		BUL E Geneva	SUI
11034,00	STANAG 4285	B=2400Bd		I F Tripoli	LBY
11034,70	SITOR A	B=100Bd S=170Hz		MFA Cairo	EGY
11037,70	2FSK	B=300Bd S=850Hz			
11039,00	*BAUDOT*	*B=50Bd S=425Hz*	DDH9	*DWD TX Pinneberg*	*D*
11041,00	CW			RUS HFDF net	RUS
11042,20	STANAG 4285	B=2400Bd	PBC	NLD N Goeree Island	HOL
11043,00	USB			BGD N net	BGD
11045,00	MIL 188-110A SER	B=2400Bd		AUS MHFCS North West Cape	AUS
11050,20	MIL 188-110B SER	B=2400Bd		S MIL	S
11051,00	USB			UNOSOM	SOM
11053,00	CODAN CHIRP	B=80Bd Ch=30 ChS=81Hz	7016**		
11053,00	CODAN CHIRP	B=80Bd Ch=30 ChS=81Hz	7019**		ETH

Frequency	Mode	Mode Parameter	Callsign	User	Country
11053,00	CODAN	B=2400Bd Ch=16 ChS=112,5Hz	7019**		
11053,00	CODAN	B=2400Bd Ch=16 ChS=112,5Hz	7016**		
11055,00	STANAG 4481	B=75Bd S=850Hz	NPG	USA N Dixon , CA	USA
11055,00	ALE USB	B=125Bd Ch=8 ChS=250Hz		USA N net	USA
11056,20	2FSK	B=100Bd S=170Hz		Naples	I
11056,70	SITOR A	B=100Bd S=170Hz		MFA Cairo	EGY
11059,50	ALE USB	B=125Bd Ch=8 ChS=250Hz		UKR net	UKR
11064,00	*CIS 12*	*B=1440Bd Ch=12 ChS=200Hz*	*RIT*	*RUS N HQ Severomorsk*	*RUS*
11064,90	MIL 188-110B APP C	B=2400Bd			
11069,00	CIS OFDM 121	B=2541Bd Ch=121 ChS=25Hz		MFA Moscow	RUS
11070,00	8PSK BURST	B=2800Bd			
11073,00	CW			RUS HFDF net	RUS
11073,50	ALE USB	B=125Bd Ch=8 ChS=250Hz		USA FBI net	USA
11078,00	ALE USB	B=125Bd Ch=8 ChS=250Hz		SNG N net	SNG
11082,00	CW			RUS HFDF net	RUS
11084,20	*STANAG 4285*	*B=2400Bd*		*G MIL Akrotiri*	*CYP*
11085,50	USB		MVU	G AF VOLMET TX Inskip	G
11086,50	FAX 120/576		GYA	G N London	G
11087,00	STANAG 4197	B=1800Bd Ch=16/39 ChS=112/56Hz			
11088,00	*T600*	*B=50Bd S=200Hz*	*RCV*	*RUS N Sevastopol*	*UKR*
11090,00	MFSK17	B=125Bd Ch=17 ChS=125Hz			
11090,00	FAX 120/576		KVM70	USA CG Honululu	HWA
11094,20	*STANAG 4285*	*B=2400Bd*		*G MIL TX Akrotiri*	*CYP*
11096,00	CW			RUS air defence net	RUS
11096,00	CW		RJS	RUS NHQ Vladivostok	RUS
11098,50	ALE USB	B=125Bd Ch=8 ChS=250Hz		USA AF MARS	USA
11100,00	USB			EGY CG net	EGY
11102,00	ISR N HYBRID MODEM	B=2400Bd	4XZ	ISR N Haifa	ISR
11104,00	SSB			NASA launch support ships	USA
11104,50	STANAG 4197	B=1800Bd Ch=16/39		GRC AF net	GRC
11104,50	ALE USB	B=125Bd Ch=8 ChS=250Hz		GRC AF net	GRC
11108,00	ALE USB	B=125Bd Ch=8 ChS=250Hz		CHN net	CHN
11108,00	ALE USB	B=125Bd Ch=8 ChS=250Hz		USA FEMA net	USA
11108,00	ALE USB	B=125Bd Ch=8 ChS=250Hz		USA SHARES net	USA
11110,00	SSB			USA AF Strategic Command	USA
11110,00	STANAG 4285	B=2400Bd		MOD Rome	I
11111,00	PACTOR II	B=100Bd S=200Hz	STAT152**	MFA Tunis	TUN
11111,00	PACTOR II	B=100Bd S=200Hz	STAT22**	TUN E Rabat	MRC
11111,00	ALE USB	B=125Bd Ch=8 ChS=250Hz		F N net	
11113,00	USB			EGY CG net	EGY
11115,00	*CW*		*RIT*	*RUS N HQ Severomorsk*	*RUS*
11115,00	CHP200 ALE	B=250Bd S=170Hz			
11116,00	*T600*	*B=50Bd S=200Hz*		*RUS N Moscow*	*RUS*

Frequency	Mode	Mode Parameter	Callsign	User	Country
11116,20	CIS 16 TONE	B=2400Bd Ch=16 ChS=150Hz		RUS E	
11120,00	T600	B=50Bd S=250Hz		RUS N	RUS
11121,00	USB			USA AF MARS	USA
11123,00	CIS OFDM 45	B=2400Bd Ch=45 ChS=62,5Hz			
11125,00	USB			EGY N net	EGY
11129,00	*CIS 12*	*B=1440Bd Ch=12 ChS=200Hz*		*RUS MIL*	*RUS*
11132,00	STANAG 4538	B=2400Bd			
11132,00	PACTOR-I	B=200Bd S=170Hz	AOULEF	Arorae P	KIR
11132,50	ALE USB	B=125Bd Ch=8 ChS=250Hz		CAN F net	CAN
11133,00	ALE USB	B=125Bd Ch=8 ChS=250Hz		IRN net	IRN
11133,50	CHN 4+4	B=75Bd Ch=8 ChS=300/450Hz		CHN N	
11135,00	CLOVER 2000	B=500Bd Ch=8	BASE31	GANOB net	EGY
11135,00	CLOVER 2000	B=500Bd Ch=8	HALAEB1		
11135,00	MIL 188-110B SER	B=2400Bd	BXM73	CHN diplo	
11135,00	ALE USB	B=125Bd Ch=8 ChS=250Hz		F N net	F
11135,00	CLOVER 2000	B=500Bd Ch=8 ChS=250Hz	GANOB3		EGY
11135,00	CLOVER 2000	B=500Bd Ch=8 ChS=250Hz	HQ4		EGY
11135,00	ALE USB	B=125Bd Ch=8 ChS=250Hz		GANOB net	
11136,00	USB			EGY	
11136,00	STANAG 4197	B=1800Bd Ch=16/39 ChS=112/56Hz		EGY N net	EGY
11136,00	USB			EGY N net	EGY
11137,00	ALE USB	B=125Bd Ch=8 ChS=250Hz			
11138,00	ALE LSB	B=125Bd Ch=8 ChS=250Hz			
11139,00	T600	B=50Bd S=250Hz		RUS N	
11145,00	STANAG 4285	B=2400Bd		AUS MHFCS Humpty Doo	AUS
11146,00	*CIS 12*	*B=1440Bd Ch=12 ChS=200Hz*		*RUS N HQ Moscow*	*RUS*
11149,00	*STANAG 4285*	*B=2400Bd*		*G MIL Akrotiri*	*CYP*
11149,00	CCIR 493-4	B=100Bd S=170Hz			
11150,00	RUS HYBRID MFSK OFDM	B=40Bd Ch=16		Praha	CZE
11150,00	RUS HYBRID MFSK OFDM	B=40Bd Ch=16		Moscow	RUS
11150,00	CCIR 493-4	B=100Bd S=170Hz			
11150,00	CCIR 493-4	B=100Bd S=170Hz			
11150,00	CCIR 493-4	B=100Bd S=170Hz			
11153,00	ALE USB	B=125Bd Ch=8 ChS=250Hz		UKR net	UKR
11153,80	RUS HYBRID MFSK OFDM	B=40Bd Ch=16			
11154,50	MIL 188-110B SER	B=2400Bd		USA CG	USA
11155,00	*CW*		*RIT*	*RUS N HQ Severomorsk*	*RUS*
11155,00	CW		RJP58	RUS F	RUS
11156,00	CW		RIW	RUS N HQ Moscow	RUS
11156,00	ALE USB	B=125Bd Ch=8 ChS=250Hz		ALG MIL net	ALG
11159,00	*STANAG 4481*	*B=75Bd S=850Hz*	*NSY*	*USA N Niscemy*	*I*
11159,00	CCIR 493-4	B=100Bd S=170Hz			

Frequency	Mode	Mode Parameter	Callsign	User	Country
11163,00	ALE USB	B=125Bd Ch=8 ChS=250Hz		ALG AF net	ALG
11163,00	CCIR 493-4	B=100Bd S=170Hz			
11165,00	SSB			Kiev ACC	UKR
11166,00	CIS 3000	B=3000Bd		RUS diplo	RUS
11168,00	CCIR 493-4	B=100Bd S=170Hz			
11168,60	ALE USB	B=125Bd Ch=8 ChS=250Hz		USA diplo net	USA
11171,00	ALE USB	B=125Bd Ch=8 ChS=250Hz		ALG MIL net	ALG
11173,50	CIS 12	B=1440Bd Ch=12 ChS=200Hz	RCV	RUS N HQ Sevastopol	UKR
11174,00	PACTOR III	B=100Bd S=200Hz	V8V2222	Sailmail Brunai Bay	BRU
11175,00	SSB			USA AF Croughton	G
11175,00	SSB			USA AF Puerto Rico	PTR
11175,00	USB			USA AF Sigonella	I
11175,00	SSB			USA AF Diego Garcia	DGA
11175,00	SSB		AFA	USA AF Andrews AFB , Camp Springs , MD	USA
11175,00	SSB			USA AF Ascension AFB	ASC
11175,00	SSB		AFS	USA AF Offutt AFB , Omaha , NE	USA
11175,00	SSB		AIF80	USA AF Yokota AB	J
11175,00	TE-204 MODEM	B=150Bd S=880Hz Ch=4 ChS=440Hz			
11175,00	USB		AIF80	USA AF Yokota AB	J
11175,00	ALE USB	B=125Bd Ch=8 ChS=250Hz		USA AF net	USA
11175,00	USB			USA AF Guam	GUM
11175,00	USB			USA AF Hickam , HWA	USA
11175,00	USB			USA AF McClellan	USA
11175,00	*BUL MFSK*	*B=240Bd Ch=8 ChS=240Hz*		*MFA Sofia*	*BUL*
11177,00	*STANAG 4481*	*B=75Bd S=850Hz*		*USA MIL Diega Garcia*	*DGA*
11177,00	STANAG 4481	B=50Bd S=850Hz	NPM	USA N Lualualei	HWA
11178,00	ALE USB	B=125Bd Ch=8 ChS=250Hz		G AF	
11178,00	USB			POL AF net	POL
11178,00	*PSK*	*B=8000Bd*		*Niscemi*	*I*
11181,00	ALE USB	B=125Bd Ch=8 ChS=250Hz		USA AF net	
11183,00	STANAG 4197	B=1800Bd Ch=16/39 ChS=112/56Hz			
11184,00	HFDL	B=1800Bd	H03**	Reykjavik	ISL
11190,00	USB			CAN AF net	CAN
11193,00	USB			RUS AF net	RUS
11196,00	ALE USB	B=125Bd Ch=8 ChS=250Hz		USA COTHEN net	USA
11199,00	ALE USB	B=125Bd Ch=8 ChS=250Hz		USA AF NIPR net	USA
11205,00	STANAG 4197	B=1800Bd Ch=16/39 ChS=112/56Hz			
11211,00	STANAG 4285	B=2400Bd			
11213,00	*STANAG 4481*	*B=75Bd S=850Hz*	*MKL*	*G N Inskip*	*G*
11215,00	ALE USB	B=125Bd Ch=8 ChS=250Hz		MTN net	MTN
11215,50	ALE USB	B=125Bd Ch=8 ChS=250Hz			
11217,00	ALE USB	B=125Bd Ch=8 ChS=250Hz		USA MIL net	
11217,00	USB		DHM91	D AF Münster	D
11217,00	ALE USB	B=125Bd Ch=8 ChS=250Hz		G MIL net	G

Frequency	Mode	Mode Parameter	Callsign	User	Country
11217,00	ALE USB	B=125Bd Ch=8 ChS=250Hz		USA SHARES net	USA
11217,60	ALE USB	B=125Bd Ch=8 ChS=250Hz		USA diplo net	
11218,00	ALE LSB	B=125Bd Ch=8 ChS=250Hz			
11220,00	SSB			USA AF Strategic Command	USA
11220,00	TE-204 MODEM	B=150Bd S=880Hz Ch=4 ChS=440Hz		USA AF	USA
11222,00	*STANAG 4481*	*B=75Bd S=850Hz*	*NSS*	*USA N TX Davidsonville , MD*	*USA*
11223,00	ALE USB	B=125Bd Ch=8 ChS=250Hz		G MIL Net	G
11226,00	SSB		CMI	CUBANA Habana	CUB
11226,00	ALE USB	B=125Bd Ch=8 ChS=250Hz		AUS MIL net	AUS
11232,00	USB		CHR	CAN AF Trenton	CAN
11232,00	SSB			CAN AF Alert	CAN
11232,00	SSB			CAN AF Chatham Island	CAN
11232,00	SSB			CAN AF Sydney	CAN
11232,00	SSB			CAN AF Vancouver	CAN
11232,00	SSB		CJX	CAN AF St. John`s	CAN
11232,00	USB		PAER	NLD N ship HNLMS DE RUYTER	NLD
11235,00	*ALE USB*	*B=125Bd Ch=8 ChS=250Hz*		*I AF net*	*I*
11239,00	STANAG 4197	B=1800Bd Ch=16/39 ChS=112/56Hz		EGY N net	EGY
11246,00	*ALE USB*	*B=125Bd Ch=8 ChS=250Hz*		*ISR AF net*	
11247,00	SSB			G AF Gibraltar	GIB
11250,00	ALE USB	B=125Bd Ch=8 ChS=250Hz		USA AF net	
11252,00	SSB			NASA launch support ships	USA
11253,00	*USB*		*MKL*	*G AF VOLMET Inskip*	*G*
11256,00	USB		NPX	USARP Amundson-ScottSouth Pole Station	ANT
11261,00	ALE LSB	B=125Bd Ch=8 ChS=250Hz		TWN N Net	TWN
11263,00	USB			RUS AF net	RUS
11265,00	USB		CHR	CAN AF Trenton	CAN
11270,00	ALE USB	B=125Bd Ch=8 ChS=250Hz		ROU diplo net	ROU
11271,00	USB		CHR	CAN AF Trenton	CAN
11273,00	CHN DATALINK	B=2400Bd Ch=30 ChS=75Hz		CHN N	CHN
11279,00	USB		VFG	Gander ACC	CAN
11279,00	SSB			Aktyubinsk VOLMET	KAZ
11279,00	SSB			Baku VOLMET	AZE
11279,00	SSB			Bodoe ACC	NOR
11279,00	SSB			Iceland ACC , Reykjavik	ISL
11279,00	SSB			Karaganda VOLMET	KAZ
11279,00	SSB			MWARA NAT	NAT
11279,00	SSB			RDARA 2B	
11279,00	SSB			RDARA 6F	
11279,00	SSB			RDARA 9C	
11279,00	SSB			Tbilisi VOLMET	GEO
11279,00	USB		EIP	Shanwik ACC	IRL
11282,00	SSB			MWARA CEP	CEP
11282,00	SSB			RDARA 13H	
11282,00	SSB			RDARA 4A	
11282,00	SSB			RDARA 6G	
11282,00	SSB			San Francisco ACC , CA	USA
11285,00	SSB			RDARA 2A	
11285,00	SSB			RDARA 3B	

Frequency	Mode	Mode Parameter	Callsign	User	Country
11285,00	SSB			RDARA 7	
11285,00	USB			Colombo ACC	CLN
11288,00	SSB			RDARA 11B	
11288,00	SSB			RDARA 5A	
11288,00	SSB			RDARA 6G	
11288,00	USB		BOING EVERETT	Boing Seattle	USA
11288,00	USB		BOING SEATTLE	Boing Seattle	USA
11291,00	USB			Cayenne ACC	GUF
11291,00	SSB			Canarias ACC , Las Palmas	CNR
11291,00	SSB			Dakar ACC	SEN
11291,00	SSB			MWARA SAT	SAT
11291,00	SSB			Paramaribo ACC	SUR
11291,00	SSB			RDARA 3B	
11291,00	SSB			RDARA 3C	
11291,00	SSB			Recife ACC	B
11291,00	SSB			Sal ACC	CPV
11291,00	USB			Manaus ACC	B
11291,00	USB			Johannesburg ACC	AFS
11291,00	USB			Rio de Janerio ACC	B
11291,00	SSB			Canarias ACC , Las Palmas	CNR
11294,00	SSB			RDARA 2A	
11294,00	SSB			RDARA 6G	
11294,00	SSB			RDARA 7C	
11297,00	SSB			Kiev VOLMET	UKR
11297,00	SSB			RDARA 12F	
11297,00	SSB			RDARA 2	
11297,00	SSB		RLAP	Rostov VOLMET	RUS
11297,00	SSB		UHD	ST. Petersburg VOLMET	RUS
11300,00	SSB			Benghazi ACC	LBY
11300,00	SSB			Aden ACC	YEM
11300,00	SSB			Asmara ACC	ERT
11300,00	SSB			Mumbai ACC	IND
11300,00	SSB			Bujumbura ACC	BDI
11300,00	SSB			Cairo ACC	EGY
11300,00	SSB			Dar es Salaam ACC	TZA
11300,00	SSB			Entebbe ACC	UGA
11300,00	SSB			Khartoum ACC	SDN
11300,00	SSB			Kigali ACC	RRW
11300,00	SSB			Mogadishu ACC	SOM
11300,00	SSB			MWARA AFI	AFI
11300,00	SSB			Port Sudan ACC	SDN
11300,00	SSB			RDARA 13H	
11300,00	SSB			RDARA 6G	
11300,00	SSB			Sanaa ACC	YEM
11300,00	SSB			Seychelles ACC , Mahe	SEY
11300,00	SSB			Tripoli ACC	LBY
11300,00	SSB		5YD	Nairobi ACC	KEN
11300,00	SSB		DJR	Djibouti ACC	DJI
11300,00	SSB		ETD3	Addis Abbeba ACC	ETH
11300,00	USB			Bahrain ACC	BHR
11300,00	USB			Comoros ACC	COM
11300,00	USB			Hargeisa ACC	SOM
11300,00	USB			Jeddah ACC	ARS
11300,00	USB			Kisimayu ACC	SOM
11300,00	USB			Male ACC	MLD
11300,00	USB			Abadan ACC	IRN
11300,00	USB			Almaty ACC	KAZ
11300,00	USB			Ashkabad ACC	TKM
11300,00	USB			Bishkek ACC	KGZ
11300,00	USB			New Delhi ACC	IND

Frequency	Mode	Mode Parameter	Callsign	User	Country
11300,00	USB			Dushanbe ACC	TJK
11300,00	USB			Kabul ACC	AFG
11300,00	USB			Karachi ACC	PAK
11300,00	USB			Kathmandu ACC	NPL
11300,00	USB			Kuwait ACC	KWT
11300,00	USB			Lahore ACC	PAK
11300,00	USB			Male ACC	MLD
11300,00	USB			Muscat ACC	OMA
11300,00	USB			Odessa ACC	UKR
11300,00	USB			Samarkhand ACC	UZB
11300,00	USB			Seychelles ACC	SEY
11300,00	USB			Tashkent ACC	UZB
11300,00	USB			Tehran ACC	IRN
11300,00	USB			Tbilisi ACC	GEO
11300,00	USB			Urumqi ACC	CHN
11300,00	USB			Yerevan ACC	ARM
11303,00	SSB			RDARA 13E	
11303,00	SSB			RDARA 3C	
11306,00	HFDL	B=1800Bd	H16**	Agana	GUM
11306,00	SSB			Lima ACC	PRU
11306,00	SSB			RDARA 11B	
11306,00	SSB			RDARA 6G	
11306,00	SSB			RDARA 7E	
11306,00	USB		BOING SEATTLE	Boing Seattle	USA
11306,00	USB		BOING EVERETT	Boing Seattle	USA
11309,00	SSB			MWARA NAT	NAT
11309,00	SSB		KEA5	New York ACC	USA
11309,00	SSB			RDARA 3A	
11309,00	SSB			RDARA 6D	
11309,00	SSB		CSY	Santa Maria ACC	AZR
11312,00	SSB			RDARA 5	
11312,00	SSB			RDARA 9C	
11312,00	HFDL	B=1800Bd	H02**	Molokai , HI	HWA
11315,00	ALE USB	B=125Bd Ch=8 ChS=250Hz		MTN MOI net	MTN
11315,00	SSB			RDARA 6G	
11315,00	SSB			VOLMET CAR	CAR
11318,00	SSB			Jekaterinenburg VOLMET	RUS
11318,00	SSB		UNNN	Novosibirsk VOLMET	RUS
11318,00	SSB			RDARA 13D	
11318,00	SSB			RDARA 3	
11318,00	SSB			RDARA 4A	
11318,00	SSB		UBB2	Syktyvkar VOLMET	RUS
11318,00	SSB		RVPE	Tyumen VOLMET	RUS
11318,00	HFDL	B=1800Bd	H13**	Santa Cruz	BOL
11318,00	CW			RUS Strategic AF net	RUS
11320,00	*DPRK ARQ*	*B=600Bd S=600Hz*		*DPRK E Moscow*	*RUS*
11321,00	SSB			RDARA 13F	
11321,00	SSB			RDARA 6A	
11321,00	HFDL	B=1800Bd	H08**	Johannesburg	AFS
11324,00	SSB			RDARA 12C	
11324,00	SSB			RDARA 3A	
11324,00	SSB			RDARA 3C	
11324,00	SSB			RDARA 4B	
11325,00	CHN 4+4	B=75Bd Ch=8 ChS=300/450Hz			
11327,00	HFDL	B=1800Bd	H01**	San Francisco , CA	USA
11327,00	SSB			MWARA SP	SP
11327,00	SSB			RDARA 13C	
11327,00	SSB			RDARA 3B	

Frequency	Mode	Mode Parameter	Callsign	User	Country
11327,00	SSB			RDARA 5	
11330,00	SSB		KEA5	New York ACC	USA
11330,00	SSB			MWARA AFI	AFI
11330,00	SSB			MWARA NP	NP
11330,00	SSB			RDARA 13F	
11330,00	SSB			RDARA 3A	
11330,00	USB			Barranquilla ACC	CLM
11330,00	USB			Boyeros ACC	CUB
11330,00	USB			Cayenne ACC	GUF
11330,00	USB			Georgetown ACC	GUY
11330,00	USB			Maiquetia ACC	VEN
11330,00	USB			Panama ACC	PNR
11330,00	USB			Paramaribo ACC	SUR
11330,00	USB			Piarco ACC	TRD
11333,00	SSB			RDARA 10	
11333,00	SSB			RDARA 2B	
11333,00	SSB			RDARA 2C	
11336,00	USB			Iceland ACC , Reyklavik	ISL
11336,00	SSB			MWARA NAT	NAT
11336,00	SSB			RDARA 3	
11336,00	SSB		VFG	Gander ACC	CAN
11336,00	USB		EIP	Shanwik ACC	IRL
11339,00	SSB			RDARA 13K	
11339,00	SSB			RDARA 2B	
11339,00	SSB			RDARA 6B	
11339,00	SSB			RDARA 9	
11340,00	ALE USB	B=125Bd Ch=8 ChS=250Hz		MTN net	MTN
11342,00	SSB			W II	WW
11342,00	SSB			W III	WW
11342,00	USB			San Francisco ACC , CA	USA
11345,00	SSB			W I	WW
11345,00	SSB			W IV	WW
11345,00	USB		SDJ	Stockholm R	S
11348,00	SSB			W II	WW
11348,00	SSB			W V	WW
11348,00	HFDL	B=1800Bd	H02**	Molokai , HI	HWA
11348,00	HFDL	B=1800Bd	H17**	Telde , Gran Canaria	E
11350,00	CCIR 493-4	B=100Bd S=170Hz			
11350,00	CCIR 493-4	B=100Bd S=170Hz			
11350,00	ALE USB	B=125Bd Ch=8 ChS=250Hz		GANOB net	EGY
11351,00	SSB			W I	WW
11351,00	SSB			W III	WW
11353,00	ALE USB	B=125Bd Ch=8 ChS=250Hz		IRN net	IRN
11354,00	USB		NOVATOR	RUS N AF	RUS
11354,00	SSB			W II	WW
11354,00	SSB			W V	WW
11354,00	USB		PRIBOY	RUS N AF Moscow	RUS
11354,00	*USB WITH FULL CARRIER*		*KROCKET*	*Moscow N AF*	*RUS*
11354,00	HFDL	B=1800Bd	H09**	Utqiagvik , AK	ALS
11354,00	USB			Airbus Toulouse Technique	F
11357,00	SSB			RDARA 10A	
11357,00	SSB			RDARA 6A	
11357,00	SSB			RDARA 6E	
11360,00	USB		DAVLENIE	RUS AF Taganrog	RUS
11360,00	USB			RUS AF net	RUS
11360,00	USB			Barranquilla ACC	CLM
11360,00	USB			Bogota ACC	CLM
11360,00	USB			Maiquetia ACC	VEN

Frequency	Mode	Mode Parameter	Callsign	User	Country
11360,00	USB			Lima ACC	PRU
11360,00	USB			Antofagasta ACC	CHL
11360,00	USB			Buenos Aires ACC	ARG
11360,00	USB			Cordoba ACC	ARG
11360,00	SSB			Asuncion ACC	PRG
11360,00	SSB			MWARA SAM	SAM
11360,00	SSB			Quito ACC	EQA
11360,00	SSB			RDARA 14	
11360,00	SSB			RDARA 2	
11360,00	SSB			RDARA 3	
11360,00	USB		KORSAR	RUS AF Moscow	RUS
11360,00	USB			Easter Island ACC	CHL
11360,00	USB		KADRIL	RUS AF Ryazan	RUS
11360,00	USB		BUNTAR	RUS AF Orenburg	RUS
11360,00	USB			La Paz ACC	BOL
11360,00	USB			Lima ACC	PRU
11360,00	USB		PROSELOK	RUS AF Pskov	RUS
11360,00	USB		KLARNETIST	RUS AF Migalovo/Tver	RUS
11360,00	USB			Puerto Montt ACC	CHL
11360,00	USB			Punta Arenas ACC	CHL
11360,00	USB			Santa Cruz ACC	BOL
11360,00	USB			Santiago ACC	CHL
11360,00	USB			Talara ACC	BOL
11360,00	USB			Ushuaia ACC	ARG
11363,00	SSB			RDARA 10A	
11363,00	SSB			RDARA 6E	
11366,00	SSB			RDARA 13K	
11366,00	SSB			RDARA 1C	
11366,00	SSB			RDARA 6B	
11366,00	SSB			RDARA 6F	
11369,00	USB		LWB	Ezeiza VOLMET	ARG
11369,00	SSB			RDARA 13G	
11369,00	SSB			RDARA 6G	
11370,00	USB			RUS AF	
11372,00	SSB			RDARA 2C	
11372,00	SSB			RDARA 3B	
11372,00	SSB			RDARA 6D	
11375,00	SSB			MWARA MID	MID
11375,00	SSB			RDARA 10A	
11375,00	SSB			RDARA 13C	
11375,00	USB			Aden ACC	YEM
11375,00	USB			Amman ACC	JOR
11375,00	USB			Ankara ACC	TUR
11375,00	USB			Beirut ACC	LBN
11375,00	USB			Cairo ACC	EGY
11375,00	USB			Damascus ACC	SYR
11375,00	USB			Jeddah ACC	ARS
11375,00	USB			Kuwait ACC	KWT
11375,00	USB			Manama ACC	BHR
11375,00	USB			Odessa ACC	UKR
11375,00	USB			Simferopol ACC	UKR
11375,00	USB			Sanaa ACC	YEM
11375,00	USB			Tehran ACC	IRN
11375,00	USB			Tbilisi ACC	GEO
11375,00	USB			Yerevan ACC	ARM
11375,00	USB			Aktyubinsk ACC	KAZ
11375,00	USB			Almaty ACC	KAZ
11375,00	USB			Bishkek ACC	KGZ
11375,00	USB			Dushanbe ACC	TJK
11375,00	USB			Kuybyshev ACC	RUS

Frequency	Mode	Mode Parameter	Callsign	User	Country
11375,00	USB			Moscow ACC	RUS
11375,00	USB			Kzyl-Orda ACC	KAZ
11375,00	USB			Samarkhand ACC	UZB
11375,00	USB			Tashkent ACC	UZB
11375,00	USB			Uralsk ACC	KAZ
11378,00	SSB			RDARA 13M	
11378,00	SSB			RDARA 3C	
11378,00	SSB			VOLMET EUR	EUR
11381,00	SSB			RDARA 12E	
11381,00	SSB			RDARA 12J	
11381,00	SSB			RDARA 6	
11384,00	SSB			MWARA CWP	CWP
11384,00	SSB			RDARA 12J	
11384,00	SSB			RDARA 1D	
11384,00	HFDL	B=1800Bd	H07**	Shannon	IRL
11384,00	USB			Hong Kong ACC	HKG
11384,00	USB			Naha ACC	J
11384,00	USB			Port Moresby ACC	PNG
11384,00	USB			San Francisco ACC , CA	USA
11384,00	USB			Seoul ACC	KOR
11384,00	USB			Taipei ACC	TWN
11384,00	USB			Tokyo ACC	J
11387,00	USB		AXQ421	Australian Volmet	AUS
11387,00	SSB		HSD	Bangkok VOLMET	THA
11387,00	SSB		AWB	Mumbai VOLMET	IND
11387,00	SSB		AWC	Calcutta VOLMET	IND
11387,00	SSB		ARA	Karachi VOLMET	PAK
11387,00	SSB			MWARA CAR	CAR
11387,00	SSB		9VA43	Singapore VOLMET	SNG
11387,00	SSB			AUS VOLMET Brisbane	AUS
11387,00	SSB			VOLMET SEA	SEA
11387,00	HFDL	B=1800Bd	H04**	Riverhead , NY	USA
11390,00	SSB			RDARA 10	
11390,00	SSB			RDARA 2	
11393,00	SSB			RDARA 12E	
11393,00	SSB			RDARA 9B	
11393,00	SSB			VOLMET MID	MID
11395,00	ALE USB	B=125Bd Ch=8 ChS=250Hz			
11396,00	USB			Brisbane ACC	AUS
11396,00	SSB			Bali ACC	INS
11396,00	SSB			Bangkok ACC	THA
11396,00	SSB			Boyeros ACC	CUB
11396,00	SSB			Ho Chi Min Ville ACC	VTN
11396,00	SSB			Hongkong ACC	HKG
11396,00	SSB			Jakarta ACC	INS
11396,00	SSB			Manila ACC	PHL
11396,00	SSB			Merida ACC	MEX
11396,00	SSB			MWARA CAR	CAR
11396,00	SSB			MWARA EA	EA
11396,00	SSB			MWARA SEA	SEA
11396,00	SSB		KEA5	New York ACC	USA
11396,00	SSB			Panama ACC	PNR
11396,00	SSB			Piarco ACC	TRD
11396,00	SSB			Singapore ACC	SNG
11396,00	SSB			Ujung Pandang ACC	INS
11396,00	USB			Barranquilla ACC	CLM
11396,00	USB			Guangzhou ACC	CHN
11396,00	USB			Irkutsk ACC	RUS
11396,00	USB			Pyongyang ACC	KRE
11396,00	USB			Ulaanbaatar ACC	MNG
11396,00	USB			Hanoi ACC	VTN

Frequency	Mode	Mode Parameter	Callsign	User	Country
11396,00	USB			Male ACC	MLD
11396,00	USB			Kuala Lumpur ACC	MLA
11396,00	USB			Manila ACC	PHL
11396,00	USB			Kota Kinabalu ACC	MLA
11396,00	USB			Seoul ACC	KOR
11396,00	USB			Tokyo ACC	J
11396,00	USB			Vientianne ACC	LAO
11400,50	ALE USB	B=125Bd Ch=8 ChS=250Hz		AUT MIL net	
11402,00	ALE USB	B=125Bd Ch=8 ChS=250Hz		USA Civil Air Patrol net	USA
11402,00	CHN 4+4	B=75Bd Ch=8 ChS=300/450Hz		CHN MIL	
11403,00	USB		PEGASO	E Combat Air Command Torrejon de Ardoz	E
11405,00	ALE USB	B=125Bd Ch=8 ChS=250Hz		S FRO net	S
11405,70	PACTOR FEC	B=100Bd S=200Hz		PAK N	PAK
11406,00	CIS OFDM 93	B=2400Bd Ch=93 ChS=31,25Hz		CIS diplo	RUS
11407,00	SSB			NASA rocket booster recovery	USA
11408,00	SSB			USA AF Strategic Command	USA
11409,00	CIS MFSK 20	B=20Bd Ch=20 ChS=40Hz		RUS INTEL	RUS
11409,70	SITOR A	B=100Bd S=170Hz		MFA Cairo	EGY
11410,00	ALE USB	B=125Bd Ch=8 ChS=250Hz		G MIL net	G
11411,00	CIS 300-500	B=300Bd S=500Hz		RUS AF	RUS
11414,00	USB			USA AF Cape Radio	USA
11414,00	CIS 12	B=1440Bd Ch=12 ChS=200Hz	RCV	RUS N HQ Sevastopol	UKR
11415,00	CIS 50-500 BAUDOT	B=50Bd S=500Hz		RUS	RUS
11418,00	CW		RMP	RUS N HQ Kaliningrad	RUS
11420,00	ALE USB	B=125Bd Ch=8 ChS=250Hz		ALG MIL net	ALG
11423,00	CIS 4FSK	B=150Bd Ch=4 ChS=4000Hz			
11424,00	CHN 4+4	B=75Bd Ch=8 ChS=300/450Hz		CHN MIL	
11425,00	ALE USB	B=125Bd Ch=8 ChS=250Hz		ROU net	
11426,00	ALE USB	B=125Bd Ch=8 ChS=250Hz		ALG MIL net	ALG
11426,50	ALE USB	B=125Bd Ch=8 ChS=250Hz		USA SHARES net	USA
11429,00	ALE USB	B=125Bd Ch=8 ChS=250Hz		CHL N net	CHL
11430,00	STANAG 4538	B=2400Bd			
11430,00	MIL 188-110C APP D	B=2400Bd			
11430,00	ALE USB	B=125Bd Ch=8 ChS=250Hz		USA FEMA net	USA
11430,00	ALE USB	B=125Bd Ch=8 ChS=250Hz		MTN net	MTN
11430,00	MIL 188-110B 39TONE	B=2400Bd Ch=39 ChS=56,25Hz			
11434,00	ALE USB	B=125Bd Ch=8 ChS=250Hz		I AF net	I
11434,70	ALE USB	B=125Bd Ch=8 ChS=250Hz		CHN net	CHN
11435,00	WINDRM	B=1912,5Bd Ch=51 ChS=46,875Hz		CUB Intelligence Directorate	CUB
11435,00	ALE USB	B=125Bd Ch=8 ChS=250Hz		Queensland Deptartment of Community Safety	AUS

Frequency	Mode	Mode Parameter	Callsign	User	Country
11437,50	ALE USB	B=125Bd Ch=8 ChS=250Hz			
11437,90	PACTOR	B=200Bd S=170Hz	EURI	ALG customs	ALG
11439,00	PACKET RADIO	B=1250Bd S=1250Hz		INS N net	INS
11439,00	CHN 4+4	B=75Bd Ch=8 ChS=300/450Hz		CHN MIL	
11440,00	ALE USB	B=125Bd Ch=8 ChS=250Hz		SVK AF net	SVK
11441,00	MIL 188-110A SER	B=2400Bd		AUS MHFCS North West Cape	AUS
11443,50	DPRK FSK	B=600Bd S=600Hz		DPRK E	
11444,00	MIL 188-110B 39TONE	B=2400Bd Ch=39 ChS=56,25Hz		AUS MIL	AUS
11444,70	SITOR A	B=100Bd S=170Hz		MFA Cairo	EGY
11445,00	ALE USB	B=125Bd Ch=8 ChS=250Hz		CHN net	CHN
11446,00	MIL 188-110B SER	B=2400Bd		S MIL	S
11448,00	ALE USB	B=125Bd Ch=8 ChS=250Hz		USA FEMA net	USA
11448,00	ALE USB	B=125Bd Ch=8 ChS=250Hz			
11450,00	2FSK	B=1200Bd S=1200Hz		I MIL	I
11450,00	*ALE LSB*	*B=125Bd Ch=8 ChS=250Hz*		*MOZ gas project*	*MOZ*
11450,00	USB			HFoZ/Radtel net	AUS
11450,00	CIS MFSK 68	B=50Bd Ch=68 ChS=47Hz		RUS diplo	RUS
11451,00	ALE USB	B=125Bd Ch=8 ChS=250Hz			USA
11454,00	CIS 3000	B=3000Bd		RUS diplo	RUS
11455,00	ALE USB	B=125Bd Ch=8 ChS=250Hz		ALG MOI net	ALG
11455,00	ALE USB	B=125Bd Ch=8 ChS=250Hz		CLM N	CLM
11455,00	DPRK ARQ	B=1200Bd S=600Hz		DPRK E Sofia	BUL
11455,20	*STANAG 4285*	*B=2400Bd*		*NOR N Oslo*	*NOR*
11456,00	USB		XVT	Da Nang R	VTN
11456,00	USB		XVG	Hai Phong R	VTN
11456,00	USB		XVS	Ho Chi Minh R	VTN
11458,00	*ALE USB*	*B=125Bd Ch=8 ChS=250Hz*		*ROU diplo net*	*ROU*
11460,20	LINK 22	B=2250Bd			
11462,00	WINDRM	B=1912,5Bd Ch=51 ChS=46,875Hz		CUB Intelligence Directorate	CUB
11468,00	T600	B=50Bd S=100Hz		RUS N Murmansk	RUS
11468,00	*T600*	*B=50Bd S=250Hz*	*RDL*	*RUS N TX Kaliningrad*	*RUS*
11470,00	CIS 50-500	B=50Bd S=500Hz	RGY2	RUS GOV	RUS
11470,00	*STANAG 4285*	*B=2400Bd*		*G MIL St. Eval*	*G*
11472,60	ALE USB	B=125Bd Ch=8 ChS=250Hz		USA diplo net	
11480,00	MIL 188-110B SER	B=2400Bd		ALB F net	ALB
11480,00	ALE USB	B=125Bd Ch=8 ChS=250Hz		ALB F net	ALB
11481,00	AUS ISB MODEM	B=600Bd S=345Hz		AUS MIL Lyndoch	AUS
11484,00	CW			RUS HFDF net	RUS
11484,00	USB		AAR9EN	USA MARS	USA
11487,00	USB			HFoZ/Radtel net	AUS
11487,00	USB		VKE237	AUS HF Radio Club	AUS
11488,00	ALE LSB	B=125Bd Ch=8 ChS=250Hz		Public health net , TX	USA
11490,00	USB		VNJ	Casey Base , AUS	ANT

Frequency	Mode	Mode Parameter	Callsign	User	Country
11490,00	USB			AUS Davis Base	ANT
11490,00	USB			Macquarie Island , AUS	ANT
11490,00	USB			Mawson Base , AUS	ANT
11490,00	ALE USB	B=125Bd Ch=8 ChS=250Hz		S FRO net	S
11490,00	*CIS 50-500*	*B=50Bd S=500Hz*		*Moscow*	*RUS*
11490,00	ALE USB	B=125Bd Ch=8 ChS=250Hz		S net	S
11491,00	ALE USB	B=125Bd Ch=8 ChS=250Hz		USA FBI net	USA
11494,00	ALE USB	B=125Bd Ch=8 ChS=250Hz		USA Department of Agriculture	USA
11494,00	ALE USB	B=125Bd Ch=8 ChS=250Hz		USA CG net	USA
11494,00	ALE USB	B=125Bd Ch=8 ChS=250Hz		USA COTHEN net	USA
11500,00	*CIS 100-500*	*B=100Bd S=500Hz*		*RUS N Murmansk*	*RUS*
11500,00	*CIS 100-500*	*B=100Bd S=500Hz*		*RUS N Moscow*	*RUS*
11500,00	ALE USB	B=125Bd Ch=8 ChS=250Hz		TUR Disaster and Emergency net	TUR
11500,00	CIS ARQ	B=100Bd S=500Hz		RUS PTT	RUS
11500,00	PACTOR III	B=100Bd S=200Hz		MFA Praha	CZE
11505,00	SITOR B	B=100Bd S=170Hz	PWZ33	B N Rio de Janeiro	B
11506,00	ALE USB	B=125Bd Ch=8 ChS=250Hz		POL MIL net	
11507,00	MIL 188-110B SER	B=2400Bd		CZR E Kiev	UKR
11523,00	*STANAG 4285*	*B=2400Bd*		*G F St. Eval*	*G*
11524,00	*T600*	*B=50Bd S=200Hz*		*RUS N Moscow*	*RUS*
11524,00	T600	B=50Bd S=100Hz		RUS N Murmansk	RUS
11527,50	CHN 4+4	B=75Bd Ch=8 ChS=300/450Hz		CHN N	
11530,00	WINDRM	B=1912,5Bd Ch=51 ChS=46,875Hz		CUB Intelligence Directorate	CUB
11530,00	CODAN CHIRP	B=80Bd Ch=30 ChS=250Hz		BGD N net	BGD
11530,00	CODAN	B=2400Bd Ch=16 ChS=112,5Hz		BGD N net	BGD
11530,00	USB			BGD N net	BGD
11532,00	RUS DBPSK 16FSK	B=1866Bd Ch=16 ChS=175Hz		RUS INTEL	RUS
11534,00	T600	B=50Bd S=200Hz		RUS N	RUS
11536,00	*RUS DBPSK 16FSK*	*B=1866Bd Ch=16 ChS=175Hz*		*RUS INTEL*	*RUS*
11538,20	*STANAG 4285*	*B=2400Bd*	*DHJ58*	*D N Glücksburg TX Neuharlingersiel*	*D*
11540,00	ALE USB	B=125Bd Ch=8 ChS=250Hz		F N net	
11543,60	ALE USB	B=125Bd Ch=8 ChS=250Hz		F N net	F
11544,00	CIS 3000	B=3000Bd		RUS diplo	RUS
11545,00	WINDRM	B=1912,5Bd Ch=51 ChS=46,875Hz		CUB Intelligence Directorate	CUB
11547,00	ALE USB	B=125Bd Ch=8 ChS=250Hz		USA AF net	USA
11550,00	*ALE USB*	*B=125Bd Ch=8 ChS=250Hz*		*F Airbus net*	*F*
11550,00	ALE USB	B=125Bd Ch=8 ChS=250Hz		G MIL net	G
11552,00	STANAG 4481	B=75Bd S=850Hz	NSS	USA N TX Davidsonville , MD	USA
11553,00	USB			McMurdo Ops	
11553,00	USB		NHG	USARP Palmer Station	ANT
11555,00	ALE USB	B=125Bd Ch=8 ChS=250Hz		USA Army NG net Utah	USA
11557,00	MIL 188-110A SER	B=2400Bd		MFA Bucharest	ROU

Frequency	Mode	Mode Parameter	Callsign	User	Country
11557,00	MIL 188-110A SER	B=2400Bd		ROU E Stockholm	S
11573,00	MIL 188-110B SER	B=2400Bd		ROU E	
11576,50	ALE USB	B=125Bd Ch=8 ChS=250Hz		USA Defence Logistic Agency	USA
11578,20	STANAG 4285	B=2400Bd	DHJ58	D N Glücksburg TX Marlow	D
11584,00	ALE USB	B=125Bd Ch=8 ChS=250Hz		G MIL net TX Ascension Island	ASC
11598,00	*CIS 50-200*	*B=50Bd S=200Hz*	*RMP*	*RUS N HQ Kaliningrad*	*RUS*
11612,00	USB		VMS469	Reids Radiodata net	AUS
11612,00	SSB		VKS737	AUS 4WD net	AUS
11620,00	ALE USB	B=125Bd Ch=8 ChS=250Hz		CHL SENAPRED net	CHL
11621,00	SSB			NASA rocket booster recovery	USA
11635,00	WINDRM	B=1912,5Bd Ch=51 ChS=46,875Hz		CUB Intelligence Directorate	CUB
11635,00	WINDRM	B=1912,5Bd Ch=51 ChS=46,875Hz		CUB Intelligence Directorate	CUB
11637,00	ALE USB	B=125Bd Ch=8 ChS=250Hz		USA FAA	USA
11638,00	*BAUDOT*	*B=50Bd S=425Hz*	*DDK7*	*DWD TX Pinneberg*	*D*
11650,00	ALE USB	B=125Bd Ch=8 ChS=250Hz		ARG CG net	ARG
11680,00	ALE USB	B=125Bd Ch=8 ChS=250Hz		ROU diplo net	ROU
11688,00	STANAG 4481	B=50Bd S=850Hz	NAU	USA N San Juan	PTR
11688,00	STANAG 4481	B=50Bd S=850Hz	NSS	USA N TX Davidsonville , MD	USA
11717,00	*CIS 12*	*B=1440Bd Ch=12 ChS=200Hz*		*RUS MIL Moscow*	*RUS*
11780,00	MFSK9	B=100Bd Ch=9 ChS=5000Hz			
11820,00	DPRK ARQ	B=600Bd S=600Hz		DPRK E Sofia	BUL
11836,00	CIS 12	B=1440Bd Ch=12 ChS=200Hz	RCV	RUS N HQ Sevastopol	UKR
11848,00	*CIS 50-1000*	*B=50Bd S=1000Hz*		*RUS MIL Moscow*	*RUS*
11892,00	T600	B=50Bd S=200Hz		RUS N	
11903,10	FAX 120/576			Swedish Meteorological and Hydrological Institute	S
11953,50	ALE USB	B=125Bd Ch=8 ChS=250Hz		USA FBI net	USA
11957,00	ALE USB	B=125Bd Ch=8 ChS=250Hz		USA FEMA net	USA
11962,00	PACTOR II	B=100Bd S=200Hz		PAK N	
11980,60	STANAG 4481	B=75Bd S=850Hz	AJE	USA AF AWS Croughton	G
11980,70	STANAG 4481	B=50Bd S=850Hz	NAU	USA N San Juan	PTR
11981,00	STANAG 4481	B=75Bd S=850Hz	NAU	USA N San Juan	PTR
11986,70	*STANAG 4481*	*B=75Bd S=850Hz*		*G F St. Eval*	*G*
11988,50	USB			USA Army MARS	
12000,00	FMCW			Pendeen	G
12000,00	FMCW			Perranporth	G
12015,00	STANAG 4285	B=2400Bd	NSS	USA N TX Davidsonville , MD	USA
12015,00	STANAG 4481	B=50Bd S=850Hz	NAU	USA N San Juan	PTR
12015,00	STANAG 4481	B=50Bd S=850Hz	NSS	USA N TX Davidsonville , MD	USA
12022,00	ALE USB	B=125Bd Ch=8 ChS=250Hz		TWN N net	TWN
12024,00	CHN DATALINK	B=2400Bd Ch=30 ChS=75Hz		Wuming area	CHN
12043,00	STANAG 4285	B=2400Bd		G F St. Eval	G
12044,00	*CIS 12*	*B=1440Bd Ch=12 ChS=200Hz*		*RUS MIL Smolensk*	*RUS*
12056,00	CW		RIW	RUS N HQ Moscow	RUS
12062,00	CW		K4MT**		
12062,00	CW		NT9P**		RUS

Frequency	Mode	Mode Parameter	Callsign	User	Country
12068,50	ALE USB	B=125Bd Ch=8 ChS=250Hz		USA Defence Logistic Agency	USA
12070,00	ALE USB	B=125Bd Ch=8 ChS=250Hz		USA ACE	
12070,00	ALE USB	B=125Bd Ch=8 ChS=250Hz		B SIVAM net	B
12076,00	ALE USB	B=125Bd Ch=8 ChS=250Hz			
12076,00	ALE USB	B=125Bd Ch=8 ChS=250Hz		USA SAC net	USA
12078,50	*STANAG 4285*	*B=2400Bd*		*G MIL Crimond*	*G*
12081,00	ALE USB	B=125Bd Ch=8 ChS=250Hz		USA Civil Air Patrol net	USA
12081,00	CIS 3000	B=3000Bd		RUS diplo	RUS
12090,00	*ALE USB*	*B=125Bd Ch=8 ChS=250Hz*		*F Airbus net*	*F*
12100,00	ALE USB	B=125Bd Ch=8 ChS=250Hz		USA CAP net	USA
12101,40	STANAG 4285	B=2400Bd		AUS MHFCS Humpty Doo	
12102,00	RUS HYBRID MODEM MFSK-PSK	B=40Bd Ch=16 ChS=40Hz		Moscow	RUS
12104,00	*CIS 75-200*	*B=75Bd S=200Hz*		*RUS MIL Moscow*	*RUS*
12105,00	STANAG 4285	B=2400Bd		AUS MHFCS Humpty Doo	
12105,00	CIS 68TONE	B=50Bd Ch=68 ChS=47Hz		RUS diplo	RUS
12107,00	*CIS MFSK 11*	*B=125Bd Ch=11 ChS=250Hz*		*Moscow*	*RUS*
12107,50	ALE USB	B=125Bd Ch=8 ChS=250Hz			
12109,00	ALE USB	B=125Bd Ch=8 ChS=250Hz			
12115,00	CIS 50-500	B=50Bd S=500Hz		RUS	RUS
12117,00	ALE USB	B=125Bd Ch=8 ChS=250Hz		CHN MIL net	CHN
12117,00	MIL 188-110b 39TONE	B=2400Bd Ch=39 ChS=56,25Hz		CHN MIL	CHN
12120,00	*STANAG 4481*	*B=75Bd S=850Hz*	*NAU*	*USA N San Juan*	*PTR*
12121,00	BAUDOT	B=50Bd S=500Hz Ch=1		RUS	RUS
12122,00	ALE USB	B=125Bd Ch=8 ChS=250Hz		USA MIL net	USA
12127,00	PACTOR III	B=100Bd S=200Hz	WHX	Droop Mountain , WV	USA
12129,00	ALE USB	B=125Bd Ch=8 ChS=250Hz		USA FEMA net	USA
12129,50	CIS 50-1000	B=50Bd S=1000Hz		RUS	RUS
12131,00	CW		WI2XER	Skycast Services	USA
12132,50	ALE USB	B=125Bd Ch=8 ChS=250Hz		F N net	
12132,60	STANAG 4285	B=2400Bd	FUJ	F N Noumea	NCL
12135,00	BAUDOOT 1,5 STB	B=50Bd S=500Hz	K4MT**		
12136,00	CW		RDL	RUS N Moscow	RUS
12138,50	ALE USB	B=125Bd Ch=8 ChS=250Hz		USA FBI net	USA
12140,00	ALE USB	B=125Bd Ch=8 ChS=250Hz		SVK AF net	SVK
12140,00	ALE USB	B=125Bd Ch=8 ChS=250Hz		TUN diplo net	TUN
12140,00	PACTOR II	B=100Bd S=200Hz	STAT25**	TUN E Brussels	BEL
12140,00	PACTOR II	B=100Bd S=200Hz	STAT152**	MFA Tunis	TUN
12140,00	ALE USB	B=125Bd Ch=8 ChS=250Hz		TUN diplo net	TUN
12140,00	CIS ARQ	B=100Bd S=500Hz		RUS PTT	RUS

Frequency	Mode	Mode Parameter	Callsign	User	Country
12140,00	CW		WI2XER	Skycast Services	USA
12142,00	CODAN CHIRP	B=80Bd Ch=30 ChS=250Hz		BGD N net	BGD
12142,00	CODAN	B=2400Bd Ch=16 ChS=112,5Hz		BGD N net	BGD
12142,00	USB			BGD N net	BGD
12142,00	CIS 3000	B=3000Bd		RUS diplo	RUS
12142,00	*STANAG 4285*	*B=2400Bd*		*G MIL St Eval*	*G*
12143,00	CIS OFDM 60	B=2400Bd Ch=60 ChS=44,5Hz		RUS	RUS
12144,70	STANAG 4285	B=2400Bd	FUJ	F N Noumea	NCL
12148,20	USB		VNJ	Casey Base , AUS	ANT
12148,20	USB			AUS Davis Base	ANT
12148,20	USB			Macquarie Island , AUS	ANT
12148,20	USB			Mawson Base , AUS	ANT
12150,00	FMCW			FallBack22, Pt. Sal, CA	USA
12151,00	ALE USB	B=125Bd Ch=8 ChS=250Hz		ALG Mil net	ALG
12151,00	CIS OFDM 128	B=2666Bd Ch=128 ChS=23,5Hz		RUS diplo	RUS
12152,00	*STANAG 4285*	*B=2400Bd*		*G MIL Akrotiri*	*CYP*
12155,00	ALE USB	B=125Bd Ch=8 ChS=250Hz		ALG MIL net	ALG
12155,00	CIS 4FSK	B=150Bd Ch=4 ChS=4000Hz			
12157,00	FMCW			Bodega Marine Laboratory, CA	USA
12157,00	FMCW			Slide Ranch, CA	USA
12157,00	FMCW			Fort Stevens, OR	USA
12158,00	MIL 188-110B SER	B=2400Bd		TUR UEKAE	TUR
12158,00	CW		RGT77	RUS MIL Moscow, RVSN	RUS
12160,00	ALE USB	B=125Bd Ch=8 ChS=250Hz		MRC Mil net	MRC
12160,50	BAUDOT	B=50Bd S=170Hz	8WB2	IND E Kabul	AFG
12162,00	*RUS HYBRID MODEM MFSK-PSK*	*B=40Bd Ch=16*		*Moscow*	*RUS*
12162,00	*RUS HYBRID MODEM PSK-PSK*	*B=125Bd Ch=2 ChS=250Hz*		*Moscow*	*RUS*
12164,00	CIS OFDM 45	B=2400Bd Ch=45 ChS=62,5Hz		RUS diplo	RUS
12165,00	*LINK 11 ISB*	*B=2250Bd Ch=16 ChS=330/110/550Hz*	*PBB*	*NLD N Den Helder*	*HOL*
12165,00	*CIS ARQ*	*B=100Bd S=500Hz*		*N Smolensk*	*RUS*
12166,00	*CIS 12*	*B=1440Bd Ch=12 ChS=200Hz*		*RUS MIL Moscow*	*RUS*
12167,00	ALE LSB	B=125Bd Ch=8 ChS=250Hz		Public health net , TX	USA
12168,00	CIS MFSK 20	B=20Bd Ch=20 ChS=40Hz		RUS INTEL	RUS
12170,00	ALE USB	B=125Bd Ch=8 ChS=250Hz		USA Army NG net Utah	USA
12170,50	ALE USB	B=125Bd Ch=8 ChS=250Hz			
12177,00	ALG AF BELL-103 MODEM	B=300Bd S=200Hz		ALG AF	ALG
12180,00	WINDRM	B=1912,5Bd Ch=51 ChS=46,875Hz		CUB Intelligence Directorate	CUB
12182,00	*STANAG 4481*	*B=75Bd S=850Hz*		*G MIL St. Eval*	*G*
12184,00	*CIS 50-500*	*B=50Bd S=500Hz*	*RVQ4*	*RUS GOV Moscow*	*RUS*
12184,50	ALE USB	B=125Bd Ch=8 ChS=250Hz			

Frequency	Mode	Mode Parameter	Callsign	User	Country
12185,00	ALE USB	B=125Bd Ch=8 ChS=250Hz		ALG AF net	ALG
12188,50	PACTOR III	B=100Bd S=200Hz	AAN3DCA	USA MARS Station, Potomac Heights , MD	USA
12188,50	PACTOR III	B=100Bd S=200Hz	AAN4TYS	USA MARS	USA
12188,50	PACTOR III	B=100Bd S=200Hz	AAN5TNC	USA MARS	USA
12190,00	*STANAG 4285*	*B=2400Bd*		*S MIL*	*S*
12192,70	STANAG 4285	B=2400Bd	NSY	USA N Niscemi , Sicily	I
12197,00	USB			HFoZ/Radtel net	AUS
12197,00	USB		VKE237	AUS HF Radio Club	AUS
12200,00	ALE USB	B=125Bd Ch=8 ChS=250Hz		USA CAP net	USA
12200,00	FMCW			Refugio State Beach	USA
12200,00	USB			EGY N net	EGY
12200,00	USB			EGY N	EGY
12202,20	*STANAG 4285*	*B=2400Bd*		*G MIL Inskip*	*G*
12204,00	ALE USB	B=125Bd Ch=8 ChS=250Hz		SNG N net	SNG
12207,00	MIL 188-110B SER	B=2400Bd		CZE E	
12209,00	ALE USB	B=125Bd Ch=8 ChS=250Hz		TUR AFAD net	TUR
12210,00	CIS 50-500	B=50Bd S=500Hz	RWB5	RUS PTT	RUS
12211,50	USB		VMS469	Reids Radiodata net	AUS
12212,00	ALE USB	B=125Bd Ch=8 ChS=250Hz		USA NG net	USA
12214,00	*RUS DBPSK 16FSK*	*B=1866Bd Ch=16 ChS=175Hz*		*RUS INTEL*	*RUS*
12214,00	DBPSK	B=62,5Bd		RUS	RUS
12215,00	BUL MFSK	B=240Bd Ch=8 ChS=240Hz		MFA Sofia	BUL
12215,00	*CIS 4FSK*	*B=150Bd Ch=4 ChS=4000Hz*		*Kaliningard*	*RUS*
12215,50	ALE USB	B=125Bd Ch=8 ChS=250Hz		USA FBI net	USA
12216,00	ALE USB	B=125Bd Ch=8 ChS=250Hz		USA FEMA net	USA
12216,00	USB			HFoZ/Radtel net	AUS
12218,00	ALE USB	B=125Bd Ch=8 ChS=250Hz		USA Civial Air Patrol net	USA
12220,00	*BUL MFSK*	*B=240Bd Ch=8 ChS=240Hz*		*MFA Sofia*	*BUL*
12220,00	RUS HYBRID MFSK OFDM	B=40Bd Ch=16		Moscow	RUS
12220,00	RUS HYBRID MFSK OFDM	B=40Bd Ch=16		Praha	CZE
12220,00	*RFSM8000*	*B=2400Bd*		*MFA Sofia*	*BUL*
12221,00	RFSM8000	B=2400Bd		BUL E	
12221,00	RFSM8000	B=2400Bd		BUL E Amman	SYR
12221,00	RFSM8000	B=2400Bd		BUL E London	G
12221,00	RFSM8000	B=2400Bd		BUL E Tunis	TUN
12222,00	ALE USB	B=125Bd Ch=8 ChS=250Hz		USA CAP net	USA
12222,00	DPRK ARQ	B=600Bd S=600Hz		DPRK E Moscow	RUS
12222,00	RUS DBPSK 16FSK	B=1866Bd Ch=16 ChS=175Hz		RUS INTEL	RUS
12222,00	ALE USB	B=125Bd Ch=8 ChS=250Hz		USA COTHEN net	USA
12222,70	SITOR A	B=100Bd S=170Hz		MFA Cairo	EGY
12226,00	RFSM8000	B=2400Bd		MFA Sofia	BUL
12226,00	RFSM8000	B=2400Bd		MFA Sofia	BUL
12227,00	BAUDOT	B=50Bd S=500Hz			
12227,00	CIS 50-500 BAUDOT	B=50Bd S=500Hz			
12227,00	CW		RGT77	RUS MIL Moscow, RVSN	RUS

Frequency	Mode	Mode Parameter	Callsign	User	Country
12228,00	RFSM8000	B=2400Bd		MFA Sofia	BUL
12228,00	RFSM8000	B=2400Bd		MFA Sofia	BUL
12230,00	SSB			SS Ch 1201	
12230,00	*ALE USB*	*B=125Bd Ch=8 ChS=250Hz*		*G MIL net*	*G*
12230,00	*CIS 50-500*	*B=50Bd S=500Hz*		*RUS MIL Ekaterinburg*	*RUS*
12230,00	CLOVER 2000	B=500Bd Ch=8 ChS=250Hz	CN1EC		B
12231,00	ALE USB	B=125Bd Ch=8 ChS=250Hz		ROU diplo net	ROU
12232,00	FMCW			Waldport, OR	USA
12233,00	SSB			SS Ch 1202	
12235,00	STANAG 4285	B=2400Bd Ch=2	CTA	POR N Lisbon	POR
12236,00	SSB			SS Ch 1203	
12239,00	SSB			SS Ch 1204	
12242,00	SSB			SS Ch 1205	
12245,00	ALE USB	B=125Bd Ch=8 ChS=250Hz		G MIL net	G
12245,00	SSB			SS Ch 1206	
12248,00	ALE USB	B=125Bd Ch=8 ChS=250Hz		USA N net	USA
12248,00	SSB			SS Ch 1207	
12250,00	ALE USB	B=125Bd Ch=8 ChS=250Hz		F N net	F
12250,00	USB			EGY N net	EGY
12251,00	*ALE USB*	*B=125Bd Ch=8 ChS=250Hz*		*ISR AF net*	*ISR*
12251,00	SSB			SS Ch 1208	
12254,00	SSB			SS Ch 1209	
12257,00	SSB			SS Ch 1210	
12260,00	SSB			SS Ch 1211	
12260,00	CW		RCRE	RUS N ship FEODOR GOLOVIN	RUS
12260,00	USB			UN MONUSCO net	UGA
12260,00	USB		RCRE	RUS N ship FEODOR GOLOVIN	RUS
12263,00	SSB			SS Ch 1212	
12266,00	SSB			SS Ch 1213	
12269,00	SSB			SS Ch 1214	
12270,00	ALE USB	B=125Bd Ch=8 ChS=250Hz		USA FEMA net	USA
12272,00	SSB			SS Ch 1215	
12275,00	SSB			SS Ch 1216	
12278,00	SSB			SS Ch 1217	
12280,20	*STANAG 4285*	*B=2400Bd*	*DHJ58*	*D N Glücksburg*	*D*
12281,00	SSB			SS Ch 1218	
12284,00	SSB			SS Ch 1219	
12287,00	SSB			SS Ch 1220	
12290,00	USB			Cape Town R	AFS
12290,00	USB			Bilbao R via Madrid	E
12290,00	USB		CWS	URG N Montevideo	URG
12290,00	USB		NMF	USA CG Boston , MA	USA
12290,00	USB		CWF	Punta Carretas R	URG
12290,00	USB		ZLM	Taupo Maritime R	NZL
12290,00	USB		HCY	Ayora R , Santa Cruz	EQA
12290,00	USB		JNA	J CG HQ Tokyo	J
12290,00	USB		VFF	Iqaluit R	CAN
12290,00	USB		VRC	Hong Kong R	CHN
12290,00	USB		NMA	USA CG Miami , FL	USA
12290,00	USB			USA CG Honolulu	HWA
12290,00	USB			USA CG Point Reyes	USA
12290,00	USB		PNK	Jayapura R , Irian Jaya	INS
12290,00	USB		PKM	Bitung R , Sulawesi	INS
12290,00	SSB			SS Ch 1221	
12290,00	USB		P2M	Port Moresbay R	PNG

Frequency	Mode	Mode Parameter	Callsign	User	Country
12293,00	SSB			SS Ch 1222	
12296,00	SSB			SS Ch 1223	
12299,00	SSB			SS Ch 1224	
12302,00	SSB			SS Ch 1225	
12303,20	STANAG 4285	B=2400Bd	IDN	I N Napoli	I
12305,00	SSB			SS Ch 1226	
12308,00	SSB			SS Ch 1227	
12311,00	SSB			SS Ch 1228	
12311,00	ALE USB	B=125Bd Ch=8 ChS=250Hz		F AF net	F
12313,00	ALE USB	B=125Bd Ch=8 ChS=250Hz		MTN MOI net	MTN
12313,00	ALE USB	B=125Bd Ch=8 ChS=250Hz		G MIL net	G
12313,20	STANAG 4285	B=2400Bd	IDN	I N Naples	I
12314,00	SSB			SS Ch 1229	
12315,00	ALE USB	B=125Bd Ch=8 ChS=250Hz		MLT N net	MLT
12317,00	SSB			SS Ch 1230	
12320,00	SSB			SS Ch 1231	
12323,00	SSB			SS Ch 1232	
12323,20	*STANAG 4285*	*B=2400Bd*		*G MIL TX Akrotiri*	*CYP*
12323,20	*STANAG 4285*	*B=2400Bd*		*G MIL TX Inskip*	*G*
12326,00	SSB			SS Ch 1233	
12329,00	SSB			SS Ch 1234	
12331,10	BAUDOT	B=75Bd S=850Hz			
12332,00	SSB			SS Ch 1235	
12333,00	DPRK ARQ	B=600Bd S=600Hz		DPRK E Moscow	RUS
12333,10	*BAUDOT*	*B=75Bd S=850Hz*		*Port Said*	*EGY*
12335,00	SSB			SS Ch 1236	
12337,00	CW		REA4	RUS AF Moscow	RUS
12338,00	SSB			SS Ch 1237	
12341,00	SSB			SS Ch 1238	
12344,00	SSB			SS Ch 1239	
12345,00	DPRK ARQ	B=600Bd S=600Hz		DPRK E Moscow	RUS
12347,00	SSB			SS Ch 1240	
12350,00	SSB			SS Ch 1241	
12350,00	USB			Trans Atlantic Cruisers net	
12350,00	USB		KPK	SSCA Trans-Atlantic Cruisers Net	USA
12353,00	USB		L2A	ARG N Buenos Aires	ARG
12353,00	USB		L2A	ARG N Mar del Plata	ARG
12353,00	USB		XVQ	Hon Gai R	VTN
12353,00	SSB			simplex frequency ship - ship	
12353,00	USB		WQCN860	Ocean-Pro R Naples , FL	USA
12354,40	ALE USB	B=125Bd Ch=8 ChS=250Hz		TUN MIL net	TUN
12356,00	USB		XVG	Hai Phong R	VTN
12356,00	PACTOR III	B=100Bd S=200Hz	PYB45	Sao Paulo	B
12356,00	SSB			simplex frequency ship - ship	
12356,00	USB		ZLM	Taupo Maritime R	NZL
12359,00	ALE USB	B=125Bd Ch=8 ChS=250Hz		G MIL net	G
12359,00	USB			Cape Town R	AFS
12359,00	USB		CBV	Valpariso R	CHL
12359,00	USB		XVT	Da Nang R	VTN
12359,00	USB		XVS	Ho Chi Minh R	VTN
12359,00	USB		PNK	Jayapura R , Irian Jaya	INS
12359,00	USB		PKM	Bitung R , Sulawesi	INS
12359,00	SSB			simplex frequency ship - ship	
12359,00	USB		CBM	Magallanes Zonal R	CHL
12359,00	USB			Yachtsmen Atlantic net	
12359,00	USB		VAX498	South Bound II R Burlington , ONT	CAN

Frequency	Mode	Mode Parameter	Callsign	User	Country
12362,00	USB		VMW	Wiluna M	AUS
12362,00	SSB			simplex frequency ship - ship	
12364,00	*STANAG 4285*	*B=2400Bd*	*TBB*	*TUR N Ankara*	*TUR*
12365,00	USB		VMC	Charleville M	AUS
12365,00	USB		XVT	Da Nang R	VTN
12365,00	SSB			simplex frequency ship - ship	
12365,00	PACTOR III	B=100Bd S=200Hz	PYB45	Sao Paulo	B
12368,20	STANAG 4285	B=2400Bd			
12370,00	SITOR A	B=100Bd S=170Hz		J CG Tokyo	J
12370,00	CW		RFX56	RUS N destroyer ADMIRAL PANTELEYEV	RUS
12370,00	CIS 12	B=1440Bd Ch=12 ChS=200Hz	RAL 2	RUS N HQ Astrakhan	RUS
12372,00	*RS-ARQ*	*B=228,5Bd S=170Hz*		*Stockholm*	*S*
12373,50	PACTOR III	B=100Bd S=200Hz	ZKN2SM	Sailmail Niue	NIU
12374,00	*CIS 12*	*B=1440Bd Ch=12 ChS=200Hz*		*RUS N ship*	
12375,00	STANAG 4285	B=2400Bd	CTA	POR N Lisbon TX Azores	POR
12377,00	FMCW			Pointe de Brezellec	F
12382,00	FAX 120/576		XSG	Shanghai R	CHN
12383,00	ALE USB	B=125Bd Ch=8 ChS=250Hz		G MIL net	G
12384,00	J OFDM 30+2	B=1500Bd Ch=32 ChS=25Hz		J N broadcast	J
12390,20	*STANAG 4285*	*B=2400Bd*		*G MIL TX Akrotiri*	*CYP*
12394,00	*STANAG 4285*	*B=2400Bd*		*G MIL St. Eval*	*G*
12398,00	CIS 12	B=1440Bd Ch=12 ChS=200Hz	RIT	RUS N HQ Severomorsk	RUS
12398,00	CIS 12	B=1440Bd Ch=12 ChS=200Hz	RLD69	RUS N ship GEORGY POBEDONOSETS BDK 45	RUS
12400,00	FMCW			Pointe de Garchine	F
12400,00	STANAG 4285	B=2400Bd	EBBO	E N ship CENTINELA P-7	E
12410,00	CHN DBPSK	B=2400Bd Ch=1		CHN AF net	CHN
12412,60	FAX 120/576		NOJ	USA CG Kodiak	ALS
12414,00	*CIS 12*	*B=1440Bd Ch=12 ChS=200Hz*		*RUS MIL*	*RUS*
12414,00	*STANAG 4285*	*B=2400Bd*		*G MIL TX Inskip*	*G*
12417,50	ALE USB	B=125Bd Ch=8 ChS=250Hz			
12418,00	STANAG 4285	B=2400Bd		G MIL Inskip	G
12420,00	CLOVER 2000	B=500Bd Ch=8			
12424,80	FSK2	B=200Bd S=400Hz			
12430,00	FMCW			Sao Juliao	POR
12431,00	ALE USB	B=125Bd Ch=8 ChS=250Hz		B N net	B
12431,00	ALE USB	B=125Bd Ch=8 ChS=250Hz		I Guarda di Finanza net	I
12435,00	FMCW			Formby radar site	G
12444,00	DPRK ARQ	B=600Bd S=600Hz		DPRK E Madrid	E
12449,00	BPSK	B=4800Bd		KNL Networks	FNL
12450,00	FMCW			Formby radar site	G
12456,00	*STANAG 4285*	*B=2400Bd*	*TBB*	*TUR N Bafa*	*TUR*
12459,00	ALE USB	B=125Bd Ch=8 ChS=250Hz		G MIL net	G
12459,00	ALE USB	B=125Bd Ch=8 ChS=250Hz			
12460,00	BPSK	B=2400Bd			
12464,00	CW		REO	RUS N	RUS
12464,00	CW		RFF78	RUS N	RUS
12464,00	CW		RFK76	RUS N	RUS
12464,00	CW		RGN90	RUS N	RUS
12464,00	CW		RGR78	RUS N	RUS
12464,00	CW		RHM81	RUS N ship KOROLEV	RUS

Frequency	Mode	Mode Parameter	Callsign	User	Country
12464,00	CW		RHO62	RUS N	RUS
12464,00	CW		RHV42	RUS N ship YAMAL	RUS
12464,00	CW		RIP90	RUS N	RUS
12464,00	CW		RIR96	RUS N ship AZOV	RUS
12464,00	CW		RIR98	RUS N	RUS
12464,00	CW		RAL65	RUS N ship	RUS
12464,00	CW		RHY73	RUS N	RUS
12464,00	CW		RKO81	RUS N ship LENA	RUS
12464,00	CW		RBES	RUS N ship SLIVA	RUS
12464,00	CW		RCIV	RUS N ship SARATOV	RUS
12464,00	CW		RMRV	RUS N ship SMOLNYY	RUS
12464,00	CW		RLX92	RUS N ship YAMAL	RUS
12464,00	CW		RHN85	RUS N ship MINSK	RUS
12464,00	CW		RMC99	RUS N ship YEVGENIY KHVOROV	RUS
12464,00	CW		RMMA	RUS N ship VIKTOR LEONOV	RUS
12464,00	CW		RCQ47	RUS Nship	RUS
12464,00	CW		RLD69	RUS N ship GEORGY POBEDONOSETS BDK 45	RUS
12464,00	CW		RHI99	RUS N ship ALEXANDER SHABALIN	RUS
12464,00	CW		RJS81	RUS N ship Amur Class PM-82	RUS
12464,00	CW		RJV33	RUS N	RUS
12464,00	CW		UCTA5	RUS N ship YELYNA	RUS
12464,00	CW		RMKW	RUS N tug SERGEI BALK	RUS
12464,00	CW		RBCS	RUS N ship	RUS
12464,00	CW		RMAK	RUS N ship	RUS
12464,00	CW		RFV99	RUS N	RUS
12464,00	CW		RFE70	RUS N	RUS
12464,00	CW		RMEV	RUS N ship EVEGENY CHUROV	RUS
12464,00	CW		RMGB	RUS N ship NIKOLAY FILCHENKOV	RUS
12464,00	CW		RGR70	RUS N	RUS
12464,00	CW		RMGB	RUS N ship IMAN	RUS
12464,00	CW		RHC86	RUS N	RUS
12464,00	CW		RFH70	RUS N ship SMETLIVYY	RUS
12464,00	CW		RMIK	RUS N ship ADMIRAL VLADIMIRSKIY	RUS
12464,00	CW		RGZ59	RUS landing ship NIKOLA FILCHENKOV	RUS
12464,00	CW		RJD31	RUS N	RUS
12464,00	CW		RMWT	RUS N ship GIGROMETR	RUS
12464,00	CW		RKH80	RUS N	RUS
12464,00	CW		RBIZ	RUS N repair ship PM-138 Amur Class	RUS
12464,00	CW		RKB91	RUS N ship KOLA	RUS
12464,00	CW		RCRE	RUS N ship	RUS
12464,00	CW		RAL48	RUS N ship NIKOLAI CHIKER	RUS
12465,00	FMCW			Formby radar site	G
12468,00	HARRIS SELCAL				
12469,80	FMCW			Vila real de Santo	POR
12471,70	ALE USB	B=125Bd Ch=8 ChS=250Hz		TUN N net	TUN
12477,00	.			SS Ch 1	
12477,50	.			SS Ch 2	
12478,00	.			SS Ch 3	
12478,50	.			SS Ch 4	
12479,00	.			SS Ch 5	
12479,50	SITOR A	B=100Bd S=170Hz	ZLM	Taupo Maritime R	NZL
12479,50	.			SS Ch 6	
12480,00	.			SS Ch 7	
12480,50	.			SS Ch 8	
12481,00	.			SS Ch 9	
12481,50	.			SS Ch 10	
12482,00	.			SS Ch 11	
12482,50	.			SS Ch 12	
12483,00	.			SS Ch 13	
12483,50	.			SS Ch 14	

Frequency	Mode	Mode Parameter	Callsign	User	Country
12484,00	.			SS Ch 15	
12484,50	.			SS Ch 16	
12485,00	.			SS Ch 17	
12485,50	.			SS Ch 18	
12486,00	.			SS Ch 19	
12486,50	.			SS Ch 20	
12487,00	.			SS Ch 21	
12487,50	.			SS Ch 22	
12488,00	.			SS Ch 23	
12488,50	.			SS Ch 24	
12489,00	.			SS Ch 25	
12489,50	.			SS Ch 26	
12490,00	.			SS Ch 27	
12490,50	.			SS Ch 28	
12491,00	.			SS Ch 29	
12491,50	.			SS Ch 30	
12492,00	.			SS Ch 31	
12492,50	.			SS Ch 32	
12493,00	.			SS Ch 33	
12493,50	.			SS Ch 34	
12494,00	BPSK	B=25Bd			
12494,00	.			SS Ch 35	
12494,50	.			SS Ch 36	
12495,00	.			SS Ch 37	
12495,50	.			SS Ch 38	
12496,00	.			SS Ch 39	
12496,50	.			SS Ch 40	
12497,00	.			SS Ch 41	
12497,00	CODAN	B=2400Bd Ch=16 ChS=112,5Hz		IRN net	IRN
12497,00	CODAN CHIRP	B=80Bd Ch=30 ChS=81Hz		IRN net	IRN
12497,00	*ALE USB*	*B=125Bd Ch=8 ChS=250Hz*		*IRN net*	*IRN*
12497,50	.			SS Ch 42	
12498,00	.			SS Ch 43	
12498,50	.			SS Ch 44	
12499,00	.			SS Ch 45	
12499,50	.			SS Ch 46	
12500,00	PACTOR III	B=100Bd S=200Hz	KLT	Austin , TX	USA
12500,00	.			SS Ch 47	
12500,00	FMCW			Sylt radar site	D
12500,00	FMCW			Wangerooge	D
12500,00	FMCW			Büsum radar site	NLD
12500,50	.			SS Ch 48	
12501,00	.			SS Ch 49	
12501,50	.			SS Ch 50	
12502,00	T600	B=50Bd S=200Hz	RIT	RUS N HQ Severomorsk	RUS
12502,00	.			SS Ch 51	
12502,50	.			SS Ch 52	
12503,00	.			SS Ch 53	
12503,50	.			SS Ch 54	
12503,60	ALE USB	B=125Bd Ch=8 ChS=250Hz		USA diplo net	USA
12504,00	.			SS Ch 55	
12504,50	.			SS Ch 56	
12505,00	.			SS Ch 57	
12505,50	.			SS Ch 58	
12506,00	.			SS Ch 59	
12506,50	.			SS Ch 60	
12507,00	.			SS Ch 61	
12507,50	.			SS Ch 62	

Frequency	Mode	Mode Parameter	Callsign	User	Country
12508,00	.			SS Ch 63	
12508,50	.			SS Ch 64	
12509,00	.			SS Ch 65	
12509,50	.			SS Ch 66	
12510,00	.			SS Ch 67	
12510,50	.			SS Ch 68	
12511,00	.			SS Ch 69	
12511,50	.			SS Ch 70	
12512,00	.			SS Ch 71	
12512,50	.			SS Ch 72	
12513,00	.			SS Ch 73	
12513,50	.			SS Ch 74	
12514,00	.			SS Ch 75	
12514,50	.			SS Ch 76	
12515,00	.			SS Ch 77	
12515,50	.			SS Ch 78	
12516,00	.			SS Ch 79	
12516,50	.			SS Ch 80	
12517,00	.			SS Ch 81	
12517,50	.			SS Ch 82	
12518,00	.			SS Ch 83	
12518,50	.			SS Ch 84	
12519,00	.			SS Ch 85	
12519,50	.			SS Ch 86	
12520,00	SITOR , CW	B=100Bd S=170Hz	JNA	J CG HQ Tokyo	J
12520,00	SITOR , CW	B=100Bd S=170Hz	VRC	Hong Kong R	CHN
12520,00	SITOR B	B=100Bd S=170Hz	ZLM	Taupo Maritime R	NZL
12520,00	SITOR , CW	B=100Bd S=170Hz	PKD	Surabaya R , Djawa	INS
12520,00	SITOR , CW	B=100Bd S=170Hz	PNK	Jayapura R , Irian Jaya	INS
12520,00	SITOR , CW	B=100Bd S=170Hz	PKM	Bitung R , Sulawesi	INS
12520,00	SITOR , CW	B=100Bd S=170Hz	PKP	Dumai R , Sumatera	INS
12520,00	.			Distress/safety frequency	WW
12520,50	.			SS Ch 88	
12521,00	CW		RHQ33	RUS N	RUS
12521,00	.			SS Ch 89	
12521,50	.			SS Ch 90	
12522,00	.			SS Ch 91	
12522,50	.			SS Ch 92	
12523,00	.			SS Ch 93	
12523,50	.			SS Ch 94	
12524,00	.			SS Ch 95	
12524,50	.			SS Ch 96	
12525,00	.			SS Ch 97	
12525,50	.			SS Ch 98	
12526,00	.			SS Ch 99	
12526,50	.			SS Ch 100	
12527,00	.			SS Ch 101	
12527,50	.			SS Ch 102	
12528,00	.			SS Ch 103	
12528,50	MIL 188-110B 39TONE	B=2400Bd Ch=39 ChS=56,25Hz		VEN N	VEN
12528,50	.			SS Ch 104	
12529,00	.			SS Ch 105	
12529,50	.			SS Ch 106	
12530,00	.			SS Ch 107	
12530,50	.			SS Ch 108	
12531,00	.			SS Ch 109	
12531,50	.			SS Ch 110	
12532,00	.			SS Ch 111	
12532,50	.			SS Ch 112	
12533,00	.			SS Ch 113	
12533,50	.			SS Ch 114	

Frequency	Mode	Mode Parameter	Callsign	User	Country
12534,00	.			SS Ch 115	
12534,50	.			SS Ch 116	
12535,00	.			SS Ch 117	
12535,50	.			SS Ch 118	
12536,00	.			SS Ch 119	
12536,50	.			SS Ch 120	
12537,00	.			SS Ch 121	
12537,50	.			SS Ch 122	
12538,00	.			SS Ch 123	
12538,50	ALE USB	B=125Bd Ch=8 ChS=250Hz			
12538,50	.			SS Ch 124	
12539,00	.			SS Ch 125	
12539,00	CW		HLW	Seoul R	KOR
12539,50	.			SS Ch 126	
12540,00	.			SS Ch 127	
12540,50	.			SS Ch 128	
12541,00	.			SS Ch 129	
12541,50	.			SS Ch 130	
12542,00	.			SS Ch 131	
12542,50	.			SS Ch 132	
12543,00	.			SS Ch 133	
12543,50	.			SS Ch 134	
12544,00	.			SS Ch 135	
12544,50	.			SS Ch 136	
12545,00	.			SS Ch 137	
12545,50	.			SS Ch 138	
12546,00	.			SS Ch 139	
12546,00	ALE USB	B=125Bd Ch=8 ChS=250Hz		VEN N	VEN
12546,50	.			SS Ch 140	
12547,00	.			SS Ch 141	
12547,50	.			SS Ch 142	
12548,00	.			SS Ch 143	
12548,50	.			SS Ch 144	
12549,00	.			SS Ch 145	
12549,50	.			SS Ch 146	
12550,00	CW			SS calling frequency Ch 1	
12550,50	CW			SS calling frequency Ch 2	
12551,00	CW			SS calling frequency Ch 3	
12551,50	CW			SS calling frequency Ch 4	
12552,00	CW			SS calling frequency Ch 5	
12552,50	CW			SS calling frequency Ch 6	
12553,00	CW			SS calling frequency Ch 7	
12553,50	CW			SS calling frequency Ch 8	
12554,00	CW			SS calling frequency Ch 9	
12554,50	CW			SS calling frequency Ch 10	
12555,00	DPRK ARQ	B=600Bd S=600Hz		DPRK E Moscow	RUS
12555,00	.			SS Ch 147	
12555,50	.			SS Ch 148	
12556,00	.			SS Ch 149	
12556,00	CW		RIW	RUS N HQ Moscow	RUS
12556,50	.			SS Ch 150	
12557,00	.			SS Ch 151	
12557,50	.			SS Ch 152	
12558,00	.			SS Ch 153	
12558,50	.			SS Ch 154	
12559,00	.			SS Ch 155	
12559,50	.			SS Ch 156	
12560,00	.			SS Ch 1	
12560,50	.			SS Ch 2	
12561,00	.			SS Ch 3	

Frequency	Mode	Mode Parameter	Callsign	User	Country
12561,50	.			SS Ch 4	
12562,00	.			SS Ch 5	
12562,50	.			SS Ch 6	
12563,00	.			SS Ch 7	
12563,50	.			SS Ch 8	
12564,00	*ALE USB*	*B=125Bd Ch=8 ChS=250Hz*		*ROU MIL net*	*ROU*
12564,00	.			SS Ch 9	
12564,50	.			SS Ch 10	
12565,00	.			SS Ch 11	
12565,50	.			SS Ch 12	
12566,00	.			SS Ch 13	
12566,50	.			SS Ch 14	
12566,70	PACTOR I	B=100Bd S=170Hz	PWGI	B N patrol ship GUAIBA , P-41	B
12566,70	PACTOR I	B=100Bd S=170Hz	PWN33	B N Natal	B
12567,00	.			SS Ch 15	
12567,50	.			SS Ch 16	
12568,00	.			SS Ch 17	
12568,20	STANAG 4285	B=2400Bd	PBC	NLD N Goeree Island	HOL
12568,50	.			SS Ch 18	
12569,00	.			SS Ch 19	
12569,50	.			SS Ch 20	
12570,00	.			SS Ch 21	
12570,50	.			SS Ch 22	
12571,00	.			SS Ch 23	
12571,50	.			SS Ch 24	
12572,00	.			SS Ch 25	
12572,00	2FSK	B=300Bd S=500Hz			
12572,50	ALE USB	B=125Bd Ch=8 ChS=250Hz		G MIL net	G
12572,50	.			SS Ch 26	
12573,00	.			SS Ch 27	
12573,10	ALE3G	B=2400Bd			
12573,50	.			SS Ch 28	
12574,00	.			SS Ch 29	
12574,50	.			SS Ch 30	
12575,00	.			SS Ch 31	
12575,50	.			SS Ch 32	
12576,00	.			SS Ch 33	
12576,00	PACTOR III	B=100Bd S=200Hz	FOHXM	Sailmail Manihi Atoll	OCE
12576,50	.			SS Ch 34	
12577,00	DSC	B=100Bd S=170Hz	JNA	J CG HQ Tokyo	J
12577,00	USB		VFF	Iqaluit R	CAN
12577,00	.			DSC distress/safety frequency	WW
12577,00	*GMDSS*	*B=100Bd S=170Hz*			*I*
12577,50	.			SS DSC frequency	WW
12578,00	.			SS DSC frequency	WW
12578,00	CW			RUS N calling frequency	RUS
12578,50	.			SS DSC frequency	WW
12579,00	SITOR-B	B=100Bd S=170Hz	UAT	Moscow R	RUS
12579,00	.			CS Ch 1	
12579,00	SITOR B	B=100Bd S=170Hz	UFZ	Vladivostok R	RUS
12579,00	SITOR B	B=100Bd S=170Hz	NMF	USA CG Boston , MA	USA
12579,00	SITOR B	B=100Bd S=170Hz	NMO	USA CG Honolulu	HWA
12579,00	SITOR B	B=100Bd S=170Hz	NRV	USA CG Guam	GUM
12579,50	.			CS Ch 2	
12580,00	SITOR , CW	B=100Bd S=170Hz	PKD	Surabaya R , Djawa	INS
12580,00	SITOR , CW	B=100Bd S=170Hz	PKP	Dumai R , Sumatera	INS
12580,00	.			CS Ch 3	
12580,00	SITOR B	B=100Bd S=170Hz			ARG
12580,50	PACTOR III	B=200Bd S=200Hz	OSY	Sailmail Brugge	BEL
12580,50	.			CS Ch 4	

Frequency	Mode	Mode Parameter	Callsign	User	Country
12581,00	.			CS Ch 5	
12581,00	SITOR , CW	B=100Bd S=170Hz	JCS	Choshi R	J
12581,50	SITOR , CW	B=100Bd S=170Hz	XSV	Tianjin R	CHN
12581,50	.			CS Ch 6	
12582,00	SITOR , CW	B=100Bd S=170Hz	PKM	Bitung R , Sulawesi	INS
12582,00	.			CS Ch 7	
12582,50	SITOR A	B=100Bd S=170Hz	SUP	Port Said R	EGY
12582,50	.			CS Ch 8	
12583,00	.			CS Ch 9	
12583,50	.			CS Ch 10	
12584,00	.			CS Ch 11	
12584,50	.			CS Ch 12	
12585,00	.			CS Ch 13	
12585,50	BAUDOT	B=45,45Bd S=170Hz	KPH	San Francisco R , CA (Bolinas/Port Reyes)	USA
12585,50	.			CS Ch 14	
12585,50	SITOR A	B=100Bd S=170Hz	KPH	San Francisco R , CA (Bolinas/Port Reyes)	USA
12586,00	SITOR , CW	B=100Bd S=170Hz	UFM3	Nevelsk R	RUS
12586,00	.			CS Ch 15	
12586,00	SITOR , CW	B=100Bd S=170Hz	4PB	Colombo R	CLN
12586,00	SITOR , CW	B=100Bd S=170Hz	UDK2	Murmansk R	RUS
12586,50	.			CS Ch 16	
12587,00	.			CS Ch 17	
12587,50	.			CS Ch 18	
12587,50	SITOR , CW	B=100Bd S=170Hz	SVO	Olympia R	GRC
12588,00	SITOR , CW	B=100Bd S=170Hz	PNK	Jayapura R , Irian Jaya	INS
12588,00	.			CS Ch 19	
12588,50	.			CS Ch 20	
12589,00	.			CS Ch 21	
12589,50	.			CS Ch 22	
12590,00	SITOR , CW	B=100Bd S=170Hz	JNA	J CG HQ Tokyo	J
12590,00	*T600*	*B=50Bd S=200Hz*		*RUS N Moscow*	*RUS*
12590,00	.			CS Ch 23	
12590,50	.			CS Ch 24	
12590,50	SITOR , CW	B=100Bd S=170Hz	KLB	Seattle R	USA
12590,50	SITOR , CW	B=100Bd S=170Hz	SVO58	Olympia R	GRC
12591,00	.			CS Ch 25	
12591,50	SITOR , CW	B=100Bd S=170Hz	UFL	Vladivostok R	RUS
12591,50	.			CS Ch 26	
12592,00	*T600*	*B=50Bd S=200Hz*		*RUS N Moscow*	*RUS*
12592,00	.			CS Ch 27	
12592,00	*T600*	*B=50Bd S=200Hz*		*RUS MIL Kazan*	*RUS*
12592,50	.			CS Ch 28	
12593,00	.			CS Ch 29	
12593,00	CW		RGT77	RUS MIL Moscow, RVSN	RUS
12593,50	.			CS Ch 30	
12594,00	.			CS Ch 31	
12594,50	SITOR , CW	B=100Bd S=170Hz	A9M	Bahrain R	BHR
12594,50	.			CS Ch 32	
12595,00	.			CS Ch 33	
12595,50	.			CS Ch 34	
12596,00	.			CS Ch 35	
12596,00	SITOR B	B=100Bd S=170Hz	RLK7	Arkchangelsk R	RUS
12596,50	.			CS Ch 36	
12597,00	.			CS Ch 37	
12597,50	.			CS Ch 38	
12598,00	.			CS Ch 39	
12598,50	.			CS Ch 40	
12599,00	.			CS Ch 41	
12599,50	.			CS Ch 42	
12599,50	SITOR , CW	B=100Bd S=170Hz	UAT	Moscow R	RUS
12600,00	.			CS Ch 43	

Frequency	Mode	Mode Parameter	Callsign	User	Country
12600,00	USB			EGY N net	EGY
12600,50	.			CS Ch 44	
12601,00	.			CS Ch 45	
12601,50	.			CS Ch 46	
12602,00	.			CS Ch 47	
12602,50	.			CS Ch 48	
12603,00	.			CS Ch 49	
12603,50	.			CS Ch 50	
12603,50	SITOR , CW	B=100Bd S=170Hz	SVO5	Olympia R	GRC
12604,00	.			CS Ch 51	
12604,50	.			CS Ch 52	
12605,00	.			CS Ch 53	
12605,50	.			CS Ch 54	
12606,00	.			CS Ch 55	
12606,00	SITOR , CW	B=100Bd S=170Hz	UIW	Kaliningrad R	RUS
12606,50	.			CS Ch 56	
12607,00	.			CS Ch 57	
12607,50	SITOR , CW	B=100Bd S=170Hz	UFL	Vladivostok R	RUS
12607,50	.			CS Ch 58	
12608,00	.			CS Ch 59	
12608,50	.			CS Ch 60	
12609,00	.			CS Ch 61	
12609,00	SITOR , CW	B=100Bd S=170Hz	4PB	Colombo R	CLN
12609,50	.			CS Ch 62	
12610,00	SITOR , CW	B=100Bd S=170Hz	XSA2	Nanjing R	CHN
12610,00	.			CS Ch 63	
12610,00	T600	B=50Bd S=200Hz	RIT	RUS N HQ Severomorsk	RUS
12610,50	.			CS Ch 64	
12611,00	.			CS Ch 65	
12611,50	.			CS Ch 66	
12612,00	.			CS Ch 67	
12612,50	.			CS Ch 68	
12613,00	.			CS Ch 69	
12613,00	SITOR , CW	B=100Bd S=170Hz	XSQ	Guangzhou R	CHN
12613,50	.			CS Ch 70	
12614,00	.			CS Ch 71	
12614,50	.			CS Ch 72	
12615,00	.			CS Ch 73	
12615,50	.			CS Ch 74	
12616,00	.			CS Ch 75	
12616,50	.			CS Ch 76	
12617,00	.			CS Ch 77	
12617,50	.			CS Ch 78	
12618,00	.			CS Ch 79	
12618,50	.			CS Ch 80	
12619,00	SITOR , CW	B=100Bd S=170Hz	EQI	Abbas R	IRN
12619,00	.			CS Ch 81	
12619,50	.			CS Ch 82	
12620,00	.			CS Ch 83	
12620,50	.			CS Ch 84	
12621,50	.			CS Ch 86	
12621,50	SITOR , CW	B=100Bd S=170Hz	VZG420	Townsville R	AUS
12622,00	.			CS Ch 87	
12622,50	.			CS Ch 88	
12622,50	SITOR , CW	B=100Bd S=170Hz	XSQ	Guangzhou R	CHN
12623,00	.			CS Ch 89	
12623,50	.			CS Ch 90	
12624,00	.			CS Ch 91	
12624,50	.			CS Ch 92	
12625,00	.			CS Ch 93	
12625,50	.			CS Ch 94	
12626,00	.			CS Ch 95	

Frequency	Mode	Mode Parameter	Callsign	User	Country
12626,50	.			CS Ch 96	
12627,00	.			CS Ch 97	
12627,50	.			CS Ch 98	
12628,00	.			CS Ch 99	
12628,50	.			CS Ch 100	
12629,00	SITOR , CW	B=100Bd S=170Hz	VJS	Perth R	AUS
12629,00	SITOR , CW	B=100Bd S=170Hz	TAH	Istanbul R	TUR
12629,00	.			CS Ch 101	
12629,25	FAX 120/576		XSQ	Guangzhou R	CHN
12629,50	.			CS Ch 102	
12630,00	.			CS Ch 103	
12630,50	.			CS Ch 104	
12631,00	.			CS Ch 105	
12631,00	*T600*	*B=50Bd S=200Hz*	*RMP*	*RUS N Kaliningrad*	*RUS*
12631,50	.			CS Ch 106	
12632,00	.			CS Ch 107	
12632,00	SITOR , CW	B=100Bd S=170Hz	WOO	Ocean Gate R , NJ	USA
12632,50	.			CS Ch 108	
12633,00	.			CS Ch 109	
12633,50	.			CS Ch 110	
12634,00	.			CS Ch 111	
12634,50	.			CS Ch 112	
12634,50	SITOR , CW	B=100Bd S=170Hz	TAH	Istanbul R	TUR
12635,00	.			CS Ch 113	
12635,50	.			CS Ch 114	
12636,00	.			CS Ch 115	
12636,50	.			CS Ch 116	
12637,00	.			CS Ch 117	
12637,50	.			CS Ch 118	
12637,50	SITOR , CW	B=100Bd S=170Hz	XSG	Shanghai R	CHN
12638,00	.			CS Ch 119	
12638,50	.			CS Ch 120	
12639,00	.			CS Ch 121	
12639,50	.			CS Ch 122	
12640,00	.			CS Ch 123	
12640,50	.			CS Ch 124	
12641,00	PACTOR , SITOR	B=100Bd S=170Hz	SAB	Göteborg R	S
12641,00	.			CS Ch 125	
12641,50	.			CS Ch 126	
12642,00	.			CS Ch 127	
12642,50	.			CS Ch 128	
12643,00	.			CS Ch 129	
12643,50	.			CS Ch 130	
12644,00	.			CS Ch 131	
12644,50	.			CS Ch 132	
12645,00	.			CS Ch 133	
12645,50	.			CS Ch 134	
12645,50	SITOR , CW	B=100Bd S=170Hz	XSQ	Guangzhou R	CHN
12646,00	.			CS Ch 135	
12646,50	.			CS Ch 136	
12647,00	.			CS Ch 137	
12647,50	.			CS Ch 138	
12648,00	.			CS Ch 139	
12648,50	.			CS Ch 140	
12648,50	SITOR , CW	B=100Bd S=170Hz	XSQ	Guangzhou R	CHN
12649,00	.			CS Ch 141	
12649,50	.			CS Ch 142	
12649,50	SITOR , CW	B=100Bd S=170Hz	XSG	Shanghai R	CHN
12650,00	.			CS Ch 143	
12650,50	.			CS Ch 144	
12651,00	.			CS Ch 145	
12651,50	.			CS Ch 146	

Frequency	Mode	Mode Parameter	Callsign	User	Country
12652,00	.			CS Ch 147	
12652,50	.			CS Ch 148	
12653,00	.			CS Ch 149	
12653,50	.			CS Ch 150	
12654,00	.			CS Ch 151	
12654,00	SITOR , CW	B=100Bd S=170Hz	TAH	Istanbul R	TUR
12654,50	.			CS Ch 152	
12655,00	.			CS Ch 153	
12655,00	FMCW			Venice, FL	USA
12655,50	.			CS Ch 154	
12656,00	.			CS Ch 155	
12656,50	.			CS Ch 156	
12657,00	DSC	B=100Bd S=170Hz	JNA	J CG HQ Tokyo	J
12657,00	DSC	B=100Bd S=170Hz	PKX	Jakarta R , Jawa	INS
12657,00	DSC	B=100Bd S=170Hz	PNK	Jayapura R , Irian Jaya	INS
12657,00	DSC	B=100Bd S=170Hz	PKE	Amboina R , Ceram	INS
12657,00	DSC	B=100Bd S=170Hz	PKM	Bitung R , Sulawesi	INS
12657,00	DSC	B=100Bd S=170Hz	PKF	Makassar R , Sulawesi	INS
12657,00	DSC	B=100Bd S=170Hz	PKB	Belawan R , Sumatera	INS
12657,00	DSC	B=100Bd S=170Hz	PKP	Dumai R , Sumatera	INS
12657,50	SITOR , CW	B=100Bd S=170Hz	UFL	Vladivostok R	RUS
12658,20	STANAG 4285	B=2400Bd	EBA	E N Madrid	E
12660,00	ALE LSB	B=125Bd Ch=8 ChS=250Hz		VEN N net	VEN
12663,10	PACTOR III	B=100Bd S=200Hz	WNU	Austin , TX	USA
12664,00	*STANAG 4285*	*B=2400Bd*	*FUO*	*F N Toulon*	*F*
12664,50	STANAG 4285	B=2400Bd	FUG	F N Saissac	F
12664,50	STANAG 4285	B=2400Bd	FUM	F N Papeete , Tahiti	OCE
12665,00	FAX 120/576		PWZ33	B N Rio de Janeiro	B
12666,50	*BAUDOT*	*B=150Bd S=850Hz*	*FUG*	*F N Saissac*	*F*
12666,50	STANAG 4285	B=2400Bd	FUG8	F N La Regine	F
12666,50	*STANAG 4285*	*B=2400Bd*	*FUG*	*F N Saissac*	*F*
12666,70	*STANAG 4285*	*B=2400Bd*	*FUG12*	*F N La Regine*	*F*
12671,80	ALE USB	B=125Bd Ch=8 ChS=250Hz		CHN net	CHN
12672,00	STANAG 4285	B=2400Bd		AUS MHFCS Townsville	AUS
12673,50	CW		JOU	Nagasaki R	J
12673,70	STANAG 4285	B=2400Bd	IDR	I N Rome	I
12675,50	CW		A4M	Muscat R	OMA
12678,00	ALE USB	B=125Bd Ch=8 ChS=250Hz		MTN MOI net	MTN
12680,00	PACTOR II	B=200Bd S=200Hz	VZX	Sailmail Darawank	AUS
12682,50	CW		ODR4	Beyrouth R	LBN
12682,50	CW		PKE	Amboina R , Ceram	INS
12682,50	CW		PKF	Makassar R , Sulawesi	INS
12682,50	CW		PKP	Dumai R , Sumatera	INS
12682,50	CW		PNK	Jayapura R , Irian Jaya	INS
12682,50	CW		TNA12	Pointe Noir R	COG
12685,50	*STANAG 4285*	*B=2400Bd*		*G MIL TX Akrotiri*	*CYP*
12688,00	CW		RGT77	RUS MIL Moscow, RVSN	RUS
12689,00	CW		LZL5	Bourgas R	BUL
12689,00	PACTOR III	B=100Bd S=200Hz	RC01	Sailmail Maputo	MOZ
12690,00	CW		PKG	Bandjarmasin R , Kalimantan	INS
12691,20	STANAG 4285	B=2400Bd	FUX	F F Le Port	REU
12692,00	CW		RAA	RUS N HQ Moscow	RUS
12694,00	CW		ZRH	AFS N Capetown	AFS
12694,00	PACTOR III	B=100Bd S=200Hz	RC01	Sailmail Maputo	MOZ
12695,50	CW		KFS	San Francisco R , CA (Bolinas/Port Reyes)	USA
12700,00	STANAG 4285	B=2400Bd	FUG	F N Saissac	F
12700,00	FMCW			Dania Beach, FL	USA
12701,00	*F N FSK*	*B=50Bd S=850Hz Ch=1*	*FUG*	*F N Saissac*	*F*

Frequency	Mode	Mode Parameter	Callsign	User	Country
12701,00	STANAG 4285	B=2400Bd	CTA	POR N Lisbon	POR
12702,50	PACTOR I	B=100Bd S=170Hz	PWZ33	B N Rio de Janeiro	B
12704,50	CW		PKD	Surabaya R , Djawa	INS
12704,50	STANAG 4285	B=2400Bd	CTA	POR N Lisbon	POR
12704,50	CW		JFX	Kagoshima R	J
12704,50	CW		PKM	Bitung R , Sulawesi	INS
12705,00	CW		RGT77	RUS MIL Moscow, RVSN	RUS
12705,50	STANAG 4285 LSB	B=2400Bd	CTA12	POR N Lisbon	F
12711,00	PACTOR I	B=100Bd S=200Hz	PWZ33	B N Rio de Janeiro	B
12713,00	STANAG 4285	B=2400Bd	FUV	F N Djibouti	DJI
12720,00	CW		RIW	RUS N HQ Moscow	RUS
12720,00	CW		RJF94	RUS N AF HQ Moscow	RUS
12724,00	STANAG 4285	B=2400Bd	FUM	F N Papeete , Tahiti	OCE
12730,20	STANAG 4285	B=2400Bd	FUG	F N Saissac	F
12731,20	PACTOR I FEC	B=100Bd S=200Hz	PWN33	B N Natal	B
12735,00	STANAG 4285	B=2400Bd Ch=2	CTA	POR N Lisbon	POR
12735,00	USB			RUS AF net	RUS
12741,00	STANAG 4285	B=2400Bd		IND N	IND
12741,00	T600	B=50Bd S=200Hz	RDL	RUS N Murmansk	RUS
12750,00	USB		6YX	JMC CG Jamaica	JMC
12750,00	FAX 120/576		NMF	USA CG Boston , MA	USA
12752,00	CW		C6N	Nassau R	BAH
12752,00	CW		RIF4	RUS N CHMS Severomorsk	RUS
12752,00	CIS 12	B=1440Bd Ch=12 ChS=200Hz		RUS N ship	
12752,50	ALE USB	B=125Bd Ch=8 ChS=250Hz		G MIL net	G
12753,00	CW		RIT	RUS N HQ Severomorsk	RUS
12757,50	ALE USB	B=125Bd Ch=8 ChS=250Hz		G MIL net	G
12763,00	CHN4+4	B=75Bd Ch=8 ChS=300/450Hz		CHN N net	CHN
12770,00	CIS 100-500 BURST	B=100Bd S=500Hz		Moscow	RUS
12770,00	CIS ARQ	B=100Bd S=500Hz	RGE31	RUS PTT	RUS
12776,00	T600	B=50Bd S=200Hz	RJD56	RUS N HQ Murmansk	RUS
12783,00	STANAG 4285	B=2400Bd		G MIL St. Eval	G
12783,50	BAUDOT	B=50Bd S=850Hz	9MR	MLA N Penang	MLA
12786,00	FAX 120/576		NMC	USA CG Pt. Reyes , CA	USA
12788,00	USB		NMG	USA CG New Orleans , LA	USA
12789,90	FAX 120/576		NMG	USA CG New Orleans , LA	USA
12793,50	MIL 188-110A SER	B=2400Bd		AUS MHFCS North West Cape	AUS
12800,00	MIL 188-110B SER	B=2400Bd		EGY N	EGY
12800,00	STANAG 4285	B=2400Bd		AUS MIL Riverina	AUS
12800,00	USB			EGY N net	EGY
12804,50	STANAG 4285	B=2400Bd	CKN	CAN MIL Aldergrove	CAN
12806,00	ALE USB	B=125Bd Ch=8 ChS=250Hz			
12808,50	USB		USI	Kherson R	UKR
12808,50	CW		VTG7	IND N Mumbai	IND
12808,50	CW		KPH	San Francisco R , CA (Bolinas/Port Reyes)	USA
12811,20	STANAG 4285	B=2400Bd	CFH	CAN N Halifax	CAN
12812,00	MIL 188-110A SER	B=2400Bd		AUS MHFCS North West Cape	AUS
12825,00	STANAG 4285	B=2400Bd	NSS	USA N TX Davidsonville , MD	USA
12825,00	SITOR , CW	B=100Bd S=170Hz	UFH	Petrapavlovsk-Kamchatskii R	RUS
12826,00	STANAG 4285	B=2400Bd	NSS	USA N TX Davidsonville , MD	USA
12829,20	STANAG 4285	B=2400Bd		CAN MIL Aldergrove	CAN
12832,00	T600	B=50Bd S=250Hz	RDL	RUS N Moscow	RUS
12835,00	STANAG 4285	B=2400Bd	NSS	USA N TX Davidsonville , MD	USA
12840,00	CW		VTP7	IND N Vishakhapatnam	IND
12840,50	BAUDOT	B=75Bd S=850Hz	PBB	NLD N Den Helder	HOL
12840,50	BAUDOT	B=75Bd S=850Hz	PBC	HOL N Goeree Island	HOL

Frequency	Mode	Mode Parameter	Callsign	User	Country
12843,00	CW		HLO	Seoul R	KOR
12847,20	*STANAG 4285*	*B=2400Bd*	*CFH*	*CAN N Halifax*	*CAN*
12850,00	*DPRK ARQ*	*B=600Bd S=600Hz*		*DPRK E Moscow*	*RUS*
12850,00	USB			EGY N net	EGY
12856,00	CW		XSG	Shanghai R	CHN
12857,00	*STANAG 4285*	*B=2400Bd*	*6WW*	*F N Dakar*	*SEN*
12861,00	*STANAG 4285*	*B=2400Bd*		*AUS MHFCS Humpty Doo*	*AUS*
12862,00	CHN4+4	B=75Bd Ch=8 ChS=300/450Hz		CHN N net	CHN
12870,00	STANAG 4285	B=2400Bd		G MIL Ascension	SHN
12871,50	CW		XSG	Shanghai R	CHN
12872,00	CHN4+4	B=75Bd Ch=8 ChS=300/450Hz		CHN N net	CHN
12874,00	*STANAG 4285*	*B=2400Bd*	*FUG*	*F N Saissac*	*F*
12876,00	ALE USB	B=125Bd Ch=8 ChS=250Hz		USA FEMA net	USA
12876,00	*STANAG 4285*	*B=2400Bd*		*AUS MHFCS Humpty Doo*	*AUS*
12877,50	BAUDOT	B=50Bd S=170Hz	UGK5	Kaliningrad R	RUS
12883,50	*STANAG 4285*	*B=2400Bd*	*FUO*	*F N Toulon*	*F*
12892,00	STANAG 4285	B=2400Bd	CKN	CAN MIL Aldergrove	CAN
12896,00	ALE USB	B=125Bd Ch=8 ChS=250Hz		G MIL net	G
12912,00	CW		PKR	Semarang R , Djawa	INS
12912,20	*STANAG 4285*	*B=2400Bd*	*CFH*	*CAN N Halifax*	*CAN*
12916,50	CW		HLF	Seoul R	KOR
12922,00	ALE USB	B=125Bd Ch=8 ChS=250Hz		USA FEMA net	USA
12923,00	CW		HLW2	Seoul R	KOR
12924,00	J OFDM 30+2	B=1500Bd Ch=32 ChS=25Hz		J N broadcast	J
12930,80	*STANAG 4285*	*B=2400Bd*	*EBA*	*E N Madrid*	*E*
12932,60	STANAG 4285	B=2400Bd	EBA	E N Madrid	E
12935,00	CW		HLG	Seoul R	KOR
12942,00	STANAG 4285	B=2400Bd		AUS MHFCS Humpty Doo	AUS
12942,00	MIL 188-110A SER	B=2400Bd		AUS MHFCS North West Cape	AUS
12950,00	FMCW			Granite Canyon, CA	USA
12958,20	*STANAG 4285*	*B=2400Bd*	*PBC*	*NLD N Goeree Island*	*HOL*
12969,00	CW		XSV	Tianjin R	CHN
12970,50	CW		PKB	Belawan R , Sumatera	INS
12970,50	CW		PKX	Jakarta R , Jawa	INS
12970,50	CW		PKY4	Sorong R , Irian Jaya	INS
12971,00	ALE USB	B=125Bd Ch=8 ChS=250Hz		IRQ Border Guard net	IRQ
12973,50	ALE USB	B=125Bd Ch=8 ChS=250Hz		AUS MIL net	AUS
12980,00	CW		PKI2	Jakarta R	INS
12980,00	CW		PKC2	Plaju R , Sumatera	INS
12980,00	FMCW			Ft. De Soto, FL	USA
12981,00	CW		RGT77	RUS MIL Moscow, RVSN	RUS
12982,00	MHF-50 MODEM	B=54,3Bd Ch=32 ChS=64,5Hz	ZSJ	AFS N Capetown	AFS
12982,20	STANAG 4285	B=2400Bd	CFH	CAN N Halifax	CAN
12990,00	*STANAG 4285*	*B=2400Bd*	*CFH*	*CAN N Halifax*	*CAN*
12998,00	MIL 188-110C APP D	B=2400Bd			
13002,50	CIS1200 SDPSK	B=1200Bd		RUS MIL	RUS
13008,00	ALE USB	B=125Bd Ch=8 ChS=250Hz		USA FEMA net	USA
13011,00	CW		AQP6	PAK N Karachi	PAK
13016,00	MIL 188-110C APP D	B=2400Bd			
13021,00	CW			RUS HFDF net	RUS

Frequency	Mode	Mode Parameter	Callsign	User	Country
13025,00	MIL 188-110C APP D	B=2400Bd			
13028,00	CIS 8181	B=81Bd S=500Hz		RUS MIL	RUS
13031,20	*STANAG 4285*	*B=2400Bd*	*FUF*	*F N Fort de France*	*MRT*
13032,00	T600	B=50Bd S=200Hz	RJD56	RUS N Murmansk	RUS
13033,50	PACTOR III	B=100Bd S=200Hz	WHL23	Saint Augustine R	USA
13035,00	CW		RJS	RUS N HQ Vladivostok	RUS
13042,00	CW		RCV	RUS N HQ Sevastopol	UKR
13042,50	STANAG 4285	B=2400Bd	FUF	F N Fort de France	MRT
13042,50	STANAG 4285	B=2400Bd	FUV	F N Djibouti	DJI
13050,00	BAUDOT	B=50Bd S=170Hz	UDK2	Murmansk R	RUS
13057,60	*STANAG 4285*	*B=2400Bd*	*EBA*	*E N Madrid*	*E*
13066,00	CW			RUS HFDF net	RUS
13068,00	T600	B=50Bd S=250Hz		RUS N Vladivostok	RUS
13072,00	*STANAG 4285*	*B=2400Bd*	*PJK*	*HOL N Willemstad*	*ATN*
13073,20	*STANAG 4285*	*B=2400Bd*	*EBA*	*E N Madrid*	*E*
13073,80	AUS MIL FSK MODEM	B=300Bd S=345Hz		AUS MIL	AUS
13074,00	FAX 120/576		JFC	Misaki R	J
13074,00	FAX 120/576		JFW	Fukushima R	J
13077,00	USB		V5W	Walvis Bay R	NMB
13077,00	USB			Bilbao R via Madrid	E
13077,00	USB		VFF	Iqaluit R	CAN
13077,00	USB		XSL	Fuzhou R	CHN
13077,00	USB		UFL	Vladivostok R	RUS
13077,00	USB		XSU	Yantai R	CHN
13077,00	USB		XVS	Ho Chi Minh R	VTN
13077,00	USB		XSA2	Nanjing R	CHN
13077,00	USB		XSZ	Dalian R	CHN
13077,00	SSB			CS Ch 1201	
13077,00	USB		UAT	Moscow R	RUS
13078,50	PACTOR III	B=100Bd S=200Hz	DZO33	Manila R	PHL
13080,00	USB		CBV	Valpariso R	CHL
13080,00	SSB			CS Ch 1202	
13080,00	USB		3DP	Suva R	FJI
13080,00	SSB		SVN51	Olympia R	GRC
13083,00	USB			Lisbon R	POR
13083,00	USB		SDJ	Stockholm R	S
13083,00	USB		XVS	Ho Chi Minh R	VTN
13083,00	SSB			CS Ch 1203	
13083,00	SSB		D4A	S. Vincente de Cabo R	CPV
13083,00	ALE USB	B=125Bd Ch=8 ChS=250Hz		ARINC Urgent Link net	USA
13086,00	USB		HSA	Bangkok R	THA
13086,00	SSB			CS Ch 1204	
13089,00	FAX 120/576		NMN	USA CG Chesapeake , VA	USA
13089,00	USB		YKM7	Lattakia R	SYR
13089,00	CW		USI	Kherson R	UKR
13089,00	USB		NRV	USA CG Guam	GUM
13089,00	USB		LSD836	Argentina R	ARG
13089,00	USB		NMN	USA CG Portsmouth , VA	USA
13089,00	USB		NMA	USA CG Miami , FL	USA
13089,00	USB			USA CG Honolulu	HWA
13089,00	SSB		TUA	Abidjan R	CTI
13089,00	SSB			CS Ch 1205	
13089,00	USB		NMC	USA CG Pt. Reynes , CA	USA
13089,00	USB		NMO	USA CG Honolulu	HWA
13092,00	USB		XSV	Tianjin R	CHN
13092,00	USB		XVG	Hai Phong R	VTN
13092,00	SSB		EKA	Yerevan R	ARM
13092,00	SSB			CS Ch 1206	
13095,00	USB		EQI	Abbas R	IRN

Frequency	Mode	Mode Parameter	Callsign	User	Country
13095,00	USB			San Lorenzo R	EQA
13095,00	USB		HCY	Ayora R , Santa Cruz	EQA
13095,00	USB			Cristobal R	EQA
13095,00	USB			Floreana R , Santa Maria	EQA
13095,00	USB			Seymour R , Baltra	EQA
13095,00	USB			Villamil R , Isabela	EQA
13095,00	SSB			CS Ch 1207	
13095,00	SSB		D4A	S. Vincente de Cabo R	CPV
13095,00	SSB		OSU51	Oostende R	BEL
13095,00	USB		HCG	Guayaquil R	EQA
13098,00	SSB			CS Ch 1208	
13098,00	USB		5BA	Cyprus R	CYP
13098,00	USB		UIW	Kaliningrad R	RUS
13101,00	USB		ESA	Tallin R	EST
13101,00	USB		A9M	Bahrain R	BHR
13101,00	USB			Cape Town R	AFS
13101,00	USB		HCY	Ayora R , Santa Cruz	EQA
13101,00	USB		PPR	Rio R	B
13101,00	USB		KLB	Seattle R , WA	USA
13101,00	CW		TRA	Libreville R	GAB
13101,00	SSB			CS Ch 1209	
13104,00	USB		J2A	Djibouti R	DJI
13104,00	USB		YQI	Constanta R	ROU
13104,00	USB		CWF	Punta Carretas R	URG
13104,00	USB		CBV	Valpariso R	CHL
13104,00	USB		PKM	Bitung R , Sulawesi	INS
13104,00	SSB			CS Ch 1210	
13104,00	USB		PKE	Amboina R , Ceram	INS
13107,00	USB		PPR	Rio R	B
13107,00	USB		9MG	Penang R	MLA
13107,00	USB		XST	Qingdao R	CHN
13107,00	SSB			CS Ch 1211	
13107,00	USB		XSQ	Guangzhou R	CHN
13110,00	USB		3BM	Mauritius R	MAU
13110,00	STANAG 4538	B=2400Bd			
13110,00	SSB			CS Ch 1212	
13110,00	USB		PKG	Bandjarmasin R , Kalimantan	INS
13111,00	ALE USB	B=125Bd Ch=8 ChS=250Hz		IRQ Border Guard net	IRQ
13113,00	USB		EQJ	Chahbahar R	IRN
13113,00	USB		HLS	Seoul R	KOR
13113,00	SSB			CS Ch 1213	
13116,00	USB		EQI	Abbas R	IRN
13116,00	USB		PPR	Rio R	B
13116,00	USB		XVN	Nha Trang R	VTN
13116,00	SSB			CS Ch 1214	
13116,00	USB		VFA	Inuvik R	CAN
13118,00	ALE USB	B=125Bd Ch=8 ChS=250Hz		MLT N net	MLT
13119,00	USB		A9M	Bahrain R	BHR
13119,00	USB		SDJ	Stockholm R	S
13119,00	SSB			CS Ch 1215	
13119,00	SSB		S7Q	Seychelles R	SEY
13122,00	USB		SUH	Alexandria R	EGY
13122,00	USB		EQJ	Chahbahar R	IRN
13122,00	SSB			CS Ch 1216	
13125,00	USB		LZW	Varna R	BUL
13125,00	USB			Saint Helena R	SHN
13125,00	USB		OYR	Aasiaat R	GRL
13125,00	SSB			CS Ch 1217	
13125,00	*CIS 12*	*B=1440Bd Ch=12 ChS=200Hz*	*RIT*	*RUS N HQ Severomorsk*	*RUS*

Frequency	Mode	Mode Parameter	Callsign	User	Country
13128,00	USB		EQC	Amirabad R	IRN
13128,00	USB		CBV	Valpariso R	CHL
13128,00	SSB			CS Ch 1218	
13128,00	USB		TAH	Istanbul R	TUR
13128,70	USB		PKX	Jakarta R , Jawa	INS
13131,00	USB		LZW	Varna R	BUL
13131,00	USB		UFM3	Nevelsk R	RUS
13131,00	USB		SDJ	Stockholm R	S
13131,00	USB		PPR	Rio R	B
13131,00	USB		UUT	Odessa R	UKR
13131,00	SSB			CS Ch 1219	
13134,00	USB		EQI	Abbas R	IRN
13134,00	USB		STP	Port Sudan R	SDN
13134,00	USB		ZBR	Bermuda Harbour R	BER
13134,00	SSB		EQM	Bushehr R	IRN
13134,00	SSB			CS Ch 1220	
13134,00	*USB*		*SVO54*	*Olympia R*	*GRC*
13134,90	USB		PKD	Surabaya R , Djawa	INS
13137,00	USB		PPR	Rio R	B
13137,00	USB		9MG	Penang R	MLA
13137,00	USB		UFH	Petrapavlovsk-Kamchatskii R	RUS
13137,00	SSB			CS Ch 1221	
13137,00	SSB		4PB	Colombo R	CLN
13137,00	SSB		A7D	Doha R	QAT
13140,00	USB		SUP	Port Said R	EGY
13140,00	USB		L3A	ARG N Comodoro Rivadavia	ARG
13140,00	SSB		EQN	Khomeini R	IRN
13140,00	SSB			CS Ch 1222	
13140,00	SSB		TAH	Istanbul R	TUR
13140,00	USB		E5R	Rarotonga R	CKH
13140,00	USB		WCC	Chatham R , MA	USA
13142,00	USB		XVN	Nha Trang R	VTN
13143,00	SSB			CS Ch 1223	
13143,00	SSB		CNP	Casablanca R	MRC
13146,00	USB			Cape Town R	AFS
13146,00	USB		CBV	Valpariso R	CHL
13146,00	USB		XVG	Hai Phong R	VTN
13146,00	SSB			CS Ch 1224	
13146,00	USB		3AC	Monaco R	MCO
13146,00	SSB		4PB	Colombo R	CLN
13147,30	SSB		ODR8	Beyrouth R	LBN
13149,00	USB		UDK2	Murmansk R	RUS
13149,00	USB		P2M	Port Moresbay R	PNG
13149,00	SSB			CS Ch 1225	
13149,00	USB		XSQ	Guangzhou R	CHN
13149,00	USB		KLN	South Bend R , IN	USA
13150,00	FMCW			Yaquina Head South, OR	USA
13152,00	USB		SDJ	Stockholm R	S
13152,00	USB		9WH	Kota Kinabalu R	MLA
13152,00	USB		9MG	Penang R	MLA
13152,00	USB		CBV	Valpariso R	CHL
13152,00	USB		9WW	Kuching R	MLA
13152,00	CW		EQM	Bushehr R	IRN
13152,00	SSB			CS Ch 1226	
13152,00	ALE USB	B=125Bd Ch=8 ChS=250Hz		ARINC Urgent Link net	USA
13155,00	SSB			CS Ch 1227	
13155,00	SSB		TJC	Douala R	CME
13158,00	USB		PPR	Rio R	B
13158,00	USB		CBV	Valpariso R	CHL
13158,00	USB		XVS	Ho Chi Minh R	VTN
13158,00	SSB			CS Ch 1228	

Frequency	Mode	Mode Parameter	Callsign	User	Country
13158,00	USB		KPH	Rio Vista R , CA	USA
13160,00	CHN 4+4	B=75Bd Ch=8 ChS=300/450Hz		CHN MIL	CHN
13161,00	USB		LSD836	Argentina R	ARG
13161,00	USB		HLS	Seoul R	KOR
13161,00	SSB			CS Ch 1229	
13162,80	USB		XFL	Mazatlan P	MEX
13164,00	SSB			CS Ch 1230	
13166,00		B=75Bd S=975Hz			
13167,00	USB		XVG	Hai Phong R	VTN
13167,00	SSB			CS Ch 1231	
13167,00	SSB		EQN	Khomeini R	IRN
13167,00	SSB		ZPC	Asuncion R	PRG
13170,00	USB		LSD836	Argentina R	ARG
13170,00	SSB			CS Ch 1232	
13170,00	SSB		SVO55	Olympia R	GRC
13173,00	USB		SXE	Aspropyrgos Attikis R	GRC
13173,00	USB			Peiraias CG R	GRC
13173,00	USB		JNA	J CG HQ Tokyo	J
13173,00	USB		PPR	Rio R	B
13173,00	USB		UFL	Vladivostok R	RUS
13173,00	USB		XVG	Hai Phong R	VTN
13173,00	SSB			CS Ch 1233	
13173,00	SSB		TAH	Istanbul R	TUR
13175,20	USB		XFL	Mazatlan P	MEX
13176,00	DSC	B=100Bd S=170Hz	HLS	Seoul R	KOR
13176,00	USB		A4M	Muscat R	OMA
13176,00	USB		XSG	Shanghai R	CHN
13176,00	SSB			CS Ch 1234	
13179,00	USB		JNA	J CG HQ Tokyo	J
13179,00	USB		LSD836	Argentina R	ARG
13179,00	USB		HSA	Bangkok R	THA
13179,00	SSB			CS Ch 1235	
13182,00	SSB			CS Ch 1236	
13182,00	USB		XSQ	Guangzhou R	CHN
13185,00	USB		XVG	Hai Phong R	VTN
13185,00	SSB			CS Ch 1237	
13188,00	USB		CBV	Valpariso R	CHL
13188,00	SSB			CS Ch 1238	
13188,00	USB		XSG	Shanghai R	CHN
13188,00	MIL 188-110B SER	B=2400Bd			
13191,00	USB		LZW	Varna R	BUL
13191,00	USB		XVG	Hai Phong R	VTN
13191,00	SSB			CS Ch 1239	
13191,00	SSB		TAH	Istanbul R	TUR
13194,00	USB		LSD836	Argentina R	ARG
13194,00	USB		XSQ	Guangzhou R	CHN
13194,00	SSB			CS Ch 1240	
13197,00	ALE USB	B=125Bd Ch=8 ChS=250Hz		CHN net	
13197,00	SSB			CS Ch 1241	
13197,00	USB		UUT	Odessa R	UKR
13200,00	USB			USA AF Croughton	G
13200,00	USB			USA AF Sigonella	I
13200,00	USB			USA AF Lajes AFB	AZR
13200,00	USB			USA AF Puerto Rico	PTR
13200,00	USB			USA AF Diego Garcia	DGA
13200,00	USB			USA AF Guam	GUM
13200,00	USB			USA AF Hickam , HWA	USA
13200,00	USB		AFS	USA AF Offutt AFB , Omaha , NE	USA
13200,00	ALE USB	B=125Bd Ch=8 ChS=250Hz		USA AF net	USA

Frequency	Mode	Mode Parameter	Callsign	User	Country
13200,00	USB			USA AF Ascension AFB	ASC
13200,00	SSB		AIF80	USA AF Yokota AB	J
13200,00	SSB		AKA	USA AF Elmendorf AFB , Anchorage	ALS
13200,00	DPRK ARQ	B=600Bd S=600Hz		DPRK E Moscow	RUS
13200,00	DPRK ARQ	B=600Bd S=600Hz		DPRK E Rome	I
13200,00	DPRK ARQ	B=600Bd S=600Hz		DPRK E Warsaw	POL
13200,00	DPRK ARQ	B=600Bd S=600Hz		DPRK E Sofia	BIL
13200,00	FMCW			Upper Trestles, San Diego, CA	USA
13200,00	USB			USA AF Andrews	USA
13202,00	STANAG 4481	B=75Bd S=850Hz	NPG	USA N Dixon , CA	USA
13205,00	SSB			USA AF Strategic Command	USA
13212,50	ALE USB	B=125Bd Ch=8 ChS=250Hz			
13215,00	ALE USB	B=125Bd Ch=8 ChS=250Hz		USA AF net	
13217,00	ALE USB	B=125Bd Ch=8 ChS=250Hz		USA AF net	USA
13220,00	ALE USB	B=125Bd Ch=8 ChS=250Hz		I AF net	I
13221,00	ALE USB	B=125Bd Ch=8 ChS=250Hz		USA CG net	USA
13222,00	DPRK ARQ	B=600Bd S=600Hz		DPRK E Moscow	RUS
13227,00	SSB			NASA launch support aircraft	USA
13227,00	STANAG 4481	B=50Bd S=850Hz	NAU	USA N San Juan	PTR
13229,00	STANAG 4481	B=50Bd S=850Hz	NAU	USA N San Juan	PTR
13233,00	USB		DHM91	D AF Münster	D
13236,00	USB		CIRCUS VERT	F AF HQ Villacoublay	F
13241,00	SSB			USAF Strategic Command	USA
13242,00	ALE USB	B=125Bd Ch=8 ChS=250Hz		USA SHARES net	USA
13257,00	USB		CHR	CAN AF Trenton	CAN
13257,90	CW		S**	RUS N Severomorsk	RUS
13258,00	CW		C**	RUS N Moscow	RUS
13258,40	CW		M**	RUS N Magadan	RUS
13260,00	SSB			Carribean Bridgetown	BRB
13261,00	SSB			Brisbane ACC	AUS
13261,00	SSB			Auckland ACC	NZL
13261,00	SSB			Nadi ACC	FJI
13261,00	SSB			VOLMET AFI	AFI
13261,00	USB			San Francisco ACC , CA	USA
13261,00	USB			Easter Island ACC	CHL
13261,00	USB			Tahiti ACC	OCE
13261,00	USB			Port Vila ACC	VUT
13261,00	USB			Rarotonga ACC	CKH
13261,00	USB			Wallis ACC	WAL
13264,00	SSB			RDARA 14	
13264,00	SSB			VOLMET EUR	EUR
13264,00	USB		EIP	Shannon VOLMET	IRL
13264,00	HFDL	B=1800Bd	H11**	Albrook	PNR
13267,00	SSB			Irkutsk VOLMET	RUS
13267,00	SSB			Jakutsk VOLMET	RUS
13267,00	SSB		UGEF	Khabarovsk VOLMET	RUS
13267,00	SSB			Magadan VOLMET	RUS
13267,00	SSB			NOAA Hurricane Center Miami , FL	USA
13267,00	SSB			RDARA 13H	
13267,00	SSB			RDARA 3	
13268,00	PACTOR	B=100Bd S=200Hz	EN8CT		
13268,00	PACTOR	B=100Bd S=200Hz	VL4RA		
13268,00	PACTOR	B=100Bd S=200Hz	SF1QE		
13270,00	SSB			RDARA 6G	
13270,00	SSB			VOLMET NAT	NAT
13270,00	USB		VFG	Gander VOLMET	CAN

Frequency	Mode	Mode Parameter	Callsign	User	Country
13270,00	USB		WSY70	New York VOLMET	USA
13270,00	HFDL	B=1800Bd	H06**	Hat Yai	THA
13270,00	*ALE USB*	*B=125Bd Ch=8 ChS=250Hz*		*TUR Disaster and Emergency net*	*TUR*
13273,00	USB			Kano ACC	NIG
13273,00	USB			Gao ACC	MLI
13273,00	USB			Ndjamena ACC	NGR
13273,00	SSB			Alger ACC	ALG
13273,00	SSB			MWARA AFI	AFI
13273,00	SSB			Niamey ACC	NGR
13273,00	USB			Tamanrasset ACC	ALG
13273,00	USB			Timimoun ACC	ALG
13273,00	USB			Tripoli ACC	LBY
13273,00	USB			Tunis ACC	TUN
13275,00	ALE USB	B=125Bd Ch=8 ChS=250Hz			
13276,00	HFDL	B=1800Bd	H01**	San Francisco , CA	USA
13276,00	HFDL	B=1800Bd	H02**	Molokai , HI	HWA
13276,00	SSB			RDARA 6G	
13276,00	SSB			VOLMET NAT	NAT
13276,00	HFDL	B=1800Bd	H04**	Riverhead , NY	USA
13279,00	SSB			VOLMET NCA	NCA
13279,00	SSB			VOLMET SAM	SAM
13282,00	USB		ZKAK	Auckland VOLMET	NZL
13282,00	USB		VRK	Honkong VOLMET	HKG
13282,00	USB		JIA	Tokyo VOLMET	J
13282,00	SSB			VOLMET PAC	PAC
13282,00	USB		KVM70	Honolulu VOLMET	HWA
13285,00	USB		3UW33	Guangzhou VOLMET	CHN
13285,00	SSB		BSQ	Bejing VOLMET	CHN
13285,00	SSB			RDARA 10	
13285,00	SSB			VOLMET SEA	SEA
13288,00	SSB			Aden ACC	YEM
13288,00	SSB			Alma Ata ACC	KAZ
13288,00	SSB			Bahrain ACC	BHR
13288,00	SSB			Mumbai ACC	IND
13288,00	SSB			Cairo ACC	EGY
13288,00	SSB			Dushanbe ACC	TJK
13288,00	SSB			Kabul ACC	AFG
13288,00	SSB			Karachi ACC	PAK
13288,00	SSB			Khartoum ACC	SDN
13288,00	SSB			Kigali ACC	RRW
13288,00	SSB			Male ACC	MLD
13288,00	SSB			MWARA AFI	AFI
13288,00	SSB			MWARA EUR	EUR
13288,00	SSB			MWARA MID	MID
13288,00	SSB			Tashkent ACC	UZB
13288,00	SSB			Tehran ACC	IRN
13288,00	SSB		ETD3	Addis Abbeba ACC	ETH
13288,00	USB			Arkhangelsk ACC	RUS
13288,00	USB			Asmara ACC	ERI
13288,00	USB			Benghazi ACC	LBY
13288,00	USB			Bujumbura ACC	BDI
13288,00	USB			Comoros ACC	COM
13288,00	USB			Dar es Salaam ACC	TZA
13288,00	USB			Entebbe ACC	UGA
13288,00	USB			Hargeisa ACC	SOM
13288,00	USB			Djibouti ACC	DJI
13288,00	USB			Jeddah ACC	ARS
13288,00	USB			Kisimayu ACC	SOM
13288,00	USB			Mogadishu ACC	SOM
13288,00	USB			Nairobi ACC	KEN

Frequency	Mode	Mode Parameter	Callsign	User	Country
13288,00	USB			Port Sudan ACC	SDN
13288,00	USB			Sanaa ACC	YEM
13288,00	USB			Seychelles ACC	SEY
13288,00	USB			Tripoli ACC	LBY
13288,00	USB			Ashkabad ACC	TKM
13288,00	USB			Bishkek ACC	KGZ
13288,00	USB			Mumbai ACC	IND
13288,00	USB			New Delhi ACC	IND
13288,00	USB			Kathmandu ACC	NPL
13288,00	USB			Kuwait ACC	KWT
13288,00	USB			Lahore ACC	PAK
13288,00	USB			Musact ACC	OMA
13288,00	USB			Simferopol ACC	UKR
13288,00	USB			Samarkhand ACC	UZB
13288,00	USB			Tbilisi ACC	GEO
13288,00	USB			Urumqi ACC	CHN
13288,00	USB			Yerevan ACC	ARM
13288,00	USB			Beirut ACC	LBN
13288,00	USB			Berlin ACC	D
13288,00	USB			Klev ACC	UKR
13288,00	USB			Lvov ACC	UKR
13288,00	USB			Minsk ACC	BLR
13288,00	USB			Moscow ACC	RUS
13288,00	USB			Murmansk ACC	RUS
13288,00	USB			Odessa ACC	UKR
13288,00	USB			Riga ACC	LTU
13288,00	USB			Sofia ACC	BUL
13288,00	USB			Syktyvkar ACC	RUS
13288,00	USB			Velikiye ACC	RUS
13288,00	USB			Vologda ACC	RUS
13288,00	USB			Tunis ACC	TUN
13288,00	USB			Vilnius ACC	LT
13288,00	USB			San Francisco ACC , CA	USA
13291,00	USB			Bodo ACC	NOR
13291,00	SSB			Iceland ACC , Reykjavik	ISL
13291,00	SSB			MWARA NAT	NAT
13291,00	SSB		KEA5	New York ACC	USA
13291,00	SSB			RDARA 6	
13291,00	SSB		CSY	Santa Maria ACC	AZR
13291,00	USB		EIP	Shanwik ACC	IRL
13291,00	SSB		VFG	Gander ACC	CAN
13294,00	SSB			Accra ACC	GHA
13294,00	SSB			Kano ACC	NIG
13294,00	SSB			Kinshasa ACC	ZAI
13294,00	SSB			Luanda ACC	AGL
13294,00	SSB			MWARA AFI	AFI
13294,00	SSB			Niamey ACC	NGR
13294,00	SSB			N`djamena ACC	TCD
13294,00	SSB			Yaounde ACC	CME
13294,00	USB			San Francisco ACC , CA	USA
13294,00	USB			Tokyo ACC	J
13294,00	USB			Bangui ACC	CAF
13294,00	USB			Douala ACC	CME
13294,00	USB			Entebbe ACC	UGA
13294,00	USB			Franceville ACC	GAB
13294,00	USB			Garoua ACC	CME
13294,00	USB			Goma ACC	COG
13294,00	USB			Harare ACC	ZWE
13294,00	USB			Kisangani ACC	COG
13294,00	USB			Lagos ACC	NIG
13294,00	USB			Libreville ACC	GAB
13294,00	USB			Lubumbashi ACC	COG

Frequency	Mode	Mode Parameter	Callsign	User	Country
13294,00	USB			Lusaka ACC	ZMB
13294,00	USB			Maiduguri ACC	NIG
13294,00	USB			Maroua ACC	CME
13294,00	USB			Niamtougou ACC	TGO
13294,00	USB			Pointe Noire ACC	KGO
13294,00	USB			Port Gentil ACC	GAB
13294,00	USB			Roberts ACC	LBR
13294,00	USB			Sao Tome ACC	CAF
13294,00	USB			Windhoek ACC	NMB
13297,00	SSB			Bogota ACC	CLM
13297,00	SSB			Boyeros ACC	CUB
13297,00	SSB			Maiquetia ACC , Caracas	VEN
13297,00	SSB			MWARA CAR	CAR
13297,00	SSB			MWARA EA	EA
13297,00	SSB			MWARA SAM	SAM
13297,00	SSB		KEA5	New York ACC	USA
13297,00	USB			Piarco ACC	TRD
13297,00	SSB			Santa Cruz ACC	BOL
13297,00	USB			Barranquilla ACC	CLM
13297,00	USB			Merida ACC	MEX
13297,00	USB			Panama ACC	PNR
13297,00	USB			Belem ACC	B
13297,00	USB			Brasilia ACC	B
13297,00	USB			Iquitos ACC	PRU
13297,00	USB			Leticia ACC	CLM
13297,00	USB			Manaus ACC	B
13297,00	USB			Porto Velho ACC	B
13297,00	USB			Rio de Janeiro ACC	B
13297,00	USB			Cayenne ACC	GUF
13297,00	USB			Georgetown ACC	GUY
13297,00	USB			Paramaribo ACC	SUR
13297,00	USB			Recife ACC	B
13297,00	USB			Asuncion ACC	PRG
13297,00	USB			Buenos Aires ACC	ARG
13297,00	USB			Campo Grande ACC	B
13297,00	USB			La Paz ACC	BOL
13297,00	USB			Montevideo ACC	URG
13297,00	USB			Lima ACC	PRU
13297,00	USB			Porto Alegre ACC	B
13297,00	USB			Port Velho ACC	B
13297,00	USB			Salvador ACC	B
13300,00	SSB			Hongkong ACC	HKG
13300,00	SSB			Manila ACC	PHL
13300,00	SSB			Mataveri ACC	PAQ
13300,00	SSB			MWARA CEP	CEP
13300,00	SSB			MWARA CWP	CWP
13300,00	SSB			MWARA NP	NP
13300,00	SSB			MWARA SP	SP
13300,00	SSB			Naha ACC	J
13300,00	SSB			Port Moresby ACC	PNG
13300,00	SSB			RDARA 4	
13300,00	SSB			San Francisco ACC , CA	USA
13300,00	USB			Seoul ACC	KOR
13300,00	USB			Taipei ACC	TWN
13300,00	USB			Tokyoi ACC	J
13303,00	HFDL	B=1800Bd	H17**	Telde , Gran Canaria	E
13303,00	SSB			MWARA EA	EA
13303,00	SSB			MWARA NCA	NCA
13303,00	USB			Chita ACC	RUS
13303,00	USB			Chulman ACC	RUS
13303,00	USB			Ekimchan ACC	RUS
13303,00	USB			Irkutsk ACC	RUS

Frequency	Mode	Mode Parameter	Callsign	User	Country
13303,00	USB			Kirensk ACC	RUS
13303,00	USB			Khabarovsk ACC	RUS
13303,00	USB			Pyongyang ACC	KRE
13303,00	USB			Ulaanbaatar ACC	MNG
13303,00	USB			Ulan Ude ACC	RUS
13306,00	SSB			Mumbai ACC	IND
13306,00	SSB			Brisbane ACC	AUS
13306,00	SSB			Canarias ACC , Las Palmas	CNR
13306,00	SSB			Cocos Island ACC	ICO
13306,00	SSB			Colombo ACC	CLN
13306,00	SSB			Dar es Salaam ACC	TZA
13306,00	SSB			Iceland ACC , Reykjavik	ISL
13306,00	USB			Jeddah ACC	ARS
13306,00	SSB			Male ACC	MLD
13306,00	SSB			Mauritiua ACC , Plaisance	MAU
13306,00	SSB			MWARA INO	INO
13306,00	SSB			MWARA NAT	NAT
13306,00	SSB		KEA5	New York ACC	USA
13306,00	SSB			Kigali ACC	RRW
13306,00	SSB		CSY	Santa Maria ACC	AZR
13306,00	SSB			Seychelles ACC , Mahe	SEY
13306,00	USB		EIP	Shanwik ACC	IRL
13306,00	SSB		5YD	Nairobi ACC	KEN
13306,00	SSB		VFG	Gander ACC	CAN
13306,00	USB			Paramaribo ACC	SUR
13306,00	USB			Piarco ACC	TRD
13306,00	USB		BOING EVERETT	Boing Seattle	USA
13306,00	USB		BOING SEATTLE	Boing Seattle	USA
13306,00	USB			Antananarivo ACC	MDG
13306,00	USB			Beira ACC	MOZ
13306,00	USB			Harare ACC	ZWE
13306,00	USB			Lilongwe ACC	MWI
13306,00	USB			Lusaka ACC	ZMB
13306,00	USB			Madras ACC	IND
13306,00	USB			Mahajanga ACC	MDG
13306,00	USB			Moroni ACC	COM
13306,00	USB			Perth ACC	AUS
13306,00	USB			St.Denis ACC	REU
13306,00	USB			Toamasina ACC	MDG
13309,00	SSB			Bangkok ACC	THA
13309,00	SSB			Hongkong ACC	HKG
13309,00	SSB			Jakarta ACC	INS
13309,00	SSB			Manila ACC	PHL
13309,00	SSB			MWARA EA	EA
13309,00	SSB			MWARA SEA	SEA
13309,00	SSB			RDARA 13C	
13309,00	SSB			RDARA 13K	
13309,00	SSB			Singapore ACC	SNG
13309,00	USB			Guangzhou ACC	CHN
13309,00	USB			Irkutsk ACC	RUS
13309,00	USB			Pyongyang ACC	KRE
13309,00	USB			Ulaanbaatar ACC	MNG
13309,00	USB			Bali ACC	INS
13309,00	USB			Hanoi ACC	VTN
13309,00	USB			Ho Chi Minh ACC	VTN
13309,00	USB			Kuala Lumpur ACC	MLA
13309,00	USB			Kota Kinabalu ACC	MLA
13309,00	USB			Seoul ACC	KOR
13309,00	USB			Tokyo ACC	J
13309,00	USB			Vientianne ACC	LAO

Frequency	Mode	Mode Parameter	Callsign	User	Country
13312,00	HFDL	B=1800Bd	H16**	Agana	GUM
13312,00	HFDL	B=1800Bd	H02**	Molokai , HI	HWA
13312,00	SSB			MWARA MID	MID
13312,00	SSB			RDARA 11B	
13312,00	SSB			USA Customs	USA
13312,00	USB		BOING SEATTLE	Boing Seattle	USA
13312,00	USB		BOING EVERETT	Boing Seattle	USA
13312,00	ALE USB	B=125Bd Ch=8 ChS=250Hz		USA COTHEN net	USA
13315,00	SSB			Canarias ACC , Las Palmas	CNR
13315,00	SSB			Dakar ACC	SEN
13315,00	USB			Cayenne ACC	GUF
13315,00	USB			Manaus ACC	B
13315,00	USB			Paramaribo ACC	SUR
13315,00	USB			Rio de Janerio ACC	B
13315,00	USB			Sal Island ACC	CPV
13315,00	USB			Ivdel ACC	RUS
13315,00	USB			Khanty-Mansiysk ACC	RUS
13315,00	USB			Moscow ACC	RUS
13315,00	SSB			Johannesburg ACC	AFS
13315,00	SSB			MWARA NCA	NCA
13315,00	SSB			MWARA SAT	SAT
13315,00	SSB			Recife ACC	B
13315,00	HFDL	B=1800Bd	H13**	Santa Cruz	BOL
13315,00	USB			Syktyvkar ACC	RUS
13315,00	USB			Vologda ACC	RUS
13318,00	SSB			Calcutta ACC	IND
13318,00	SSB			Colombo ACC	CLN
13318,00	SSB			Jakarta ACC	INS
13318,00	SSB			Kuala Lumpur ACC	MLA
13318,00	SSB			Kunming ACC	CHN
13318,00	SSB			Madras ACC	IND
13318,00	SSB			Male ACC	MLD
13318,00	SSB			MWARA SEA	SEA
13318,00	SSB			RDARA 13	
13318,00	SSB			Singapore ACC	SNG
13318,00	SSB			Yangon ACC	BRM
13318,00	USB			Bali ACC	INS
13318,00	USB			Bangkok ACC	THA
13318,00	USB			Dhaka ACC	BGD
13318,00	USB			Guangzhou ACC	CHN
13318,00	USB			Kathmandu ACC	NPL
13318,00	USB			Brisbane ACC	AUS
13318,00	USB			Male ACC	MLD
13318,00	USB			Ujung Pandang ACC	INS
13321,00	SSB			RDARA 2	
13321,00	SSB			RDARA 3	
13321,00	HFDL	B=1800Bd	H08**	Johannesburg	AFS
13321,00	HFDL	B=1800Bd	H14**	Krasnoyarsk	RUS
13322,00	DPRK ARQ	B=1200Bd S=600Hz		DPRK E Madrid	E
13324,00	HFDL	B=1800Bd	H02**	Molokai , HI	HWA
13324,00	SSB			W I	WW
13324,00	SSB			W III	WW
13327,00	SSB			W I	WW
13327,00	SSB			W IV	WW
13327,00	USB			IBERIA Madrid ACC	E
13330,00	SSB		5YD	Nairobi ACC	KEN
13330,00	SSB			W II	WW
13330,00	SSB			W V	WW

Frequency	Mode	Mode Parameter	Callsign	User	Country
13330,00	USB			Kenya ACC	KEN
13333,00	DPRK ARQ	B=600Bd S=600Hz		DPRK E Moscow	RUS
13333,00	SSB			W I	WW
13333,00	SSB			W III	WW
13336,00	SSB			W I	WW
13336,00	SSB			W IV	WW
13339,00	USB			San Francisca ACC , CA	USA
13339,00	USB			Tokyo ACC	J
13339,00	SSB			W II	WW
13339,00	SSB			W V	WW
13339,00	USB			San Francisco ACC , CA	USA
13342,00	HFDL	B=1800Bd	H10**	Muan	KOR
13342,00	USB		SDJ	Stockholm R	S
13342,00	SSB			W I	WW
13342,00	SSB			W III	WW
13345,00	SSB			W I	WW
13345,00	SSB			W IV	WW
13348,00	SSB			W II	WW
13348,00	SSB			W V	WW
13348,00	USB			San Francisco ACC , CA	USA
13351,00	SSB			W I	WW
13351,00	SSB			W III	WW
13351,00	HFDL	B=1800Bd	H05**	Auckland	NZL
13354,00	USB			New York ACC	USA
13354,00	USB			Santa Maria ACC	AZR
13354,00	SSB			RDARA 5	
13354,00	SSB			RDARA 7	
13354,00	SSB			San Francisco ACC , CA	USA
13354,00	HFDL	B=1800Bd	H15**	Al Muharraq	BHR
13356,00	ALE USB	B=125Bd Ch=8 ChS=250Hz		CHL MOI net	CHL
13357,00	SSB			Canarias ACC , Las Palmas	CNR
13357,00	SSB			Dakar ACC	SEN
13357,00	SSB			MWARA SAT	SAT
13357,00	SSB			RDARA 2	
13357,00	SSB			Recife ACC	B
13357,00	SSB			Sal ACC	CPV
13357,00	USB			Brasilia ACC	B
13357,00	USB			Cayenne ACC	GUF
13357,00	USB			Brasilia ACC	B
13357,00	USB			Manaus ACC	B
13357,00	USB			Paramaribo ACC	SUR
13357,00	USB			Rio de Janerio ACC	B
13357,00	USB			Abidjan ACC	CTI
13357,00	USB			Bamako ACC	MLI
13357,00	USB			Bangui ACC	CAF
13357,00	USB			Bissau ACC	GNB
13357,00	USB			Bouake ACC	CTI
13357,00	USB			Casablanca ACC	MRC
13357,00	USB			Conakry ACC	GUI
13357,00	USB			Dakar ACC	SEN
13357,00	USB			Freetown ACC	SRL
13357,00	USB			Johannesburg ACC	AFS
13357,00	USB			Kano ACC	NIG
13357,00	USB			Niamey ACC	NGR
13357,00	USB			Nouadhibou ACC	MTN
13357,00	USB			Nouakchott ACC	MTN
13357,00	USB			Ouagadougou ACC	BFA
13357,00	USB			Roberts ACC	LBR
13365,00	RFSM	B=2400Bd		MFA Sofia	BUL
13366,00	CIS 100-500	B=100Bd S=500Hz		CIS	CIS

Frequency	Mode	Mode Parameter	Callsign	User	Country
13368,00	CIS MFSK 20	B=20Bd Ch=20 ChS=40Hz		RUS INTEL	RUS
13370,00	*STANAG 4285*	*B=2400Bd*		*G F Inskip*	*G*
13376,00	RUS HYBRID MODEM MFSK-PSK	B=40Bd Ch=16 ChS=40Hz		Moscow	RUS
13377,00	ALE USB	B=2400Bd ChS=250Hz		ALG MIL net	ALG
13378,00	MIL 188-110A SER	B=2400Bd			
13381,00	RUS DBPSK 16FSK	B=1866Bd Ch=16 ChS=175Hz		RUS INTEL	RUS
13384,00	CIS 50-500 BAUDOT	B=50Bd S=500Hz		RUS	RUS
13386,00	CIS MFSK 17 WIDEBAND	B=566Bd Ch=17 ChS=1200Hz		RUS diplo	RUS
13393,00	ALE USB	B=125Bd Ch=8 ChS=250Hz		ALG AF net	ALG
13395,00	FMCW			Llandulas radar site	G
13396,00	CW		RGT77	RUS MIL Moscow, RVSN	RUS
13397,00	CIS 100-500	B=100Bd S=500Hz		RUS	RUS
13404,00	CW			RUS MIL net	RUS
13408,50	T600	B=50Bd S=200Hz	RIT	RUS N HQ Severomorsk	RUS
13410,00	*STANAG 4285*	*B=2400Bd*	*6WW*	*F N Dakar*	*SEN*
13411,60	*STANAG 4285*	*B=2400Bd*	*FUE*	*F N Brest*	*F*
13411,70	STANAG 4285	B=2400Bd	6WW	F N Dakar	SEN
13412,00	FMCW			Point Reyes, CA	USA
13415,00	CODAN	B=2400Bd Ch=16 ChS=112,5Hz			
13415,00	ALE USB	B=125Bd Ch=8 ChS=250Hz			
13416,00	USB		XVS	Ho Chi Minh R	VTN
13416,00	ALE USB	B=125Bd Ch=8 ChS=250Hz		Bren Ferren Corp	USA
13417,00	ALE USB	B=125Bd Ch=8 ChS=250Hz		AUS Police WA	AUS
13418,50	ALE USB	B=125Bd Ch=8 ChS=250Hz		AUS Police WA	AUS
13419,70	STANAG 4285	B=2400Bd	FUE	F N Brest	F
13419,80	*F N FSK*	*B=50Bd S=850Hz Ch=1*	*FUG*	*F N Saissac*	*F*
13420,00	RFSM8000	B=2400Bd		MFA Sofia	BUL
13423,50	*ALE USB*	*B=125Bd Ch=8 ChS=250Hz*		*G MIL net*	*G*
13426,00	PACTOR II	B=200Bd S=200Hz	V8V2222	Sailmail Brunai Bay	BRU
13426,00	CW			RUS HFDF net	RUS
13430,00	*STANAG 4285*	*B=2400Bd*		*G MIL Akrotiri*	*CYP*
13430,00	FMCW			Llandulas radar site	G
13431,00	CCIR 493-4	B=100Bd S=170Hz			
13435,00	WINDRM	B=1912,5Bd Ch=51 ChS=46,875Hz		CUB Intelligence Directorate	CUB
13435,00	CW			RUS HFDF net	RUS
13436,00	CW			RUS HFDF net	RUS
13437,00	T600	B=50Bd S=250Hz		RUS N Vladivostok	RUS
13439,00	FMCW			Santa Cruz Island, CA	USA
13440,00	FMCW			Coal Oil Pt.	USA
13440,00	FMCW			Mandalay Generating Station	USA
13440,00	FMCW			Nicholas Canyon, CA	USA
13440,00	STANAG 4285	B=2400Bd		AUS MHFCS Humpty Doo	AUS
13440,00	CHN 4+4	B=75Bd Ch=8 ChS=300/450Hz		CHN MIL	
13440,00	FMCW			Point Mugu, CA	USA
13440,00	FMCW			San Nicolas Island, CA	USA
13440,00	MIL 188-110B SER	B=2400Bd			

Frequency	Mode	Mode Parameter	Callsign	User	Country
13440,20	*STANAG 4285*	*B=2400Bd*		*DNK N Frederickshavn*	*DNK*
13443,00	CW		RGT77	RUS MIL Moscow, RVSN	RUS
13445,00	USB			G AF Air Cadets	G
13446,00	ALE USB	B=125Bd Ch=8 ChS=250Hz		USA FEMA net	USA
13446,00	USB			HFoZ/Radtel net	AUS
13450,00	FMCW			Dyfamed	F
13450,00	FMCW			Fort Bragg, CA	USA
13450,00	FMCW			Point Conception, CA	USA
13450,00	ALE LSB	B=125Bd Ch=8 ChS=250Hz		MOZ gas project	MOZ
13450,00	FMCW			Bradley Beach, NJ	USA
13450,00	FMCW			Brigantine, NJ	USA
13450,00	FMCW			Brant Beach, NJ	USA
13450,00	FMCW			Anasco, Puerto Rico	USA
13450,00	FMCW			Strathmere, NJ	USA
13450,00	FMCW			Sea Bright, NJ	USA
13450,00	FMCW			Seaside Park, NJ	USA
13450,00	FMCW			North Wildwood, NJ	USA
13451,00	ALE USB	B=125Bd Ch=8 ChS=250Hz		USA FEMA net	USA
13452,00	CCIR 493-4	B=100Bd S=170Hz			
13457,00	ALE USB	B=125Bd Ch=8 ChS=250Hz		USA FAA	USA
13460,00	MIL 188-110B SER	B=2400Bd	CENTR6**	MFA Bucharest	ROU
13460,00	MIL 188-110B SER	B=2400Bd		ROU E Tripoli	LBY
13465,00	FMCW			Llandulas radar site	G
13467,00	FMCW			Point Pinos, Pacific Grove, CA	USA
13468,00	ALE USB	B=125Bd Ch=8 ChS=250Hz		ROU diplo	
13468,00	RUS HYBRID MODEM MFSK-PSK	B=40Bd Ch=16 ChS=40Hz		Moscow	RUS
13469,00	CW		RJE56	RUS N	RUS
13473,00	*STANAG 4285*	*B=2400Bd*		*G MIL St. Eval*	*G*
13474,00	SYSTEME 3000	B=2400Bd			
13475,00	ALE USB	B=125Bd Ch=8 ChS=250Hz		VEN N	VEN
13475,00	FMCW			St. Catherine's, GA	USA
13475,00	FMCW			Cabo Rojo, Puerto Rico	USA
13475,00	FMCW			Crandon, FL	USA
13475,00	FMCW			Cedar Island, VA	USA
13475,00	FMCW			Clam Harbour, Nova Scotia	USA
13478,00	ALE USB	B=125Bd Ch=8 ChS=250Hz		USA MARS net	USA
13479,00	CW		REA4	RUS AF HQ Moscow	RUS
13479,00	*CIS 50-1000*	*B=50Bd S=1000Hz*	*REA4*	*RUS AF HQ Moscow*	*RUS*
13486,00	*RUS HYBRID MODEM MFSK-PSK*	*B=40Bd Ch=16*		*Moscow*	*RUS*
13486,00	*RUS HYBRID MODEM PSK-PSK*	*B=125Bd Ch=2 ChS=250Hz*		*Moscow*	*RUS*
13487,20	*STANAG 4285*	*B=2400Bd*	*DHJ58*	*D N Glücksburg TX Staberhuk*	*D*
13488,00	ALE USB	B=125Bd Ch=8 ChS=250Hz		Public health net , TX	USA
13490,00	ALE USB	B=125Bd Ch=8 ChS=250Hz			
13494,00	CW		S**	RUS N Severomorsk	RUS
13499,71	FMCW			Camp Varnum, RI	USA
13500,00	FMCW			Torungen	NOR
13500,00	FMCW			Inish Orr	IRL
13500,00	FMCW			Loop Head	IRL
13500,00	FMCW			Diablo Canyon Standard Range, CA	USA

Frequency	Mode	Mode Parameter	Callsign	User	Country
13500,00	FMCW			Point Estero, CA	USA
13500,00	FMCW			MIO Laboratory Toulon area	F
13500,00	CODAR			Tuscany	I
13500,00	FMCW			Dockweiler Headquarters, CA	USA
13500,00	FMCW			Santa Cruz, CA	USA
13500,00	FMCW			Summerland Sanitary District	USA
13500,00	FMCW			Eastern Point Retreat House, MA	USA
13500,00	FMCW			Fourth Cliff Recreation Area, MA	USA
13500,00	FMCW			Grand Manan, NB, CA	USA
13500,00	FMCW			Gulfport Harbor Pier, MS	USA
13500,00	FMCW			Green's Island, Maine, USA	USA
13500,00	FMCW			Provincetown, MA	USA
13500,00	FMCW			Barkat	MLT
13500,00	FMCW			Form	E
13500,00	FMCW			Salou	E
13500,00	FMCW			Alfcada	E
13500,00	FMCW			Alfanzina	POR
13500,00	FMCW			Torungen	NOR
13500,00	*CIS VFT: M/S 1.0*	*B=100Bd S=150Hz Ch=6 ChS=490Hz*		*Kazan area*	*RUS*
13500,00	FMCW			Poseidon System Limnos Island	GRC
13500,00	FMCW			Livorno Accademia	I
13500,00	FMCW			Marina di San Vincenzo	I
13500,00	FMCW			Barkat	MLT
13500,00	FMCW			Sopu	MLT
13500,00	FMCW			Galf	E
13500,00	FMCW			Salou	E
13500,00	FMCW			Alfcada	E
13500,00	FMCW			Vinaroz	E
13500,00	FMCW			Mazagon	E
13500,00	FMCW			Sagres	POR
13500,00	FMCW			Alfanzina	POR
13503,60	ALE USB	B=125Bd Ch=8 ChS=250Hz		USA diplo net	
13509,00	CIS MFSK 20	B=20Bd Ch=20 ChS=40Hz		RUS INTEL	RUS
13510,00	BAUDOT	B=75Bd S=720Hz	CFH	CAN F Halifax	CAN
13513,80	PACTOR II	B=200Bd S=200Hz	VZX	Sailmail Darawank	AUS
13515,00	ALE USB	B=125Bd Ch=8 ChS=250Hz		USA AF net	
13518,00	CIS 3000	B=3000Bd		Moscow	RUS
13520,00	*MIL 188-110B SER*	*B=2400Bd*		*G MIL Inskip*	*G*
13525,00	FMCW			Fort Funston, CA	USA
13525,00	FMCW			Haulover Inlet, FL	USA
13526,70	SITOR A	B=100Bd S=170Hz		EGY E Tripoli	LBY
13527,50	CW		D**	RUS N Sevastopol	UKR
13527,80	CW		P**	RUS N Kaliningrad	RUS
13528,00	CW		S**	RUS N Severomorsk	RUS
13528,00	CW		C**	RUS N Moscow	RUS
13528,00	CW		F**	RUS N Vladivostok	RUS
13528,10	CW		A**	RUS N Astrakhan	
13528,30	CW		K**	RUS N	RUS
13528,40	CW		M**	RUS N Magadan	RUS
13528,50	CW		U**		RUS
13530,00	ALE USB	B=125Bd Ch=8 ChS=250Hz		ALG AF net	ALG
13533,00	FMCW			Gerstle Cove, Salt Point State Park, CA	USA
13536,10	MHF-50 MODEM	B=54,3Bd Ch=32 ChS=64,5Hz	ZSJ	AFS N Capetown	AFS
13537,00	FMCW			Washburne, OR	USA
13537,00	FMCW			Hillsboro, FL	USA
13538,00	FAX 120/576		ZSJ	AFS N Cape Naval	AFS

Frequency	Mode	Mode Parameter	Callsign	User	Country
13538,00	CIS MFSK14	B=7,8Bd Ch=14 ChS=15Hz		RUS INTEL	RUS
13541,00	T600	B=50Bd S=250Hz	RJD56	RUS N HQ Murmansk	RUS
13545,00	FMCW			Plaka radar site	GRC
13545,00	FMCW			Fissini radar site	GRC
13545,00	FMCW			Limnos island radar site	GRC
13545,00	USB			G AF Air Cadets	G
13545,00	*ALE USB*	*B=125Bd Ch=8 ChS=250Hz*		*IRL MIL*	*IRL*
13548,00	PACTOR III	B=100Bd S=200Hz	9Z4DH	Sailmail Chaguaramas	TRD
13552,00	RFSM8000	B=2400Bd		BUL diplo	BUL
13552,00	RFSM8000	B=2400Bd		BUL E Cairo	EGY
13552,00	ALE USB	B=125Bd Ch=8 ChS=250Hz		G MIL net	G
13553,90	WSPR	B=1,46Bd Ch=4 ChS=1,46Hz		WSPR net	
13555,00	DPRK ARQ	B=600Bd S=600Hz		DPRK E Moscow	RUS
13555,00	DPRK ARQ	B=1200Bd S=600Hz		DPRK E Tripoli	LBY
13555,00	CW		P		USA
13556,00	ALE USB	B=125Bd Ch=8 ChS=250Hz			
13557,00	*RUS HYBRID MODEM MFSK-PSK*	*B=40Bd Ch=16*		*Moscow*	*RUS*
13560,00	CW	B=2000Bd		RFID	
13561,00	CW		RGT77	RUS MIL Moscow, RVSN	RUS
13562,40	CW		TDV		USA
13564,90	4FSK	B=195,24Bd Ch=4 ChS=195Hz			
13566,00	KNL MODEM	B=1500Bd Ch=1		Maritme Nesh Network Management	
13568,00	CW		RGT77	RUS MIL Moscow, RVSN	RUS
13570,00	FAX 120/576		HLL2	Seoul M	KOR
13573,00	CIS 68TONE	B=50Bd Ch=68 ChS=47Hz		RUS diplo	RUS
13574,00	STANAG 4285	B=2400Bd	NSS	USA N TX Davidsonville , MD	USA
13575,00	*POL MIL FSK*	*B=100Bd S=775Hz*		*POL MIL Warszawa*	*POL*
13580,00	CW		RDL	RUS N Moscow	RUS
13580,00	ALE USB	B=125Bd Ch=8 ChS=250Hz		PAK N net	PAK
13580,70	PACTOR	B=100Bd S=200Hz	DEK2811	DRK Mücke-Merlan	D
13580,70	PACTOR	B=100Bd S=200Hz	DEK	German Red Cross	D
13600,00	OOK				
13614,00	SSB			USA AF MARS	J
13630,00	ALE USB	B=125Bd Ch=8 ChS=250Hz		USA FAA	USA
13636,00	CW		C**	RUS N Moscow	RUS
13636,00	CW		F**	RUS N Vladivostok	RUS
13636,00	CW		S**	RUS N Severomorsk	RUS
13636,00	CW		RJS	RUS N HQ Vladivostok	RUS
13676,70	STANAG 4481	B=75Bd S=850Hz			
13683,60	PACTOR III	B=100Bd S=200Hz	KDS	Ontario	CAN
13712,00	ALE USB	B=125Bd Ch=8 ChS=250Hz		Queensland Deptartment of Community Safety	AUS
13722,00	PACTOR	B=100Bd S=200Hz	NNF9CA		USA
13741,00	USB			USA Army MARS	
13849,20	*8PSK*	*B=3000Bd*		*RUS MIL Moscow*	*RUS*
13852,00	CW		RGT77	RUS MIL Moscow, RVSN	RUS
13854,70	BAUDOT	B=50Bd S=170Hz	BYN1		
13854,70	BAUDOT	B=50Bd S=170Hz	BQR3		
13858,00	FSK2	B=1200Bd S=1200Hz		G MIL Inskip	G
13861,50	PACTOR III	B=100Bd S=200Hz	CEV773	Sailmail Los Lagos	CHL

Frequency	Mode	Mode Parameter	Callsign	User	Country
13870,00	CIS 50-500	B=50Bd S=500Hz	RWB3	RUS PTT	D
13870,00	ALE USB	B=125Bd Ch=8 ChS=250Hz		ROU diplo net	ROU
13870,00	STANAG 4481	B=50Bd S=850Hz	NPG	USA N Dixon , CA	USA
13874,00	PACTOR III	B=100Bd S=200Hz	WQAB964	Sailmail San Diego , CA	USA
13874,60	MI 188-110B SER	B=2400Bd			
13875,00	PACTOR III	B=100Bd S=200Hz	CEV773	Sailmail Los Lagos	CHL
13876,00	CW		RGT77	RUS MIL Moscow, RVSN	RUS
13876,70	SITOR A	B=100Bd S=170Hz		EGY E Belgrade	SER
13877,00	ALE USB	B=125Bd Ch=8 ChS=250Hz		G MIL net	G
13878,00	SSB			NASA launch support aircraft	USA
13880,00	PACTOR II	B=200Bd S=200Hz	HPPM1	Sailmail Chiriqui	PNR
13880,00	PACTOR III	B=100Bd S=200Hz	HPPM3	Sailmail Panama	PNR
13880,00	*STANAG 4285*	*B=2400Bd*		*G MIL Akrotiri*	*CYP*
13882,50	*FAX 120/288/576*		*DDK6*	*DWD TX Pinneberg*	*D*
13884,00	CIS MFSK14	B=7,8Bd Ch=14 ChS=15Hz		RUS INTEL	RUS
13890,00	CIS 4FSK	B=150Bd Ch=4 ChS=4000Hz			RUS
13893,70	*STANAG 4285*	*B=2400Bd*	*IDN*	*I N Napoli*	*I*
13894,00	ALE USB	B=125Bd Ch=8 ChS=250Hz		USA FEMA net	USA
13899,50	CIS 50-1000	B=50Bd S=1000Hz		RUS	RUS
13900,00	VFT: CIS 100-1440	B=100Bd S=1440Hz Ch=3 ChS=480Hz		RUS	RUS
13906,40	PACTOR III	B=200Bd S=200Hz	WPTG385	Sailmail Corpus Christi , TX	USA
13907,00	SSB			USAF Strategic Command	USA
13907,00	ALE USB	B=125Bd Ch=8 ChS=250Hz		USA COTHEN net	USA
13907,20	CW		WI2XER	Skycast Services	USA
13909,20	PACTOR I	B=100Bd S=200Hz	PACMRGF		
13909,50	CIS 100-2000	B=100Bd S=2000Hz Ch=1		RUS MIL	RUS
13910,00	USB		VMD750	Austravel Safety Net	AUS
13910,00	*CIS VFT: M/S 1.0*	*B=100Bd S=150Hz Ch=6 ChS=490Hz*		*Moscow*	*RUS*
13913,00	SKYOFDM	B=64Bd Ch=28 ChS=86Hz			
13914,00	SYSTEME 3000	B=2400Bd			F
13915,00	*CIS 50-500*	*B=50Bd S=500Hz*		*RUS N Kaliningrad*	*RUS*
13915,00	CIS ARQ	B=100Bd S=500Hz			RUS
13915,00	PACTOR II	B=200Bd S=200Hz	WHV861	Sailmail San Luis Obispo , CA	USA
13915,00	*CIS 100-500 BURST*	*B=100Bd S=500Hz*		*Jekaterinenburg*	*RUS*
13915,00	CIS ARQ	B=100Bd S=500Hz	UMN3	RUS PTT Serpukhov	RUS
13920,00	FAX 120/576		VMC	Charleville M	AUS
13921,00	ALE USB	B=125Bd Ch=8 ChS=250Hz		USA AF net	USA
13921,40	PACTOR II	B=200Bd S=200Hz	WPUC469	Sailmail South Daytona , FL	USA
13922,20	*STANAG 4285*	*B=2400Bd*	*PBC*	*NLD N Goeree Island*	*HOL*
13926,40	PACTOR II	B=200Bd S=200Hz	WPTG385	Sailmail Corpus Christi , TX	USA
13927,00	USB		AFA2CM	USA AF MARS , NY	USA
13927,00	USB		AFA1QW	USA AF MARS	USA
13927,00	USB		AFA6PF	USA AF MARS Los Angeles , CA	USA
13927,00	USB		AFA1WP	USA AF MARS	USA
13927,00	USB		AFA1RE	USA AF MARS, ME	USA
13927,00	USB		AFA3HS	USA AF MARS Kansas City , KS	USA
13927,00	USB		AFA2MH	USA AF MARS , GA	USA
13927,00	USB		AFA2HS	USA AF MARS Florida	USA
13927,00	SSB			USA AF MARS	
13927,00	USB		AFA5QW	USA AF MARS Greenwood , IN	USA

Frequency	Mode	Mode Parameter	Callsign	User	Country
13927,00	USB		AFA1RT	USA AF MARS New Hampshire	USA
13930,00	PACTOR II	B=200Bd S=200Hz	KUZ533	Sailmail Honululu	HWA
13930,00	PACTOR III	B=100Bd S=200Hz	RC01	Sailmail Maputo	MOZ
13935,00	ALE USB	B=125Bd Ch=8 ChS=250Hz		USA FEMA net	USA
13937,00	PACTOR II	B=200Bd S=200Hz	XJN714	Sailmail Lunenburg , NS	CAN
13938,00	CIS 12	B=1440Bd Ch=12 ChS=200Hz		RUS MIL	RUS
13940,00	PACTOR III	B=100Bd S=200Hz	WHV382	Sailmail Friday Habor , WA	USA
13944,00	*CIS 12*	*B=1440Bd Ch=12 ChS=200Hz*		*RUS MIL St. Petersburg*	*RUS*
13945,00	ALE USB	B=125Bd Ch=8 ChS=250Hz		TUN diplo net	TUN
13946,00	PACTOR II	B=200Bd S=200Hz	WHV861	Sailmail San Luis Obispo , CA	USA
13948,00	CIS 68TONE	B=50Bd Ch=68 ChS=47Hz		RUS diplo	RUS
13950,00	BUL MFSK	B=240Bd Ch=8 ChS=240Hz		MFA Sofia	BUL
13955,00	ALE USB	B=125Bd Ch=8 ChS=250Hz		SUI F net	
13958,00	USB			USA Army MARS	
13958,00	*MIL 188-110B SER*	*B=2400Bd*		*SUI MIL Banja Luca*	*BHI*
13963,50	USB			USA AF MARS	
13965,00	PACTOR III	B=100Bd S=200Hz			
13965,00	USB			G AF Air Cadets	G
13970,00	USB			F yacht net	
13971,00	PACTOR II	B=200Bd S=200Hz	WRD719	Sailmail Palo Alto	USA
13971,00	PACTOR III	B=100Bd S=200Hz	WQLI952	Sailmail Watsonville , CA	USA
13972,00	ALE USB	B=125Bd Ch=8 ChS=250Hz		B AF net	B
13972,00	CIS OFDM 45	B=2400Bd Ch=45 ChS=62,5Hz			RUS
13972,00	CIS OFDM 93	B=2400Bd Ch=93 ChS=31,25Hz			RUS
13973,00	CW		RAL2	RUS N HQ Astrakhan	RUS
13973,00	PACTOR II	B=200Bd S=170Hz	RC2KAB**	ICRC Kabul	AFG
13973,00	PACTOR II	B=200Bd S=170Hz	RC2MAZ**	ICRC Mostar	BIH
13975,00	CW		RFH2	RUS N	RUS
13975,00	CW		RHW2	RUS N	RUS
13975,00	CW		RBL71	RUS N	RUS
13975,00	CW		RAL2	RUS N HQ Astrakhan	RUS
13975,00	CW		RDU2	RUS N	RUS
13977,50	*STANAG 4285*	*B=2400Bd Ch=1*	*HWN*	*F N Paris*	*F*
13978,00	ALE USB	B=125Bd Ch=8 ChS=250Hz		B AF net	B
13979,00	T600	B=50Bd S=500Hz		RUS N	
13980,00	PACTOR II	B=200Bd S=200Hz	HPPM2	Sailmail Chiriqui	PNR
13980,00	ALE USB	B=125Bd Ch=8 ChS=250Hz		ROU diplo net	ROU
13982,00	MHF-50 MODEM	B=54,3Bd Ch=32 ChS=64,5Hz	ZSJ	AFS N Capetown	AFS
13984,00	ALE USB	B=125Bd Ch=8 ChS=250Hz		ALG MIL net	ALG
13984,00	PACTOR IV	B=100Bd S=200Hz			
13985,00	CW		RAL2	RUS N HQ Astrakhan	RUS
13985,00	BAUDOT	B=50Bd S=500Hz		Moscow	RUS
13985,00	CIS 50-500	B=50Bd S=500Hz	RRQ6	RUS PPT	RUS
13985,00	CW		RNM6	RUS	RUS
13985,00	CIS 50-17	B=50Bd S=500Hz			
13986,00	PACTOR III	B=100Bd S=200Hz	WQLI952	Sailmail Watsonville , CA	USA
13986,00	PACTOR II	B=200Bd S=200Hz	WRD719	Sailmail Palo Alto	USA
13987,00	USB			E Civil Protection net REMER	E
13988,50	FAX 120/576		JMH	Tokyo M	J

Frequency	Mode	Mode Parameter	Callsign	User	Country
13988,50	*4FSK*	*B=100Bd Ch=4 ChS=500Hz*		*RUS MIL Moscow*	*RUS*
13989,30	*CIS 100-1000*	*B=100Bd S=1000Hz*		*RUS MIL Moscow*	*RUS*
13989,80	*CIS 100-1000*	*B=100Bd S=1000Hz*		*RUS MIL Moscow*	*RUS*
13990,30	*CIS 100-1000*	*B=100Bd S=1000Hz*		*RUS MIL Moscow*	*RUS*
13992,00	PACTOR III	B=100Bd S=200Hz	RC01	Sailmail Maputo	MOZ
13993,00	USB		AAR4IG	USA MARS	USA
13993,00	USB		AAR9EN	USA MARS	USA
13993,00	USB		AAR4MN	USA MARS	USA
13993,00	USB		AAR1BX	USA MARS	USA
13993,00	USB		AFG30	USA MARS	USA
13993,00	USB		AAR9HP	USA MARS	USA
13993,00	USB		AAR0CS	USA MARS	
13993,00	USB			USA AF MARS	D
13993,00	USB		AFI1D	USA MARS net	USA
13994,30	*CIS 100-1000*	*B=100Bd S=1000Hz*		*RUS MIL Moscow*	*RUS*
13994,80	*CIS 100-1000*	*B=100Bd S=1000Hz*		*RUS MIL Novosibirsk*	*RUS*
13996,00	ALE USB	B=125Bd Ch=8 ChS=250Hz		F N net	F
13996,00	USB			USA AF MARS	
13996,20	*CIS 100-1000*	*B=100Bd S=1000Hz*		*RUS MIL Vladiwostok*	*RUS*
13998,00	PACTOR II	B=200Bd S=200Hz	KZN508	Sailmail Rockhill , SC	USA
13998,00	PACTOR II	B=100Bd S=200Hz		IFRC net	
14002,00	STANAG 4481	B=50Bd S=850Hz	NDT	USA N Yokosuka	J
14008,00	*T600*	*B=50Bd S=500Hz*		*RUS N Moscow*	*RUS*
14024,00	*CIS 12*	*B=1440Bd Ch=12 ChS=200Hz*	*RMP*	*RUS N HQ Kaliningrad*	*RUS*
14038,50	DPRK FSK	B=600Bd S=600Hz		DPRK E Tripoli	LBY
14047,50	CW		W1AW	ARRL morse training	USA
14060,00	ALE USB	B=125Bd Ch=8 ChS=250Hz		ISR AF net	ISR
14070,00	PSK31	B=31Bd		PSK31 channel	
14072,50	MFSK			MFSK modes user	
14074,00	FT8	B=5,86Bd Ch=8 ChS=5,86Hz		FT8 channel	
14076,00	JT65	B=2,69Bd Ch=65 ChS=2,69Hz		JT65 channel	
14078,00	JT9	B=1,736Bd Ch=9 ChS=1,736Hz		JT9 channel	
14078,00	JS8	B=6,25Bd Ch=8 ChS=6,25Hz		JS8 user	
14080,00	FT4	B=5,86Bd Ch=8 ChS=5,86Hz		FT4 channel	
14081,00	*CIS 12*	*B=1440Bd Ch=12 ChS=200Hz*		*RUS N HQ Moscow*	*RUS*
14083,00	BAUDOT	B=45,45Bd S=170Hz		RTTY user	
14095,00	LONGCHAT	B=40Bd Ch=1		LongChat net	
14095,60	WSPR	B=1,46Bd Ch=4 ChS=1,46Hz		WSPR net	
14096,00	CIS 12	B=1440Bd Ch=12 ChS=200Hz	RJS	RUS NHQ Vladivostok	RUS
14100,00	CW		4U1UN	UN New York	USA
14100,00	CW		4X6TU	Tel Aviv	ISR
14100,00	CW		OH2B	Espoo	FNL
14100,00	CW		YV5B	Caracas	VEN
14100,00	CW		VE8AT	Inuvik, NT	CAN

Frequency	Mode	Mode Parameter	Callsign	User	Country
14100,00	CW		W6WX	Mt. Umunhum	USA
14100,00	CW		KH6RS	Maui	HWA
14100,00	CW		ZL6B	Masterton	NZL
14100,00	CW		VK6RBP	Rolystone	AUS
14100,00	CW		JA2IGY	Mt. Asama	J
14100,00	CW		RR9O	Novosibirsk	RUS
14100,00	CW		VR2B	Hong Kong	HKG
14100,00	CW		4S7B	Colombo	CLN
14100,00	CW		ZS6DN	Pretoria	AFS
14100,00	CW		5Z4B	Kariobangi	KEN
14100,00	CW		CS3B	São Jorge	MDR
14100,00	CW		LU4AA	Buenos Aires	ARG
14100,00	CW		OA4B	Lima	PRU
14103,00	ROBUST PACKET RADIO	B=200Bd Ch=8 ChS=60Hz		Robust Packet Radio net, HF-APRS	
14103,00	ROS USB	B=1Bd Ch=144 ChS=15,625Hz		ROS frequency	
14105,00	VARA	Ch=1		VARA chat frequency	
14107,00	HFPAGER	B=5,86Bd Ch=18		RUS HFPager net	RUS
14108,00	CW			RUS MIL net	RUS
14109,00	ALE USB	B=125Bd Ch=8 ChS=250Hz		HFLINK network	
14116,00	*T600*	*B=50Bd S=250Hz*		*RUS N Moscow*	*RUS*
14125,00	USB			Wireless Institute Civil Emergency Service WICEN	AUS
14145,00	USB			Mawson Base , AUS	ANT
14160,00	CIS 75-250	B=75Bd S=250Hz		RUS N Moscow	RUS
14192,00	T600	B=50Bd S=500Hz	RMP	RUS N Kaliningrad	RUS
14200,00	USB			EGY N net	EGY
14236,00	FREEDV	B=1900Bd Ch=14		FreeDV net	
14236,00	ARD9800 MODEM	B=2250Bd Ch=36 ChS=62,5Hz		AOR Digital Voice net	USA
14240,00	*CIS 12*	*B=1440Bd Ch=12 ChS=200Hz*		*RUS MIL Moscow*	*RUS*
14250,00	USB			EGY N net	EGY
14259,00	CIS OFDM 60	B=2400Bd Ch=60 ChS=44,5Hz		RUS diplo	RUS
14291,00	PSK4A	B=2400Bd		CHN MIL	CHN
14292,00	CW		REA4	RUS AF HQ Moscow	RUS
14297,00	USB			European maritime mobile net	
14300,00	USB			Ham radio emergency frequency	R1
14300,00	DPRK ARQ	B=600Bd S=600Hz		DPRK E Moscow	RUS
14300,00	DPRK ARQ	B=600Bd S=600Hz		DPRK E Berlin	D
14300,00	DPRK ARQ	B=600Bd S=600Hz		DPRK E Vienna	AUT
14300,00	USB			Pacific Seafarers net	
14300,00	USB			Maritime Mobile Service net	
14303,00	USB			UK maritime mobile net	
14313,00	USB			INTERMAR net	D
14325,00	USB		WX4NHC	Huricane Watch net Miami , FL	USA
14333,00	*T600*	*B=50Bd S=200Hz*	*RMP*	*RUS N Kaliningrad*	*RUS*
14343,00	CHN N MIL 188-110 MOD	B=2400Bd		CHN N	CHN
14346,00	ALE USB	B=125Bd Ch=8 ChS=250Hz		HFLINK network	
14350,00	HFD+VL73	B=2160Bd Ch=73 ChS=37,5Hz		University Las Palmas de Gran Canaria	E
14354,00	*STANAG 4285*	*B=2400Bd*		*G MIL St. Eval*	*G*
14355,70	*STANAG 4481*	*B=75Bd S=850Hz*		*G MIL St. Eval*	*G*
14356,00	*STANAG 4285*	*B=2400Bd*		*G MIL St. Eval*	*G*
14357,00	ALE USB	B=125Bd Ch=8 ChS=250Hz		USA Civil Air Patrol net	USA
14360,00	ALE LSB	B=125Bd Ch=8 ChS=250Hz		TWN N Net	TWN

Frequency	Mode	Mode Parameter	Callsign	User	Country
14360,00	MIL 188-110A SER	B=2400Bd		TWN N net	TWN
14364,00	LINK 11 CLEW	B=2250Bd Ch=16 ChS=330/110/550Hz		USA N Kitsap , WA	USA
14369,60	*STANAG 4285*	*B=2400Bd*		*NOR N Kristiansand*	*NOR*
14369,70	*STANAG 4285*	*B=2400Bd*		*DNK N TX Fredrerickshavn*	*DNK*
14375,00	WINDRM	B=1912,5Bd Ch=51 ChS=46,875Hz		CUB Intelligence Directorate	CUB
14376,00	CW			RUS MIL net	RUS
14376,20	*STANAG 4285*	*B=2400Bd Ch=1*	*IDN*	*I N Napoli*	*I*
14379,00	CIS OFDM 45	B=2400Bd Ch=45 ChS=62,5Hz		RUS diplo	RUS
14380,00	ALE USB	B=125Bd Ch=8 ChS=250Hz		CHN MIL net	CHN
14380,00	MIL 188-110A SER	B=2400Bd		CHN MIL net	CHN
14385,00	STANAG 4285	B=2400Bd		AUS MHFCS Humpty Doo	AUS
14385,00	ALE USB	B=125Bd Ch=8 ChS=250Hz		AUS MIL net	AUS
14389,00	USB			USA AF MARS	
14390,00	CIS OFDM 73TONE	B=2190Bd Ch=73 ChS=37,5Hz		RUS diplo	RUS
14392,00	USB			USA AF MARS	
14396,50	ALE USB	B=125Bd Ch=8 ChS=250Hz		USA SHARES net	USA
14398,00	*RUS HYBRID MODEM MFSK-PSK*	*B=40Bd Ch=16*		*Moscow*	*RUS*
14402,00	CW		RGT77	RUS MIL Moscow, RVSN	RUS
14402,00	*STANAG 4285*	*B=2400Bd*		*G MIL St. Eval*	*G*
14403,50	ALE USB	B=125Bd Ch=8 ChS=250Hz		G MIL net	G
14405,00	SSB			USA AF MARS	
14406,00	ALE USB	B=125Bd Ch=8 ChS=250Hz		MFA Bucharest	ROU
14406,00	MIL 188-110A SER	B=2400Bd Ch=1		MFA Bucharest	ROU
14406,00	*STANAG 4285*	*B=2400Bd*		*G F St. Eval*	*G*
14408,00	SSB			USA AF MARS	
14411,00	CW		RDL	RUS N Moscow	RUS
14411,00	*T600*	*B=50Bd S=200Hz*	*RDL*	*RUS N Moscow*	*RUS*
14411,00	*T600*	*B=50Bd S=200Hz*		*RUS N Smolensk*	*RUS*
14415,00	USB		VNJ	Casey Base , AUS	ANT
14415,00	USB			AUS Davis Base	ANT
14415,00	USB			Macquarie Island , AUS	ANT
14415,00	USB			Mawson Base , AUS	ANT
14415,00	BPSK	B=100Bd		POL INT	POL
14415,70	STANAG 4285	B=2400Bd	IGJ	I N Augusta	I
14416,00	CIS 50-500	B=50Bd S=500Hz		RUS	RUS
14420,00	ALE LSB	B=125Bd Ch=8 ChS=250Hz		TWN N net	TWN
14422,00	USB			Yacht net	I
14428,00	*CIS 75-200*	*B=75Bd S=200Hz*		*RUS MIL Moscow*	*RUS*
14430,00	OQPSK	B=48000Bd		HF Trading	USA
14434,00	STANAG 4481	B=50Bd S=850Hz	NPM	USA N Lualualei	HWA
14436,00	ALE USB	B=125Bd Ch=8 ChS=250Hz		CHN MIL net	CHN
14436,20	PACTOR II	B=200Bd S=200Hz	VZX	Sailmail Darawank	AUS
14436,20	PACTOR III	B=100Bd S=200Hz	XJN714	Sailmail Lunenburg , NSC	CAN
14441,00	STANAG 4285	B=2400Bd		AUS MHFCS Humpty Doo	AUS
14442,50	*DPRK ARQ*	*B=600Bd S=600Hz*		*DPRK E Warszawa*	*POL*
14444,00	CODAN CHIRP	B=80Bd Ch=30 ChS=81Hz		MFA Cairo	EGY
14444,00	*STANAG 4285*	*B=2400Bd*		*G F St. Eval*	*G*

Frequency	Mode	Mode Parameter	Callsign	User	Country
14444,00	CODAN	B=2400Bd Ch=16 ChS=112,5Hz		MFA Cairo	EGY
14444,00	*STANAG 4285*	*B=2400Bd*		*G MIL Akrotiri*	*CYP*
14444,00	DPRK ARQ	B=600Bd S=600Hz		DPRK E Habana	CUB
14444,00	PSK4A	B=2400Bd		CHN MIL	CHN
14455,00	*DPRK ARQ*	*B=600Bd S=600Hz*		*DPRK E Madrid*	*E*
14456,00	CW		RAL65	RUS N ship	RUS
14457,00	ALE USB	B=125Bd Ch=8 ChS=250Hz		ALG AF net	ALG
14458,40	ALE USB	B=125Bd Ch=8 ChS=250Hz		CAN CFARS net	CAN
14458,50	USB		CIW681	CAN MIL	CAN
14458,50	ALE USB	B=125Bd Ch=8 ChS=250Hz		USA FBI net	USA
14460,00	*STANAG 4285*	*B=2400Bd*		*G MIL Akrotiri*	*CYP*
14460,00	MIL 188-110A SER	B=2400Bd		AUS MHFCS North West Cape	AUS
14462,00	STANAG 4285	B=2400Bd		AUS MHFCS Humpty Doo	AUS
14462,00	*STANAG 4285*	*B=2400Bd*		*AUS MHFCS Humpty Doo*	*AUS*
14463,50				USA MARS net	USA
14467,30	*BAUDOT*	*B=50Bd S=450Hz*	*DDH8*	*DWD TX Pinneberg*	*D*
14468,50	6CH PSK	Ch=6 ChS=400Hz			
14473,00	CIS 50-500 BAUDOT	B=50Bd S=500Hz		RUS	
14484,00	USB		AAR9JL	USA Army MARS	USA
14484,00	USB		AAR0QM	USA Army MARS	USA
14484,00	USB			USA Army MARS net	USA
14485,00	CIS 50-500	B=50Bd S=500Hz	RTK	RUS PTT	RUS
14485,50	*ALE USB*	*B=125Bd Ch=8 ChS=250Hz*		*G MIL net*	*G*
14493,00	T600	B=50Bd S=200Hz	RIT	RUS N HQ Severomorsk	RUS
14493,50	ALE USB	B=125Bd Ch=8 ChS=250Hz		USA FBI net	USA
14493,50	USB			USA FBI net	USA
14494,00	*STANAG 4285*	*B=2400Bd*		*G MIL St. Eval*	*G*
14500,00	USB			G AF Air Cadets	G
14500,00	CIS 50-500	B=50Bd S=500Hz	RRR32	RUS PTT Smolensk	RUS
14502,00	STANAG 4285	B=2400Bd	NSS	USA N TX Davidsonville , MD	USA
14508,00	*CODAN*	*B=2400Bd Ch=16 ChS=112,5Hz*		*IRN net Teheran*	*IRN*
14508,00	*ALE USB*	*B=125Bd Ch=8 ChS=250Hz*		*IRN net*	*IRN*
14508,00	*CODAN CHIRP*	*B=80Bd Ch=30 ChS=81Hz*		*IRN net*	*IRN*
14508,50	*ALE USB*	*B=125Bd Ch=8 ChS=250Hz*		*G MIL net*	*G*
14520,00	DPRK ARQ	B=600Bd S=600Hz		DPRK E	
14520,00	AUS ISB MODEM	B=600Bd S=345Hz		AUS MIL	AUS
14530,00	DPRK ARQ	B=600Bd S=600Hz		DPRK E Moscow	RUS
14532,50	ALE USB	B=125Bd Ch=8 ChS=250Hz		USA FBI net	USA
14535,50	STANAG 4285	B=2400Bd		G N Crimond	G
14538,00	STANAG 4285	B=2400Bd		G MIL Akrotiri	CYP
14538,00	CIS MFSK14	B=7,8Bd Ch=14 ChS=15Hz		RUS INTEL	RUS
14540,00	DPRK ARQ	B=600Bd S=600Hz		DPRK E Addis Abeba	ETH
14545,00	DPRK ARQ	B=600Bd S=600Hz		DPRK E	
14548,20	*STANAG 4285*	*B=2400Bd*		*G MIL TX Akrotiri*	*CYP*
14550,00	ALE USB	B=125Bd Ch=8 ChS=250Hz		MRC MIL net	MRC
14550,00	*ALE USB*	*B=125Bd Ch=8 ChS=250Hz*		*F Airbus net*	*F*
14552,00	CW			RUS N	RUS
14552,00	CW		RGT77	RUS MIL Moscow, RVSN	RUS

Frequency	Mode	Mode Parameter	Callsign	User	Country
14556,00	*CW*		*RIW*	*RUS N HQ Moscow*	*RUS*
14556,00	CW		RAA	RUS N	RUS
14556,00	*CW*		*RDL*	*RUS N Moscow*	*RUS*
14557,00	CIS 100-1000	B=100Bd S=1000Hz			
14557,00	CIS OFDM 60	B=2400Bd Ch=60 ChS=44,5Hz		RUS diplo	RUS
14562,00	CIS 50-500	B=50Bd S=500Hz		RUS	RUS
14567,00	ALE USB	B=125Bd Ch=8 ChS=250Hz		USA FEMA net	USA
14570,00	ALE USB	B=125Bd Ch=8 ChS=250Hz		F N net	F
14572,00	ALE USB	B=125Bd Ch=8 ChS=250Hz		AUS Police NSW	AUS
14572,00	ALE USB	B=125Bd Ch=8 ChS=250Hz		AUS Police NT	AUS
14572,00	ALE USB	B=125Bd Ch=8 ChS=250Hz		AUS Police QLD	AUS
14572,00	ALE USB	B=125Bd Ch=8 ChS=250Hz		AUS Police WA	AUS
14572,00	ALE USB	B=125Bd Ch=8 ChS=250Hz		SAPOL Communication Infrastructure	AUS
14580,20	STANAG 4285	B=2400Bd		G MIL St. Eval	G
14581,00	*T600*	*B=50Bd S=200Hz*		*RUS N Moscow*	*RUS*
14581,10	*T600*	*B=50Bd S=200Hz*		*RUS N Smolensk*	*RUS*
14581,50	STANAG 5030	B=50Bd S=50Hz			
14582,00	ALE USB	B=125Bd Ch=8 ChS=250Hz		USA COTHEN net	USA
14584,20	*STANAG 4285*	*B=2400Bd*		*NOR N TX Stavanger*	*NOR*
14585,00	PACKET RADIO	B=1200Bd S=1000Hz		INS N net	INS
14587,00	CW		RAA	RUS N HQ Moscow	RUS
14588,00	PACTOR III	B=100Bd S=200Hz	RC01	Sailmail Maputo	MOZ
14590,00	HFT MODEM		WI2XNX	10Band LLC	USA
14592,00	CW		CGA984	Essex County, ON	CAN
14593,00	T600	B=50Bd S=250Hz	RJS	RUS N HQ Vladivostok	RUS
14600,00	MIL 188-110A SER	B=2400Bd		EGY N net	EGY
14600,00	*CIS 12*	*B=1440Bd Ch=12 ChS=200Hz*	*RCV*	*RUS N HQ Sevastopol*	*UKR*
14600,00	USB			EGY N net	EGY
14605,00	*STANAG 4285*	*B=2400Bd*		*AUS MHFCS Humpty Doo*	*AUS*
14606,00	USB			USA AF MARS	AZR
14609,00	CIS MFSK 20	B=20Bd Ch=20 ChS=40Hz		RUS INTEL	RUS
14610,00	USB		PWBL	B N training ship BRASIL U-27	B
14610,00	ALE USB	B=125Bd Ch=8 ChS=250Hz		MRC net	MRC
14631,00	STANAG 4285	B=2400Bd	CTA	POR N Lisbon	POR
14631,00	STANAG 4285	B=2400Bd	CTA	POR N Lisbon	POR
14643,00	CIS OFDM 60	B=2400Bd Ch=60 ChS=44,5Hz		RUS diplo	RUS
14650,00	DPRK ARQ	B=600Bd S=600Hz		DPRK E Moscow	RUS
14655,00	T600	B=50Bd S=200Hz	RJS	RUS N HQ Vladivostok	RUS
14658,00	USB			USA Army MARS	
14661,00	BPSK	B=2400Bd		FIN diplo	
14662,00	USB		VKE237	AUS HF Radio Club	AUS
14664,00	T600	B=50Bd S=200Hz	RDL	RUS N Moscow	RUS
14664,00	CW		RDL	RUS N Moscow	RUS
14669,00	CHN DBPSK	B=2400Bd Ch=1		CHN AF net	CHN
14670,00	CW , SSB		CHU	Ottawa TS	CAN
14674,00	CIS 12	B=1440Bd Ch=12 ChS=200Hz	RCV	RUS N HQ Sevastopol	UKR

Frequency	Mode	Mode Parameter	Callsign	User	Country
14681,50	ALE USB	B=125Bd Ch=8 ChS=250Hz		USA Army NG net Utah	USA
14693,00	CHN DBPSK	B=2400Bd Ch=1		CHN AF net	CHN
14693,00	MEROD	B=150Bd S=800Hz Ch=1		MFA Helsinki	FIN
14693,00	SKYOFDM	B=1792Bd Ch=28 ChS=86Hz		MFA Helsinki	FIN
14693,00	NOKIA MSG TERMINAL	B=301,1Bd S=760Hz		MFA Helsinki	FIN
14693,00	BPSK	B=2400Bd		FIN diplo	
14704,00	T600	B=50Bd S=250Hz	RDL	RUS N Moscow	RUS
14705,00	*MIL 188-110A SER*	*B=2400Bd*		*AUS MHFCS North West Cape*	*AUS*
14706,00	*STANAG 4285*	*B=2400Bd*		*AUS MHFCS Humpty Doo*	*AUS*
14710,00	SKYOFDM	B=64Bd Ch=22 ChS=86Hz		MFA Helsinki	FIN
14710,00	ALE USB	B=125Bd Ch=8 ChS=250Hz		FIN diplo net	FIN
14710,00	USB			GRC net	GRC
14711,00	IRN N QPSK ADAPTIVE	B=468Bd		IRN N	IRN
14711,00	IRN N QPSK ADAPTIVE	B=468Bd		IRN N	IRN
14711,00	IRN N QPSK ADAPTIVE	B=468Bd		IRN N	IRN
14716,00	CW			RUS HFDF net	RUS
14718,60	STANAG 4285	B=2400Bd	FUX	F N Le Port	REU
14721,00	*DPRK ARQ*	*B=1200Bd S=600Hz*		*DPRK E Caracas*	*VEN*
14722,20	*STANAG 4285*	*B=2400Bd*	*FUG*	*F N Saissac*	*F*
14723,70	STANAG 4285	B=2400Bd	FUX	F N Le Port	REU
14723,70	*STANAG 4285*	*B=2400Bd*		*G MIL TX Akrotiri*	*CYP*
14724,00	ALE USB	B=125Bd Ch=8 ChS=250Hz		TUN diplo net	TUN
14730,00	OQPSK	B=48000Bd		HF Trading	USA
14738,00	CIS MFSK14	B=7,8Bd Ch=14 ChS=15Hz		RUS INTEL	RUS
14741,50	*DPRK ARQ*	*B=600Bd S=600Hz*		*MFA Pyongyang*	*KRE*
14745,00	ALE USB	B=125Bd Ch=8 ChS=250Hz		Queensland Deptartment of Community Safety	AUS
14746,00	CW		UAL	MFA Moscow	RUS
14747,50	*STANAG 4481*	*B=75Bd S=850Hz*	*CFH*	*CAN N Halifax*	*CAN*
14754,50	MIL 188-110B SER	B=2400Bd		S MIL	S
14756,00	RFSM8000	B=2400Bd		BUL E Moscow	RUS
14757,00	ALE USB	B=125Bd Ch=8 ChS=250Hz		SNG N net	SNG
14766,00	T600	B=50Bd S=200Hz		RUS N Bishkek	KGZ
14776,00	ALE USB	B=125Bd Ch=8 ChS=250Hz		USA FEMA net	USA
14776,00	RFSM8000	B=2400Bd		MFA Sofia	BUL
14780,00	BAUDOT	B=50Bd S=500Hz		Ministry of railway Moscow	RUS
14780,00	OQPSK	B=48000Bd		HF Trading	USA
14781,00	CIS 150-500	B=150Bd S=500Hz		RUS AF	RUS
14782,00	CHN N MIL 188-110 MOD	B=2400Bd		CHN N	CHN
14783,20	*STANAG 4285*	*B=2400Bd*		*G MIL Akrotiri*	*CYP*
14786,00	RFSM8000	B=2400Bd		BUL diplo	BUL
14786,00	RFSM8000	B=2400Bd		MFA Sofia	BUL
14790,00	ALE LSB	B=125Bd Ch=8 ChS=250Hz		VEN N net	VEN
14794,00	CIS 12	B=1440Bd Ch=12 ChS=200Hz		RUS MIL	RUS
14796,00	RFSM8000	B=2400Bd		BUL E Pairs	F
14796,00	RFSM8000	B=2400Bd		BUL E Madrid	E

Frequency	Mode	Mode Parameter	Callsign	User	Country
14800,00	USB			EGY N net	EGY
14807,50	ALE 3G	B=2400Bd		UI MIL net	
14828,60	HFD+VL73	B=2160Bd Ch=73 ChS=37,5Hz		University Las Palmas de Gran Canaria	E
14830,00	*T600*	*B=50Bd S=500Hz*		*RUS N Vladivostok*	*RUS*
14830,00	HFT MODEM		WI2XNX	10Band LLC	USA
14830,00	OQPSK	B=48000Bd		HF Trading	USA
14836,00	ALE USB	B=125Bd Ch=8 ChS=250Hz		USA FEMA net	USA
14841,00	CW		RGT77	RUS MIL Moscow, RVSN	RUS
14848,00	*STANAG 4285*	*B=2400Bd*		*G MIL TX St. Eval*	*G*
14850,00	USB			EGY N net	EGY
14851,00	DPRK ARQ	B=600Bd S=600Hz		DPRK E Caracas	VEN
14851,80	PACTOR II FEC	B=100Bd S=200Hz		PAK N Karachi	PAK
14855,50	THOR22	B=21,533Bd Ch=18 ChS=21,533Hz	NNX3XA		USA
14862,00	ALE USB	B=125Bd Ch=8 ChS=250Hz		TWN N net	TWN
14865,00	USB		VMD750	AUSTRAVEL safety net	AUS
14870,40	STANAG 4285	B=2400Bd		AUS MHFCS Humpty Doo	AUS
14873,00	STANAG 4285	B=2400Bd		AUS MHFCS Townsville	AUS
14874,00	STANAG 4285	B=2400Bd		AUS MHFCS Humpty Doo	AUS
14875,00	ALE USB	B=125Bd Ch=8 ChS=250Hz		G MIL net	G
14880,00	*DPRK ARQ*	*B=600Bd S=600Hz*		*DPRK E Moecow*	*RUS*
14880,00	OQPSK	B=48000Bd		HF Trading	USA
14885,00	ALE USB	B=125Bd Ch=8 ChS=250Hz		USA FEMA net	USA
14885,00	DPRK ARQ	B=600Bd S=600Hz		DPRK E Sofia	BUL
14887,00	STANAG 4481	B=75Bd S=850Hz	NAU	USA N San Juan	PTR
14892,00	ALE USB	B=125Bd Ch=8 ChS=250Hz		UN MINUSCA net	
14896,00	CIS 75-200	B=75Bd S=200Hz		RUS N	RUS
14899,00	ALE USB	B=125Bd Ch=8 ChS=250Hz		USA FEMA net	USA
14901,20	*STANAG 4285*	*B=2400Bd*		*G F Akrotiri*	*CYP*
14908,50	6CH PSK	Ch=6 ChS=400Hz			
14914,00	ALE USB	B=125Bd Ch=8 ChS=250Hz		USA Civil Air Patrol net	
14925,00	ALE USB	B=125Bd Ch=8 ChS=250Hz		FIN diplo net	FIN
14928,50	ALE USB	B=125Bd Ch=8 ChS=250Hz		USA AHRES net	USA
14935,00	ALE USB	B=125Bd Ch=8 ChS=250Hz		USA MARS net	USA
14968,00	ALE USB	B=125Bd Ch=8 ChS=250Hz		G MIL net	G
14977,00	USB		VMS469	Reids Radiodata net	AUS
14977,00	USB		VKS737	AUS 4WD net	AUS
14987,00	PACTOR II	B=200Bd S=200Hz	V8V2222	Sailmail Brunai Bay	BRU
14987,00	ALE USB	B=125Bd Ch=8 ChS=250Hz		CHN net	CHN
14996,00	*CW*		*RWM*	*Moscow TS*	*RUS*
14998,00	MHF-50 MODEM	B=54,3Bd Ch=32 ChS=64,5Hz	ZSJ	AFS N Capetown	AFS
15000,00	STANAG 4285	B=2400Bd		G MIL Cromond	G
15000,00	AM		LOL	Buenos Aires TS	ARG
15000,00	ALE USB	B=125Bd Ch=8 ChS=250Hz		USA NG net	USA
15000,00	AM		WWV	Fort Collins TS , CO	USA
15000,00	AM		WWVH	Kekaha Kauai TS	HWA
15000,00	ALE USB	B=125Bd Ch=8 ChS=250Hz		TUN F Net	

Frequency	Mode	Mode Parameter	Callsign	User	Country
15000,00	CW		BPM	Lingtong TS	CHN

15000 – 20000 kHz

Frequency	Mode	Mode Parameter	Callsign	User	Country
15000,00	STANAG 4285	B=2400Bd		G MIL Cromond	G
15000,00	AM		LOL	Buenos Aires TS	ARG
15000,00	ALE USB	B=125Bd Ch=8 ChS=250Hz		USA NG net	USA
15000,00	AM		WWV	Fort Collins TS , CO	USA
15000,00	AM		WWVH	Kekaha Kauai TS	HWA
15000,00	ALE USB	B=125Bd Ch=8 ChS=250Hz		TUN F Net	
15000,00	CW		BPM	Lingtong TS	CHN
15006,00	AM		EBC	San Fernando TS	E
15016,00	USB			USA AF Croughton	G
15016,00	USB			USA AF Sigonella	I
15016,00	USB			USA AF Lajes AFB	AZR
15016,00	USB			USA AF Diego Garcia	DGA
15016,00	USB			USA AF Puerto Rico	PTR
15016,00	USB		AFS	USA AF Offutt AFB , Omaha , NE	USA
15016,00	USB		AIE	USA AF Guam	GUM
15016,00	ALE USB	B=125Bd Ch=8 ChS=250Hz		USA AF net	
15016,00	USB			USA AF Hickam , HWA	USA
15016,00	USB			USA AF Ascension	ASC
15016,00	USB		AKA	USA AF Elmendorf AFB , Anchorage	ALS
15016,00	USB		AFA	USA AF Andrews AFB , Camp Springs , MD	USA
15016,00	USB		AIF80	USA AF Yokota AB	J
15018,00	STANAG 4481	B=50Bd S=850Hz		USA N Diega Garcia	DGA
15018,00	STANAG 4481	B=50Bd S=850Hz		USA N Crougthon	G
15025,00	HFDL	B=1800Bd	H03**	Reykjavik	ISL
15031,00	USB		CHR	CAN AF Trenton	CAN
15034,00	USB		CHR	CAN AF Trenton	CAN
15040,00	ALE USB	B=125Bd Ch=8 ChS=250Hz		G MIL net	G
15042,00	ALE USB	B=125Bd Ch=8 ChS=250Hz		I AF net	I
15043,00	ALE USB	B=125Bd Ch=8 ChS=250Hz		USA AF net	
15043,00	*ALE USB*	*B=125Bd Ch=8 ChS=250Hz*		*Task Force Vortex Camp Bondsteel*	
15047,00	MIL 188-110C	B=2400Bd		MEX N Net	MEX
15049,50	ALE USB	B=125Bd Ch=8 ChS=250Hz			
15080,00	NOKIA MSG TERMINAL	B=301,1Bd S=760Hz		MFA Helsinki	
15080,00	CW		RGT77	RUS MIL Moscow, RVSN	RUS
15091,00	MIL 188-110A SER	B=2400Bd		G MIL Akrotiri	CYP
15091,00	ALE USB	B=125Bd Ch=8 ChS=250Hz		USA AF net	USA
15094,00	ALE USB	B=125Bd Ch=8 ChS=250Hz		USA FEMA net	USA
15094,00	ALE USB	B=125Bd Ch=8 ChS=250Hz		USA SHARES net	USA
15200,00	USB			EGY N net	EGY
15377,00	*STANAG 4285*	*B=2400Bd*		*G MIL St. Eval*	*G*
15448,00	USB			RUS MIL Rostov-na-Donu	RUS
15494,20	*STANAG 4285*	*B=2400Bd*		*G MIL TX Akrotiri*	*CYP*
15500,00	*STANAG 4285*	*B=2400Bd*	*CFH*	*CAN N Halifax*	*CAN*

Frequency	Mode	Mode Parameter	Callsign	User	Country
15500,00	DPRK ARQ	B=600Bd S=600Hz		DPRK E Pretoria	AFS
15520,00	CIS 50-125	B=50Bd S=125Hz		RUS N	RUS
15520,12	T600	B=50Bd S=200Hz		RUS N Murmansk	RUS
15530,00	DPRK ARQ	B=600Bd S=600Hz		DPRK E Addis Ababa	ETH
15532,00	*CIS 12*	*B=1440Bd Ch=12 ChS=200Hz*		*RUS N HQ Moscow*	*RUS*
15555,00	DPRK ARQ	B=600Bd S=600Hz		DPRK E Moscow	RUS
15566,00	ALE USB	B=125Bd Ch=8 ChS=250Hz		USA FEMA net	USA
15586,00	CW		RCV	RUS N HQ Sevastopol	UKR
15586,00	CW		RMP	RUS N HQ Kaliningrad	RUS
15601,50	*STANAG 4481*	*B=75Bd S=850Hz*	*NPN*	*USA N Guam*	*GUM*
15602,00	ALE USB	B=125Bd Ch=8 ChS=250Hz		USA Civil Air Patrol net	USA
15602,00	CIS 3000	B=3000Bd		RUS diplo	RUS
15615,00	FAX 120/576		VMW	Wiluna M	AUS
15625,00	ALE USB	B=125Bd Ch=8 ChS=250Hz		TUN diplo net	TUN
15628,50	ALE USB	B=125Bd Ch=8 ChS=250Hz		USA SHARES net	USA
15658,00	ALE USB	B=125Bd Ch=8 ChS=250Hz		USA FEMA net	USA
15661,00	ALE LSB	B=125Bd Ch=8 ChS=250Hz		Public health net , TX	USA
15670,00	MFSK32	B=31,25Bd Ch=16 ChS=39,375Hz		VOA Radiogram , TX Edward R. Murrow , NC	USA
15670,00	MFSK8	B=7,8125Bd Ch=32 ChS=9,875Hz		VOA Radiogram , TX Edward R. Murrow , NC	USA
15708,00	ALE USB	B=125Bd Ch=8 ChS=250Hz		USA FEMA net	USA
15712,00	PACTOR III	B=100Bd S=200Hz	AFA0AM	USA MARS	USA
15723,00	ALE USB	B=125Bd Ch=8 ChS=250Hz		CHN MIL net	CHN
15723,00	CHN 8FSK	B=125Bd Ch=8 ChS=250Hz		CHN MIL	CHN
15735,50	AUSB	B=600Bd S=345Hz		POL MIL net	POL
15741,00	ALE USB	B=125Bd Ch=8 ChS=250Hz		ROU diplo net	ROU
15741,00	MIL 188-110A SER	B=2400Bd		ROU E Baku	AZE
15741,00	MIL 188-110A SER	B=2400Bd		MFA Bucharest	ROU
15778,00	T600	B=50Bd S=100Hz		RUS N	RUS
15778,00	*T600*	*B=50Bd S=200Hz*	*RCV*	*RUS N Sevastopol*	*UKR*
15803,00	*STANAG 4285*	*B=2400Bd*		*G MIL Akrotiri*	*CYP*
15805,00	ALE LSB	B=125Bd Ch=8 ChS=250Hz		CHL N net	CHL
15808,00	CW		RGT77	RUS MIL Moscow, RVSN	RUS
15812,00	*STANAG 4285*	*B=2400Bd*		*G MIL Akrotiri*	*CYP*
15812,00	CW		RIT	RUS N HQ Severomorsk	RUS
15814,00	CIS MFSK14	B=7,8Bd Ch=14 ChS=15Hz		RUS INTEL	RUS
15816,00	CIS 8181	B=81Bd S=500Hz		RUS MIL	RUS
15818,20	STANAG 4285	B=2400Bd	CFH	CAN N Halifax	CAN
15824,70	*STANAG 4481*	*B=75Bd S=850Hz*		*G MIL St. Eval*	*G*
15832,00	ALE USB	B=125Bd Ch=8 ChS=250Hz		G MIL net	G
15834,00	ALE USB	B=125Bd Ch=8 ChS=250Hz		ALG AF net	ALG
15840,00	ALE USB	B=125Bd Ch=8 ChS=250Hz		USA FEMA	USA
15845,00	USB		VNJ	Casey Base , AUS	ANT
15845,00	USB			AUS Davis Base	ANT
15845,00	USB			Macquarie Island , AUS	ANT
15845,00	USB			Mawson Base , AUS	ANT

Frequency	Mode	Mode Parameter	Callsign	User	Country
15864,00	*CODAN CHIRP*	*B=80Bd Ch=30 ChS=81Hz*		*IRN net*	*IRN*
15864,00	*CODAN*	*B=2400Bd Ch=16 ChS=112,5Hz*		*IRN net*	*IRN*
15864,00	ALE USB	B=125Bd Ch=8 ChS=250Hz		IRN net	IRN
15867,00	ALE USB	B=125Bd Ch=8 ChS=250Hz		USA COTHEN net	USA
15870,00	CIS 50-500	B=50Bd S=500Hz	REA9	RUS GOV	RUS
15877,50	ALE USB	B=125Bd Ch=8 ChS=250Hz		B N net	B
15890,50	USB			HFoZ/Radtel net	AUS
15893,00	ALE USB	B=125Bd Ch=8 ChS=250Hz		SUI F net	
15896,50	PACKET RADIO	B=1200Bd S=1000Hz	CE4TST		
15896,50	PACKET RADIO	B=1200Bd S=1000Hz	HQ2CEN		
15909,00	ALE USB	B=125Bd Ch=8 ChS=250Hz		LVA F net	
15913,00	CIS MFSK14	B=7,8Bd Ch=14 ChS=15Hz		RUS INTEL	RUS
15917,00	BAUDOT	B=50Bd S=170Hz	8WB1	IND E Belgrade	SER
15918,20	*STANAG 4285*	*B=2400Bd*	*CFH*	*CAN N Halifax*	*CAN*
15920,00	BAUDOT	B=75Bd S=850Hz	CFH	CAN F Halifax	CAN
15929,00	STANAG 4285	B=2400Bd	NSS	USA N TX Davidsonville , MD	USA
15937,00	BR 6028	B=75Bd S=170Hz Ch=7 ChS=340Hz	ZLO29	NZL N Irirangi	NZL
15938,00	STANAG 4285	B=2400Bd		G MIL St. Eval	G
15940,00	CIS 50-500	B=50Bd S=500Hz	RPD2	RUS GOV	RUS
15950,00	STANAG 4197	B=1800Bd Ch=16/39 ChS=112/56Hz		EGY N net	EGY
15950,00	USB			EGY N net	EGY
15953,50	ALE USB	B=125Bd Ch=8 ChS=250Hz		USA FBI net	USA
15959,00	*STANAG 4481*	*B=75Bd S=850Hz*	*NSS*	*USA N TX Davidsonville , MD*	*USA*
15968,00	USB		VMS469	Reids Radiodata net	AUS
15972,00	USB		VKE237	AUS HF Radio Club	AUS
15975,00	ALE USB	B=125Bd Ch=8 ChS=250Hz		USA NG net	USA
15978,00	RUS MFSK 66	B=3000Bd Ch=66 ChS=46,875Hz		MFA Moscow	RUS
15986,00	CHN OFDM 30TONE	B=1800Bd Ch=30 ChS=60Hz		CHN MIL	CHN
15992,50	USB		CAPITOLE	F AF Taverny	F
16010,00	SITOR A	B=100Bd S=170Hz		MFA Cairo	EGY
16010,00	SITOR A	B=100Bd S=170Hz		EGY E Washington	USA
16010,50	ALE USB	B=125Bd Ch=8 ChS=250Hz		AUS MIL net	AUS
16010,50	ALE USB	B=125Bd Ch=8 ChS=250Hz		POL MIL net	POL
16011,00	ALE USB	B=125Bd Ch=8 ChS=250Hz		USA DACN net	USA
16015,00	ALE USB	B=125Bd Ch=8 ChS=250Hz		CHL police net	CHL
16022,00	CODAN	B=2400Bd Ch=16 ChS=112,5Hz		MFA Cairo	EGY
16023,70	SITOR A	B=100Bd S=170Hz		EGY E Kuala Lumpur	MLA
16023,70	SITOR A	B=100Bd S=170Hz		MFA Cairo	EGY
16035,00	FAX 120/576 60/288		9VF252	KYODO Singapore	SNG
16039,50	ALE USB	B=125Bd Ch=8 ChS=250Hz		USA MARS net	USA

Frequency	Mode	Mode Parameter	Callsign	User	Country
16046,00	FMCW			Kalaeloa, HI	USA
16047,00	STANAG 4481	B=75Bd S=850Hz	NPM	USA N Lualualei	HWA
16050,00	FMCW			Jekyll Island, GA, USA	USA
16055,00	STANAG 4481	B=75Bd S=850Hz	NDT	USN Yokosuka	J
16075,00	USB			AUS	
16075,00	CCIR 493-4	B=100Bd S=170Hz		AUS	
16075,00	CCIR 493-4	B=100Bd S=170Hz		AUS	
16076,00	USB		CYRANO	F AF E-3F airplane	F
16077,00	ALE USB	B=125Bd Ch=8 ChS=250Hz		USA ACE net	USA
16078,50	ALE USB	B=125Bd Ch=8 ChS=250Hz		USA Defence Logistic Agency net	
16078,60	STANAG 4197	B=1800Bd Ch=16/39 ChS=112/56Hz			
16083,00	CIS 4FSK	B=150Bd Ch=4 ChS=4000Hz			RUS
16087,00	USB			USA N Barbers Point , HI	USA
16090,00	ALE USB	B=125Bd Ch=8 ChS=250Hz		USA FBI net	USA
16096,00	ALE USB	B=125Bd Ch=8 ChS=250Hz		MTN net	MTN
16100,00	FMCW			Ter Heijde radar site	HOL
16100,00	IRN QPSK	B=468Bd		IRN N	IRN
16104,50	USB			HFoZ/Radtel net	AUS
16105,00	ALE USB	B=125Bd Ch=8 ChS=250Hz		POL MIL net	
16107,70	*STANAG 4285*	*B=2400Bd*		*G MIL Akrotiri*	*CYP*
16110,00	STANAG	B=2400Bd	VWGZ	IND N Vishakapatnam	IND
16112,00	*T600*	*B=50Bd S=250Hz*	*RJD56*	*RUS N HQ Murmansk*	*RUS*
16112,00	ALE USB	B=125Bd Ch=8 ChS=250Hz		MTN MOI net	MTN
16112,00	FMCW			Long Point Wildlife Refuge, MA	USA
16112,00	CW		RJS	RUS NHQ Vladivostok	RUS
16114,00	CIS MFSK14	B=7,8Bd Ch=14 ChS=15Hz		RUS INTEL	RUS
16117,50	ALE USB	B=125Bd Ch=8 ChS=250Hz		USA SAC net	USA
16118,00	CIS OFDM 45	B=2400Bd Ch=45 ChS=62,5Hz		RUS diplo	RUS
16120,00	*DPRK ARQ*	*B=600Bd S=600Hz*		*DPRK E Geneva*	*SUI*
16121,00	BR 6028	B=75Bd S=170Hz Ch=8 ChS=340Hz			
16122,20	*STANAG 4285*	*B=2400Bd*	*CFH*	*CAN N Halifax*	*CAN*
16123,00	T600	B=50Bd S=250Hz	RDL	RUS N Moscow	RUS
16123,00	CIS OFDM 45	B=2400Bd Ch=45 ChS=62,5Hz		RUS	RUS
16123,00	*STANAG 4285*	*B=2400Bd*	*NSS*	*USA N TX Davidsonville , MD*	*USA*
16124,00	*CIS OFDM 45*	*B=2400Bd Ch=45 ChS=62,5Hz*		*Moscow*	*RUS*
16124,30	STANAG 4481	B=50Bd S=850Hz	NAU	USA N San Juan	PTR
16125,00	ALE USB	B=125Bd Ch=8 ChS=250Hz		TUN diplo net	TUN
16126,00	*STANAG 4285*	*B=2400Bd*		*G MIL Akrotiri*	*CYP*
16126,20	STANAG 4285	B=2400Bd	DHJ58	D N Glücksburg TX Marlow	D
16126,70	PACTOR II	B=100Bd S=200Hz	STAT16**	TUN E Riyadh	ARS
16126,70	PACTOR II	B=100Bd S=200Hz	STAT151**	MFA Tunis	TUN
16127,50	PACTOR III	B=100Bd S=200Hz		EQA N	EQA
16133,00	ALE USB	B=125Bd Ch=8 ChS=250Hz		SUI MIL net	SUI
16133,00	MIL 188-110A	B=2400Bd		SUI MIL net	SUI
16135,00	FAX 120/576		KVM70	USA CG Honululu	HWA

Frequency	Mode	Mode Parameter	Callsign	User	Country
16138,00	CIS MFSK 13	B=31,25Bd Ch=13 ChS=125Hz		MFA Moscow	RUS
16140,00	CODAR			SAKMEO Thuwal	ARS
16140,00	CODAR			SAKMEO Rabigh	ARS
16145,00	*STANAG 4285*	*B=2400Bd*		*G MIL TX Falkland*	*FLK*
16148,00	ALE USB	B=125Bd Ch=8 ChS=250Hz		G MIL net	G
16148,50	ALE USB	B=125Bd Ch=8 ChS=250Hz		CHN net	
16150,00	FMCW			MIO Laboratory Nice area	F
16150,00	FMCW			Jordan River, BC, Canada	USA
16150,00	FMCW			Race Rocks, Canada	USA
16150,00	FMCW			Lewes, DE	USA
16150,00	FMCW			Myrtle Beach State Park, NC	USA
16153,00	*CIS OFDM 128*	*B=2666Bd Ch=128 ChS=23,5Hz*		*Moscow*	*RUS*
16156,00	CW		UAL	MFA Moscow	RUS
16156,00	CIS 75-500	B=75Bd S=500Hz		MFA Moscow	RUS
16161,00	*CODAN*	*B=2400Bd Ch=16 ChS=112,5Hz*		*IRN net*	*IRN*
16161,00	*CODAN CHIRP*	*B=80Bd Ch=30 ChS=81Hz*		*IRN net*	*IRN*
16161,00	ALE USB	B=125Bd Ch=8 ChS=250Hz		IRN net	IRN
16167,00	CIS MFSK14	B=7,8Bd Ch=14 ChS=15Hz		RUS INTEL	RUS
16173,00	FMCW			Ouddorp	NLD
16173,50	ALE USB	B=125Bd Ch=8 ChS=250Hz		USA FBI net	USA
16175,00	*VFT: CIS 100-1440*	*B=100Bd S=1440Hz Ch=3 ChS=480Hz*		*near Samara*	*RUS*
16175,00	VFT: CIS 100-1440	B=100Bd S=1440Hz Ch=3 ChS=480Hz		Moscow	RUS
16175,00	FMCW			Antares	F
16180,00	WINDRM	B=1912,5Bd Ch=51 ChS=46,875Hz		CUB Intelligence Directorate	CUB
16180,00	CIS 50-500	B=50Bd S=500Hz	RNM6	RUS PTT	RUS
16183,00	ALE USB	B=125Bd Ch=8 ChS=250Hz		AUS police net	AUS
16185,00	PACTOR III	B=100Bd S=200Hz	WHX	Droop Mountain , WV	USA
16188,00	FMCW			Horseneck Beach State Reserve, MA	USA
16188,00	FMCW			Nantucket, MA	USA
16190,00	*CIS OFDM 128*	*B=2666Bd Ch=128 ChS=23,5Hz*		*Moscow*	*RUS*
16193,50	STANAG 4285	B=2400Bd		G MIL Akrotiri	CYP
16194,00	ALE USB	B=125Bd Ch=8 ChS=250Hz		UNAMA net	AFG
16200,00	FMCW			MVCO Meteorological Mast, MA	USA
16200,00	USB			EGY N net	EGY
16201,00	ALE USB	B=125Bd Ch=8 ChS=250Hz		USA FEMA net	USA
16207,00	*T600*	*B=50Bd S=200Hz*	*RJD56*	*RUS N HQ Murmansk*	*RUS*
16210,00	RFSM8000	B=2400Bd		BUL E	
16215,00	BPSK QPSK	B=1333Bd			
16215,00	BPSK QPSK	B=2000Bd			
16223,00	CW		RGT77	RUS MIL Moscow RVSN	RUS
16228,00	STANAG 4481	B=75Bd S=850Hz	NPG	USA N Dixon , CA	USA
16231,00	ALE USB	B=125Bd Ch=8 ChS=250Hz		HFoZ/Radtel net	AUS
16231,00	USB			HFoZ/Radtel net	AUS
16233,00	CW		RGT77	RUS MIL Moscow, RVSN	RUS

Frequency	Mode	Mode Parameter	Callsign	User	Country
16234,00	CHN 4+4	B=75Bd Ch=8 ChS=300/450Hz		CHN N	CHN
16234,00	*T600*	*B=50Bd S=500Hz*		*RUS N Smolensk*	*RUS*
16234,00	*T600*	*B=50Bd S=200Hz*		*RUS N Moscow*	*RUS*
16234,00	T600	B=50Bd S=200Hz		RUS N	RUS
16238,00	ALE USB	B=125Bd Ch=8 ChS=250Hz		USA FEMA net	USA
16240,00	USB			HFoZ/Radtel net	AUS
16243,00	USB			USA Army MARS	
16246,00	OQPSK	B=6500Bd			
16252,00	DPRK ARQ	B=600Bd S=600Hz		DPRK E Moscow	RUS
16252,00	DPRK ARQ	B=600Bd S=600Hz		DPRK E Rome	I
16252,00	DPRK ARQ	B=600Bd S=600Hz		DPRK E Geneve	SUI
16266,50	QPSK	B=1200Bd			RUS
16268,50	*STANAG 4481*	*B=75Bd S=850Hz*	*NPG*	*USA N Dixon , CA*	*USA*
16269,00	*CIS 12*	*B=1440Bd Ch=12 ChS=200Hz*	*RCV*	*RUS N HQ Sevastopol*	*UKR*
16271,50	CIS 12	B=1440Bd Ch=12 ChS=200Hz		RUS N ship	
16273,00	MAHRS	B=2400Bd			
16274,20	STANAG 4285	B=2400Bd		G MIL Akrotiri	CYP
16280,00	ALE USB	B=125Bd Ch=8 ChS=250Hz		TUN diplo net	TUN
16283,60	ALE USB	B=125Bd Ch=8 ChS=250Hz		USA diplo net	USA
16285,00	ALE USB	B=125Bd Ch=8 ChS=250Hz		TUN diplo net	TUN
16287,00	*STANAG 4285*	*B=2400Bd*		*G MIL TX Ascension*	*SHN*
16290,00	ALE USB	B=125Bd Ch=8 ChS=250Hz		TUN diplo net	TUN
16293,00	ALE LSB	B=125Bd Ch=8 ChS=250Hz		CHN N net	CHN
16297,00	RFSM8000	B=2400Bd		MFA Sofia	BUL
16298,50	USB			USA AF MARS	USA
16300,00	HFT MODEM		WI2XNX	10Band LLC	USA
16301,50	CW		CGA984	Essex County, ON	CAN
16316,00	*T600*	*B=50Bd S=200Hz*	*RMP*	*RUS N Kaliningrad*	*RUS*
16317,00	CODAN CHIRP	B=80Bd Ch=30 ChS=81Hz		MFA Cairo	EGY
16317,00	CODAN	B=2400Bd Ch=16 ChS=112,5Hz		MFA Cairo	EGY
16318,00	ALE USB	B=125Bd Ch=8 ChS=250Hz		ROU diplo net	ROU
16318,50	CHN 4+4	B=75Bd Ch=8 ChS=300/450Hz		CHN N	CHN
16321,00	ALE USB	B=125Bd Ch=8 ChS=250Hz		G MIL net	G
16325,00	DPRK ARQ	B=600Bd S=600Hz		DPRK E	
16325,50	6CH PSK	Ch=6 ChS=400Hz		Sao Paulo	B
16331,40	CW		M**	RUS N Magadan	RUS
16331,70	*CW*		*D****	*RUS N Sevastopol*	*UKR*
16331,80	CW		P**	RUS N Kaliningrad	RUS
16331,90	CW		S**	RUS N Severomorsk	RUS
16332,00	CW		C**	RUS N Moscow	RUS
16332,10	CW		A**	RUS N Astrakhan	
16332,20	CW		F**	RUS N Vladivostok	RUS
16332,30	CW		K**	RUS N Petropavlovsk	RUS
16332,90	CW		M**	RUS N Magadan	RUS
16333,00	CIS 3000	B=3000Bd			
16338,50	*CIS 12*	*B=1440Bd Ch=12 ChS=200Hz*		*RUS N HQ Moscow*	*RUS*
16340,00	CODAN	B=2400Bd Ch=16 ChS=112,5Hz		MFA Cairo	EGY

Frequency	Mode	Mode Parameter	Callsign	User	Country
16341,20	ALE USB	B=125Bd Ch=8 ChS=250Hz		USA FBI net	USA
16342,00	*STANAG 4285*	*B=2400Bd*		*G F Akrotiri*	*CYP*
16345,00	ALE USB	B=125Bd Ch=8 ChS=250Hz		ROU diplo net	ROU
16345,00	MIL 188-110A SER	B=2400Bd		ROU E Alger	ALG
16345,00	MIL 188-110A SER	B=2400Bd		MFA Bucharest	ROU
16346,70	SITOR A	B=100Bd S=170Hz		MFA Cairo	EGY
16348,00	ALE USB	B=125Bd Ch=8 ChS=250Hz		USA FAA net	USA
16350,00	MIL 188-110B	B=2400Bd		G	
16350,00	*ALE USB*	*B=125Bd Ch=8 ChS=250Hz*		*G MIL Net*	*G*
16357,00	*STANAG 4285*	*B=2400Bd*		*G MIL Akrotiri*	*CYP*
16357,00	ALE LSB	B=125Bd Ch=8 ChS=250Hz		ARG CG net	ARG
16357,00	ALE USB	B=125Bd Ch=8 ChS=250Hz		G MIL net	
16357,70	*2FSK*	*B=300Bd S=850Hz*		*Akrotiri*	*CYP*
16358,60	ALE USB	B=125Bd Ch=8 ChS=250Hz		USA diplo net	
16360,00	SSB			SS Ch 1601	
16363,00	SSB			SS Ch 1602	
16366,00	SSB			SS Ch 1603	
16366,00	SSB			SS Ch 1603	
16368,00	T600	B=50Bd S=100Hz			
16369,00	SSB			SS Ch 1604	
16372,00	SSB			SS Ch 1605	
16373,00	BAUDOT	B=50Bd S=400Hz	8WB4	IND E Teheran	IRN
16375,00	SSB			SS Ch 1606	
16378,00	SSB			SS Ch 1607	
16381,00	SSB			SS Ch 1608	
16384,00	SSB			SS Ch 1609	
16387,00	SSB			SS Ch 1610	
16390,00	SSB			SS Ch 1611	
16393,00	SSB			SS Ch 1612	
16396,00	SSB			SS Ch 1613	
16399,00	SSB			SS Ch 1614	
16402,00	ALE USB	B=125Bd Ch=8 ChS=250Hz		MLT N net	MLT
16402,00	SSB			SS Ch 1615	
16405,00	SSB			SS Ch 1616	
16408,00	SSB			SS Ch 1617	
16408,00	SITOR B	B=100Bd S=170Hz	SVO	Olympia R	GRC
16411,00	SSB			SS Ch 1618	
16414,00	SSB			SS Ch 1619	
16414,20	BAUDOT	B=50Bd S=230Hz	8WB4	IND E Teheran	IRN
16417,00	SSB			SS Ch 1620	
16420,00	USB			Cape Town R	AFS
16420,00	USB		CWS	URG N Montevideo	URG
16420,00	USB		NMF	USA CG Boston , MA	USA
16420,00	USB		CWF	Punta Carretas R	URG
16420,00	USB		ZLM	Taupo Maritime R	NZL
16420,00	USB		HCY	Ayora R , Santa Cruz	EQA
16420,00	USB		JNA	J CG HQ Tokyo	J
16420,00	USB		VFF	Iqaluit R	CAN
16420,00	USB		VRC	Hong Kong R	CHN
16420,00	USB		NMA	USA CG Miami , FL	USA
16420,00	USB			USA CG Honolulu	HWA
16420,00	USB			USA CG Point Reyes	USA
16420,00	USB		PNK	Jayapura R , Irian Jaya	INS
16420,00	USB		PKM	Bitung R , Sulawesi	INS
16420,00	SSB			SS Ch 1621	

Frequency	Mode	Mode Parameter	Callsign	User	Country
16423,00	SSB			SS Ch 1622	
16426,00	SSB			SS Ch 1623	
16429,00	SSB			SS Ch 1624	
16432,00	SSB			SS Ch 1625	
16433,40	*STANAG 4285*	*B=2400Bd*		*DNK N TX Frederickshavn*	*DNK*
16435,00	SSB			SS Ch 1626	
16437,60	*STANAG 4285*	*B=2400Bd*	*FUG*	*F N Saissac*	*F*
16438,00	SSB			SS Ch 1627	
16441,00	CHN 4+4	B=75Bd Ch=8 ChS=300/450Hz			
16441,00	SSB			SS Ch 1628	
16444,00	SSB			SS Ch 1629	
16447,00	SSB			SS Ch 1630	
16448,20	STANAG 4285	B=2400Bd	IDN	I N Napoli	I
16450,00	DPRK ARQ	B=600Bd S=600Hz		DPRK E Moscow	RUS
16450,00	SSB			SS Ch 1631	
16450,00	STANAG 4285	B=2400Bd	IDN21	I N Naples	I
16453,00	SSB			SS Ch 1632	
16456,00	SSB			SS Ch 1633	
16459,00	SSB			SS Ch 1634	
16462,00	SSB			SS Ch 1635	
16465,00	SSB			SS Ch 1636	
16468,00	SSB			SS Ch 1637	
16471,00	SSB			SS Ch 1638	
16474,00	SSB			SS Ch 1639	
16477,00	SSB			SS Ch 1640	
16478,00	THALES SKYMASTER ALE	Ch=8			
16480,00	SSB			SS Ch 1641	
16483,00	SSB			SS Ch 1642	
16486,00	SSB			SS Ch 1643	
16487,00	*STANAG 4285*	*B=2400Bd*	*JWT*	*NOR N Stavanger*	*NOR*
16489,00	SSB			SS Ch 1644	
16490,00	ALE USB	B=125Bd Ch=8 ChS=250Hz		ALG police net	ALG
16492,00	SSB			SS Ch 1645	
16495,00	SSB			SS Ch 1646	
16498,00	SSB			SS Ch 1647	
16500,00	STANAG 4197	B=1800Bd Ch=16/39 ChS=112/56Hz		EGY N net	EGY
16500,00	USB			EGY N net	EGY
16501,00	SSB			SS Ch 1648	
16504,00	SSB			SS Ch 1649	
16506,00	CHN 4+4	B=75Bd Ch=8 ChS=300/450Hz		CHN MIL	CHN
16507,00	SSB			SS Ch 1650	
16510,00	STANAG 4197	B=1800Bd Ch=16/39 ChS=112/56Hz			
16510,00	SSB			SS Ch 1651	
16513,00	SSB			SS Ch 1652	
16513,20	*STANAG 4285*	*B=2400Bd*		*NOR N Oslo*	*NOR*
16513,20	*STANAG 4285*	*B=2400Bd*		*NOR N TX Stavanger*	*NOR*
16516,00	SSB			SS Ch 1653	
16519,00	SSB			SS Ch 1654	
16520,00	ALE USB	B=125Bd Ch=8 ChS=250Hz		ALG MIL net	ALG
16522,00	SSB			SS Ch 1655	
16522,00	STANAG 4481	B=600Bd S=600Hz		AUS MIL	AUS
16525,00	SSB			SS Ch 1656	
16528,00	USB		VMW	Wiluna M	AUS
16528,00	USB		L2A	ARG N Buenos Aires	ARG

Frequency	Mode	Mode Parameter	Callsign	User	Country
16528,00	USB		L3A	ARG N Comodoro Rivadavia	ARG
16528,00	USB		L2A	ARG N Mar del Plata	ARG
16528,00	PACTOR III	B=100Bd S=200Hz	PYB45	Sao Paulo	B
16528,00	SSB			simplex frequency ship - ship	
16528,00	USB		WQCN860	Ocean-Pro R Naples , FL	USA
16529,40	ALE USB	B=125Bd Ch=8 ChS=250Hz		TUN N net	TUN
16531,00	SSB			simplex frequency ship - ship	
16531,00	USB		ZLM	Taupo Maritime R	NZL
16531,00	USB		VAX498	South Bound II R Burlington , ONT	CAN
16534,00	SSB			simplex frequency ship - ship	
16537,00	SSB			Liepaja R	LVA
16537,00	USB		CBV	Valpariso R	CHL
16537,00	USB		PNK	Jayapura R , Irian Jaya	INS
16537,00	USB		PKM	Bitung R , Sulawesi	INS
16537,00	SSB			simplex frequency ship - ship	
16540,00	SSB			simplex frequency ship - ship	
16543,00	SSB			simplex frequency ship - ship	
16543,00	USB			F sailor net	
16544,00	ALE USB	B=125Bd Ch=8 ChS=250Hz		CLM N net	CLM
16546,00	USB		VMC	Charleville M	AUS
16546,00	SSB			simplex frequency ship - ship	
16548,00	PACTOR III	B=100Bd S=200Hz	PYB45	Sao Paulo	B
16549,20	*STANAG 4285*	*B=2400Bd*		*G MIL TX Akrotiri*	*CYP*
16552,50	MIL 188-110A SER	B=2400Bd		AUS MHFCS North West Cape	AUS
16553,00	J OFDM 30+2	B=1500Bd Ch=32 ChS=25Hz		J N broadcast	J
16553,20	*STANAG 4481*	*B=100Bd S=850Hz*		*GRC MIL Crete*	*GRC*
16554,00	KNL MODEM	B=1500Bd Ch=1		Maritme Nesh Network Management	
16555,00	*MIL 188-110B SER*	*B=2400Bd*		*POL MIL*	*AFG*
16557,10	FAX 120/576			Beijing M	CHN
16557,20	*STANAG 4285*	*B=2400Bd*	EBA	*E N Madrid*	*E*
16559,00	FAX 120/576		XSG	Shanghai R	CHN
16560,70	*STANAG 4481*	*B=75Bd S=850Hz*		*G F Inskip*	*G*
16562,50	FAX 120/576		XSQ	Guangzhou M	CHN
16563,50	PACTOR III	B=100Bd S=200Hz	ZKN2SM	Sailmail Niue	NIU
16568,20	MIL 188-110B	B=2400Bd	PBB	NLD N Den Helder TX Flevo	HOL
16575,70	STANAG 4285	B=2400Bd	IDR	I N Rome	I
16576,00	USB		PBB	NLD N Den Helder	HOL
16576,00	USB		PBC	NLD N Goeree Island	HOL
16586,00	CHN 4+4	B=75Bd Ch=8 ChS=300/450Hz		CHN MIL	
16588,00	*STANAG 4285*	*B=2400Bd*		*G MIL Akrotiri*	*CYP*
16603,00	CIS 12	B=1440Bd Ch=12 ChS=200Hz	RLD69	RUS N ship GEORGY POBEDONOSETS BDK 45	RUS
16603,00	CIS 12	B=1440Bd Ch=12 ChS=200Hz	RIT	RUS N HQ Severomorsk	RUS
16606,00	*STANAG 4285*	*B=2400Bd*		*Gdansk*	*POL*
16606,00	*MIL 188-110B SER*	*B=2400Bd*		*G N ship near Tallinn*	*EST*
16606,00	*MIL 188-110B SER*	*B=2400Bd*		*G F St. Eval*	*G*
16608,50	.			CS Ch 6	WW
16608,60	*STANAG 4285*	*B=2400Bd*		*G MIL TX Akrotiri*	*CYP*
16615,00	CIS 12	B=1440Bd Ch=12 ChS=200Hz	RLD69	RUS N ship GEORGY POBEDONOSETS BDK 45	RUS
16615,00	CIS 12	B=1440Bd Ch=12 ChS=200Hz	RIT	RUS N HQ Severomorsk	RUS
16626,00	ALE USB	B=125Bd Ch=8 ChS=250Hz		IRN net	IRN
16626,00	*CODAN*	*B=2400Bd Ch=16 ChS=112,5Hz*		*IRN net*	*IRN*
16626,00	*CODAN CHIRP*	*B=80Bd Ch=30 ChS=81Hz*		*IRN net*	*IRN*

Frequency	Mode	Mode Parameter	Callsign	User	Country
16641,00	CW		JFG	Shizuokaken Gyogyo R	J
16662,00	BPSK	B=4800Bd		KNL Networks	FNL
16662,00	BPSK	B=9600Bd		KNL Networks	FNL
16662,00	BPSK	B=19200Bd		KNL Networks	FNL
16663,00	*STANAG 4285*	*B=2400Bd*		*NOR N Bergen*	*NOR*
16666,00	DPRK ARQ	B=600Bd S=600Hz		DPRK E Havana	CUB
16668,50	DPRK ARQ	B=600Bd S=600Hz		DPRK E	
16683,50	.			SS Ch 1	
16684,00	.			SS Ch 2	
16684,50	PACTOR III	B=200Bd S=200Hz	OSY	Sailmail Brugge	BEL
16684,50	.			SS Ch 3	
16685,00	.			SS Ch 4	
16685,50	.			SS Ch 5	
16686,00	SITOR A	B=100Bd S=170Hz	ZLM	Taupo Maritime R	NZL
16686,00	.			SS Ch 6	
16686,50	.			SS Ch 7	
16687,00	.			SS Ch 8	
16687,50	.			SS Ch 9	
16688,00	.			SS Ch 10	
16688,50	.			SS Ch 11	
16689,00	.			SS Ch 12	
16689,50	.			SS Ch 13	
16690,00	.			SS Ch 14	
16690,50	.			SS Ch 15	
16691,00	.			SS Ch 16	
16691,50	.			SS Ch 17	
16692,00	.			SS Ch 18	
16692,50	.			SS Ch 19	
16693,00	.			SS Ch 20	
16693,50	.			SS Ch 21	
16694,00	.			SS Ch 22	
16694,50	.			SS Ch 23	
16695,00	SITOR , CW	B=100Bd S=170Hz	JNA	J CG HQ Tokyo	J
16695,00	SITOR , CW	B=100Bd S=170Hz	VRC	Hong Kong R	CHN
16695,00	SITOR B	B=100Bd S=170Hz	ZLM	Taupo Maritime R	NZL
16695,00	SITOR , CW	B=100Bd S=170Hz	PKD	Surabaya R , Djawa	INS
16695,00	SITOR , CW	B=100Bd S=170Hz	PNK	Jayapura R , Irian Jaya	INS
16695,00	SITOR , CW	B=100Bd S=170Hz	PKM	Bitung R , Sulawesi	INS
16695,00	SITOR , CW	B=100Bd S=170Hz	PKP	Dumai R , Sumatera	INS
16695,00	.			Distress/safety frequency	WW
16695,50	.			SS Ch 25	
16696,00	.			SS Ch 26	
16696,50	.			SS Ch 27	
16697,00	.			SS Ch 28	
16697,50	.			SS Ch 29	
16698,00	.			SS Ch 30	
16698,50	.			SS Ch 31	
16699,00	.			SS Ch 32	
16699,50	.			SS Ch 33	
16700,00	.			SS Ch 34	
16700,50	.			SS Ch 35	
16701,00	.			SS Ch 36	
16701,50	.			SS Ch 37	
16702,00	.			SS Ch 38	
16702,50	.			SS Ch 39	
16703,00	.			SS Ch 40	
16703,50	.			SS Ch 41	
16704,00	.			SS Ch 42	
16704,50	.			SS Ch 43	
16705,00	.			SS Ch 44	
16705,50	.			SS Ch 45	
16706,00	.				

Frequency	Mode	Mode Parameter	Callsign	User	Country
16706,00	.			SS Ch 46	
16706,50	.			SS Ch 47	
16707,00	.			SS Ch 48	
16707,50	.			SS Ch 49	
16708,00	.			SS Ch 50	
16708,50	.			SS Ch 51	
16709,00	.			SS Ch 52	
16709,50	.			SS Ch 53	
16710,00	.			SS Ch 54	
16710,50	.			SS Ch 55	
16711,00	.			SS Ch 56	
16711,50	.			SS Ch 57	
16712,00	.			SS Ch 58	
16712,50	.			SS Ch 59	
16713,00	.			SS Ch 60	
16713,50	.			SS Ch 61	
16714,00	.			SS Ch 62	
16714,50	.			SS Ch 63	
16715,00	.			SS Ch 64	
16715,50	.			SS Ch 65	
16716,00	ALE LSB	B=125Bd Ch=8 ChS=250Hz			
16716,00	.			SS Ch 66	
16716,00	*MIL 188-110B SER*	*B=2400Bd*		*MRC MIL Beni Ansar*	*MRC*
16716,50	.			SS Ch 67	
16717,00	.			SS Ch 68	
16717,50	.			SS Ch 69	
16718,00	.			SS Ch 70	
16718,50	.			SS Ch 71	
16719,00	.			SS Ch 72	
16719,50	.			SS Ch 73	
16720,00	.			SS Ch 74	
16720,50	.			SS Ch 75	
16721,00	.			SS Ch 76	
16721,50	.			SS Ch 77	
16722,00	.			SS Ch 78	
16722,50	.			SS Ch 79	
16723,00	.			SS Ch 80	
16723,50	.			SS Ch 81	
16724,00	.			SS Ch 82	
16724,00	ALE USB	B=125Bd Ch=8 ChS=250Hz		CHN net	CHN
16724,50	.			SS Ch 83	
16725,00	.			SS Ch 84	
16725,50	.			SS Ch 85	
16726,00	.			SS Ch 86	
16726,50	.			SS Ch 87	
16727,00	.			SS Ch 88	
16727,50	.			SS Ch 89	
16728,00	.			SS Ch 90	
16728,00	CW			RUS N calling frequency	RUS
16728,50	.			SS Ch 91	
16729,00	.			SS Ch 92	
16729,50	.			SS Ch 93	
16730,00	.			SS Ch 94	
16730,50	.			SS Ch 95	
16731,00	.			SS Ch 96	
16731,50	.			SS Ch 97	
16732,00	.			SS Ch 98	
16732,50	.			SS Ch 99	
16733,00	.			SS Ch 100	
16733,50	.			SS Ch 101	

Frequency	Mode	Mode Parameter	Callsign	User	Country
16734,00	CW			SS calling frequency Ch 1	
16734,50	CW			SS calling frequency Ch 2	
16735,00	CW			SS calling frequency Ch 3	
16735,50	CW			SS calling frequency Ch 4	
16736,00	CW			SS calling frequency Ch 5	
16736,50	CW			SS calling frequency Ch 6	
16737,00	CW			SS calling frequency Ch 7	
16737,50	CW			SS calling frequency Ch 8	
16738,00	CW			SS calling frequency Ch 9	
16738,50	CW			SS calling frequency Ch 10	
16739,00	.			SS Ch 102	
16739,50	.			SS Ch 103	
16740,00	.			SS Ch 104	
16740,50	.			SS Ch 105	
16741,00	.			SS Ch 106	
16741,50	.			SS Ch 107	
16742,00	.			SS Ch 108	
16742,50	.			SS Ch 109	
16743,00	.			SS Ch 110	
16743,50	.			SS Ch 111	
16744,00	.			SS Ch 112	
16744,50	.			SS Ch 113	
16745,00	.			SS Ch 114	
16745,50	.			SS Ch 115	
16746,00	.			SS Ch 116	
16746,50	.			SS Ch 117	
16747,00	.			SS Ch 118	
16747,50	.			SS Ch 119	
16748,00	.			SS Ch 120	
16748,20	STANAG 4285	B=2400Bd	PBC	HOL N Goroe Island	HOL
16748,50	.			SS Ch 121	
16749,00	.			SS Ch 122	
16749,50	.			SS Ch 123	
16750,00	.			SS Ch 124	
16750,50	.			SS Ch 125	
16751,00	.			SS Ch 126	
16751,50	.			SS Ch 127	
16752,00	.			SS Ch 128	
16752,50	.			SS Ch 129	
16753,00	.			SS Ch 130	
16753,50	.			SS Ch 131	
16754,00	.			SS Ch 132	
16754,50	.			SS Ch 133	
16755,00	.			SS Ch 134	
16755,50	.			SS Ch 135	
16756,00	.			SS Ch 136	
16756,50	.			SS Ch 137	
16757,00	.			SS Ch 138	
16757,50	.			SS Ch 139	
16758,00	.			SS Ch 140	
16758,50	.			SS Ch 141	
16759,00	.			SS Ch 142	
16759,50	.			SS Ch 143	
16760,00	.			SS Ch 144	
16760,50	.			SS Ch 145	
16761,00	.			SS Ch 146	
16761,50	.			SS Ch 147	
16762,00	.			SS Ch 148	
16762,50	.			SS Ch 149	
16763,00	.			SS Ch 150	
16763,50	.			SS Ch 151	

Frequency	Mode	Mode Parameter	Callsign	User	Country
16764,00	ALE USB	B=125Bd Ch=8 ChS=250Hz		F net	F
16764,00	.			SS Ch 152	
16764,50	.			SS Ch 153	
16765,00	.			SS Ch 154	
16765,50	.			SS Ch 155	
16766,00	.			SS Ch 156	
16766,50	.			SS Ch 157	
16767,00	.			SS Ch 158	
16767,50	.			SS Ch 159	
16768,00	.			SS Ch 160	
16768,50	.			SS Ch 161	
16769,00	.			SS Ch 162	
16769,50	.			SS Ch 163	
16770,00	.			SS Ch 164	
16770,50	.			SS Ch 165	
16771,00	.			SS Ch 166	
16771,50	.			SS Ch 167	
16772,00	.			SS Ch 168	
16772,50	.			SS Ch 169	
16773,00	.			SS Ch 170	
16773,50	.			SS Ch 171	
16774,00	.			SS Ch 172	
16774,50	.			SS Ch 173	
16775,00	.			SS Ch 174	
16775,50	.			SS Ch 175	
16776,00	.			SS Ch 176	
16776,50	.			SS Ch 177	
16777,00	.			SS Ch 178	
16777,50	.			SS Ch 179	
16778,00	.			SS Ch 180	
16778,50	.			SS Ch 181	
16779,00	.			SS Ch 182	
16779,50	.			SS Ch 183	
16780,00	.			SS Ch 184	
16780,50	.			SS Ch 185	
16781,00	.			SS Ch 186	
16781,50	.			SS Ch 187	
16782,00	.			SS Ch 188	
16782,50	.			SS Ch 189	
16783,00	.			SS Ch 190	
16783,50	.			SS Ch 191	
16784,00	.			SS Ch 192	
16784,50	.			SS Ch 193	
16785,00	.			SS Ch 1	
16785,50	PACTOR III	B=100Bd S=200Hz	FOHXM	Sailmail Manihi Atoll	OCE
16785,50	.			SS Ch 2	
16786,00	PACTOR II	B=200Bd S=200Hz	V8V2222	Sailmail Brunai Bay	BRU
16786,00	.			SS Ch 3	
16786,50	.			SS Ch 4	
16787,00	.			SS Ch 5	
16787,50	.			SS Ch 6	
16788,00	.			SS Ch 7	
16788,50	.			SS Ch 8	
16789,00	.			SS Ch 9	
16789,50	*MIL 188-110B 39TONE*	*B=2400Bd Ch=39 ChS=56,25Hz*		*MRC MIL Marakesch*	*MRC*
16789,50	.			SS Ch 10	
16790,00	.			SS Ch 11	
16790,50	.			SS Ch 12	
16791,00	.			SS Ch 13	
16791,50	.			SS Ch 14	

Frequency	Mode	Mode Parameter	Callsign	User	Country
16791,70	*STANAG 4285*	*B=2400Bd*	*IDR*	*I N TX Rome*	*I*
16792,00	.			SS Ch 15	
16792,50	.			SS Ch 16	
16793,00	.			SS Ch 17	
16793,50	.			SS Ch 18	
16794,00	.			SS Ch 19	
16794,50	.			SS Ch 20	
16795,00	.			SS Ch 21	
16795,50	.			SS Ch 22	
16796,00	.			SS Ch 23	
16796,00	PACTOR I FEC	B=100Bd S=200Hz		B N net	B
16796,50	.			SS Ch 24	
16797,00	.			SS Ch 25	
16797,50	.			SS Ch 26	
16798,00	.			SS Ch 27	
16798,50	.			SS Ch 28	
16799,00	.			SS Ch 29	
16799,50	.			SS Ch 30	
16800,00	.			SS Ch 31	
16800,00	USB			EGY N net	EGY
16800,50	.			SS Ch 32	
16801,00	.			SS Ch 33	
16801,50	.			SS Ch 34	
16802,00	.			SS Ch 35	
16802,50	.			SS Ch 36	
16803,00	.			SS Ch 37	
16803,50	.			SS Ch 38	
16804,00	.			SS Ch 39	
16804,00	CW		NAF	Cape Prince of Wales	ALS
16804,50	DSC	B=100Bd S=170Hz	JNA	J CG HQ Tokyo	J
16804,50	USB		VFF	Iqaluit R	CAN
16804,50	.			DSC distress/safety frequency	WW
16805,00	.			SS DSC frequency	
16806,00	.			SS DSC frequency	
16806,50	.			MSI frequency	
16806,50	SITOR B	B=100Bd S=170Hz	NMC	USA CG Pt. Reyes , CA	USA
16806,50	SITOR B	B=100Bd S=170Hz	NRV	USA CG Guam	GUM
16806,50	SITOR B	B=100Bd S=170Hz	NMF	USA CG Boston , MA	USA
16807,00	.			CS Ch 1	
16807,50	.			CS Ch 2	
16808,00	SITOR , CW	B=100Bd S=170Hz	UFM3	Nevelsk R	RUS
16808,00	SITOR , CW	B=100Bd S=170Hz	XSV	Tianjin R	CHN
16808,00	*T600*	*B=50Bd S=200Hz*	*RJD56*	*RUS N HQ Murmansk*	*RUS*
16808,00	T600	B=50Bd S=500Hz		RUS N	RUS
16808,00	*T600*	*B=50Bd S=200Hz*	*RDL*	*RUS N Moscow*	*RUS*
16808,00	.			CS Ch 3	
16808,00	T600	B=50Bd S=200Hz		RUS N Smolensk	RUS
16808,50	.			CS Ch 4	
16809,00	.			CS Ch 5	
16809,50	.			CS Ch 6	
16810,00	.			CS Ch 7	
16810,00	SITOR , CW	B=100Bd S=170Hz	JCS	Choshi R	J
16810,00	SITOR , CW	B=100Bd S=170Hz	PKD	Surabaya R , Djawa	INS
16810,50	.			CS Ch 8	
16811,00	.			CS Ch 9	
16811,50	SITOR , CW	B=100Bd S=170Hz	A9M	Bahrain R	BHR
16811,50	.			CS Ch 10	
16812,00	SITOR , CW	B=100Bd S=170Hz	JNA	J CG HQ Tokyo	J
16812,00	SITOR , CW	B=100Bd S=170Hz	PNK	Jayapura R , Irian Jaya	INS
16812,00	.			CS Ch 11	
16812,50	.			CS Ch 12	
16813,00	.			CS Ch 13	

Frequency	Mode	Mode Parameter	Callsign	User	Country
16813,00	SITOR , CW	B=100Bd S=170Hz	UAT	Moscow R	RUS
16813,20	STANAG 4285	B=2400Bd	EBA	E N Madrid	E
16813,50	.			CS Ch 14	
16814,00	SITOR , CW	B=100Bd S=170Hz	PKP	Dumai R , Sumatera	INS
16814,00	.			CS Ch 15	
16814,50	.			CS Ch 16	
16815,00	SITOR , CW	B=100Bd S=170Hz	SVO614	Olympia R	GRC
16815,00	.			CS Ch 17	
16815,50	.			CS Ch 18	
16816,00	.			CS Ch 19	
16816,50	.			CS Ch 20	
16817,00	.			CS Ch 21	
16817,50	.			CS Ch 22	
16818,00	SITOR , CW	B=100Bd S=170Hz	SVO615	Olympia R	GRC
16818,00	.			CS Ch 23	
16818,50	.			CS Ch 25	
16819,00	.			CS Ch 26	
16819,50	.			CS Ch 27	
16820,00	.			CS Ch 28	
16820,50	.			CS Ch 29	
16821,00	SITOR , CW	B=100Bd S=170Hz	PKM	Bitung R , Sulawesi	INS
16821,00	.			CS Ch 30	
16821,50	.			CS Ch 31	
16822,00	.			CS Ch 32	
16822,50	SITOR , CW	B=100Bd S=170Hz	UDK2	Murmansk R	RUS
16822,50	.			CS Ch 33	
16823,00	.			CS Ch 34	
16823,50	.			CS Ch 35	
16824,00	.			CS Ch 36	
16824,50	.			CS Ch 37	
16824,70	STANAG 4285	B=2400Bd	IDR	I N Rome	I
16825,00	.			CS Ch 38	
16825,50	.			CS Ch 39	
16826,00	.			CS Ch 40	
16826,25	FAX 120/576		XSQ	Guangzhou R	CHN
16826,50	.			CS Ch 41	
16827,00	.			CS Ch 42	
16827,50	.			CS Ch 43	
16828,00	.			CS Ch 44	
16828,50	.			CS Ch 45	
16829,00	.			CS Ch 46	
16829,50	.			CS Ch 47	
16830,00	.			CS Ch 48	
16830,50	.			CS Ch 49	
16830,50	*SITOR , CW*	*B=100Bd S=170Hz*	*SVO6*	*Olympia R*	*GRC*
16831,00	.			CS Ch 50	
16831,50	.			CS Ch 51	
16832,00	.			CS Ch 52	
16832,50	.			CS Ch 53	
16833,00	.			CS Ch 54	
16833,50	.			CS Ch 55	
16833,50	SITOR , CW	B=100Bd S=170Hz	UIT		
16833,50	SITOR , CW	B=100Bd S=170Hz	UIW	Kaliningrad R	RUS
16834,00	.			CS Ch 56	
16834,50	.			CS Ch 57	
16835,00	.			CS Ch 58	
16835,50	.			CS Ch 59	
16836,00	SITOR , CW	B=100Bd S=170Hz	UFL	Vladivostok R	RUS
16836,00	.			CS Ch 60	
16836,50	.			CS Ch 61	
16836,60	ALE USB	B=125Bd Ch=8 ChS=250Hz		USA diplo net	

Frequency	Mode	Mode Parameter	Callsign	User	Country
16837,00	.			CS Ch 62	
16837,50	.			CS Ch 63	
16838,00	.			CS Ch 64	
16838,50	.			CS Ch 65	
16839,00	.			CS Ch 66	
16839,50	.			CS Ch 67	
16840,00	.			CS Ch 68	
16840,50	.			CS Ch 69	
16841,00	.			CS Ch 70	
16841,50	.			CS Ch 71	
16842,00	.			CS Ch 72	
16842,50	.			CS Ch 73	
16843,00	.			CS Ch 74	
16843,50	.			CS Ch 75	
16844,00	.			CS Ch 76	
16844,50	.			CS Ch 77	
16845,00	.			CS Ch 78	
16845,50	.			CS Ch 79	
16846,00	.			CS Ch 80	
16846,50	.			CS Ch 81	
16847,00	.			CS Ch 82	
16847,50	.			CS Ch 83	
16848,00	.			CS Ch 84	
16848,50	.			CS Ch 85	
16849,00	.			CS Ch 86	
16849,50	.			CS Ch 87	
16850,00	SITOR B	B=100Bd S=170Hz	XSQ	Guangzhou R	CHN
16850,50	.			CS Ch 89	
16851,00	.			CS Ch 90	
16851,00	SITOR A	B=100Bd S=170Hz	XSQ	Guangzhou R	CHN
16851,50	.			CS Ch 91	
16852,00	.			CS Ch 92	
16852,50	.			CS Ch 93	
16853,00	.			CS Ch 94	
16853,50	.			CS Ch 95	
16854,00	SITOR , CW	B=100Bd S=170Hz	XSQ	Guangzhou R	CHN
16854,00	.			CS Ch 96	
16854,50	.			CS Ch 97	
16855,00	.			CS Ch 98	
16855,50	.			CS Ch 99	
16856,00	*T600*	*B=50Bd S=1000Hz*		*RUS AF Kaliningrad*	*RUS*
16856,00	.			CS Ch 100	
16856,50	.			CS Ch 101	
16857,00	.			CS Ch 102	
16857,50	.			CS Ch 103	
16858,00	.			CS Ch 104	
16858,50	.			CS Ch 105	
16859,00	.			CS Ch 106	
16859,50	.			CS Ch 107	
16860,00	.			CS Ch 108	
16860,50	.			CS Ch 109	
16861,00	.			CS Ch 110	
16861,50	.			CS Ch 111	
16861,70	CW		PKD	Surabaya R , Djawa	INS
16861,70	CW		PKB	Belawan R , Sumatera	INS
16861,70	CW		PKX	Jakarta R , Jawa	INS
16862,00	.			CS Ch 112	
16862,50	.			CS Ch 113	
16863,00	.			CS Ch 114	
16863,50	.			CS Ch 115	
16864,00	.			CS Ch 116	
16864,50	.			CS Ch 117	

Frequency	Mode	Mode Parameter	Callsign	User	Country
16865,00	.			CS Ch 118	
16865,50	.			CS Ch 119	
16866,00	.			CS Ch 120	
16866,00	CW		LZL5	Bourgas R	BUL
16866,50	.			CS Ch 121	
16867,00	.			CS Ch 122	
16867,50	.			CS Ch 123	
16868,00	.			CS Ch 124	
16868,50	.			CS Ch 125	
16869,00	.			CS Ch 126	
16869,50	.			CS Ch 127	
16870,00	.			CS Ch 128	
16870,50	.			CS Ch 129	
16871,00	.			CS Ch 130	
16871,50	.			CS Ch 131	
16872,00	.			CS Ch 132	
16872,50	.			CS Ch 133	
16873,00	.			CS Ch 134	
16873,50	.			CS Ch 135	
16874,00	.			CS Ch 136	
16874,50	.			CS Ch 137	
16875,00	.			CS Ch 138	
16875,50	.			CS Ch 139	
16876,00	.			CS Ch 140	
16876,50	.			CS Ch 141	
16877,00	.			CS Ch 142	
16877,50	.			CS Ch 143	
16878,00	.			CS Ch 144	
16878,50	.			CS Ch 145	
16879,00	.			CS Ch 146	
16879,50	.			CS Ch 147	
16880,00	.			CS Ch 148	
16880,00	SITOR , CW	B=100Bd S=170Hz	XSQ	Guangzhou R	CHN
16880,50	.			CS Ch 149	
16881,00	.			CS Ch 150	
16881,50	.			CS Ch 151	
16882,00	.			CS Ch 152	
16882,50	.			CS Ch 153	
16883,00	.			CS Ch 154	
16883,50	.			CS Ch 155	
16884,00	.			CS Ch 156	
16884,50	.			CS Ch 157	
16885,00	.			CS Ch 158	
16885,50	.			CS Ch 159	
16886,00	.			CS Ch 160	
16886,00	SITOR , CW	B=100Bd S=170Hz	TAH	Istanbul R	TUR
16886,50	.			CS Ch 161	
16887,00	.			CS Ch 162	
16887,50	.			CS Ch 163	
16888,00	.			CS Ch 164	
16888,50	.			CS Ch 165	
16889,00	.			CS Ch 166	
16889,00	DPRK ARQ	B=600Bd S=600Hz		DPRK E Caracas	VEN
16889,50	.			CS Ch 167	
16890,00	.			CS Ch 168	
16890,50	.			CS Ch 169	
16891,00	.			CS Ch 170	
16891,50	.			CS Ch 171	
16892,00	.			CS Ch 172	
16892,00	SITOR , CW	B=100Bd S=170Hz	XSG	Shanghai R	CHN
16892,50	.			CS Ch 173	
16893,00	.			CS Ch 174	

Frequency	Mode	Mode Parameter	Callsign	User	Country
16893,50	.			CS Ch 175	
16894,00	.			CS Ch 176	
16894,50	.			CS Ch 177	
16895,00	.			CS Ch 178	
16895,50	.			CS Ch 179	
16896,00	.			CS Ch 180	
16896,50	.			CS Ch 181	
16897,00	.			CS Ch 182	
16897,50	.			CS Ch 183	
16898,00	.			CS Ch 184	
16898,00	SITOR , CW	B=100Bd S=170Hz	XSG	Shanghai R	CHN
16898,50	.			CS Ch 185	
16899,00	.			CS Ch 186	
16899,50	.			CS Ch 187	
16900,00	.			CS Ch 188	
16900,00	ALE USB	B=125Bd Ch=8 ChS=250Hz		UAE net	UAE
16900,50	.			CS Ch 189	
16901,00	.			CS Ch 190	
16901,50	.			CS Ch 191	
16902,00	.			CS Ch 192	
16902,50	.			CS Ch 193	
16903,00	DSC	B=100Bd S=170Hz	JNA	J CG HQ Tokyo	J
16903,00	DSC	B=100Bd S=170Hz	PKX	Jakarta R , Jawa	INS
16903,50	SITOR , CW	B=100Bd S=170Hz	UFL	Vladivostok R	RUS
16905,00	STANAG 4285	B=2400Bd	FUV	F N Djibouti	DJI
16907,00	STANAG 4285	B=2400Bd	CTA	POR N Lisbon	POR
16907,50	FAX 120/576		JFC	Misaki R	J
16907,50	FAX 120/576		JFA	Chuo Fisheries	J
16908,00	PACTOR II	B=200Bd S=200Hz	VZX	Sailmail Darawank	AUS
16910,00	CW		HLJ	Seoul R	KOR
16910,50	ALE USB	B=125Bd Ch=8 ChS=250Hz		G MIL net	G
16910,70	*STANAG 4285*	*B=2400Bd*	*FUG*	*F N Saissac*	*F*
16912,00	T600	B=50Bd S=200Hz	RJD56	RUS N HQ Murmansk	RUS
16913,20	STANAG 4285	B=2400Bd	FUX	F N Le Port	REU
16915,00	STANAG 4285	B=2400Bd	FUX	F N Le Port	REU
16921,60	PACTOR III	B=100Bd S=200Hz	WNU	Austin , TX	USA
16922,10	STANAG 4285	B=2400Bd		AUS MHFCS Townsville	AUS
16927,00	BAUDOT	B=50Bd S=170Hz	UGK5	Kaliningrad R	RUS
16930,00	CW		OSN416	BEL N Oostende	BEL
16935,00	T600	B=50Bd S=250Hz	RDL	RUS N Moscow	RUS
16938,00	CW		VTG8	IND N Mumbai	IND
16941,00	STANAG 4285	B=2400Bd	VWGZ	IND N Vishakapatnam	IND
16950,00	CW		9MB6	MLA N Georgtown , Penang Island	MLA
16951,50	*STANAG 4285*	*B=2400Bd*	*6WW*	*F N Dakar*	*SEN*
16953,00	MIL 188-110A SER	B=2400Bd		AUS MHFCS North West Cape	AUS
16953,00	*STANAG 4285*	*B=2400Bd*		*AUS MHFCS North West Cape*	*AUS*
16953,00	*STANAG 4285*	*B=2400Bd*		*AUS MHFCS Humpty Doo*	*AUS*
16956,00	CW		PZN5	Paramaribo R	SUR
16957,80	STANAG 4285	B=2400Bd	FUJ	F N Noumea	NCL
16958,50	*CIS 12*	*B=1440Bd Ch=12 ChS=200Hz*		*Area of Smolensk*	*RUS*
16961,50	*STANAG 4285*	*B=2400Bd*	*FUF*	*F N Fort de France*	*MRT*
16966,00	*F N FSK*	*B=50Bd S=850Hz Ch=1*	*FUG*	*F N Saissac*	*F*
16966,00	CW		ZRQ	AFS N Silvermine	AFS
16966,20	*STANAG 4285*	*B=2400Bd*	*FUG*	*F N Saissac*	*F*
16971,00	FAX 60/120/576		JJC	Tokyo M	J
16976,00	PACTOR I FEC	B=100Bd S=200Hz	PWZ33	B N Rio de Janeiro	B
16977,00	CW		JFG	Shizuokaken Gyogyo R	J
16977,00	CW		JFM	Muroto R	J

Frequency	Mode	Mode Parameter	Callsign	User	Country
16984,00	PACTOR FEC	B=100Bd S=200Hz		B N	B
16986,00	*BAUDOT*	*B=75Bd S=850Hz*	*CTP*	*POR N Lisbon*	*POR*
16990,00	CW		HLO	Seoul R	KOR
17007,00	MIL 188-110A SER	B=2400Bd		AUS MHFCS North West Cape	AUS
17014,00	CW		JFW	Fukushima R	J
17015,70	CW		S**	RUS N Severomorsk	RUS
17016,00	CW		C**	RUS N Moscow	RUS
17016,80	CW		KPH	San Francisco R , CA (Bolinas/Port Reyes)	USA
17020,00	*BAUDOT*	*B=50Bd S=170Hz*	*UDK2*	*Murmansk R*	*RUS*
17021,20	STANAG 4285	B=2400Bd Ch=1	JWT	NOR N Stavanger	NOR
17024,50	*CIS 12*	*B=1440Bd Ch=12 ChS=200Hz*		*RUS N HQ Moscow*	*RUS*
17029,00	*STANAG 4285*	*B=2400Bd*		*G MIL Crimond*	*G*
17040,00	T600	B=50Bd S=200Hz	RJS	RUS N HQ Vladivostok	RUS
17045,00	SITOR , CW	B=100Bd S=170Hz	UFH	Petrapavlovsk-Kamchatskii R	RUS
17055,00	CHN 4+4	B=75Bd Ch=8 ChS=300/450Hz		CHN MIL	CHN
17055,00	ALE USB	B=125Bd Ch=8 ChS=250Hz		POL MIL net	POL
17060,60	STANAG 4285	B=2400Bd	FUX	F N Le Port	REU
17063,50	*STANAG 4285*	*B=2400Bd*	*FUG*	*F N Saissac*	*F*
17064,00	*STANAG 4285*	*B=2400Bd*	*6WW*	*F N Dakar*	*SEN*
17069,00	FAX 120/576 60/288		JJC	Tokyo M	J
17074,40	CW		PNK	Jayapura R , Irian Jaya	INS
17082,20	*STANAG 4285*	*B=2400Bd*	*CFH*	*CAN N Halifax*	*CAN*
17097,60	STANAG 4285	B=2400Bd	FUM	F N Papeete , Tahiti	OCE
17102,00	T600	B=50Bd S=250Hz	RIT	RUS N HQ Severomorsk	RUS
17103,00	PACTOR III	B=100Bd S=200Hz	9Z4DH	Sailmail Chaguaramas	TRD
17103,20	CW		XSG	Shanghai R	CHN
17106,70	STANAG 4285	B=2400Bd	FUF	F N Fort de France	MRT
17115,00	CW		LZS32	Burgas R	BUL
17116,00	*STANAG 4285*	*B=2400Bd*		*AUS MHFCS Humpty Doo*	*AUS*
17117,00	*MIL 188-110B SER*	*B=2400Bd*		*AUS MHFCS Humpty Doo*	*AUS*
17117,40	STANAG 4285	B=2400Bd	FUG	F N Saissac	F
17117,60	BAUDOT	B=75Bd S=850Hz	PBB	NLD N Den Helder	NLD
17121,00	*STANAG 4285*	*B=2400Bd*	*FUG*	*F N Saissac*	*F*
17125,20	*STANAG 4285*	*B=2400Bd*		*AUS MHFCS North West Cape*	*AUS*
17125,50	MIL 188-110A SER	B=2400Bd		AUS MHFCS North West Cape	AUS
17126,50	MIL 188-110A SER	B=2400Bd		NZL MIL Waiouru	NZL
17127,20	STANAG 4285	B=2400Bd	CTA	POR N Lisbon	POR
17130,00	CW		HLW	Seoul R	KOR
17130,00	*STANAG 4285*	*B=2400Bd*		*G F St. Eval*	*G*
17132,00	CW		ZSJ6	AFS N Capetown	AFS
17146,40	FAX 120/576		NMG	USA CG New Orleans , LA	USA
17146,40	FAX 120/576		CBV	Valpariso R	CHL
17151,20	FAX 120/576		NMC	USA CG Pt. Reyes , CA	USA
17156,00	ALE USB	B=125Bd Ch=8 ChS=250Hz		IRN net	IRN
17180,00	*STANAG 4285*	*B=2400Bd*	*FUG*	*F N Saissac*	*F*
17184,80	CW		PKA	Sabang R , We	INS
17184,80	CW		PKE	Amboina R , Ceram	INS
17184,80	CW		PKP	Dumai R , Sumatera	INS
17188,00	*STANAG 4285*	*B=2400Bd*	*FUG*	*F N Saissac*	*F*
17198,50	PACTOR III	B=100Bd S=200Hz	DZO37	Manila R	PHL
17201,70	*STANAG 4285*	*B=2400Bd*		*G F Akrotiri*	*CYP*
17217,10	STANAG 4285	B=2400Bd		POR N Lisbon	POR
17223,40	CW		A4M	Muscat R	OMA
17231,00	CW		JFC	Misaki R	J
17233,50	*STANAG 4285*	*B=2400Bd*		*AUS MHFCS Humpty Doo*	*AUS*
17237,60	STANAG 4285	B=2400Bd	EBA	E N Madrid	E
17237,60	*STANAG 4285*	*B=2400Bd*	*EBK*	*E N Las Palmas*	*CNR*

Frequency	Mode	Mode Parameter	Callsign	User	Country
17237,70	*STANAG 4285*	*B=2400Bd*	*CTA*	*POR N Lisbon*	*POR*
17242,00	USB		LSD836	Argentina R	ARG
17242,00	MHF-50 MODEM	B=54,3Bd Ch=32 ChS=64,5Hz	ZSJ	AFS N Capetown	AFS
17242,00	SSB			CS Ch 1601	
17242,00	SSB		S7Q	Seychelles R	SEY
17245,00	SSB			CS Ch 1602	
17248,00	USB		OYR	Aasiaat R	GRL
17248,00	SSB			CS Ch 1603	
17248,00	USB		5BA	Cyprus R	CYP
17251,00	SSB			CS Ch 1604	
17251,00	SSB		SVO61	Olympia R	GRC
17254,00	USB		EQO	Now Shar R	IRN
17254,00	USB		SDJ	Stockholm R	S
17254,00	SSB			CS Ch 1605	
17254,00	SSB		EQM	Bushehr R	IRN
17257,00	USB		PPR	Rio R	B
17257,00	SSB			CS Ch 1606	
17257,00	SSB		TAH	Istanbul R	TUR
17257,00	USB		UAT	Moscow R	RUS
17260,00	USB		EQC	Amirabad R	IRN
17260,00	USB		UFL	Vladivostok R	RUS
17260,00	SSB			CS Ch 1607	
17263,00	USB			Cape Town R	AFS
17263,00	USB		SDJ	Stockholm R	S
17263,00	SSB			CS Ch 1608	
17266,00	SSB			CS Ch 1609	
17266,00	SSB		SVO62	Olympia R	GRC
17269,00	USB		SUH	Alexandria R	EGY
17269,00	USB		CWF	Punta Carretas R	URG
17269,00	USB		PNK	Jayapura R , Irian Jaya	INS
17269,00	USB		PKM	Bitung R , Sulawesi	INS
17269,00	SSB			CS Ch 1610	
17272,00	USB		9WH	Kota Kinabalu R	MLA
17272,00	USB		9MG	Penang R	MLA
17272,00	SSB			CS Ch 1611	
17272,00	USB		PPR	Rio R	B
17272,00	SSB		TAH	Istanbul R	TUR
17275,00	USB		UDK2	Murmansk R	RUS
17275,00	SSB			CS Ch 1612	
17278,00	USB		EQI	Abbas R	IRN
17278,00	CW		EQN	Khomeini R	IRN
17278,00	SSB			CS Ch 1613	
17278,00	SSB		OSU63	Oostende R	BEL
17278,00	USB		PPR	Rio R	B
17281,00	USB		SDJ	Stockholm R	S
17281,00	SSB			CS Ch 1614	
17281,00	USB		XSQ	Guangzhou R	CHN
17284,00	USB		EQI	Abbas R	IRN
17284,00	USB			Lisbon R	POR
17284,00	SSB			CS Ch 1615	
17284,00	SSB		D4A	S. Vincente de Cabo R	CPV
17287,00	USB		XSQ	Guangzhou R	CHN
17287,00	SSB			CS Ch 1616	
17287,00	SSB		EQN	Khomeini R	IRN
17287,00	ALE USB	B=125Bd Ch=8 ChS=250Hz		ARINC Urgent Link net	USA
17290,00	USB		EQI	Abbas R	IRN
17290,00	SSB			CS Ch 1617	
17290,00	SSB		SVO63	Olympia R	GRC
17293,00	USB		A9M	Bahrain R	BHR
17293,00	USB		ZBR	Bermuda Harbour R	BER

Frequency	Mode	Mode Parameter	Callsign	User	Country
17293,00	SSB			CS Ch 1618	
17293,00	SSB		TAH	Istanbul R	TUR
17296,00	USB		3BM	Mauritius R	MAU
17296,00	SSB			CS Ch 1619	
17299,00	SSB			CS Ch 1620	
17302,00	USB		LSD836	Argentina R	ARG
17302,00	USB		9MG	Penang R	MLA
17302,00	SSB			CS Ch 1621	
17302,00	SSB		4PB	Colombo R	CLN
17302,00	SSB		A7D	Doha R	QAT
17302,00	USB		PPR	Rio R	B
17305,00	USB		SVO64	Olympia R	GRC
17305,00	USB		YQI	Constanta R	ROU
17305,00	SSB			CS Ch 1622	
17308,00	SSB			CS Ch 1623	
17311,00	USB		KLB	Seattle R , WA	USA
17311,00	SSB			CS Ch 1624	
17311,00	USB		UIW	Kaliningrad R	RUS
17314,00	USB		EQJ	Chahbahar R	IRN
17314,00	USB		NMC	USCG San Francisco , CA	USA
17314,00	USB		NMN	USA CG Portsmouth , VA	USA
17314,00	USB		NMA	USA CG Miami , FL	USA
17314,00	USB		NMF	USA CG Boston , MA	USA
17314,00	USB			USA CG Honolulu	HWA
17314,00	USB			USA CG Point Reyes	USA
17314,00	SSB			CS Ch 1625	
17314,00	SSB		SVO65	Olympia R	GRC
17317,00	SSB			CS Ch 1626	
17317,00	SSB		SVO66	Olympia R	GRC
17320,00	SSB			CS Ch 1627	
17323,00	USB		UFM3	Nevelsk R	RUS
17323,00	SSB			CS Ch 1628	
17326,00	USB		EQJ	Chahbahar R	IRN
17326,00	USB		STP	Port Sudan R	SDN
17326,00	SSB			CS Ch 1629	
17326,00	SSB		SVO67	Olympia R	GRC
17329,00	SSB			CS Ch 1630	
17332,00	USB		LZW	Varna R	BUL
17332,00	USB		CBV	Valpariso R	CHL
17332,00	USB		XSU	Yantai R	CHN
17332,00	SSB			CS Ch 1631	
17335,00	SSB			CS Ch 1632	
17338,00	USB			Cape Town R	AFS
17338,00	SSB			CS Ch 1633	
17341,00	USB		HLS	Seoul R	KOR
17341,00	SSB		TUA	Abidjan R	CTI
17341,00	SSB			CS Ch 1634	
17341,00	USB		SVO68	Olympia R	GRC
17344,00	USB		YKM7	Lattakia R	SYR
17344,00	SSB			CS Ch 1635	
17344,00	SSB		D4A	S. Vincente de Cabo R	CPV
17344,00	SSB		LZW	Varna R	BUL
17347,00	SSB			CS Ch 1636	
17350,00	USB		HLS	Seoul R	KOR
17350,00	SSB			CS Ch 1637	
17353,00	SSB			CS Ch 1638	
17353,00	SSB		CNP	Casablanca R	MRC
17356,00	SSB			CS Ch 1639	
17359,00	USB		ESA	Tallin R	EST
17359,00	USB		CBV	Valpariso R	CHL
17359,00	SSB			CS Ch 1640	
17359,00	SSB		SVO69	Olympia R	GRC

Frequency	Mode	Mode Parameter	Callsign	User	Country
17361,00	*CODAN CHIRP*	*B=80Bd Ch=30 ChS=81Hz*		*IRN net*	*IRN*
17361,00	*ALE USB*	*B=125Bd Ch=8 ChS=250Hz*		*IRN net*	*IRN*
17361,00	*CODAN*	*B=2400Bd Ch=16 ChS=112,5Hz*		*IRN net*	*IRN*
17362,00	USB		SDJ	Stockholm R	S
17362,00	SSB			CS Ch 1641	
17365,00	SSB			CS Ch 1642	
17368,00	SSB			CS Ch 1643	
17371,00	SSB			CS Ch 1644	
17374,00	USB		LSD836	Argentina R	ARG
17374,00	USB		A4M	Muscat R	OMA
17374,00	SSB			CS Ch 1645	
17375,00	IRN N QPSK ADAPTIVE	B=468Bd		IRN N	
17377,00	USB		XVG	Hai Phong R	VTN
17377,00	SSB			CS Ch 1646	
17380,00	IRN N PSK	B=936Bd		IRN	IRN
17380,00	SSB			CS Ch 1647	
17380,00	8PSK	B=2400Bd			
17383,00	SSB			CS Ch 1648	
17386,00	SSB			CS Ch 1649	
17386,00	SSB		LZW	Varna R	BUL
17386,00	USB		PPR	Rio R	B
17388,50	CHN 4+4	B=75Bd Ch=8 ChS=300/450Hz		CHN MIL Red Sea	
17389,00	USB		SVO611	Olympia R	GRC
17389,00	SSB			CS Ch 1650	
17392,00	USB		XSV	Tianjin R	CHN
17392,00	SSB			CS Ch 1651	
17395,00	USB		JNA	J CG HQ Tokyo	J
17395,00	SSB			CS Ch 1652	
17398,00	USB		XSQ	Guangzhou R	CHN
17398,00	SSB			CS Ch 1653	
17398,20	*STANAG 4285*	*B=2400Bd Ch=1*		*G F Akrotiri*	*CYP*
17401,00	USB		JNA	J CG HQ Tokyo	J
17401,00	SSB			CS Ch 1654	
17404,00	USB		SXE	Aspropyrgos Attikis R	GRC
17404,00	USB		SVO612	Olympia R	GRC
17404,00	USB			Peiraias CG R	GRC
17404,00	USB		CBV	Valpariso R	CHL
17404,00	USB		UUT	Odessa R	UKR
17404,00	SSB			CS Ch 1655	
17405,00	*STANAG 4285*	*B=2400Bd Ch=1*		*G MIL St. Eval*	*G*
17407,00	USB		XSG	Shanghai R	CHN
17407,00	SSB			CS Ch 1656	
17411,00	ALE USB	B=125Bd Ch=8 ChS=250Hz		CHL disaster control net	CHL
17412,00	ALE USB	B=125Bd Ch=8 ChS=250Hz		USA Civil Air Patrol net	USA
17413,50	DPRK FSK	B=600Bd S=600Hz		DPRK diplo net	
17415,00	*STANAG 4285*	*B=2400Bd*		*G MIL Akrotiri*	*CYP*
17415,00	ALE USB	B=125Bd Ch=8 ChS=250Hz		USA CAP	USA
17418,00	ALE USB	B=125Bd Ch=8 ChS=250Hz		FIN diplo net	FIN
17418,00	SKYOFDM	B=64Bd Ch=22 ChS=86Hz		MFA Helsinki	FIN
17418,00	MIL 188-110A SER	B=2400Bd		AUS MHFCS North West Cape	AUS
17420,00	SITOR A	B=100Bd S=170Hz		MFA Cairo	EGY
17420,00	SITOR A	B=100Bd S=170Hz		EGY E Dakar	SEN
17420,50	STANAG 4285	B=2400Bd Ch=1		I MIL attache net	I

Frequency	Mode	Mode Parameter	Callsign	User	Country
17423,00	4FSK	B=250Bd Ch=4 ChS=500Hz			
17423,00	4FSK	B=160Bd Ch=4 ChS=320Hz			
17423,00	*CIS 50-500*	*B=50Bd S=500Hz*		*RUS MIL Moscow*	*RUS*
17426,00	ALE USB	B=125Bd Ch=8 ChS=250Hz		CHL DC net	CHL
17430,00	FAX 120/576 60/288		9VF235	KYODO Singapore	SNG
17430,00	CIS MFSK 11	B=125Bd Ch=11 ChS=250Hz			RUS
17432,00	*STANAG 4285*	*B=2400Bd*		*G MIL Akrotiri*	*CYP*
17435,00	ALE USB	B=125Bd Ch=8 ChS=250Hz		ROU diplo net	ROU
17440,00	RFSM8000	B=2400Bd		BUL E	
17450,00	RADAR	Ch=1		CAN Meteor Orbit Radar CMOR	CAN
17450,00				CAN Meteor Orbit Radar	CAN
17455,00	RFSM8000	B=2400Bd		MFA Sofia	BUL
17458,50	ALE USB	B=125Bd Ch=8 ChS=250Hz		USA SHARES net	USA
17458,50	ALE USB	B=125Bd Ch=8 ChS=250Hz	5RDAAA	USA MARS	USA
17460,00	ALE USB	B=125Bd Ch=8 ChS=250Hz		FIN diplo net	FIN
17460,00	*T600*	*B=50Bd S=250Hz Ch=1*		*RUS N Moscow*	*RUS*
17460,00	RFSM8000	B=2400Bd		BUL E	
17462,00	CIS MFSK14	B=7,8Bd Ch=14 ChS=15Hz		RUS INTEL	RUS
17463,00	USB		VMD750	Austravel Safety Net	AUS
17467,00	ALE USB	B=125Bd Ch=8 ChS=250Hz		TUN diplo net	TUN
17475,00	CIS 50-500	B=50Bd S=500Hz	RQF	RUS GOV	RUS
17478,50	ALE USB	B=125Bd Ch=8 ChS=250Hz		USA NG net	USA
17479,20	*STANAG 4285*	*B=2400Bd*	*DHJ58*	*D N Glücksburg TX Marlow*	*D*
17480,00	USB			AUS Davis Base	ANT
17480,00	USB			Macquarie Island , AUS	ANT
17480,00	USB			Mawson Base , AUS	ANT
17480,00	WINDRM	B=1912,5Bd Ch=51 ChS=46,875Hz		CUB Intelligence Directorate	CUB
17480,00	RUS MFSK 66	B=3000Bd Ch=66 ChS=46,875Hz		RUS E Abidjan	CTI
17487,00	ALE USB	B=125Bd Ch=8 ChS=250Hz		USA SHARES net	USA
17489,00	ALE USB	B=125Bd Ch=8 ChS=250Hz		TUN diplo net	TUN
17501,50	ALE USB	B=125Bd Ch=8 ChS=250Hz		USA NG net	USA
17519,00	ALE USB	B=125Bd Ch=8 ChS=250Hz		USA FEMA net	USA
17537,00	CIS 50-500	B=50Bd S=500Hz		Moscow	RUS
17562,00	STANAG 4285	B=2400Bd	CTA	POR N Lisbon	POR
17615,00	CW		RCV	RUS N HQ Sevastopol	UKR
17682,70	*STANAG 4285*	*B=2400Bd*	*EBA*	*E N Madrid*	*E*
17685,00	*STANAG 4285*	*B=2400Bd*		*G MIL St. Eval*	*G*
17860,00	MFSK8	B=7,8125Bd Ch=32 ChS=9,875Hz		VOA Radiogram , TX Edward R. Murrow , NC	USA
17860,00	MFSK32	B=31,25Bd Ch=16 ChS=39,375Hz		VOA Radiogram	USA
17901,00	HFDL	B=1800Bd	H11**	Albrook	PNR
17901,00	USB			PanAM R	USA
17901,00	SSB			RDARA 12	

Frequency	Mode	Mode Parameter	Callsign	User	Country
17904,00	SSB			MWARA CEP	CEP
17904,00	SSB			MWARA CWP	CWP
17904,00	SSB			MWARA NP	NP
17904,00	SSB			MWARA SP	SP
17904,00	USB			Hong Kong ACC	HKG
17904,00	USB			Manila ACC	PHL
17904,00	USB			Naha ACC	J
17904,00	USB			Port Moseby ACC	PNG
17904,00	USB			Seoul ACC	KOR
17904,00	SSB			RDARA 4	
17904,00	SSB			San Francisco ACC , CA	USA
17904,00	SSB			Tokyo ACC	J
17904,00	USB			Taipei ACC	TWN
17907,00	SSB			Merida ACC	MEX
17907,00	SSB			MWARA CAR	CAR
17907,00	SSB			MWARA EA	EA
17907,00	SSB			MWARA SAM	SAM
17907,00	SSB			MWARA SEA	SEA
17907,00	SSB		KEA5	New York ACC	USA
17907,00	USB			Barranquilla	CLM
17907,00	USB			Boyeros ACC	CUB
17907,00	USB			New York ACC	USA
17907,00	USB			Panama ACC	PNR
17907,00	USB			Piarco ACC	TRD
17907,00	USB			Cayenne ACC	GUF
17907,00	USB			Georgetown ACC	GUY
17907,00	USB			Maiquetia ACC	VEN
17907,00	USB			Paramaribo ACC	SUR
17907,00	USB			Brasilia ACC	B
17907,00	USB			Bogota ACC	CLM
17907,00	USB			Iquitos ACC	PRU
17907,00	USB			Leticia ACC	CLM
17907,00	USB			Manaus ACC	B
17907,00	USB			Rio de Janeiro ACC	B
17907,00	USB			Campo Grande ACC	B
17907,00	USB			Recife ACC	B
17907,00	USB			Porto Velho ACC	B
17907,00	USB			Lima ACC	PRU
17907,00	USB			Quito ACC	EQA
17907,00	USB			Buenos Aires ACC	ARG
17907,00	USB			Asuncion ACC	PRG
17907,00	USB			La Paz ACC	BOL
17907,00	USB			Montevideo ACC	URG
17907,00	USB			Porto Alegre ACC	B
17907,00	USB			Salvador ACC	B
17907,00	USB			Santa Cruz ACC	BOL
17907,00	USB			Antofagasta ACC	CHL
17907,00	USB			Easter Island ACC	CHL
17907,00	USB			Cordoba ACC	ARG
17907,00	USB			Puerto Montt ACC	CHL
17907,00	USB			Punta Arenas ACC	CHL
17907,00	USB			Santiago ACC	CHL
17907,00	USB			Talara ACC	PRU
17907,00	USB			Ushuaia ACC	ARG
17907,00	USB			Bali ACC	INS
17907,00	USB			Bangkok ACC	THA
17907,00	USB			Jakarta ACC	INS
17907,00	USB			Guangzhou ACC	CHN
17907,00	USB			Irkutsk ACC	RUS
17907,00	USB			Pyongyang ACC	KRE
17907,00	USB			Ulaanbaatar ACC	MNG
17907,00	USB			Singapore ACC	SNG

Frequency	Mode	Mode Parameter	Callsign	User	Country
17907,00	USB			Colombo ACC	CLN
17907,00	USB			Brisbane ACC	AUS
17907,00	USB			Calcutta ACC	IND
17907,00	USB			Hong Kong ACC	HKG
17907,00	USB			Hanoi ACC	VTN
17907,00	USB			Ho Chi Minh ACC	VTN
17907,00	USB			Dhaka ACC	BGD
17907,00	USB			Male ACC	MLD
17907,00	USB			Ujung Pandang ACC	INS
17907,00	USB			Kuala Lumpur ACC	MLA
17907,00	USB			Kathmandu ACC	NPL
17907,00	USB			Tokyo ACC	J
17907,00	USB			Vientianne ACC	LAO
17907,00	USB			Kota Kinabalu ACC	MLA
17907,00	USB			Kunming ACC	CHN
17907,00	USB			Madras ACC	IND
17907,00	USB			Yangon ACC	BRM
17907,00	USB			Manila ACC	PHL
17907,00	USB			Seoul ACC	KOR
17910,00	SSB			Carribean Bridgetown	BRB
17910,00	SSB			RDARA 10	
17910,00	SSB			RDARA 3	
17912,00	HFDL	B=1800Bd	H14**	Krasnoyarsk	RUS
17913,00	SSB			RDARA 13	
17913,00	SSB			RDARA 6G	
17916,00	USB		SDJ	Stockholm R	S
17916,00	SSB			W I	WW
17916,00	SSB			W III	WW
17916,00	HFDL	B=1800Bd	H05**	Auckland	NZL
17916,00	HFDL	B=1800Bd	H13**	Santa Cruz	BOL
17919,00	HFDL	B=1800Bd	H16**	Agana	GUM
17919,00	HFDL	B=1800Bd	H01**	San Francisco , CA	USA
17919,00	HFDL	B=1800Bd	H02**	Molokai , HI	HWA
17919,00	HFDL	B=1800Bd	H04**	Riverhead , NY	USA
17919,00	SSB			W II	WW
17919,00	SSB			W IV	WW
17919,00	HFDL	B=1800Bd	H09**	Utqiagvik , AK	ALS
17922,00	SSB			W I	WW
17922,00	SSB			W III	WW
17922,00	HFDL	B=1800Bd	H08**	Johannesburg	AFS
17925,00	SSB			W V	WW
17925,00	USB			San Francisco ACC , CA	USA
17928,00	SSB			W II	WW
17928,00	SSB			W III	WW
17928,00	SSB			W IV	WW
17928,00	HFDL	B=1800Bd	H06**	Hat Yai	THA
17928,00	HFDL	B=1800Bd	H17**	Telde , Gran Canaria	E
17931,00	SSB			W I	WW
17931,00	SSB			W V	WW
17934,00	SSB			W II	WW
17934,00	SSB			W III	WW
17934,00	HFDL	B=1800Bd	H09**	Utqiagvik , AK	ALS
17937,00	SSB			W IV	WW
17937,00	SSB			W V	WW
17938,20	STANAG 4285	B=2400Bd	EBA	E N Madrid	E
17940,00	SSB			W II	WW
17940,00	SSB			W III	WW
17940,00	USB			IBERIA Madrid ACC	E
17943,00	SSB			RDARA 6	
17946,00	USB			Shanwik ACC	IRL
17946,00	USB			Santa Maria ACC	AZR
17946,00	USB			Tokyo ACC	J

Frequency	Mode	Mode Parameter	Callsign	User	Country
17946,00	USB			San Francisco ACC , CA	USA
17946,00	SSB			Iceland ACC , Reykjavik	ISL
17946,00	SSB			MWARA NAT	NAT
17946,00	SSB		KEA5	New York ACC	USA
17946,00	SSB			RDARA 14	
17946,00	SSB		VFG	Gander ACC	CAN
17949,00	SSB			RDARA 5	
17952,00	USB			New York ACC	USA
17952,00	USB			Santa Maria ACC	AZR
17952,00	SSB			RDARA 3	
17955,00	SSB			Abidjan ACC	CTI
17955,00	SSB			Canarias ACC , Las Palmas	CNR
17955,00	SSB			Dakar ACC	SEN
17955,00	SSB			Johannesburg ACC	AFS
17955,00	SSB			MWARA SAT	SAT
17955,00	SSB			RDARA 6B	
17955,00	SSB			Sal ACC	CPV
17955,00	USB			Brasilia ACC	B
17955,00	USB			Cayenne ACC	GUF
17955,00	USB			Recife ACC	B
17955,00	USB			Rio de Janerio ACC	B
17955,00	USB			Dakar ACC	SEN
17955,00	USB			Manaus ACC	B
17955,00	USB			Paramaribo ACC	SUR
17955,00	USB			Bamako ACC	MLI
17955,00	USB			Bangui ACC	CAF
17955,00	USB			Bissau ACC	GNB
17955,00	USB			Bouake ACC	CTI
17955,00	USB			Casablanca ACC	MRC
17955,00	USB			Conakry ACC	GUI
17955,00	USB			Freetown ACC	SRL
17955,00	USB			Johannesburg ACC	AFS
17955,00	USB			Kano ACC	NIG
17955,00	USB			Niamey ACC	NGR
17955,00	USB			Nouadhibou ACC	MTN
17955,00	USB			Nouakchott ACC	MTN
17955,00	USB			Ouagadougou ACC	BFA
17955,00	USB			Roberts ACC	LBR
17958,00	HFDL	B=1800Bd	H10**	Muan	KOR
17958,00	SSB			MWARA NCA	NCA
17958,00	USB			Ivdel ACC	RUS
17958,00	USB			Khanty-Mansiysk ACC	RUS
17958,00	USB			Moscow ACC	RUS
17958,00	USB			Syktyvkar ACC	RUS
17958,00	USB			Vologda ACC	RUS
17958,00	USB			Barnaul ACC	RUS
17958,00	USB			Irkutsk ACC	RUS
17958,00	USB			Kirensk ACC	RUS
17958,00	USB			Kolpashevo ACC	RUS
17958,00	USB			Krasnoyarsk ACC	RUS
17958,00	USB			Novosibirsk ACC	RUS
17958,00	USB			Podkamennaya ACC	RUS
17958,00	USB			Surgut ACC	RUS
17958,00	USB			Chita ACC	RUS
17958,00	USB			Chulman ACC	RUS
17958,00	USB			Ekimchan ACC	RUS
17958,00	USB			Khabarovsk ACC	RUS
17958,00	USB			Pyongyang ACC	KRE
17958,00	USB			Ulaanbaatar ACC	MNG
17958,00	USB			Ulan Ude ACC	RUS
17958,00	USB			Beijing ACC	CHN
17958,00	USB			Guangzhou ACC	CHN

Frequency	Mode	Mode Parameter	Callsign	User	Country
17958,00	USB			Hailar ACC	CHN
17958,00	USB			Jinan ACC	CHN
17958,00	USB			Kunming ACC	CHN
17958,00	USB			Lanzhou ACC	CHN
17958,00	USB			Shanghai ACC	CHN
17958,00	USB			Shenyang ACC	CHN
17958,00	USB			Taegu ACC	KOR
17958,00	USB			Urumqi ACC	CHN
17958,00	USB			Wuhan ACC	CHN
17958,00	USB			Zhengzhou ACC	CHN
17961,00	SSB			MWARA AFI	AFI
17961,00	SSB			MWARA EUR	EUR
17961,00	SSB			MWARA INO	INO
17961,00	SSB			MWARA MID	MID
17961,00	SSB			Perth ACC	AUS
17961,00	USB			Algiers ACC	ALG
17961,00	USB			Kano ACC	NIG
17961,00	USB			Gao ACC	MLI
17961,00	USB			Niamey ACC	NGR
17961,00	USB			Ndjamena ACC	TCD
17961,00	USB			Tamanrasset ACC	ALG
17961,00	USB			Timimoun ACC	ALG
17961,00	USB			Tunis ACC	TUN
17961,00	USB			Tripoli ACC	LBY
17961,00	USB			Addis AbabaACC	ETH
17961,00	USB			Aden ACC	YEM
17961,00	USB			Asmara ACC	ERI
17961,00	USB			Bahrain ACC	BHR
17961,00	USB			Benghazi ACC	LBY
17961,00	USB			Mumbai ACC	IND
17961,00	USB			Bujumbura ACC	BDI
17961,00	USB			Cairo ACC	EGY
17961,00	USB			Comoros ACC	COM
17961,00	USB			Entebbe ACC	UGA
17961,00	USB			Hargeisa ACC	SOM
17961,00	USB			Djibouti ACC	DJI
17961,00	USB			Jeddah ACC	ARS
17961,00	USB			Khartoum ACC	SDN
17961,00	USB			Kigali ACC	RRW
17961,00	USB			Kisimayu ACC	SOM
17961,00	USB			Mogadishu ACC	SOM
17961,00	USB			Male ACC	MLD
17961,00	USB			Nairobi ACC	KEN
17961,00	USB			Port Sudan ACC	SDN
17961,00	USB			Sanaa ACC	YEM
17961,00	USB			Seychelles ACC	SEY
17961,00	USB			Accra ACC	GHA
17961,00	USB			Bangui ACC	CAF
17961,00	USB			Douala ACC	CME
17961,00	USB			Franceville ACC	GAB
17961,00	USB			Garoua ACC	CME
17961,00	USB			Goma ACC	UGA
17961,00	USB			Harare ACC	ZWE
17961,00	USB			Kinshasa ACC	COG
17961,00	USB			Kisangani ACC	COG
17961,00	USB			Lagos ACC	NIG
17961,00	USB			Libreville ACC	GAB
17961,00	USB			Luanda ACC	ANG
17961,00	USB			Lubumbashi ACC	COG
17961,00	USB			Lusaka ACC	ZMB
17961,00	USB			Maiduguri ACC	NIG
17961,00	USB			Maroua ACC	CME

Frequency	Mode	Mode Parameter	Callsign	User	Country
17961,00	USB			Niamtougou ACC	TGO
17961,00	USB			Pointe Noire ACC	COG
17961,00	USB			Port Gentil ACC	GAB
17961,00	USB			Roberts ACC	LBR
17961,00	USB			Sao Tome ACC	STP
17961,00	USB			Windhoek ACC	NMB
17961,00	USB			Yaounde ACC	CME
17961,00	USB			Antananarivo ACC	MDG
17961,00	USB			Beira ACC	MOZ
17961,00	USB			Brisbane ACC	AUS
17961,00	USB			Cocos Islands ACC	AUS
17961,00	USB			Colombo ACC	CLN
17961,00	USB			Lilongwe ACC	MWI
17961,00	USB			Cheannai ACC	IND
17961,00	USB			Mahajanga ACC	MDG
17961,00	USB			Male ACC	MLD
17961,00	USB			Mauritius ACC	MAU
17961,00	USB			Moroni ACC	COM
17961,00	USB			St.Denis ACC	REU
17961,00	USB			Toamasina ACC	MDG
17961,00	USB			Amman ACC	JOR
17961,00	USB			Ankara ACC	TUR
17961,00	USB			Beirut ACC	LBN
17961,00	USB			Damascus ACC	SYR
17961,00	USB			Kuwait ACC	KWT
17961,00	USB			Manama ACC	BHR
17961,00	USB			Simferopol ACC	UKR
17961,00	USB			Sanaa ACC	YEM
17961,00	USB			Tehran ACC	IRN
17961,00	USB			Tbilisi ACC	GEO
17961,00	USB			Yerevan ACC	ARM
17961,00	USB			Almaty ACC	KAZ
17961,00	USB			Ashkabad ACC	TKM
17961,00	USB			Bishkek ACC	KGZ
17961,00	USB			New Delhi ACC	IND
17961,00	USB			Dushanbe ACC	TJK
17961,00	USB			Kabul ACC	AFG
17961,00	USB			Karachi ACC	PAK
17961,00	USB			Kathmandu ACC	NPL
17961,00	USB			Lahore ACC	PAK
17961,00	USB			Muscat ACC	OMA
17961,00	USB			Samarkhand ACC	UZB
17961,00	USB			Tashkent ACC	UZB
17961,00	USB			Urumqi ACC	CHN
17961,00	USB			Aktyubinsk ACC	RUS
17961,00	USB			Moscow ACC	RUS
17961,00	USB			Arkhangelsk ACC	RUS
17961,00	USB			Kuybyshev ACC	RUS
17961,00	USB			Kzyl-Orda ACC	RUS
17961,00	USB			Uralsk ACC	RUS
17961,00	USB			Berlin ACC	D
17961,00	USB			Kiev ACC	UKR
17961,00	USB			Lvov ACC	UKR
17961,00	USB			Minsk ACC	BLR
17961,00	USB			Murmansk ACC	RUS
17961,00	USB			St. Petersburg ACC	RUS
17961,00	USB			Syktyvkar ACC	RUS
17961,00	USB			Velikiye ACC	RUS
17961,00	USB			Vologda ACC	RUS
17961,00	USB			Riga ACC	LVA
17961,00	USB			Sofia ACC	BUL
17961,00	USB			Vilnius ACC	LTU

Frequency	Mode	Mode Parameter	Callsign	User	Country
17964,00	SSB			Lockheed Corp. Los Angeles , CA	USA
17964,00	SSB			McDonnell Douglas , St. Louis , MO	USA
17964,00	SSB			RDARA 11B	
17964,00	SSB			RDARA 2	
17964,00	USB		BOING EVERETT	Boing Seattle	USA
17964,00	USB		BOING SEATTLE	Boing Seattle	USA
17967,00	SSB			RDARA 13A	
17967,00	SSB			RDARA 13B	
17967,00	SSB			RDARA 13E	
17967,00	SSB			RDARA 13F	
17967,00	SSB			RDARA 5	
17967,00	HFDL	B=1800Bd	H15**	Al Muharraq	BHR
17971,00	ISR N HYBRID MODEM	B=2400Bd		ISR N	ISR
17971,00	*STANAG 4285*	*B=2400Bd*		*G MIL Akrotiri*	*CYP*
17973,00	MFSK16	B=2400Bd Ch=16 ChS=125Hz			
17973,00	ALE USB	B=125Bd Ch=8 ChS=250Hz		USA AF NIPR net	USA
17976,00	ALE USB	B=125Bd Ch=8 ChS=250Hz		USA AF net	
17976,00	MIL 188-110B SER	B=2400Bd			
17985,00	HFDL	B=1800Bd	H03**	Reykjavik	ISL
17988,00	ALE USB	B=125Bd Ch=8 ChS=250Hz		USA N net	
17992,00	SSB			USAF Strategic Command	USA
17994,00	MIL 188-110B SER	B=2400Bd			
17994,00	USB		CHR	CAN AF Trenton	CAN
18000,00	ALE USB	B=125Bd Ch=8 ChS=250Hz		USA AF net	
18001,00	CW		4XZ	ISR N Haifa	ISR
18003,00	ALE USB	B=125Bd Ch=8 ChS=250Hz		USA AF net	
18003,00	ALE USB	B=125Bd Ch=8 ChS=250Hz		USA SAC net	USA
18006,50	ALE USB	B=125Bd Ch=8 ChS=250Hz			
18009,00	LINK 11 CLEW	B=2250Bd Ch=16 ChS=330/110/550Hz		F AF	F
18009,00	SSB			NASA launch support ships	USA
18009,00	USB		CAPITOLE	F AF Taverny	F
18012,00	USB		FDE	F AF Villacoublay	F
18012,00	USB		CHR	CAN AF Trenton	CAN
18012,00	STANAG 4285	B=2400Bd	FUF	F N Fort de France	MRT
18012,00	USB		CIRCUS VERT	F AF HQ Villacoublay	F
18021,00	ALE USB	B=125Bd Ch=8 ChS=250Hz		USA Government HF net	USA
18030,00	USB		KORSAR	RUS AF Moscow	RUS
18030,00	USB		KLARNETIST	RUS AF Migalovo/Tver	RUS
18033,50	DPRK FSK	B=600Bd S=600Hz		DPRK diplo net	
18035,00	ALE LSB	B=125Bd Ch=8 ChS=250Hz		TWN N net	TWN
18040,00	ALE USB	B=125Bd Ch=8 ChS=250Hz		F N net	
18040,00	STANAG 4285	B=2400Bd	FUJ	F N Noumea	NCL
18042,00	*STANAG 4285*	*B=2400Bd*		*G MIL TX Akrotiri*	*CYP*
18046,00	SSB			USAF Strategic Command	USA
18047,00	STANAG 4481	B=75Bd S=850Hz			

Frequency	Mode	Mode Parameter	Callsign	User	Country
18050,00	PACTOR III	B=100Bd S=200Hz	KLT	Austin , TX	USA
18053,50	STANAG 4285	B=2400Bd		I MIL Attaché net	
18060,00	FAX 120/576		VMW	Wiluna M	AUS
18063,00	ALE USB	B=125Bd Ch=8 ChS=250Hz		CHN MIL net	CHN
18095,00	LONGCHAT	B=40Bd Ch=1		LongChat net	
18097,50	CW		W1AW	ARRL morse training	USA
18098,00	PSK31	B=31Bd		PSK31 channel	
18100,00	FT8	B=5,86Bd Ch=8 ChS=5,86Hz		FT8 channel	
18102,00	JT65	B=2,69Bd Ch=65 ChS=2,69Hz		JT65 channel	
18103,00	MFSK			MFSK modes user	
18104,00	JT9	B=1,736Bd Ch=9 ChS=1,736Hz		JT9 channel	
18104,00	FT4	B=5,86Bd Ch=8 ChS=5,86Hz		FT4 channel	
18104,00	JS8	B=6,25Bd Ch=8 ChS=6,25Hz		JS8 user	
18104,60	WSPR	B=1,46Bd Ch=4 ChS=1,46Hz		WSPR net	
18106,00	ALE USB	B=125Bd Ch=8 ChS=250Hz		HFLINK network	
18106,00	BAUDOT	B=45,45Bd S=170Hz		RTTY user	
18106,40	ALE USB	B=125Bd Ch=8 ChS=250Hz			
18107,00	VARA	Ch=1		VARA chat freqeuency	
18107,00	*T600*	*B=50Bd S=100Hz*	*RDL*	*RUS N Moscow*	*RUS*
18107,00	*CIS 50-250*	*B=50Bd S=200Hz*		*RUS N Moscow*	*RUS*
18108,00	ROS USB	B=1Bd Ch=144 ChS=15,625Hz		ROS frequency	
18110,00	CW		VE8AT	Inuvik, NT	CAN
18110,00	CW		4U1UN	UN New York	USA
18110,00	CW		4X6TU	Tel Aviv	ISR
18110,00	CW		CS3B	Madeira	
18110,00	CW		OH2B	Espoo	FNL
18110,00	CW		YV5B	Caracas	VEN
18110,00	CW		W6WX	Mt. Umunhum	USA
18110,00	CW		KH6RS	Maui	HWA
18110,00	CW		ZL6B	Masterton	NZL
18110,00	CW		VK6RBP	Rolystone	AUS
18110,00	CW		JA2IGY	Mt. Asama	J
18110,00	CW		RR9O	Novosibirsk	RUS
18110,00	CW		VR2B	Hong Kong	HKG
18110,00	CW		4S7B	Colombo	CLN
18110,00	CW		ZS6DN	Pretoria	AFS
18110,00	CW		5Z4B	Kariobangi	KEN
18110,00	CW		LU4AA	Buenos Aires	ARG
18110,00	CW		OA4B	Lima	PRU
18117,50	ALE USB	B=125Bd Ch=8 ChS=250Hz		HFLINK network	
18140,00	PSK4A	B=2400Bd		CHN MIL	CHN
18150,00	USB			Wireless Institute Civil Emergency Service WICEN	AUS
18158,50	DPRK FSK	B=600Bd S=600Hz		MFA Pyongyang	KRE
18160,00	USB			Ham radio emergency frequency	R1
18162,50	ARD9800 MODEM	B=2250Bd Ch=36 ChS=62,5Hz		AOR Digital Voice net	USA
18170,00	*STANAG 4285*	*B=2400Bd Ch=1*	*JWT*	*NOR N Stavanger*	*NOR*
18171,00	ALE USB	B=125Bd Ch=8 ChS=250Hz		USA FBI net	USA
18172,00	PACTOR III	B=100Bd S=200Hz	9Z4DH	Sailmail Chaguaramas	TRD

Frequency	Mode	Mode Parameter	Callsign	User	Country
18175,00	MIL 188-110A SER	B=2400Bd		AUS MHFCS North West Cape	AUS
18178,00	ALE USB	B=125Bd Ch=8 ChS=250Hz		ARINC Urgent Link net	USA
18185,00	BR 6028	B=75Bd S=170Hz Ch=7 ChS=340Hz	ZLO29	NZL N Irirangi	NZL
18190,00	CW		CGA984	Essex County, ON	CAN
18194,40	*STANAG 4285*	*B=2400Bd*	*IDR*	*I N TX Rome*	*I*
18194,70	*STANAG 4285*	*B=2400Bd*	*IDR*	*I N TX Taranto*	*I*
18200,00	USB			EGY N net	EGY
18200,00	EGY QPSK	B=2400Bd		EGY	
18206,00	CW			INS N net	INS
18210,00	ALE USB	B=125Bd Ch=8 ChS=250Hz		G MIL net	G
18213,00	CIS MFSK14	B=7,8Bd Ch=14 ChS=15Hz		RUS INTEL	RUS
18215,00	CIS 3000	B=3000Bd		RUS diplo	RUS
18215,00	CIS 68TONE	B=50Bd Ch=68 ChS=47Hz		RUS diplo	RUS
18218,00	MIL 188-110B SER	B=2400Bd		B MIL net	B
18218,00	ALE USB	B=125Bd Ch=8 ChS=250Hz		B MIL net	B
18223,50	ALE USB	B=125Bd Ch=8 ChS=250Hz		AUT MIL net	
18230,00	*STANAG 4285*	*B=2400Bd*		*G MIL Falklands*	*FLK*
18231,00	*USB*			*I MIL Rome*	*I*
18234,00	PACTOR II	B=200Bd S=200Hz	XJN714	Sailmail Lunenburg , NS	CAN
18237,80	BAUDOT	B=75Bd S=170Hz	ZSC	Capetown R	AFS
18238,00	FAX 120/576		ZSJ	AFS N Capetown Naval	AFS
18240,00	PACTOR II	B=200Bd S=200Hz	HPPM1	Sailmail Chiriqui	PNR
18240,00	PACTOR III	B=100Bd S=200Hz	HPPM3	Sailmail Panama	PNR
18248,60	ALE USB	B=125Bd Ch=8 ChS=250Hz		USA diplo net	
18250,00	OQPSK	B=48000Bd		HF Trading Aurora , IL	USA
18252,50	*STANAG 4285*	*B=2400Bd*	*JWT*	*NOR N Stavanger*	*NOR*
18255,70	PACTOR II	B=100Bd S=200Hz	SA1QC		
18255,70	PACTOR II	B=100Bd S=200Hz	AB2BC		
18258,20	PACKET RADIO	B=300Bd S=200Hz	CE4ANK	Ankara	TUR
18264,00	PACTOR II	B=200Bd S=200Hz	KUZ533	Sailmail Honululu	HWA
18264,00	PACTOR III	B=100Bd S=200Hz	RC01	Sailmail Maputo	MOZ
18264,00	RUS DBPSK 16FSK	B=1866Bd Ch=16 ChS=175Hz		RUS INTEL	RUS
18264,00	ALE USB	B=125Bd Ch=8 ChS=250Hz		IRN net	IRN
18264,00	*CODAN CHIRP*	*B=80Bd Ch=30 ChS=81Hz*		*IRN net*	*IRN*
18264,00	*CODAN*	*B=2400Bd Ch=16 ChS=112,5Hz*		*IRN net*	*IRN*
18265,00	CIS 2CH QPSK	B=125Bd Ch=2 ChS=100Hz		RUS INTEL	RUS
18267,00	ALE LSB	B=125Bd Ch=8 ChS=250Hz		Public health net , TX	USA
18270,00	MIL 188-110B SER	B=2400Bd		CZE E Brasilia	B
18270,00	MIL 188-110B SER	B=2400Bd		MFA Prague	CZE
18270,50	SITOR , CW	B=100Bd S=170Hz	WPC	Pin Oak R	USA
18273,00	CW		RJD97	RUS N ship	RUS
18275,00	ALE USB	B=125Bd Ch=8 ChS=250Hz		ROU diplo net	ROU
18276,00	CIS 75-500	B=75Bd S=500Hz		RUS E Havana	CUB
18277,00	PACTOR III	B=100Bd S=200Hz	WHV382	Sailmail Friday Habor , WA	USA
18279,20	STANAG 4285	B=2400Bd	OVK	DNK N Aarhus	DNK
18295,00	ALE USB	B=125Bd Ch=8 ChS=250Hz		TUN diplo net	TUN
18296,00	PACTOR II	B=200Bd S=200Hz	WHV861	Sailmail San Luis Obispo , CA	USA

Frequency	Mode	Mode Parameter	Callsign	User	Country
18300,00	ALE USB	B=125Bd Ch=8 ChS=250Hz		CZE diplo net	CZE
18315,00	ALE USB	B=125Bd Ch=8 ChS=250Hz		G MIL net TX Ascension	ASC
18315,20	MIL 188-110B SER	B=2400Bd		S MIL	S
18315,20	*STANAG 4285*	*B=2400Bd*	*6WW*	*F N Dakar*	*SEN*
18320,00	PACTOR III	B=100Bd S=200Hz		MFA Prague	CZE
18320,00	PACTOR III	B=100Bd S=200Hz		CZE E Islamabad	PAK
18322,00	CW		CGA984	Essex County, ON	CAN
18322,00	OFDM	B=2400Bd Ch=121 ChS=25Hz		MFA Moscow	RUS
18322,00	QPSK	B=2400Bd		MFA Moscow	RUS
18325,00	ALE USB	B=125Bd Ch=8 ChS=250Hz		F N net	
18328,00	CW		RHO62	RUS N ship ADMIRAL VLADIMIRSKIY	RUS
18352,00	CHN 4+4	B=75Bd Ch=8 ChS=300/450Hz		CHN MIL	
18359,50	CIS 100-2000	B=100Bd S=2000Hz Ch=1		RUS MIL	RUS
18365,20	*STANAG 4285*	*B=2400Bd*	*6WW*	*F N Dakar*	*SEN*
18370,00	STANAG 4481	B=75Bd S=850Hz	NPN	USN Guam	GUM
18376,40	PACTOR II	B=200Bd S=200Hz	WPTG385	Sailmail Corpus Christi , TX	USA
18377,00	CODAN	B=2400Bd Ch=16 ChS=112,5Hz		MFA Cairo	EGY
18381,40	PACTOR II	B=200Bd S=200Hz	WPUC469	Sailmail South Daytona , FL	USA
18390,00	PACTOR III	B=100Bd S=200Hz	WQAB964	Sailmail San Diego , CA	USA
18403,50	ALE USB	B=125Bd Ch=8 ChS=250Hz		G MIL net	G
18405,20	STANAG 4285	B=2400Bd		G MIL Akrotiri	CYP
18407,00	STANAG 4197	B=1800Bd Ch=16/39 ChS=112/56Hz			
18407,00	ALE USB	B=125Bd Ch=8 ChS=250Hz		G MIL net	G
18408,70	CLOVER 2000	B=500Bd Ch=8 ChS=250Hz		MFA New Delhi	IND
18411,20	*STANAG 4285*	*B=2400Bd*	*6WW*	*F N Dakar*	*SEN*
18413,50	DPRK FSK	B=600Bd S=600Hz		DPRK diplo net	
18415,00	RFSM8000	B=2400Bd		BUL E	
18423,40	MIL 188-110B SER	B=2400Bd		S N Karlskrona	S
18440,00	PACTOR II	B=200Bd S=200Hz	HPPM1	Sailmail Chiriqui	PNR
18440,00	PACTOR III	B=100Bd S=200Hz	HPPM3	Sailmail Panama	PNR
18455,00	RFSM8000	B=2400Bd		BUL E	
18457,00	STANAG 4285	B=2400Bd		G Mil Akrotiri	CYP
18493,70	*STANAG 4285*	*B=2400Bd*	*FUV*	*F MIL Djibouti*	*DJI*
18495,00	FSK	B=1200Bd S=850Hz		F MIL Djibouti	DJI
18502,00	*OQPSK*	*B=2400Bd*		*G MIL Akrotiri*	*CYP*
18506,00	STANAG 4285	B=2400Bd		G MIL Akrotiri	CYP
18509,00	ALE USB	B=125Bd Ch=8 ChS=250Hz		G MIL net	G
18513,00	ALE USB	B=125Bd Ch=8 ChS=250Hz		USA Civil Air Patrol net	USA
18515,00	RFSM8000	B=2400Bd		MFA Sofia	BUL
18516,00	ALE USB	B=125Bd Ch=8 ChS=250Hz		USA Civil Air Patrol net	USA
18520,00	RFSM8000	B=2400Bd		BUL E Washington	USA
18531,50	*STANAG 4285*	*B=2400Bd*		*G MIL Akrotiri*	*CYP*
18542,00	ALE USB	B=125Bd Ch=8 ChS=250Hz		SAPOL Communication Infrastructure	AUS
18542,00	ALE USB	B=125Bd Ch=8 ChS=250Hz		AUS Police WA	AUS

Frequency	Mode	Mode Parameter	Callsign	User	Country
18542,00	ALE USB	B=125Bd Ch=8 ChS=250Hz		AUS Police QLD	AUS
18542,00	ALE USB	B=125Bd Ch=8 ChS=250Hz		AUS Police NSW	AUS
18542,00	ALE USB	B=125Bd Ch=8 ChS=250Hz		AUS Police NT	AUS
18545,00	MIL 188-110B SER	B=2400Bd			
18552,00	*OQPSK*	*B=2400Bd*		*G MIL Akrotiri*	*CYP*
18558,00	STANAG 4481	B=75Bd S=850Hz	NPM	USA N Lualualei	HWA
18562,00	CHN 4+4	B=75Bd Ch=8 ChS=300/450Hz	NLPV**	CHN MIL	
18562,00	FSK	B=50Bd S=500Hz			CUB
18578,00	ALE USB	B=125Bd Ch=8 ChS=250Hz		F N net	
18578,00	STANAG 4285	B=2400Bd	FUJ	F N Noumea	NCL
18580,90	ALE USB	B=125Bd Ch=8 ChS=250Hz		USA FBI net	USA
18585,00	STANAG 4285	B=2400Bd		AUS MHFCS Humpty Doo	AUS
18592,50	PACTOR III	B=100Bd S=200Hz	VXZ	Darawank R , New South Wales	AUS
18594,00	PACTOR II	B=200Bd S=200Hz	VZX	Sailmail Darawank	AUS
18594,00	ALE USB	B=125Bd Ch=8 ChS=250Hz		USA COTHEN net	USA
18594,00	SSB			USAF Strategic Command	USA
18594,00	ALE USB	B=125Bd Ch=8 ChS=250Hz		USA COTHEN net	USA
18598,40	STANAG 4285	B=2400Bd	FUJ	F N Noumea	NCL
18610,00	PACTOR II	B=200Bd S=200Hz	HPPM2	Sailmail Chiriqui	PNR
18617,20	STANAG 4285	B=2400Bd	PBB	NLD N Den Helder TX Flevo	HOL
18617,20	*STANAG 4285*	*B=2400Bd*	*PBB*	*NLD N Den Helder TX Zeewolde*	*HOL*
18618,00	PACTOR II	B=200Bd S=200Hz	KZN508	Sailmail Rockhill , SC	USA
18618,00	PACTOR IV	B=100Bd S=200Hz	KZN508	Sailmail node Rock Hill , SC	USA
18624,00	PACTOR III	B=100Bd S=200Hz	WQLI952	Sailmail Watsonville , CA	USA
18630,00	PACTOR II	B=200Bd S=200Hz	KZN508	Sailmail Rockhill , SC	USA
18630,00	PACTOR III	B=100Bd S=200Hz	RC01	Sailmail Maputo	MOZ
18633,00	*STANAG 4285*	*B=2400Bd*		*G MIL Akrotiri*	*CYP*
18634,00	ALE LSB	B=125Bd Ch=8 ChS=250Hz		CHL N net	CHL
18635,00	*STANAG 4285*	*B=2400Bd*	*6WW*	*F N Dakar*	*SEN*
18658,00	CIS 68TONE	B=50Bd Ch=68 ChS=47Hz		RUS diplo	RUS
18665,00	ALE USB	B=125Bd Ch=8 ChS=250Hz		PAK N net	
18666,00	ALE USB	B=125Bd Ch=8 ChS=250Hz		USA FBI net	USA
18673,00	ALE USB	B=125Bd Ch=8 ChS=250Hz		Queensland Deptartment of Community Safety	AUS
18689,60	ALE USB	B=125Bd Ch=8 ChS=250Hz		ARINC Urgent Link net	USA
18691,50	*STANAG 4285*	*B=2400Bd*	*IDR*	*I N TX Rome*	*I*
18696,00	T600	B=50Bd S=250Hz	RJS	RUS N HQ Vladivostok	RUS
18702,50	MIL 188-110B SER	B=2400Bd		S N Karlskrona	S
18709,00	T600	B=50Bd S=250Hz	RJS	RUS N HQ Vladivostok	RUS
18720,00	T600	B=50Bd S=250Hz	RJS	RUS N HQ Vladivostok	RUS
18722,70	BAUDOT	B=50Bd S=400Hz		MFA New Delhi	IND
18724,70	BAUDOT	B=50Bd S=400Hz		MFA New Delhi	IND
18726,00	CIS100-500	B=100Bd S=500Hz		RUS E Habana	CUB
18735,00	AUS MIL ISB MODEM	B=300Bd S=345Hz		AUS MIL	AUS
18735,50	6CH PSK	Ch=6 ChS=400Hz			
18761,00	BPSK	B=2400Bd		Fin diplo	
18780,00	SSB			SS Ch 1801	
18783,00	SSB			SS Ch 1802	
18786,00	SSB			SS Ch 1803	

Frequency	Mode	Mode Parameter	Callsign	User	Country
18789,00	SSB			SS Ch 1804	
18792,00	SSB			SS Ch 1805	
18795,00	SSB			SS Ch 1806	
18798,00	SSB			SS Ch 1807	
18800,00	USB			EGY N net	EGY
18801,00	SSB			SS Ch 1808	
18804,00	SSB			SS Ch 1809	
18807,00	SSB			SS Ch 1810	
18810,00	SSB			SS Ch 1811	
18813,00	SSB			SS Ch 1812	
18816,00	SSB			SS Ch 1813	
18819,00	SSB			SS Ch 1814	
18822,00	SSB			SS Ch 1815	
18825,00	SSB			simplex frequency ship - ship	
18828,00	SSB			simplex frequency ship - ship	
18831,00	SSB			simplex frequency ship - ship	
18834,00	T600	B=50Bd S=500Hz			
18834,00	USB		XVS	Ho Chi Minh R	VTN
18834,00	SSB			simplex frequency ship - ship	
18837,00	SSB			simplex frequency ship - ship	
18840,00	USB		WQCN860	Ocean-Pro R Naples , FL	USA
18840,00	SSB			simplex frequency ship - ship	
18843,00	USB		XVG	Hai Phong R	VTN
18843,00	SSB			simplex frequency ship - ship	
18847,00	*STANAG 4285*	*B=2400Bd*		*AUS MHFCS North West Cape*	*AUS*
18850,00	ALE USB	B=125Bd Ch=8 ChS=250Hz		CHN net	CHN
18856,50	PACTOR III	B=100Bd S=200Hz	ZKN2SM	Sailmail Niue	NIU
18870,50	.			SS Ch 1	
18871,00	.			SS Ch 2	
18871,50	.			SS Ch 3	
18871,70	CW		D**	RUS N Sevastopol	UKR
18872,00	.			SS Ch 4	
18872,00	T600	B=50Bd S=200Hz		RUS N	
18872,00	CIS 12	B=1440Bd Ch=12 ChS=200Hz		RUS MIL	
18872,00	CIS 75-200	B=75Bd S=200Hz		RUS N	
18872,50	.			SS Ch 5	
18873,00	.			SS Ch 6	
18873,50	.			SS Ch 7	
18874,00	.			SS Ch 8	
18874,50	.			SS Ch 9	
18875,00	.			SS Ch 10	
18875,00	ALE USB	B=125Bd Ch=8 ChS=250Hz		G MIL net	G
18875,50	.			SS Ch 11	
18876,00	.			SS Ch 12	
18876,50	.			SS Ch 13	
18877,00	.			SS Ch 14	
18877,50	.			SS Ch 15	
18878,00	.			SS Ch 16	
18878,50	.			SS Ch 17	
18879,00	.			SS Ch 18	
18879,50	.			SS Ch 19	
18880,00	.			SS Ch 20	
18880,00	*STANAG 4285*	*B=2400Bd*		*TUR N Ankara*	*TUR*
18880,50	.			SS Ch 21	
18881,00	.			SS Ch 22	
18881,50	.			SS Ch 23	
18882,00	.			SS Ch 24	
18882,50	.			SS Ch 25	
18883,00	.			SS Ch 26	

Frequency	Mode	Mode Parameter	Callsign	User	Country
18883,50	.			SS Ch 27	
18884,00	.			SS Ch 28	
18884,50	.			SS Ch 29	
18885,00	.			SS Ch 30	
18885,50	.			SS Ch 31	
18886,00	.			SS Ch 32	
18886,50	.			SS Ch 33	
18887,00	.			SS Ch 34	
18887,50	.			SS Ch 35	
18888,00	.			SS Ch 36	
18888,50	.			SS Ch 37	
18889,00	.			SS Ch 38	
18889,50	.			SS Ch 39	
18890,00	.			SS Ch 40	
18890,50	.			SS Ch 41	
18891,00	.			SS Ch 42	
18891,50	.			SS Ch 43	
18892,00	.			SS Ch 44	
18892,50	.			SS Ch 45	
18893,00	PACTOR II	B=200Bd S=200Hz	V8V2222	Sailmail Brunai Bay	BRU
18893,00	.			SS Ch 1	
18893,50	.			SS Ch 2	
18894,00	.			SS Ch 3	
18894,50	.			SS Ch 4	
18895,00	.			SS Ch 5	
18895,50	.			SS Ch 6	
18896,00	.			SS Ch 7	
18896,50	.			SS Ch 8	
18897,00	.			SS Ch 9	
18897,50	.			SS Ch 10	
18898,00	.			SS Ch 11	
18898,50	.			SS DSC frequency	WW
18899,00	.			SS DSC frequency	WW
18899,50	.			SS DSC frequency	WW
18935,00	SSB			MOI Alger	ALG
18938,00	PACTOR III	B=100Bd S=200Hz	NCS355	USA SHARES Aurora , CO	USA
18938,00	PACTOR III	B=100Bd S=200Hz	NCS362	USA SHARES Spanaway , WA	USA
18944,60	ALE USB	B=125Bd Ch=8 ChS=250Hz		USA diplo net	
18954,00	ALE USB	B=125Bd Ch=8 ChS=250Hz		SNG N net	SNG
18954,00	USB			SNG N net	SNG
18956,00	2FSK	B=96Bd S=500Hz		SNG N RSS Resolution	SNG
18965,00	STANAG 4285	B=2400Bd	FUM	F N Papeete , Tahiti	OCE
18976,00	CW		A4I	Muscat	OMA
18980,00	CW		7CB	INS N Belawan	INS
18981,00	CHN 4+4	B=75Bd Ch=8 ChS=300/450Hz		CHN MIL	CHN
18985,00	STANAG 4285	B=2400Bd	FUM	F N Papeete , Tahiti	OCE
19047,00	RUS DBPSK 16FSK	B=1866Bd Ch=16 ChS=175Hz		RUS INTEL	RUS
19051,70	*STANAG 4285*	*B=2400Bd*		*G MIL TX Akrotiri*	*CYP*
19056,00	CIS 75-500	B=75Bd S=500Hz		RUS	RUS
19056,00	CIS 12	B=1440Bd Ch=12 ChS=200Hz		RUS MIL Yekaterinenburg	RUS
19056,00	CIS 12	B=1440Bd Ch=12 ChS=200Hz		RUS MIL	RUS
19060,00	ALE USB	B=125Bd Ch=8 ChS=250Hz		USA FEMA net	USA
19066,50	MIL 188-110A SER	B=2400Bd		AUS MHFCS North West Cape	AUS
19073,50	ALE USB	B=125Bd Ch=8 ChS=250Hz		AUS MHFCS net	AUS

Frequency	Mode	Mode Parameter	Callsign	User	Country
19079,50	ALE USB	B=125Bd Ch=8 ChS=250Hz		USA NG net	USA
19080,00	MIL 188-110B SER	B=2400Bd			
19085,00	ALE USB	B=125Bd Ch=8 ChS=250Hz		TUN diplo net	
19102,00	KOR 2FSK	B=1200Bd S=850Hz Ch=1		KOR MIL Busan	KOR
19102,00	OQPSK	B=48000Bd		HF Trading Long Island	USA
19102,00	OQPSK	B=48000Bd	WI2XER	Skycast Services	USA
19117,00	CIS 2CH QPSK	B=125Bd Ch=2 ChS=100Hz		RUS INTEL	RUS
19117,00	RUS DBPSK 16FSK	B=1866Bd Ch=16 ChS=175Hz		RUS INTEL	RUS
19119,00	BAUDOT	B=50Bd S=500Hz			
19123,00	OFDM				
19131,00	ALE USB	B=125Bd Ch=8 ChS=250Hz		USA FBI net	USA
19160,00	CIS 50-500	B=50Bd S=500Hz		Moscow	RUS
19162,00	PSK-FSK	B=800Bd			
19177,00	ALE USB	B=125Bd Ch=8 ChS=250Hz		TWN N net	TWN
19177,00	MIL 188-110A SER	B=2400Bd		TWN N net	TWN
19190,00	PACTOR II	B=100Bd S=200Hz			
19191,00	*CODAN*	*B=2400Bd Ch=16 ChS=112,5Hz*		*IRN net*	*IRN*
19191,00	*CODAN CHIRP*	*B=80Bd Ch=30 ChS=81Hz*		*IRN net*	*IRN*
19191,00	ALE USB	B=125Bd Ch=8 ChS=250Hz		IRN net	IRN
19200,00	ALE USB	B=125Bd Ch=8 ChS=250Hz		VEN N net	VEN
19200,00	ALE USB	B=125Bd Ch=8 ChS=250Hz		ALG AF net	ALG
19200,00	USB			EGY N net	EGY
19201,00	*CW*		*RCV*	*RUS N HQ Sevastopol*	*UKR*
19203,00	ALE USB	B=125Bd Ch=8 ChS=250Hz		F N net	
19210,00	ALE USB	B=125Bd Ch=8 ChS=250Hz		USA FEMA net	USA
19210,00	*T600*	*B=50Bd S=200Hz*		*RUS N Moscow*	*RUS*
19215,00	ALE USB	B=125Bd Ch=8 ChS=250Hz		F N net	
19228,20	PACTOR I	B=100Bd S=200Hz		Karachi	PAK
19241,50	*PACTOR IV*	*B=100Bd S=200Hz*	*DRA65*	*Max Planck Institute Leipzig*	*D*
19241,50	PACTOR IV	B=100Bd S=200Hz	9SD42	Max Planck field unit Congo	COG
19241,50	PACTOR III	B=100Bd S=200Hz	9SD56	Max Planck field unit Congo	COG
19246,70	SITOR A	B=100Bd S=170Hz Ch=1		MFA Cairo	EGY
19255,00	USB		VNJ	Casey Base , AUS	ANT
19255,00	USB			AUS Davis Base	ANT
19255,00	USB			Macquarie Island , AUS	ANT
19255,00	USB			Mawson Base , AUS	ANT
19255,00	ALE USB	B=125Bd Ch=8 ChS=250Hz		TUN diplo net	TUN
19255,00	ALE USB	B=125Bd Ch=8 ChS=250Hz		TUN diplo net	TUN
19256,00	*T600*	*B=50Bd S=200Hz*		*RUS N Murmansk*	*RUS*
19303,00	SSB			NASA launch support ships	USA
19305,00	CIS OFDM 45	B=2400Bd Ch=45 ChS=62,5Hz		RUS diplo	RUS
19323,00	CW		CGA984	Essex County, ON	CAN
19344,50	ALE USB	B=125Bd Ch=8 ChS=250Hz		USA FBI net	USA

Frequency	Mode	Mode Parameter	Callsign	User	Country
19350,00	*STANAG 4285*	*B=2400Bd*		*G MIL Ascension*	*ASC*
19385,00	ALE USB	B=125Bd Ch=8 ChS=250Hz		TUN diplo net	TUN
19385,00	*STANAG 4285*	*B=2400Bd*		*G MIL Ascension Island*	*ASC*
19419,50	DPRK FSK	B=600Bd S=600Hz		DPRK diplo net	
19507,20	STANAG 4285	B=2400Bd	DHJ58	D N Glücksburg TX Marlow	D
19528,20	*STANAG 4285*	*B=2400Bd*		*G MIL Crimond*	
19602,00	ALE USB	B=125Bd Ch=8 ChS=250Hz		ISR AF net	ISR
19639,00	ALE 3G	B=2400Bd		UI MIL net	
19675,00	ALE USB	B=125Bd Ch=8 ChS=250Hz		TUN diplo net	TUN
19680,50	.			MSI Frequency	WW
19680,50	SITOR B	B=100Bd S=170Hz	L2C	ARG N Buenos Aires	ARG
19681,00	.			CS Ch 1	WW
19681,50	.			CS Ch 2	WW
19682,00	.			CS Ch 3	WW
19682,50	.			CS Ch 4	WW
19683,00	.			CS Ch 5	WW
19683,50	.			CS Ch 6	WW
19684,00	.			CS Ch 7	WW
19684,50	SITOR , CW	B=100Bd S=170Hz	JNA	J CG HQ Tokyo	J
19684,50	.			CS Ch 8	WW
19685,00	.			CS Ch 9	WW
19685,50	.			CS Ch 10	WW
19686,00	.			CS Ch 11	WW
19686,50	.			CS Ch 12	WW
19687,00	.			CS Ch 13	WW
19687,50	.			CS Ch 14	WW
19688,00	.			CS Ch 15	WW
19688,00	*T600*	*B=50Bd S=200Hz*	*RIT*	*RUS N HQ Severomorsk*	*RUS*
19688,50	.			CS Ch 16	WW
19689,00	.			CS Ch 17	WW
19689,50	.			CS Ch 18	WW
19690,00	.			CS Ch 19	WW
19690,50	.			CS Ch 20	WW
19691,00	.			CS Ch 21	WW
19691,50	.			CS Ch 22	WW
19692,00	.			CS Ch 23	WW
19692,50	.			CS Ch 24	WW
19693,00	.			CS Ch 25	WW
19693,50	.			CS Ch 26	WW
19694,00	.			CS Ch 27	WW
19694,50	.			CS Ch 28	WW
19695,00	.			CS Ch 28	WW
19695,50	.			CS Ch 30	WW
19696,00	.			CS Ch 31	WW
19696,50	.			CS Ch 32	WW
19697,00	.			CS Ch 33	WW
19697,50	.			CS Ch 34	WW
19698,00	.			CS Ch 35	WW
19698,50	.			CS Ch 36	WW
19699,00	.			CS Ch 37	WW
19699,50	.			CS Ch 38	WW
19700,00	.			CS Ch 39	WW
19700,50	.			CS Ch 40	WW
19701,00	.			CS Ch 41	WW
19701,50	.			CS Ch 42	WW
19702,00	.			CS Ch 43	WW
19702,50	.			CS Ch 44	WW
19703,00	.			CS Ch 45	WW
19703,50	DSC	B=100Bd S=170Hz	JNA	J CG HQ Tokyo	J

Frequency	Mode	Mode Parameter	Callsign	User	Country
19707,00	PACTOR I FEC	B=100Bd S=200Hz	PWZ33	B N Rio de Janeiro	B
19708,00	MIL 188-110B SER	B=2400Bd		AUS MHFCS Humpty Doo	AUS
19709,00	ALE USB	B=125Bd Ch=8 ChS=250Hz		B N net	B
19729,00	MIL 188-110A SER	B=2400Bd		AUS MHFCS North West Cape	AUS
19739,40	*STANAG 4285*	*B=2400Bd*	*CTP*	*POR N Lisbon*	*POR*
19743,40	*STANAG 4285 LSB*	*B=2400Bd*	*CTA16*	*POR N Lisbon*	*POR*
19755,00	SSB			CS Ch 1801	
19755,00	ALE USB	B=125Bd Ch=8 ChS=250Hz		ARINC Urgent Link net	USA
19758,00	SSB			CS Ch 1802	
19761,00	SSB			CS Ch 1803	
19764,00	USB		SDJ	Stockholm R	S
19764,00	USB		JNA	J CG HQ Tokyo	J
19764,00	USB		UFL	Vladivostok R	RUS
19764,00	SSB			CS Ch 1804	
19767,00	USB		JNA	J CG HQ Tokyo	J
19767,00	SSB			CS Ch 1805	
19770,00	USB		XSQ	Guangzhou R	CHN
19770,00	SSB			CS Ch 1806	
19773,00	USB		LSD836	Argentina R	ARG
19773,00	SSB			CS Ch 1807	
19776,00	SSB			CS Ch 1808	
19779,00	USB		SVO613	Olympia R	GRC
19779,00	SSB			CS Ch 1809	
19782,00	SSB			CS Ch 1810	
19782,00	SSB		LZW	Varna R	BUL
19783,00	CW		AQP	PAK N Karachi	PAK
19785,00	USB		SXE	Aspropyrgos Attikis R	GRC
19785,00	USB			Peiraias CG R	GRC
19785,00	SSB			CS Ch 1811	
19788,00	USB		LSD836	Argentina R	ARG
19788,00	FAX 120/576		XSG	Shanghai R	CHN
19788,00	SSB			CS Ch 1812	
19791,00	USB		SDJ	Stockholm R	S
19791,00	SSB			CS Ch 1813	
19794,00	USB		XSQ	Guangzhou R	CHN
19794,00	SSB			CS Ch 1814	
19797,00	USB		LSD836	Argentina R	ARG
19797,00	SSB			CS Ch 1815	
19798,50	CIS OFDM 112	B=3360Bd Ch=112 ChS=32Hz		RUS	RUS
19800,00	USB			EGY N net	EGY
19800,00	EGY QPSK	B=2400Bd		EGY N	EGY
19800,00	STANAG 4197	B=1800Bd Ch=16/39 ChS=112/56Hz		EGY N net	EGY
19800,00	USB			EGY N	EGY
19814,00	ALE USB	B=125Bd Ch=8 ChS=250Hz		USA Civil Air Patrol net	USA
19880,00	PACTOR II FEC	B=100Bd S=200Hz	AQP	PAK N Karachi	PAK
19889,70	PACTOR II FEC	B=100Bd S=200Hz	AQP	PAK N Karachi	PAK
19892,00	*STANAG 4285*	*B=2400Bd*		*G MIL Akrotiri*	*CYP*
19917,00	CW		RIW	RUS N HQ Moscow	RUS
19921,00	RUS DBPSK 16FSK	B=1866Bd Ch=16 ChS=175Hz		RUS INTEL	RUS
19922,00	CIS 2CH QPSK	B=125Bd Ch=2 ChS=100Hz		RUS INTEL	RUS
19922,00	OQPSK	B=48000Bd	WH2XWU	County Information Services	USA
19936,00	*T600*	*B=50Bd S=200Hz*	*RIW*	*RUS N Moscow*	*RUS*
19936,00	T600	B=50Bd S=250Hz		RUS N Vladivostok	RUS

Frequency	Mode	Mode Parameter	Callsign	User	Country
19953,50	ALE USB	B=125Bd Ch=8 ChS=250Hz		USA FBI net	USA
19969,00	ALE USB	B=125Bd Ch=8 ChS=250Hz		USA FEMA net	USA
19980,00	*STANAG 4285*	*B=2400Bd*		*TUR MIL Ankara*	*TUR*
19984,00	T600	B=50Bd S=250Hz		RUS N Vladivostok	RUS
19993,00	SSB			MSV SAR operational frequency	WW
20000,00	AM		WWV	Fort Collins TS , CO	USA

20000 – 30000 kHz

Frequency	Mode	Mode Parameter	Callsign	User	Country
20006,50	DPRK FSK	B=600Bd S=600Hz		DPRK diplo net	
20016,70	SITOR B	B=100Bd S=170Hz		MFA Cairo	EGY
20027,00	RUS DBPSK 16FSK	B=1866Bd Ch=16 ChS=175Hz		RUS INTEL	RUS
20031,00	ALE USB	B=125Bd Ch=8 ChS=250Hz		USA AF net	
20047,60	CW		D**	RUS N Sevastopol	UKR
20047,90	CW		S**	RUS N Arkhangelsk	RUS
20048,00	CW		C**	RUS N Moscow	RUS
20048,20	CW		F**	RUS N Vladivostok	RUS
20048,30	CW		K**	RUS N Petropavlovsk	RUS
20048,40	CW		M**	RUS N Magadan	GLK
20050,00	CODAN	B=2400Bd Ch=16 ChS=112,5Hz		MFA Cairo	EGY
20050,00	CODAN	B=2400Bd Ch=16 ChS=112,5Hz		EGY E Bamako	MLI
20050,00	CODAN CHIRP	B=80Bd Ch=30 ChS=81Hz		MFA Cairo	EGY
20055,00	CODAN	B=2400Bd Ch=16 ChS=112,5Hz		MFA Cairo	EGY
20055,00	CODAN	B=2400Bd Ch=16 ChS=112,5Hz		EGY E New Delhi	IND
20055,00	CODAN	B=2400Bd Ch=16 ChS=112,5Hz		EGY E Islamabad	PAK
20055,00	CODAN	B=2400Bd Ch=16 ChS=112,5Hz		EGY E Washington	USA
20055,00	CODAN	B=2400Bd Ch=16 ChS=112,5Hz		EGY E Abidjan	CTI
20056,70	SITOR A	B=100Bd S=170Hz		MFA Cairo	EGY
20090,50	STANAG 4481	B=75Bd S=850Hz			
20096,00	T600	B=50Bd S=200Hz	RDL	RUS N	RUS
20100,00	*QPSK*	*B=1200Bd Ch=1*		*Area of Yekaterinenburg*	*RUS*
20100,00	USB			EGY N net	EGY
20106,20	STANAG 4285	B=2400Bd		AUS MHFCS Humpty Doo	AUS
20107,00	ALE USB	B=125Bd Ch=8 ChS=250Hz		USA SHARES net	USA
20109,70	STANAG 4285	B=2400Bd	OVG	DNK N Frederikshavn	DNK
20110,00	STANAG 4285	B=2400Bd		AUS MHFCS Humpty Doo	AUS
20124,00	ALE USB	B=125Bd Ch=8 ChS=250Hz		FIN diplo net	FIN
20124,00	SKYOFDM	B=64Bd Ch=22 ChS=86Hz		MFA Helsinki	FIN
20149,00	PACTOR II FEC	B=100Bd S=200Hz	AQP	PAK N Karachi	PAK
20152,90	CW		CGA984	Essex County, ON	CAN
20168,00	ALE USB	B=125Bd Ch=8 ChS=250Hz		G MIL net	G
20178,70	CODAN	B=2400Bd Ch=16 ChS=112,5Hz		MFA Cairo	EGY

Frequency	Mode	Mode Parameter	Callsign	User	Country
20178,70	CODAN CHIRP	B=80Bd Ch=30 ChS=250Hz		MFA Cairo	EGY
20183,00	ALE USB	B=125Bd Ch=8 ChS=250Hz		B MIL net	B
20186,00	SSB			NASA launch tracking net	USA
20186,00	SSB			NASA Cape R , FL	USA
20198,20	*STANAG 4285*	*B=2400Bd*		*G MIL Akrotiri*	*CYP*
20200,00	STANAG 4285	B=2400Bd		G MIL Crimond	G
20200,20	*STANAG 4285*	*B=2400Bd*	*6WW*	*F N Dakar*	*SEN*
20200,50	CW			RUS HFDF net	RUS
20238,00	CW		RDL	RUS N Moscow	RUS
20254,50	*STANAG 4285*	*B=2400Bd Ch=1*	*JWT*	*NOR N Stavanger*	*NOR*
20268,00	T600	B=50Bd S=250Hz	RMP	RUS N HQ Kaliningrad	RUS
20275,00	ALE USB	B=125Bd Ch=8 ChS=250Hz		G MIL net TX Ascension	ASC
20277,00	STANAG 4285	B=2400Bd		AUS MHFCS Humpty Doo	AUS
20287,00	*STANAG 4285*	*B=2400Bd*		*G MIL TX Akrotiri*	*CYP*
20300,00	STANAG 4197	B=1800Bd Ch=16/39 ChS=112/56Hz		EGY N net	EGY
20325,00	ALE USB	B=125Bd Ch=8 ChS=250Hz		F N net	
20325,00	STANAG 4285	B=2400Bd	FUJ	F N Noumea	NCL
20331,00	DPRK ARQ	B=600Bd S=600Hz		DPRK E	
20336,00	MAZIELKA	B=1Bd Ch=6		RUS diplo	RUS
20348,50	ALE USB	B=125Bd Ch=8 ChS=250Hz		USA FBI net	USA
20366,00	MIL 188-110A	B=2400Bd		AUS MIL Noth West Cape	AUS
20366,00	ALE USB	B=125Bd Ch=8 ChS=250Hz		CHN net	CHN
20366,00	*STANAG 4285*	*B=2400Bd*		*AUS MHFCS Humpty Doo*	*AUS*
20373,00	PACTOR II	B=200Bd S=200Hz	V8V2222	Sailmail Brunai Bay	BRU
20382,00	CHN 4+4	B=75Bd Ch=8 ChS=300/450Hz		CHN MIL	
20390,00	USB			USA AF Cape Radio	USA
20402,50	ALE USB	B=125Bd Ch=8 ChS=250Hz		USA FBI net	USA
20414,00	ALE USB	B=125Bd Ch=8 ChS=250Hz		USA FEMA net	USA
20423,50	ALE USB	B=125Bd Ch=8 ChS=250Hz		G MIL net	G
20430,00	ALE USB	B=125Bd Ch=8 ChS=250Hz		G MIL net	G
20457,00	PACTOR III	B=100Bd S=200Hz	OL1A	CZE MIL Prague	CZE
20461,00	RUS DBPSK 16FSK	B=1866Bd Ch=16 ChS=175Hz		RUS INTEL	RUS
20469,00	FAX 120/576		VMC	Charleville M	AUS
20500,00	ALE USB	B=125Bd Ch=8 ChS=250Hz			
20513,00	CIS MFSK14	B=7,8Bd Ch=14 ChS=15Hz		RUS INTEL	RUS
20520,00	STANAG 4197	B=1800Bd Ch=16/39			
20528,00	PACTOR III	B=100Bd S=200Hz	9Z4DH	Sailmail Chaguaramas	TRD
20536,00	*T600*	*B=50Bd S=200Hz*	*RCV*	*RUS N HQ Sevastopol*	*UKR*
20540,00	*STANAG 4285*	*B=2400Bd*		*G MIL Akrotiri*	*CYP*
20550,00	ALE USB	B=125Bd Ch=8 ChS=250Hz		MRC MIL net	MRC
20555,50	IRN N PSK	B=963Bd		IRN N	IRN
20558,50	DPRK FSK	B=600Bd S=600Hz		DPRK diplo net	
20584,50	DPRK FSK	B=600Bd S=600Hz		DPRK diplo net	
20597,00	ALE USB	B=125Bd Ch=8 ChS=250Hz		USA AF net	USA

Frequency	Mode	Mode Parameter	Callsign	User	Country
20602,50	ALE USB	B=125Bd Ch=8 ChS=250Hz		USA FBI net	USA
20619,00	PACTOR II	B=100Bd S=200Hz			
20631,00	SSB			USAF Strategic Command	USA
20632,00	STANAG 4285	B=2400Bd		AUS MHFCS Humpty Doo	AUS
20640,00	T600	B=50Bd S=250Hz	RJD56	RUS N HQ Murmansk	RUS
20647,00	STANAG 4481	B=75Bd S=850Hz			
20670,00	CIS OFDM 112	B=3360Bd Ch=112 ChS=32Hz		RUS diplo	RUS
20726,00	RUS DBPSK 16FSK	B=1866Bd Ch=16 ChS=175Hz		RUS INTEL	RUS
20763,00	USB			USA AF MARS	
20764,00	BAUDOT	B=75Bd S=450Hz		RUS diplo Habana	CUB
20810,60	ALE USB	B=125Bd Ch=8 ChS=250Hz		USA diplo net	USA
20840,00	T600	B=50Bd S=250Hz		RUS N St. Petersburg	RUS
20841,00	BAUDOT	B=75Bd S=500Hz		RUS diplo Habana	CUB
20841,00	CIS OFDM 45	B=2400Bd Ch=45 ChS=62,5Hz		RUS diplo Habana	CUB
20861,50	ALE USB	B=125Bd Ch=8 ChS=250Hz			
20873,00	USB			USA AF MARS	
20881,00	RUS DBPSK 16FSK	B=1866Bd Ch=16 ChS=175Hz		RUS INTEL	RUS
20890,00	CIS 3000	B=3000Bd		RUS diplo	RUS
20890,00	ALE USB	B=125Bd Ch=8 ChS=250Hz		USA COTHEN net	USA
20906,00	ALE USB	B=125Bd Ch=8 ChS=250Hz		USA National Guard net	USA
20907,50	ALE USB	B=125Bd Ch=8 ChS=250Hz		USA Defence Logistic Agency net	USA
20908,00	ALE USB	B=125Bd Ch=8 ChS=250Hz		FIN diplo net	
20908,20	*ALE USB*	*B=125Bd Ch=8 ChS=250Hz*		*Kunhing area*	*MMR*
20908,20	*MIL 188-110B SER*	*B=2400Bd*		*Kunhing area*	*MMR*
20918,00	*STANAG 4285*	*B=2400Bd*		*G MIL Ascension*	*SHN*
20946,00	PACTOR III	B=100Bd S=200Hz	STCIP		
20953,00	STANAG 4285	B=2400Bd		G MIL Crimond	G
20955,00	STANAG 4285	B=2400Bd	NPN	USA N Guam	GUM
20955,00	RFSM8000	B=2400Bd		BUL diplo net	BUL
20960,00	USB			USA AF MARS	
20971,50	STANAG 4285	B=2400Bd		G MIL Akrotiri	CYP
20976,50	ALE USB	B=125Bd Ch=8 ChS=250Hz		USA SHARES net	USA
20981,00	STANAG 4285	B=2400Bd		G MIL Akrotiri	CYP
20982,00	ALE USB	B=125Bd Ch=8 ChS=250Hz		G MIL net	G
20982,00	CHN 4+4	B=75Bd Ch=8 ChS=300/450Hz		CHN MIL	
20987,00	*STANAG 4285*	*B=2400Bd*		*G MIL Akrotiri*	*CYP*
20991,00	USB			USA AF MARS	
20991,70	CW		D**	RUS N Sevastopol	UKR
20991,70	CW		S**	RUS N Severomorsk	RUS
20991,80	CW		P**	RUS N Kaliningrad	RUS
20992,00	CW		C**	RUS N Moscow	RUS
20992,50	USB			USA AF MARS	
20994,00	PACTOR III	B=100Bd S=200Hz	AEM1WF	USA Army MARS	USA
21034,00	CHN 4+4	B=75Bd Ch=8 ChS=300/450Hz		CHN MIL	CHN
21067,50	CW		W1AW	ARRL morse training	USA
21070,00	PSK31	B=31Bd		PSK31 channel	

Frequency	Mode	Mode Parameter	Callsign	User	Country
21074,00	FT8	B=5,86Bd Ch=8 ChS=5,86Hz		FT8 channel	
21076,00	JT65	B=2,69Bd Ch=65 ChS=2,69Hz		JT65 channel	
21078,00	JT9	B=1,736Bd Ch=9 ChS=1,736Hz		JT9 channel	
21078,00	JS8	B=6,25Bd Ch=8 ChS=6,25Hz		JS8 user	
21080,00	BAUDOT	B=45,45Bd S=170Hz		RTTY user	
21086,50	MFSK			MFSK modes user	
21094,60	WSPR	B=1,46Bd Ch=4 ChS=1,46Hz		WSPR net	
21095,00	LONGCHAT	B=40Bd Ch=1		LongChat net	
21096,00	ALE USB	B=125Bd Ch=8 ChS=250Hz		HFLINK network	
21105,00	VARA	Ch=1		VARA chat freqeuency	
21122,00	ROS USB	B=1Bd Ch=144 ChS=15,625Hz		ROS frequency	
21140,00	FT4	B=5,86Bd Ch=8 ChS=5,86Hz		FT4 channel	
21145,00	ALE USB	B=125Bd Ch=8 ChS=250Hz		MRC MIL net	MRC
21150,00	CW		KH6RS	Maui	HWA
21150,00	CW		4U1UN	New York City	USA
21150,00	CW		W6WX	San Jose , CA	USA
21150,00	CW		YV5B	Caracas	VEN
21150,00	CW		5Z4B	Kilifi	CYP
21150,00	CW		CS3B	Madeira	E
21150,00	CW		4X6TU	Tel aviv	ISR
21150,00	CW		VE8AT	Inuvik, NT	CAN
21150,00	CW		ZL6B	Masterton	NZL
21150,00	CW		VK6RBP	Rolystone	AUS
21150,00	CW		JA2IGY	Mt. Asama	J
21150,00	CW		RR9O	Novosibirsk	RUS
21150,00	CW		VR2B	Hong Kong	HKG
21150,00	CW		4S7B	Colombo	CLN
21150,00	CW		ZS6DN	Pretoria	AFS
21150,00	CW		OH2B	Lohja	FNL
21150,00	CW		LU4AA	Buenos Aires	ARG
21150,00	CW		OA4B	Lima	PRU
21190,00	USB			Wireless Institute Civil Emergency Service WICEN	AUS
21360,00	USB			Ham radio emergency frequency	R1
21370,00	ARD9800 MODEM	B=2250Bd Ch=36 ChS=62,5Hz		AOR Digital Voice net	USA
21412,00	USB			Pacific Maritime Mobile net	
21432,50	ALE USB	B=125Bd Ch=8 ChS=250Hz		HFLINK network	
21438,00	CW		RCV	RUS N HQ Sevastopol	UKR
21484,00	STANAG 4285	B=2400Bd		G MIL Ascension	SHN
21764,00	T600	B=50Bd S=200Hz		RUS N Vladivostok	RUS
21784,00	T600	B=50Bd S=200Hz	RMP	RUS N HQ Kaliningrad	RUS
21855,00	CODAN	B=2400Bd Ch=16 ChS=112,5Hz		MFA Cairo	EGY
21856,70	SITOR A	B=100Bd S=170Hz		MFA Cairo	EGY
21859,00	RUS DBPSK 16FSK	B=1866Bd Ch=16 ChS=175Hz		RUS INTEL	RUS
21863,25	ALE USB	B=125Bd Ch=8 ChS=250Hz		USA Civil Air Patrol net	USA
21866,00	PACTOR II	B=200Bd S=200Hz	XJN714	Sailmail Lunenburg , NS	CAN
21866,00	ALE USB	B=125Bd Ch=8 ChS=250Hz		USA FEMA net	USA

Frequency	Mode	Mode Parameter	Callsign	User	Country
21866,50	ALE USB	B=125Bd Ch=8 ChS=250Hz		USA FEMA	USA
21882,00	ALE USB	B=125Bd Ch=8 ChS=250Hz		USA FEMA net	USA
21918,00	RUS DBPSK 16FSK	B=1866Bd Ch=16 ChS=175Hz		RUS INTEL	RUS
21919,00	ALE USB	B=125Bd Ch=8 ChS=250Hz		USA FEMA net	USA
21925,00	USB			San Francisco ACC , CA	USA
21925,00	USB			Tokyo ACC	J
21928,00	HFDL	B=1800Bd	H09**	Utqiagvik , AK	ALS
21928,00	HFDL	B=1800Bd	H16**	Agana	GUM
21931,00	HFDL	B=1800Bd	H10**	Muan	KOR
21931,00	HFDL	B=1800Bd	H04**	Riverhead , NY	USA
21931,00	USB		BOING EVERETT	Boing Seattle	USA
21931,00	USB		BOING SEATTLE	Boing Seattle	USA
21934,00	HFDL	B=1800Bd	H01**	San Francisco , CA	USA
21937,00	HFDL	B=1800Bd	H02**	Molokai , HI	HWA
21937,00	SSB			NOAA Hurricane Center Miami , FL	USA
21937,00	HFDL	B=1800Bd	H09**	Utqiagvik , AK	ALS
21940,00	SSB			W I	WW
21943,00	SSB			W V	WW
21946,00	SSB			ARINC Houston , FL	USA
21946,00	SSB			W I	WW
21949,00	HFDL	B=1800Bd	H06**	Hat Yai	THA
21949,00	SSB			W III	WW
21949,00	HFDL	B=1800Bd	H08**	Johannesburg	AFS
21952,00	SSB			W I	WW
21955,00	USB			PanAM R	USA
21955,00	SSB			W IV	WW
21955,00	HFDL	B=1800Bd	H17**	Telde , Gran Canaria	E
21958,00	SSB			W I	WW
21961,00	SSB			W V	WW
21964,00	SSB			W II	WW
21964,00	USB			San Francisco ACC , CA	USA
21967,00	SSB			W I	WW
21967,00	USB			IBERIA Madrid ACC	E
21970,00	SSB			W III	WW
21973,00	SSB			W I	WW
21976,00	SSB			W IV	WW
21979,00	SSB			W I	WW
21982,00	SSB			W V	WW
21982,00	HFDL	B=1800Bd	H15**	Al Muharraq	BHR
21985,00	SSB			W II	WW
21985,00	USB			San Francisco ACC , CA	USA
21988,00	SSB			W I	WW
21990,00	HFDL	B=1800Bd	H14**	Krasnoyarsk	RUS
21991,00	SSB			W IV	WW
21994,00	SSB			W V	WW
21997,00	SSB			W I	WW
21997,00	HFDL	B=1800Bd	H13**	Santa Cruz	BOL
22000,00	SITOR B	B=100Bd S=170Hz	SVO	Olympia R	GRC
22000,00	SSB			SS Ch 2201	
22003,00	SSB			SS Ch 2202	
22006,00	SSB			SS Ch 2203	
22009,00	SSB			SS Ch 2204	
22012,00	SSB			SS Ch 2205	
22015,00	ALE USB	B=125Bd Ch=8 ChS=250Hz		Queensland Deptartment of Community Safety	AUS
22015,00	SSB			SS Ch 2206	
22018,00	SSB			SS Ch 2207	

Frequency	Mode	Mode Parameter	Callsign	User	Country
22021,00	SSB			SS Ch 2208	
22024,00	SSB			SS Ch 2209	
22027,00	SSB			SS Ch 2210	
22030,00	SSB			SS Ch 2211	
22033,00	SSB			SS Ch 2212	
22036,00	SSB			SS Ch 2213	
22039,00	SSB			SS Ch 2214	
22042,00	SSB			SS Ch 2215	
22045,00	SSB			SS Ch 2216	
22046,50	BPSK	B=1200Bd			RUS
22048,00	SSB			SS Ch 2217	
22051,00	SSB			SS Ch 2218	
22054,00	SSB			SS Ch 2219	
22057,00	SSB			SS Ch 2220	
22060,00	SSB			SS Ch 2221	
22063,00	SSB			SS Ch 2222	
22063,00	STANAG 4197	B=1800Bd Ch=16/39 ChS=112/56Hz		USA MIL	USA
22066,00	SSB			SS Ch 2223	
22069,00	SSB			SS Ch 2224	
22072,00	SSB			SS Ch 2225	
22075,00	SSB			SS Ch 2226	
22078,00	SSB			SS Ch 2227	
22081,00	SSB			SS Ch 2228	
22084,00	SSB			SS Ch 2229	
22087,00	SSB			SS Ch 2230	
22090,00	SSB			SS Ch 2231	
22093,00	SSB			SS Ch 2232	
22096,00	SSB			SS Ch 2233	
22099,00	SSB			SS Ch 2234	
22100,00	STANAG 4197	B=1800Bd Ch=16/39 ChS=112/56Hz		EGY N net	EGY
22100,00	USB			EGY N	EGY
22102,00	SSB			SS Ch 2235	
22105,00	SSB			SS Ch 2236	
22108,00	SSB			SS Ch 2237	
22111,00	SSB			SS Ch 2238	
22114,00	SSB			SS Ch 2239	
22117,00	SSB			SS Ch 2240	
22120,00	SSB			SS Ch 2241	
22123,00	SSB			SS Ch 2242	
22126,00	SSB			SS Ch 2243	
22129,00	SSB			SS Ch 2244	
22132,00	SSB			SS Ch 2245	
22135,00	SSB			SS Ch 2246	
22138,00	SSB			SS Ch 2247	
22141,00	SSB			SS Ch 2248	
22144,00	SSB			SS Ch 2249	
22147,00	SSB			SS Ch 2250	
22150,00	SSB			SS Ch 2251	
22153,00	SSB			SS Ch 2252	
22156,00	SSB			SS Ch 2253	
22159,00	SSB			simplex frequency ship - ship	
22162,00	SSB			simplex frequency ship - ship	
22165,00	SSB			simplex frequency ship - ship	
22168,00	SSB			simplex frequency ship - ship	
22171,00	SSB			simplex frequency ship - ship	
22174,00	SSB			simplex frequency ship - ship	
22177,00	SSB			simplex frequency ship - ship	
22177,00	STANAG 4285	B=2400Bd	ICT	I N Rome TX Taranto	I

Frequency	Mode	Mode Parameter	Callsign	User	Country
22177,00	ALE USB	B=125Bd Ch=8 ChS=250Hz		G MIL net	G
22177,80	ALE USB	B=125Bd Ch=8 ChS=250Hz		G MIL net	G
22186,00	STANAG 4285	B=2400Bd		G MIL Ascension Island	ASC
22210,00	STANAG 4285	B=2400Bd	CTA	POR N Lisbon	POR
22212,00	PACTOR III	B=100Bd S=200Hz	RC01	Sailmail Maputo	MOZ
22220,20	STANAG 4285	B=2400Bd		G MIL Akrotiri	CYP
22226,00	STANAG 4285	B=2400Bd		G MIL St. Eval	G
22226,00	STANAG 4285	B=2400Bd		G MIL Inskip	G
22228,00	BAUDOT	B=75Bd S=850Hz	PBB	NLD N Den Helder	HOL
22230,00	STANAG 4285	B=2400Bd		G MIL Akrotiri	CYP
22234,50	ALE USB	B=125Bd Ch=8 ChS=250Hz		G MIL net	G
22260,00	CHN 4+4	B=75Bd Ch=8 ChS=300/450Hz		CHN ship antarctic	
22273,00	USB		SVO	Olympia R	GRC
22279,50	CW			SS calling frequency Ch 1	
22280,00	CW			SS calling frequency Ch 2	
22280,50	CW			SS calling frequency Ch 3	
22281,00	CW			SS calling frequency Ch 4	
22281,50	CW			SS calling frequency Ch 5	
22282,00	CW			SS calling frequency Ch 6	
22282,50	CW			SS calling frequency Ch 7	
22283,00	CW			SS calling frequency Ch 8	
22283,50	CW			SS calling frequency Ch 9	
22284,00	CW			SS calling frequency Ch 10	
22284,50	.			SS Ch 1	
22285,00	MHF-50 MODEM	B=54,3Bd Ch=32 ChS=64,5Hz	ZSJ	AFS N Capetown	AFS
22285,00	.			SS Ch 2	
22285,50	.			SS Ch 3	
22286,00	.			SS Ch 4	
22286,50	.			SS Ch 5	
22287,00	.			SS Ch 6	
22287,50	.			SS Ch 7	
22288,00	.			SS Ch 8	
22288,50	.			SS Ch 9	
22289,00	.			SS Ch 10	
22289,50	.			SS Ch 11	
22290,00	.			SS Ch 12	
22290,50	.			SS Ch 13	
22291,00	.			SS Ch 14	
22291,50	.			SS Ch 15	
22292,00	.			SS Ch 16	
22292,50	.			SS Ch 17	
22293,00	.			SS Ch 18	
22293,50	.			SS Ch 19	
22294,00	.			SS Ch 20	
22294,50	.			SS Ch 21	
22295,00	.			SS Ch 22	
22295,50	.			SS Ch 23	
22296,00	.			SS Ch 24	
22296,50	.			SS Ch 25	
22297,00	.			SS Ch 26	
22297,50	.			SS Ch 27	
22298,00	.			SS Ch 28	
22298,50	.			SS Ch 29	
22299,00	.			SS Ch 30	
22299,50	.			SS Ch 31	
22300,00	.			SS Ch 32	
22300,50	.			SS Ch 33	

Frequency	Mode	Mode Parameter	Callsign	User	Country
22301,00	.			SS Ch 34	
22301,50	.			SS Ch 35	
22302,00	.			SS Ch 36	
22302,50	.			SS Ch 37	
22303,00	.			SS Ch 38	
22303,50	.			SS Ch 39	
22304,00	.			SS Ch 40	
22304,50	.			SS Ch 41	
22305,00	.			SS Ch 42	
22305,50	.			SS Ch 43	
22306,00	.			SS Ch 44	
22306,50	.			SS Ch 45	
22307,00	.			SS Ch 46	
22307,50	.			SS Ch 47	
22308,00	.			SS Ch 48	
22308,50	.			SS Ch 49	
22308,70	STANAG 4285	B=2400Bd	IDR	I N Rome TX Isola di San Pietro	I
22309,00	.			SS Ch 50	
22309,50	.			SS Ch 51	
22310,00	.			SS Ch 52	
22310,50	.			SS Ch 53	
22311,00	.			SS Ch 54	
22311,50	.			SS Ch 55	
22312,00	.			SS Ch 56	
22312,50	.			SS Ch 57	
22313,00	.			SS Ch 58	
22313,50	.			SS Ch 59	
22314,00	.			SS Ch 60	
22314,50	.			SS Ch 61	
22315,00	.			SS Ch 62	
22315,50	.			SS Ch 63	
22316,00	.			SS Ch 64	
22316,50	.			SS Ch 65	
22317,00	.			SS Ch 66	
22317,50	.			SS Ch 67	
22318,00	.			SS Ch 68	
22318,50	.			SS Ch 69	
22319,00	.			SS Ch 70	
22319,50	.			SS Ch 71	
22320,00	.			SS Ch 72	
22320,50	.			SS Ch 73	
22321,00	.			SS Ch 74	
22321,50	.			SS Ch 75	
22322,00	.			SS Ch 76	
22322,50	.			SS Ch 77	
22323,00	.			SS Ch 78	
22323,50	.			SS Ch 79	
22324,00	.			SS Ch 80	
22324,50	.			SS Ch 81	
22325,00	.			SS Ch 82	
22325,50	.			SS Ch 83	
22326,00	.			SS Ch 84	
22326,50	.			SS Ch 85	
22327,00	.			SS Ch 86	
22327,50	.			SS Ch 87	
22328,00	.			SS Ch 88	
22328,50	.			SS Ch 89	
22329,00	.			SS Ch 90	
22329,50	.			SS Ch 91	
22330,00	.			SS Ch 92	
22330,50	.			SS Ch 93	
22331,00	.			SS Ch 94	

Frequency	Mode	Mode Parameter	Callsign	User	Country
22331,50	.			SS Ch 95	
22332,00	.			SS Ch 96	
22332,50	.			SS Ch 97	
22333,00	.			SS Ch 98	
22333,50	.			SS Ch 99	
22334,00	.			SS Ch 100	
22334,50	.			SS Ch 101	
22335,00	.			SS Ch 102	
22335,50	.			SS Ch 103	
22336,00	.			SS Ch 104	
22336,50	.			SS Ch 105	
22337,00	.			SS Ch 106	
22337,50	.			SS Ch 107	
22338,00	.			SS Ch 108	
22338,50	.			SS Ch 109	
22339,00	.			SS Ch 110	
22339,50	.			SS Ch 111	
22340,00	.			SS Ch 112	
22340,00	CW		LZL7	Bourgas R	BUL
22340,50	.			SS Ch 113	
22341,00	.			SS Ch 114	
22341,50	.			SS Ch 115	
22342,00	.			SS Ch 116	
22342,50	.			SS Ch 117	
22343,00	.			SS Ch 118	
22343,50	.			SS Ch 119	
22344,00	.			SS Ch 120	
22344,50	.			SS Ch 121	
22345,00	.			SS Ch 122	
22345,50	.			SS Ch 123	
22346,00	.			SS Ch 124	
22346,50	.			SS Ch 125	
22347,00	.			SS Ch 126	
22347,50	.			SS Ch 127	
22348,00	.			SS Ch 128	
22348,50	.			SS Ch 129	
22349,00	.			SS Ch 130	
22349,50	.			SS Ch 131	
22350,00	.			SS Ch 132	
22350,50	.			SS Ch 133	
22351,00	.			SS Ch 134	
22351,50	.			SS Ch 135	
22352,00	PACTOR II	B=200Bd S=200Hz	V8V2222	Sailmail Brunai Bay	BRU
22352,00	.			SS Ch 1	
22352,50	.			SS Ch 2	
22353,00	.			SS Ch 3	
22353,50	.			SS Ch 4	
22354,00	.			SS Ch 5	
22354,50	.			SS Ch 6	
22355,00	.			SS Ch 7	
22355,50	.			SS Ch 8	
22356,00	.			SS Ch 9	
22356,50	.			SS Ch 10	
22357,00	.			SS Ch 11	
22357,50	.			SS Ch 12	
22358,00	.			SS Ch 13	
22358,50	.			SS Ch 14	
22359,00	.			SS Ch 15	
22359,50	.			SS Ch 16	
22360,00	.			SS Ch 17	
22360,50	.			SS Ch 18	
22361,00	.			SS Ch 19	

Frequency	Mode	Mode Parameter	Callsign	User	Country
22361,50	.			SS Ch 20	
22362,00	.			SS Ch 21	
22362,50	.			SS Ch 22	
22363,00	.			SS Ch 23	
22363,50	.			SS Ch 24	
22364,00	.			SS Ch 25	
22364,50	.			SS Ch 26	
22365,00	.			SS Ch 27	
22365,50	.			SS Ch 28	
22366,00	.			SS Ch 29	
22366,50	.			SS Ch 30	
22367,00	.			SS Ch 31	
22367,50	.			SS Ch 32	
22368,00	.			SS Ch 33	
22368,50	.			SS Ch 34	
22369,00	.			SS Ch 35	
22369,50	.			SS Ch 36	
22370,00	.			SS Ch 37	
22370,50	.			SS Ch 38	
22371,00	.			SS Ch 39	
22371,50	.			SS Ch 40	
22372,00	.			SS Ch 41	
22372,50	.			SS Ch 42	
22373,00	.			SS Ch 43	
22373,50	.			SS Ch 44	
22374,00	.			SS Ch 45	
22374,50	.			SS DSC frequency	WW
22375,00	.			SS DSC frequency	WW
22375,00	CW		XYR24	Rangoon R	BRM
22375,50	.			SS DSC frequency	WW
22376,00	.			MSI Frequency	
22376,00	.			SS DSC frequency	WW
22376,00	SITOR B	B=100Bd S=170Hz	NRV	USA CG Guam	GUM
22376,00	SITOR B	B=100Bd S=170Hz	NMO	USA CG Honolulu	HWA
22376,50	.			CS Ch 1	
22377,00	.			CS Ch 2	
22377,50	.			CS Ch 3	
22378,00	.			CS Ch 4	
22378,50	.			CS Ch 5	
22379,00	.			CS Ch 6	
22379,50	.			CS Ch 7	
22380,00	.			CS Ch 8	
22380,50	.			CS Ch 9	
22381,00	.			CS Ch 10	
22381,50	.			CS Ch 11	
22382,00	.			CS Ch 12	
22382,50	.			CS Ch 13	
22383,00	.			CS Ch 14	
22383,50	.			CS Ch 15	
22384,00	.			CS Ch 16	
22384,50	.			CS Ch 17	
22384,50	SITOR , CW	B=100Bd S=170Hz	SVO79	Olympia R	GRC
22385,00	.			CS Ch 18	
22385,50	.			CS Ch 19	
22386,00	.			CS Ch 20	
22386,50	.			CS Ch 21	
22387,00	.			CS Ch 22	
22387,50	.			CS Ch 23	
22387,50	SITOR , CW	B=100Bd S=170Hz	SVO7	Olympia R	GRC
22388,00	.			CS Ch 24	
22388,50	.			CS Ch 25	
22389,00	.			CS Ch 26	

Frequency	Mode	Mode Parameter	Callsign	User	Country
22389,50	.			CS Ch 27	
22390,00	.			CS Ch 28	
22390,50	.			CS Ch 29	
22391,00	.			CS Ch 30	
22391,50	.			CS Ch 31	
22392,00	T600	B=50Bd S=200Hz	RDL	RUS N Moscow	RUS
22392,00	.			CS Ch 32	
22392,50	.			CS Ch 33	
22393,00	.			CS Ch 34	
22393,50	.			CS Ch 35	
22394,00	.			CS Ch 36	
22394,50	.			CS Ch 37	
22395,00	.			CS Ch 38	
22395,50	.			CS Ch 39	
22396,00	.			CS Ch 40	
22396,50	.			CS Ch 41	
22397,00	.			CS Ch 42	
22397,50	.			CS Ch 43	
22398,00	.			CS Ch 44	
22398,50	.			CS Ch 45	
22399,00	SITOR , CW	B=100Bd S=170Hz	A9M	Bahrain R	BHR
22399,00	.			CS Ch 46	
22399,50	.			CS Ch 47	
22400,00	.			CS Ch 48	
22400,50	.			CS Ch 49	
22400,50	SITOR , CW	B=100Bd S=170Hz	SVO791	Olympia R	GRC
22401,00	.			CS Ch 50	
22401,50	.			CS Ch 51	
22402,00	.			CS Ch 52	
22402,50	.			CS Ch 53	
22403,00	.			CS Ch 54	
22403,50	.			CS Ch 55	
22404,00	.			CS Ch 56	
22404,50	.			CS Ch 57	
22405,00	.			CS Ch 58	
22405,50	.			CS Ch 59	
22406,00	.			CS Ch 60	
22406,50	.			CS Ch 61	
22407,00	.			CS Ch 62	
22407,50	.			CS Ch 63	
22407,50	SITOR , CW	B=100Bd S=170Hz	UAT	Moscow R	RUS
22408,00	.			CS Ch 64	
22408,50	SITOR , CW	B=100Bd S=170Hz	UFL	Vladivostok R	RUS
22408,50	.			CS Ch 65	
22409,00	.			CS Ch 66	
22409,50	.			CS Ch 67	
22410,00	.			CS Ch 68	
22410,50	.			CS Ch 69	
22411,00	.			CS Ch 70	
22411,50	.			CS Ch 71	
22412,00	SITOR , CW	B=100Bd S=170Hz	JNA	J CG HQ Tokyo	J
22412,00	.			CS Ch 72	
22412,50	.			CS Ch 73	
22413,00	T600	B=50Bd S=200Hz	RJD56	RUS N HQ Murmansk	RUS
22413,00	.			CS Ch 74	
22413,50	.			CS Ch 75	
22414,00	.			CS Ch 76	
22414,50	.			CS Ch 77	
22415,00	.			CS Ch 78	
22415,50	.			CS Ch 79	
22416,00	.			CS Ch 80	
22416,50	.			CS Ch 81	

Frequency	Mode	Mode Parameter	Callsign	User	Country
22417,00	.			CS Ch 82	
22417,50	.			CS Ch 83	
22418,00	.			CS Ch 84	
22418,50	.			CS Ch 85	
22419,00	.			CS Ch 86	
22419,50	.			CS Ch 87	
22420,00	.			CS Ch 88	
22420,50	.			CS Ch 89	
22421,00	.			CS Ch 90	
22421,50	.			CS Ch 91	
22422,00	.			CS Ch 92	
22422,50	.			CS Ch 93	
22423,00	.			CS Ch 94	
22423,50	.			CS Ch 95	
22424,00	.			CS Ch 96	
22424,50	.			CS Ch 97	
22425,00	.			CS Ch 98	
22425,50	.			CS Ch 99	
22426,00	.			CS Ch 100	
22426,50	.			CS Ch 101	
22427,00	.			CS Ch 102	
22427,50	.			CS Ch 103	
22428,00	.			CS Ch 104	
22428,50	.			CS Ch 105	
22429,00	.			CS Ch 106	
22429,50	.			CS Ch 107	
22430,00	.			CS Ch 108	
22430,50	.			CS Ch 109	
22431,00	.			CS Ch 110	
22431,00	CW		PKX	Jakarta R , Jawa	INS
22431,50	.			CS Ch 111	
22432,00	.			CS Ch 112	
22432,50	.			CS Ch 113	
22433,00	.			CS Ch 114	
22433,50	.			CS Ch 115	
22434,00	.			CS Ch 116	
22434,50	.			CS Ch 117	
22435,00	.			CS Ch 118	
22435,50	.			CS Ch 119	
22436,00	.			CS Ch 120	
22436,50	.			CS Ch 121	
22437,00	.			CS Ch 122	
22437,50	.			CS Ch 123	
22438,00	.			CS Ch 124	
22438,50	.			CS Ch 125	
22439,00	.			CS Ch 126	
22439,50	.			CS Ch 127	
22440,00	.			CS Ch 128	
22440,50	.			CS Ch 129	
22441,00	.			CS Ch 130	
22441,50	.			CS Ch 131	
22442,00	.			CS Ch 132	
22442,50	.			CS Ch 133	
22443,00	.			CS Ch 134	
22443,50				CS Ch 135	
22444,00	DSC	B=100Bd S=170Hz	JNA	J CG HQ Tokyo	J
22444,00	DSC	B=100Bd S=170Hz	PKX	Jakarta R , Jawa	INS
22444,50	SITOR , CW	B=100Bd S=170Hz	UFL	Vladivostok R	RUS
22447,00	*STANAG 4285*	*B=2400Bd*	*FUV*	*F N Djibouti*	*DJI*
22449,00	*STANAG 4481*	*B=50Bd S=350Hz*		*AUS MIL*	*MIL*
22449,00	STANAG 4285	B=2400Bd		AUS MHFCS Humpty Doo	AUS
22456,00	USB		PPR	Rio R	B

Frequency	Mode	Mode Parameter	Callsign	User	Country
22461,00	STANAG 4285	B=2400Bd	FUJ	F N Noumea	NCL
22461,70	STANAG 4285	B=2400Bd	IDR	I N Rome	I
22471,00	*STANAG 4481*	*B=50Bd S=850Hz*	*NKW*	*USA N Diego Garcia*	*DGA*
22473,50	MIL 188-110B SER	B=2400Bd	PWZ33	B N Rio de Janeiro	B
22477,50	CW		KPH	San Francisco R , CA (Bolinas/Port Reyes)	USA
22526,00	T600	B=50Bd S=100Hz	RCV	RUS N HQ Sevastopol	UKR
22527,00	FAX 120/576		NMC	USA CG Pt. Reyes , CA	USA
22542,00	FAX 120/576 60/288		JJC	Tokyo M	J
22544,60	STANAG 4285	B=2400Bd	FUM	F N Papeete , Tahiti	OCE
22559,60	FAX 120/576		JFW	Fukushima R	J
22560,00	FAX 120/576		JFC	Misaki R	J
22567,50	PACTOR III	B=100Bd S=200Hz	DZO25	Manila R	PHL
22571,00	*ALE USB*	*B=125Bd Ch=8 ChS=250Hz*		*G MIL net*	*G*
22583,00	STANAG 4285	B=2400Bd	FUX	F N Le Port	REU
22594,70	STANAG 4285	B=2400Bd	EBK	E N Las Palmas	CNR
22594,70	*STANAG 4285*	*B=2400Bd*	*EBA*	*E N Madrid*	*E*
22606,50	CW		JFM	Muroto R	J
22611,50	CW		HLF	Seoul R	KOR
22620,00	STANAG 4285	B=2400Bd		AUS MHFCS Riverina	AUS
22620,00	MIL 188-110A SER	B=2400Bd		AUS MHFCS North West Cape	AUS
22642,00	CW		JFX	Kagoshima R	J
22643,00	PACTOR II	B=200Bd S=200Hz	HPPM2	Sailmail Chiriqui	PNR
22648,00	T600	B=50Bd S=250Hz	RCV	RUS NHQ Sevastopol	UKR
22649,00	PACTOR II	B=200Bd S=200Hz	VZX	Sailmail Darawank	AUS
22682,70	*STANAG 4285*	*B=2400Bd*	*EBK*	*E N Madrid TX Almatriche*	*CNR*
22696,00	SSB			CS Ch 2201	
22698,30	USB		PKX	Jakarta R , Jawa	INS
22699,00	USB		UFL	Vladivostok R	RUS
22699,00	SSB			CS Ch 2202	
22702,00	USB		EQI	Abbas R	IRN
22702,00	USB		SDJ	Stockholm R	S
22702,00	USB		OYR	Aasiaat R	GRL
22702,00	SSB			CS Ch 2203	
22702,00	SSB		LZW	Varna R	BUL
22705,00	USB		LSD836	Argentina R	ARG
22705,00	SSB			CS Ch 2204	
22708,00	SSB		EQM	Bushehr R	IRN
22708,00	SSB			CS Ch 2205	
22711,00	USB		A9M	Bahrain R	BHR
22711,00	SSB			CS Ch 2206	
22714,00	USB			Lisbon R	POR
22714,00	SSB			CS Ch 2207	
22714,00	SSB		D4A	S. Vincente de Cabo R	CPV
22714,00	USB		UAT	Moscow R	RUS
22717,00	USB		L2A	ARG N Buenos Aires	ARG
22717,00	USB		L2A	ARG N Mar del Plata	ARG
22717,00	USB		LZW	Varna R	BUL
22717,00	SSB			CS Ch 2208	
22720,00	*USB*		*SVO*	*Olympia R*	*GRC*
22720,00	DSC	B=100Bd S=170Hz	HLS	Seoul R	KOR
22720,00	SSB			CS Ch 2209	
22723,00	SSB			CS Ch 2210	
22723,00	SSB		SVO72	Olympia R	GRC
22726,00	USB		SDJ	Stockholm R	S
22726,00	SSB			CS Ch 2211	
22729,00	SSB			CS Ch 2212	
22732,00	USB		SDJ	Stockholm R	S
22732,00	SSB			CS Ch 2213	
22735,00	USB		EQN	Khomeini R	IRN
22735,00	SSB			CS Ch 2214	

Frequency	Mode	Mode Parameter	Callsign	User	Country
22735,00	SSB		TAH	Istanbul R	TUR
22735,00	USB		XSQ	Guangzhou R	CHN
22738,00	SSB			CS Ch 2215	
22738,00	ALE USB	B=125Bd Ch=8 ChS=250Hz		ARINC Urgent Link net	USA
22741,00	SSB			CS Ch 2216	
22744,00	USB		EQJ	Chahbahar R	IRN
22744,00	SSB			CS Ch 2217	
22744,00	SSB		SVO73	Olympia R	GRC
22747,00	SSB			CS Ch 2218	
22750,00	SSB			CS Ch 2219	
22750,00	SSB		SVO74	Olympia R	GRC
22753,00	USB		XSQ	Guangzhou R	CHN
22753,00	SSB			CS Ch 2220	
22753,00	CHN 4+4	B=75Bd Ch=8 ChS=300/450Hz		CHN N	CHN
22756,00	USB		LSD836	Argentina R	ARG
22756,00	USB		CBV	Valpariso R	CHL
22756,00	SSB			CS Ch 2221	
22756,00	SSB		4PB	Colombo R	CLN
22759,00	SSB			CS Ch 2222	
22759,00	SSB		D4A	S. Vincente de Cabo R	CPV
22762,00	SSB			CS Ch 2223	
22762,00	SSB		TAH	Istanbul R	TUR
22765,00	SSB			CS Ch 2224	
22765,00	SSB		SVO75	Olympia R	GRC
22768,00	USB		CBV	Valpariso R	CHL
22768,00	CW		TUA	Abidjan R	CTI
22768,00	SSB			CS Ch 2225	
22771,00	USB		SUH	Alexandria R	EGY
22771,00	SSB			CS Ch 2226	
22774,00	SSB			CS Ch 2227	
22777,00	USB		SDJ	Stockholm R	S
22777,00	SSB			CS Ch 2228	
22779,00	THALES SKYMASTER ALE				
22780,00	USB		HSA	Bangkok R	THA
22780,00	SSB			CS Ch 2229	
22783,00	USB		SDJ	Stockholm R	S
22783,00	SSB			CS Ch 2230	
22783,00	SSB		TAH	Istanbul R	TUR
22786,00	SSB			CS Ch 2231	
22786,00	SSB		SVO76	Olympia R	GRC
22789,00	SSB			CS Ch 2232	
22790,00	STANAG 4285	B=2400Bd		G MIL Akrotiri	CYP
22792,00	USB		YQI	Constanta R	ROU
22792,00	SSB			CS Ch 2233	
22792,00	SSB		SVO77	Olympia R	GRC
22795,00	SSB			CS Ch 2234	
22798,00	USB		EQI	Abbas R	IRN
22798,00	SSB			CS Ch 2235	
22801,00	SSB			CS Ch 2236	
22804,00	SSB			CS Ch 2237	
22806,00	*STANAG 4285*	*B=2400Bd*		*G MIL St. Eval*	*G*
22807,00	USB		SVO78	Olympia R	GRC
22807,00	SSB			CS Ch 2238	
22807,00	USB		PPR	Rio R	B
22810,00	USB		EQC	Amirabad R	IRN
22810,00	SSB			CS Ch 2239	
22813,00	USB		CBV	Valpariso R	CHL
22813,00	SSB			CS Ch 2240	
22816,00	SSB			CS Ch 2241	

Frequency	Mode	Mode Parameter	Callsign	User	Country
22819,00	SSB			CS Ch 2242	
22822,00	USB		LZW	Varna R	BUL
22822,00	USB		JNA	J CG HQ Tokyo	J
22822,00	SSB			CS Ch 2243	
22822,00	USB		PPR	Rio R	B
22825,00	USB		JNA	J CG HQ Tokyo	J
22825,00	SSB			CS Ch 2244	
22828,00	SSB			CS Ch 2245	
22831,00	SSB			CS Ch 2246	
22834,00	SSB			CS Ch 2247	
22837,00	SSB			CS Ch 2248	
22840,00	SSB			CS Ch 2249	
22843,00	SSB			CS Ch 2250	
22843,00	USB		PPR	Rio R	B
22846,00	SSB			CS Ch 2251	
22846,00	USB		PPR	Rio R	B
22849,00	SSB			CS Ch 2252	
22852,00	USB		SXE	Aspropyrgos Attikis R	GRC
22852,00	USB			Peiraias CG R	GRC
22852,00	STANAG 4285	B=2400Bd		AUS MHFCS Humpty Doo	AUS
22852,00	SSB			CS Ch 2253	
22855,00	CODAN CHIRP	B=80Bd Ch=30 ChS=250Hz		EGY E Washington	USA
22855,00	CODAN	B=2400Bd Ch=16 ChS=112,5Hz		EGY E Washington	USA
22856,70	SITOR A	B=100Bd S=170Hz		EGY E Washington	USA
22860,00	ALE USB	B=125Bd Ch=8 ChS=250Hz		MRC MIL net	MRC
22864,00	*T600*	*B=50Bd S=200Hz*		*RUS N Moscow*	*RUS*
22881,40	PACTOR II	B=200Bd S=200Hz	WPTG385	Sailmail Corpus Christi , TX	USA
22890,00	ALE USB	B=125Bd Ch=8 ChS=250Hz		G MIL net TX Ascension Island	ASC
22910,00	*STANAG 4481*	*B=75Bd S=850Hz*	*NKW*	*USA N Diego Garcia*	*DGA*
22917,40	PACTOR III	B=100Bd S=200Hz	WQLI952	Sailmail Watsonville , CA	USA
22918,50	STANAG 4285	B=2400Bd	CFH	CAN N Halifax	CAN
22921,60	ALE USB	B=125Bd Ch=8 ChS=250Hz		US DVA net	
22941,20	*STANAG 4285*	*B=2400Bd*		*G MIL Akrotiri*	*CYP*
22947,00	USB			USA AF MARS	
22961,40	PACTOR II	B=200Bd S=200Hz	WPUC469	Sailmail South Daytona , FL	USA
22964,00	CODAN CHIRP	B=80Bd Ch=30 ChS=81Hz		MFA Cairo	EGY
22964,00	CODAN	B=2400Bd Ch=16 ChS=112,5Hz		MFA Cairo	EGY
22965,70	SITOR A	B=100Bd S=170Hz		MFA Cairo	EGY
22983,00	ALE USB	B=125Bd Ch=8 ChS=250Hz		USA FEMA net	USA
23006,00	ALE USB	B=125Bd Ch=8 ChS=250Hz		USA Civil Air Patrol net	USA
23043,00	T600	B=50Bd S=250Hz	RJS	RUS N HQ Vladivostok	RUS
23050,00	PACTOR II	B=200Bd S=200Hz	HPPM1	Sailmail Chiriqui	PNR
23050,00	PACTOR III	B=100Bd S=200Hz	HPPM3	Sailmail Panama	PNR
23060,00	STANAG 4285	B=2400Bd		AUS MHFCS Humpty Doo	AUS
23060,00	PACTOR III	B=100Bd S=200Hz	WQAB964	Sailmail San Diego , CA	USA
23062,50	STANAG 4285	B=2400Bd		AUS MHFCS North West Cape	AUS
23120,00	CIS 75-1000	B=75Bd S=1000Hz		RUS	RUS
23197,00	STANAG 4285	B=2400Bd		AUS MHFCS Humpty Doo	AUS
23197,00	MIL 188-110A SER	B=2400Bd		AUS MHFCS North West Cape	AUS
23210,00	USB		SDJ	Stockholm R	S
23220,00	SSB			USAF Strategic Command	USA
23250,00	USB		CHR	CAN AF Trenton	CAN
23254,00	USB		CIRCUS VERT	F AF HQ Villacoublay	F

Frequency	Mode	Mode Parameter	Callsign	User	Country
23337,00	ALE USB	B=125Bd Ch=8 ChS=250Hz		USA AF net	
23357,00	SSB			USAF Strategic Command	USA
23405,00	STANAG 4285	B=2400Bd		G MIL Ascension	
23452,00	T600	B=50Bd S=200Hz		RUS N Vladivostok	RUS
23500,00	ALE USB	B=125Bd Ch=8 ChS=250Hz		CHN net	
23500,00	CHN HYBRID MODEM 8FSK-PSK	B=2400Bd		CHN	
23514,00	STANAG 4538	B=2400Bd		I MIL net	I
23540,00	ALE USB	B=125Bd Ch=8 ChS=250Hz		TUN diplo net	
23550,00	ALE USB	B=125Bd Ch=8 ChS=250Hz		USA FEMA net	USA
23831,20	STANAG 4285	B=2400Bd	PBB	NLD N Den Helder TX Flevo	HOL
23861,50	CW		CGA984	Essex County, ON	CAN
23866,35	ALE USB	B=125Bd Ch=8 ChS=250Hz			
24008,00	ALE USB	B=125Bd Ch=8 ChS=250Hz		USA FEMA net	USA
24080,00	ALE USB	B=125Bd Ch=8 ChS=250Hz		TUN diplo net	TUN
24100,00	FMCW			Point Bonita, CA	USA
24127,50	DPRK FSK	B=600Bd S=600Hz		DPRK diplo net	
24170,00	ALE USB	B=125Bd Ch=8 ChS=250Hz		TUN diplo net	TUN
24192,00	ROS USB	B=1Bd Ch=144 ChS=15,625Hz		ROS frequency	
24210,00	AUS MIL FSK MODEM	B=600Bd S=345Hz		AUS MIL	AUS
24274,00	SSB		AFI	USAF McClellan AFB , Sacramento , CA	USA
24331,50	ALE USB	B=125Bd Ch=8 ChS=250Hz		G MIL net	G
24400,00	FMCW			San Diego, Coronado Islands	USA
24400,00	FMCW			Westshore Coal Terminal, Tsawwassen	USA
24400,00	FMCW			Ridley Island, Canada	USA
24448,50	USB			USA AF MARS	
24450,00	ALE USB	B=125Bd Ch=8 ChS=250Hz		USA CG net	USA
24500,00	FMCW			Naval Postgraduate School, CA	USA
24526,00	ALE USB	B=125Bd Ch=8 ChS=250Hz		USA FEMA net	USA
24550,00	FMCW			North Key Largo, FL	USA
24550,00	FMCW			Old Bridge Waterfront Park, NJ	USA
24550,00	FMCW			Port Monmouth, NJ	USA
24550,00	FMCW			Staten Island, NY	USA
24563,50	USB			SRL mission Tabda	SOM
24600,00	FMCW			Ocracoke Island, NC	USA
24600,00	FMCW			Portici radar site	I
24600,00	FMCW			Castellammare di Stabia	I
24700,00	STANAG 4285	B=2400Bd	PBB	NLD N Den Helder TX Flevo	NLD
24799,00	FMCW			San Diego, Point Loma	USA
24825,00	ALE USB	B=125Bd Ch=8 ChS=250Hz		G	
24838,50	ALE USB	B=125Bd Ch=8 ChS=250Hz		USA COTHEN net	USA
24883,60	ALE USB	B=125Bd Ch=8 ChS=250Hz		UI ALE net	
24883,60	ALE USB	B=125Bd Ch=8 ChS=250Hz		USA diplo net	
24900,00	THALES SKYMASTER ALE				

Frequency	Mode	Mode Parameter	Callsign	User	Country
24915,00	FT8	B=5,86Bd Ch=8 ChS=5,86Hz		FT8 channel	
24917,00	JT65	B=2,69Bd Ch=65 ChS=2,69Hz		JT65 channel	
24919,00	JT9	B=1,736Bd Ch=9 ChS=1,736Hz		JT9 channel	
24919,00	FT4	B=5,86Bd Ch=8 ChS=5,86Hz		FT4 channel	
24920,00	PSK31	B=31Bd		PSK31 channel	
24921,50	MFSK			MFSK modes user	
24922,00	JS8	B=6,25Bd Ch=8 ChS=6,25Hz		JS8 user	
24924,60	WSPR	B=1,46Bd Ch=4 ChS=1,46Hz		WSPR net	
24925,00	BAUDOT	B=45,45Bd S=170Hz		RTTY user	
24926,00	ALE USB	B=125Bd Ch=8 ChS=250Hz		HFLINK network	
24927,00	VARA	Ch=1		VARA chat freqeuency	
24930,00	CW		4U1UN	UN New York City	USA
24930,00	CW		VE8AT	Inuvik, NT	CAN
24930,00	CW		W6WX	Mt. Umunhum	USA
24930,00	CW		KH6RS	Maui	HWA
24930,00	CW		ZL6B	Masterton	NZL
24930,00	CW		VK6RBP	Rolystone	AUS
24930,00	CW		JA2IGY	Mt. Asama	J
24930,00	CW		RR9O	Novosibirsk	RUS
24930,00	CW		VR2B	Hong Kong	HKG
24930,00	CW		4S7B	Colombo	CLN
24930,00	CW		ZS6DN	Pretoria	AFS
24930,00	CW		5Z4B	Kariobangi	KEN
24930,00	CW		4X6TU	Tel Aviv	ISR
24930,00	CW		OH2B	Lohja	FNL
24930,00	CW		CS3B	São Jorge	MDR
24930,00	CW		LU4AA	Buenos Aires	ARG
24930,00	CW		OA4B	Lima	PRU
24930,00	CW		YV5B	Caracas	VEN
24932,00	ALE USB	B=125Bd Ch=8 ChS=250Hz		HFLINK network	
24950,00	USB			Wireless Institute Civil Emergency Service WICEN	AUS
24954,00	CHN 4+4	B=75Bd Ch=8 ChS=300/450Hz		CHN MIL	CHN
25000,00	FMCW			Mutton Island	IRL
25000,00	FMCW			Spiddle	IRL
25000,00	AM		WWV	Fort Collins TS , CO	USA
25000,00	CW			MIKES Otaniemi TS	FIN
25000,00	FMCW			Piran 1	SLO
25070,00	SSB			SS Ch 2501	
25073,00	SSB			SS Ch 2502	
25076,00	SSB			SS Ch 2503	
25079,00	SSB			SS Ch 2504	
25082,00	SSB			SS Ch 2505	
25085,00	SSB			SS Ch 2506	
25088,00	SSB			SS Ch 2507	
25091,00	SSB			SS Ch 2508	
25094,00	SSB			SS Ch 2509	
25097,00	SSB			SS Ch 2510	
25100,00	FMCW			Newport Beach, CA	USA
25100,00	SSB			simplex frequency ship - ship	
25103,00	SSB			simplex frequency ship - ship	
25106,00	SSB			simplex frequency ship - ship	
25109,00	SSB			simplex frequency ship - ship	

Frequency	Mode	Mode Parameter	Callsign	User	Country
25112,00	SSB			simplex frequency ship - ship	
25115,00	SSB			simplex frequency ship - ship	
25118,00	SSB			simplex frequency ship - ship	
25127,00	ALE USB	B=125Bd Ch=8 ChS=250Hz		G MIL net	G
25171,50	.			SS calling frequency Ch A	
25172,00	.			SS calling frequency Ch C	
25172,50	.			SS calling frequency Ch B	
25173,00	.			SS Ch 1	
25173,50	.			SS Ch 2	
25174,00	.			SS Ch 3	
25174,50	.			SS Ch 4	
25175,00	.			SS Ch 5	
25175,50	.			SS Ch 6	
25176,00	.			SS Ch 7	
25176,50	.			SS Ch 8	
25177,00	.			SS Ch 9	
25177,50	.			SS Ch 10	
25178,00	.			SS Ch 11	
25178,50	.			SS Ch 12	
25179,00	.			SS Ch 13	
25179,50	.			SS Ch 14	
25180,00	.			SS Ch 15	
25180,50	.			SS Ch 16	
25181,00	.			SS Ch 17	
25181,50	.			SS Ch 18	
25182,00	.			SS Ch 19	
25182,50	.			SS Ch 20	
25183,00	.			SS Ch 21	
25183,50	.			SS Ch 22	
25184,00	.			SS Ch 23	
25184,50	.			SS Ch 24	
25185,00	.			SS Ch 25	
25185,50	.			SS Ch 26	
25186,00	.			SS Ch 27	
25186,50	.			SS Ch 28	
25187,00	.			SS Ch 29	
25187,50	.			SS Ch 30	
25188,00	.			SS Ch 31	
25188,50	.			SS Ch 32	
25189,00	.			SS Ch 33	
25189,50	.			SS Ch 34	
25190,00	.			SS Ch 35	
25190,50	.			SS Ch 36	
25191,00	.			SS Ch 37	
25191,50	.			SS Ch 38	
25192,00	.			SS Ch 39	
25192,50	.			SS Ch 40	
25193,00	.			SS Ch 1	
25193,50	.			SS Ch 2	
25194,00	.			SS Ch 3	
25194,50	.			SS Ch 4	
25195,00	.			SS Ch 5	
25195,50	.			SS Ch 6	
25196,00	.			SS Ch 7	
25196,50	.			SS Ch 8	
25197,00	.			SS Ch 9	
25197,50	.			SS Ch 10	
25198,00	.			SS Ch 11	
25198,50	.			SS Ch 12	
25199,00	.			SS Ch 13	
25199,50	.			SS Ch 14	

Frequency	Mode	Mode Parameter	Callsign	User	Country
25200,00	.			SS Ch 15	
25200,00	FMCW			Sorrento radar site	I
25200,50	.			SS Ch 16	
25201,00	.			SS Ch 17	
25201,50	.			SS Ch 18	
25202,00	.			SS Ch 19	
25202,50	.			SS Ch 20	
25203,00	.			SS Ch 21	
25203,50	.			SS Ch 22	
25204,00	.			SS Ch 23	
25204,50	.			SS Ch 24	
25205,00	.			SS Ch 25	
25205,50	.			SS Ch 26	
25206,00	.			SS Ch 27	
25206,50	.			SS Ch 28	
25207,00	.			SS Ch 29	
25207,50	.			SS Ch 30	
25208,00	.			SS Ch 31	
25208,50	.			SS DSC frequency	WW
25209,00	.			SS DSC frequency	WW
25209,50	.			SS DSC frequency	WW
25250,00	FMCW			Cape Henlopen, DE	USA
25296,00	HFT MODEM			Frankfurt	D
25300,00	FMCW			Great Captain Island, CT	USA
25324,00	STANAG 4285	B=2400Bd		AUS MHFCS Humpty Doo	AUS
25350,00	ALE USB	B=125Bd Ch=8 ChS=250Hz		USA MIL net	USA
25350,00	ALE USB	B=125Bd Ch=8 ChS=250Hz		USA	
25354,00	ALE USB	B=125Bd Ch=8 ChS=250Hz		USA civial air patrol net	USA
25359,00	THALES SKYMASTER ALE	B=125Bd Ch=8 ChS=250Hz		ALG AF	ALG
25388,00	FMCW			San Diego, Border Park	USA
25400,00	FMCW			Moss Landing Marine Laboratories, CA	USA
25400,00	FMCW			Dan Blocker, CA	USA
25400,00	FMCW			Point Fermin, CA	USA
25400,00	FMCW			Dana Point, CA	USA
25400,00	FMCW			Point Loma Waste Water Treatment Plant, CA	USA
25400,00	FMCW			Digby Island, Prince Rupert	USA
25400,00	FMCW			Georgina Point, Canada	USA
25400,00	FMCW			Iona Wastewater Treatment Plant, Vancouver	USA
25400,00	FMCW			Sandy Cove, Nova Scotia	USA
25400,00	FMCW			Silver Slipper Casino, MS	USA
25547,00	ALE USB	B=125Bd Ch=8 ChS=250Hz		USA NG net	USA
25600,00	FMCW			San Elijo State Beach, CA	USA
25600,00	FMCW			Block Island, RI	USA
25600,00	FMCW			First Landing State Park, VA	USA
25600,00	FMCW			Misquamicut, RI	USA
25600,00	FMCW			Montauk, NY	USA
25600,00	FMCW			Pass Christian Yacht Club, MS	USA
25600,00	FMCW			Squibnocket Farms, MA	USA
25600,00	FMCW			Sunset Beach Resort, Cape Charles, VA	USA
25800,00	FMCW			Rutgers Cape Shore Lab, NJ	USA
26100,50	.			MSI frequency	WW
26101,00	.			CS Ch1	
26101,50	.			CS Ch2	
26102,00	.			CS Ch3	
26102,50	.			CS Ch4	
26103,00	.			CS Ch5	

Frequency	Mode	Mode Parameter	Callsign	User	Country
26103,50	.			CS Ch6	
26104,00	.			CS Ch7	
26104,50	.			CS Ch8	
26105,00	.			CS Ch9	
26105,50	.			CS Ch10	
26106,00	.			CS Ch11	
26106,50	.			CS Ch12	
26107,00	.			CS Ch13	
26107,50	.			CS Ch14	
26108,00	.			CS Ch15	
26108,50	.			CS Ch16	
26109,00	.			CS Ch17	
26109,50	.			CS Ch18	
26110,00	.			CS Ch19	
26110,50	.			CS Ch20	
26111,00	.			CS Ch21	
26111,50	.			CS Ch22	
26112,00	.			CS Ch23	
26112,50	.			CS Ch24	
26113,00	.			CS Ch25	
26113,50	.			CS Ch26	
26114,00	.			CS Ch27	
26114,50	.			CS Ch28	
26115,00	.			CS Ch29	
26115,50	.			CS Ch30	
26116,00	.			CS Ch31	
26116,50	.			CS Ch32	
26117,00	.			CS Ch33	
26117,50	.			CS Ch34	
26118,00	.			CS Ch35	
26118,50	.			CS Ch36	
26119,00	.			CS Ch37	
26119,50	.			CS Ch38	
26120,00	.			CS Ch39	
26120,50	.			CS Ch40	
26137,50	STANAG 4285	B=2400Bd		AUS MHFCS Humpty Doo	AUS
26137,50	MIL 188-110A	B=2400Bd		AUS MHFCS North West Cape	AUS
26145,00	SSB			CS Ch 2501	
26148,00	SSB			CS Ch 2502	
26151,00	USB		LZW	Varna R	BUL
26151,00	SSB			CS Ch 2503	
26154,00	SSB			CS Ch 2504	
26157,00	SSB			CS Ch 2505	
26160,00	USB		SDJ	Stockholm R	S
26160,00	SSB			CS Ch 2506	
26163,00	USB		SXE	Aspropyrgos Attikis R	GRC
26163,00	USB		SVO81	Olympia R	GRC
26163,00	SSB			CS Ch 2507	
26166,00	USB		SVO82	Olympia R	GRC
26166,00	SSB			CS Ch 2508	
26169,00	USB		SDJ	Stockholm R	S
26169,00	SSB			CS Ch 2509	
26172,00	SSB			CS Ch 2510	
26190,00	FMCW			Stehli Beach, NY	USA
26250,00	FMCW			Ocean View Beach, VA	USA
26275,00	FMCW			Mont	I
26275,00	FMCW			Razanj	HRV
26275,00	FMCW			Stonica	HRV
26275,00	FMCW			Tarifa	E
26280,00	FMCW			Tino	I
26280,00	FMCW			Ceuta	E
26280,00	FMCW			Punta Carnero	E

Frequency	Mode	Mode Parameter	Callsign	User	Country
26300,00	FMCW			Virginia Key, FL	USA
26300,00	FMCW			Point Barrow, AK	USA
26300,00	FMCW			Sandspit, BC, Canada	USA
26310,00	FMCW			Point Atkinson, West Vancouver	USA
26310,00	FMCW			Cape Simpson, AK	USA
26370,00	ALE USB	B=125Bd Ch=8 ChS=250Hz		G MIL net	G
26530,50	ALE USB	B=125Bd Ch=8 ChS=250Hz			
26608,00	STANAG 4285	B=2400Bd		AUS MHFCS Townsville	AUS
26796,60	ALE USB	B=125Bd Ch=8 ChS=250Hz		ARINC Urgent Link net	USA
26812,00	ALE USB	B=125Bd Ch=8 ChS=250Hz		USA SHARES net	USA
26813,20	STANAG 4285	B=2400Bd	PBB	NLD N Den Helder	HOL
26890,00	ALE USB	B=125Bd Ch=8 ChS=250Hz		TUN diplo net	TUN
26965,00	AM , FM			CB Ch 1	
26975,00	AM , FM			CB Ch 2	
26985,00	AM , FM			CB Ch 3	
27005,00	AM , FM			CB Ch 4	
27015,00	AM , FM			CB Ch 5	
27025,00	AM , FM			CB Ch 6	
27035,00	AM , FM			CB Ch 7	
27055,00	AM , FM			CB Ch 8	
27065,00	AM , FM			CB Ch 9	
27075,00	AM , FM			CB Ch 10	
27085,00	AM , FM			CB Ch 11	
27095,00	CW			ETCS Eurobalise RX frequency	
27105,00	AM , FM			CB Ch 12	
27115,00	AM , FM			CB Ch 13	
27125,00	AM , FM			CB Ch 14	
27135,00	AM , FM			CB Ch 15	
27155,00	AM , FM			CB Ch 16	
27165,00	AM , FM			CB Ch 17	
27175,00	AM , FM			CB Ch 18	
27185,00	AM , FM			CB Ch 19	
27205,00	AM , FM			CB Ch 20	
27215,00	AM , FM			CB Ch 21	
27225,00	AM , FM			CB Ch 22	
27235,00	AM , FM			CB Ch 23	
27245,00	JS8	B=6,25Bd Ch=8 ChS=6,25Hz		JS8 user	
27245,00	AM , FM			CB Ch 24	
27255,00	AM , FM			CB Ch 25	
27265,00	AM , FM			CB Ch 26	
27275,00	AM , FM			CB Ch 27	
27285,00	AM , FM			CB Ch 28	
27295,00	AM , FM			CB Ch 29	
27300,00	FMCW			Kakaako, HI	USA
27305,00	AM , FM			CB Ch 30	
27315,00	AM , FM			CB Ch 31	
27325,00	AM , FM			CB Ch 32	
27335,00	AM , FM			CB Ch 33	
27345,00	AM , FM			CB Ch 34	
27355,00	AM , FM			CB Ch 35	
27365,00	AM , FM			CB Ch 36	
27375,00	AM , FM			CB Ch 37	
27385,00	AM , FM			CB Ch 38	
27395,00	AM , FM			CB Ch 39	
27404,00	AM , FM			CB Ch 40	
27560,00	ALE USB	B=125Bd Ch=8 ChS=250Hz		USA CG net	USA

Frequency	Mode	Mode Parameter	Callsign	User	Country
27590,00	T600	B=50Bd S=200Hz		RUS N	RUS
27635,00	ROS USB	B=1Bd Ch=144 ChS=15,625Hz		ROS freeband channel	
27860,00	USB		VMR360	Melbourne R	AUS
27860,00	USB		VMR361	Westernport R	AUS
27860,00	USB		VMR375	Mallacoota R	AUS
27860,00	USB		VMR502	Adelaide R	AUS
27860,00	USB		VMR555	Tumb Bay R	AUS
27860,00	USB		VMR232	Hobart R	AUS
27865,00	STANAG 4285	B=2400Bd		G MIL Akrotiri	CYP
27870,00	ALE USB	B=125Bd Ch=8 ChS=250Hz		USA AF net	
27880,00	USB		VMR300	Whyalla R	AUS
27880,00	USB		VMR360	Melbourne R	AUS
27880,00	USB		VMR361	Westernport R	AUS
27880,00	USB		VMR555	Tumb Bay R	AUS
27880,00	USB		VMR2233	Portland R	AUS
27910,00	USB		VMR300	Whyalla R	AUS
27920,50	ALE USB	B=125Bd Ch=8 ChS=250Hz			
27940,00	USB		VMR2233	Portland R	AUS
28067,50	CW		W1AW	ARRL morse training	USA
28070,00	PSK31	B=31Bd		PSK31 channel	
28072,50	MFSK			MFSK modes user	
28074,00	FT8	B=5,86Bd Ch=8 ChS=5,86Hz		FT8 channel	
28076,00	JT65	B=2,69Bd Ch=65 ChS=2,69Hz		JT65 channel	
28078,00	JT9	B=1,736Bd Ch=9 ChS=1,736Hz		JT9 channel	
28078,00	JS8	B=6,25Bd Ch=8 ChS=6,25Hz		JS8 user	
28080,00	BAUDOT	B=45,45Bd S=170Hz		RTTY user	
28095,00	LONGCHAT	B=40Bd Ch=1		LongChat net	
28105,00	VARA	Ch=1		VARA chat frequency	
28120,00	PSK31	B=31Bd		PSK31 channel	
28124,60	WSPR	B=1,46Bd Ch=4 ChS=1,46Hz		WSPR net	
28140,00	ROS USB	B=1Bd Ch=144 ChS=15,625Hz		ROS frequency	
28146,00	ALE USB	B=125Bd Ch=8 ChS=250Hz		HFLINK network	
28180,00	FT4	B=5,86Bd Ch=8 ChS=5,86Hz		FT4 channel	
28189,00	MSK144	B=2000Bd S=1000Hz	N4WLO	Theodore , AL	USA
28200,00	CW		4U1UN	UN New York City	USA
28200,00	CW		VE8AT	Inuvik, NT	CAN
28200,00	CW		W6WX	Mt. Umunhum	USA
28200,00	CW		KH6RS	Maui	HWA
28200,00	CW		ZL6B	Masterton	NZL
28200,00	CW		VK6RBP	Rolystone	AUS
28200,00	CW		JA2IGY	Mt. Asama	J
28200,00	CW		RR9O	Novosibirsk	RUS
28200,00	CW		VR2B	Hong Kong	HKG
28200,00	CW		4S7B	Colombo	CLN
28200,00	CW		ZS6DN	Pretoria	AFS
28200,00	CW		5Z4B	Kariobangi	KEN
28200,00	CW		4X6TU	Tel Aviv	ISR
28200,00	CW		OH2B	Lohja	FNL
28200,00	CW		CS3B	São Jorge	MDR
28200,00	CW		LU4AA	Buenos Aires	ARG

Frequency	Mode	Mode Parameter	Callsign	User	Country
28200,00	CW		OA4B	Lima	PRU
28200,00	CW		YV5B	Caracas	VEN
28312,50	ALE USB	B=125Bd Ch=8 ChS=250Hz		HFLINK network	
28360,00	USB			GTARC net	USA
28450,00	USB			Wireless Institute Civil Emergency Service WICEN	AUS
29300,00	ALE USB	B=125Bd Ch=8 ChS=250Hz			
29850,00	RADAR	Ch=1		CAN Meteor Orbit Radar CMOR	CAN
29894,00	ALE USB	B=125Bd Ch=8 ChS=250Hz		USA CAP net	USA
29895,00	T600	B=50Bd S=250Hz	RDL	RUS N Moscow	RUS
29981,00	ARD9800 MODEM	B=2250Bd Ch=36 ChS=62,5Hz		UI net	

Tables

Allocation of International Call Signs

Call Sign Serie	Allocation
AAA - ALZ	United States of America
AMA - AOZ	Spain
APA - ASZ	Pakistan
ATA - AWZ	India
AXA - AXZ	Australia
AYA - AZZ	Argentinia
A2A - A2Z	Botswana
A3A - A3Z	Tonga
A4A - A4Z	Oman
A5A - A5Z	Bhutan
A6A - A6Z	United Arab Emirates
A7A - A7Z	Qatar
A8A - A8Z	Liberia
A9A - A9Z	Bahrain
BAA - BZZ	China
CAA - CEZ	Chile
CFA - CKZ	Canada
CLA - CMZ	Cuba
CNA - CNZ	Morocco
COA - COZ	Cuba
CPA - CPZ	Bolivia
CQA - CUZ	Portugal
CVA - CXZ	Uruguay
CYA - CZZ	Canada
C2A - C2Z	Nauru
C3A - C3Z	Andorra
C4A - C4Z	Cyprus
C5A - C5Z	Gambia
C6A - C6Z	Bahamas
C7A - C7Z	World Meteorological Organization
C8A - C9Z	Mocambique
DAA - DRZ	Germany
DSA - DTZ	Republic of Korea
DUA - DZZ	Philippines
D2A - D3Z	Angola
D4A - D4Z	Cape Verde
D5A - D5Z	Liberia
D6A - D6Z	Comoros
D7A - D9Z	Republic of Korea
EAA - EHZ	Spain
EIA - EJZ	Ireland

Call Sign Serie	Allocation
EKA - EKZ	Armenia
ELA - ELZ	Liberia
EMA - EOZ	Ukraine
EPA - EQZ	Iran
ERA - ESZ	Moldova
ETA - ETZ	Ethopia
EUA - EWZ	Byelorussian Soviet Socialist Republic
EXA - EXZ	Kyrgyz Republic
EYA - EYZ	Tajikistan
EZA - EZZ	Turkmenistan
E2A - E2Z	Thailand
E3A - E3Z	Eritrea
FAA - FZZ	France
GAA - GZZ	Great Britain and Northern Ireland
HAA - HAZ	Hungary
HBA - HBZ	Switzerland
HCA - HDZ	Ecuador
HEA - HEZ	Switzerland
HFA - HFZ	Poland
HGA - HGZ	Hungary
HHA - HHZ	Haiti
HIA - HIZ	Dominican Republic
HJA - HKZ	Columbia
HLA - HLZ	Republic of Korea
HMA - HMZ	Democratic People's Republic of Korea
HNA - HNZ	Iraq
HOA - HPZ	Panama
HQA - HRZ	Honduras
HSA - HSZ	Thailand
HTA - HTZ	El Salvador
HVA - HVZ	Vatican City
HWA - HYZ	France
HZA - HZZ	Saudi Arabia
H2A - H2Z	Cyprus
H3A - H3Z	Panama
H4A - H4Z	Solomon Island
H6A - H7Z	Nicaragua
H8A - H9Z	Panama
IAA - IZZ	Italy

Call Sign Serie	Allocation	Call Sign Serie	Allocation
JAA - JSZ	Japan	SSA - SSZ	Egypt
JTA - JVZ	Mongolian People's Republic	SVA - SZZ	Greece
JWA - JXZ	Norway	S2A - S3Z	Bangladesh
JYA - JYZ	Jordan	S6A - S6Z	Singapore
JZA - JZZ	Indonesia	S7A - S7Z	Seychelles
J2A - J2Z	Djibouti	S9A - S9Z	Sao Tome and Principe
J3A - J3Z	Grenada	TAA - TCZ	Turkey
J4A - J4Z	Greece	TDA - TDZ	Guatemala
J5A - J5Z	Guinea - Bissau	TEA - TEZ	Costa Rica
J6A - J6Z	Saint Lucia	TFA - TFZ	Iceland
J7A - J7Z	Dominica	TGA - TGZ	Guatemala
J8A - J8Z	St. Vincent and the Grenadines	THA - THZ	France
KAA - KZZ	United States of America	TIA - TIZ	Costa Rica
LAA - LNZ	Norway	TJA - TJZ	Cameroon
LOA - LWZ	Argentina	TKA - TKZ	France
LXA - LXZ	Luxembourg	TLA - TLZ	Central African Republic
LYA - LYZ	Union of Soviet Socialist Republics	TMA - TMZ	France
LZA - LZZ	Bulgaria	TNA - TNZ	Congo
L2A - L9Z	Argentina	TOA -TQZ	France
MAA - MZZ	Great Britain and Northern Ireland	TRA - TRZ	Gabon
NAA - NZZ	United States of America	TSA - TSZ	Tunisia
OAA - OAZ	Peru	TTA - TTZ	Chad
ODA - ODZ	Lebanon	TUA - TUZ	Ivory Coast
OEA - OEZ	Austria	TVA - TVZ	France
OFA - OJZ	Finland	TYA - TYZ	Benin
OKA - OLZ	Czech Republic	TZA - TZZ	Mali
OMA - OMZ	Slovak Republic	T2A - T2Z	Tuvula
ONA - OTZ	Belgium	T3A - T3Z	Kiribati
OUA - OZZ	Denmark	T4A - T4Z	Cuba
PAA - PIZ	Netherland	T5A - T5Z	Somalia
PJA - PJZ	Netherland Antilles	T6A - T6Z	Afghanistan
PKA - POZ	Indonesia	T7A - T7Z	San Marino
PPA - PYZ	Brazil	T9A - T9Z	Bosnia and Herzegovina
PZA - PZZ	Suriname	UAA - UIZ	Union of Soviet Socialist Republics
P2A - P2Z	Papua New Guinea	UJA - UMZ	Uzbekistan
P3A - P3Z	Cyprus	UNA - UQZ	Kazakhstan
P4A - P4Z	Aruba	UUA - UZZ	Ukraine
P5A - P9Z	Democratic People's Republic of Korea	URA - UTZ	Ukrainian Soviet Socialist Republics
QAA - QZZ	Service abbreviations	UUA - UZZ	Union of Soviet Socialist Republics
RAA - RZZ	Union of Soviet Socialist Republics	VAA - VGZ	Canada
SAA - SMZ	Sweden	VHA - VNZ	Australia
SNA - SRZ	Poland	VOA - VOZ	Canada
		VPA - VSZ	Great Britain and Northern Ireland

Call Sign Serie	Allocation	Call Sign Serie	Allocation
VTA - VWZ	India	ZVA - ZZZ	Brazil
VXA - VYZ	Canada	Z2A - Z2Z	Zimbabwe
VZA - VZZ	Australia	Z3A - Z3Z	Macedonia
V2A - V2Z	Antigua and Barbuda	2AA - 2ZZ	Great Britain and Northern Ireland
V3A - V3Z	Belize	3AA - 3AZ	Monaco
V4A - V4Z	St. Christopher and Nevis	3BA - 3BZ	Mauritius
V5A - V5Z	Namibia	3CA - 3CZ	Equatorial Guinea
V6A - V6Z	Micronesia	3DA - 3DM	Swaziland
V7A - V7Z	Marshall Islands	3DN - 3DZ	Fiji
V8A - V8Z	Brunei	3EA - 3FZ	Panama
WAA - WZZ	United States of America	3GA - 3GZ	Chile
XAA - XIZ	Mexico	3HA - 3UZ	China
XJA - XOZ	Canada	3VA - 3VZ	Tunisia
XPA - XPZ	Denmark	3WA - 3WZ	Viet Nam
XQA - XRZ	Chile	3XA - 3XZ	Guinea
XSA - XSZ	China	3YA - 3YZ	Norway
XTA - XTZ	Burkina Faso	3ZA - 3ZZ	Poland
XUA - XUZ	Kampuchea	4AA - 4CA	Mexico
XVA - XVZ	Viet Nam	4DA - 4IZ	Philippines
XWA - XWZ	Laos	4JA - 4JZ	Union of Soviet Socialist Republics
XXA - XXZ	Portugal	4LA - 4LZ	Georgia
XYA - XZZ	Myanmar	4MA - 4MZ	Venezuela
YAA - YAZ	Afghanistan	4NA - 4OZ	Yugoslavia
YBA - YHZ	Indonesia	4PA - 4SZ	Sri Lanka
YIA - YIZ	Iraq	4TA - 4TZ	Peru
YJA - YJZ	New Hebrides	4UA - 4UZ	United Nations Organization
YKA - YKZ	Syria	4VA - 4VZ	Haiti
YLA - YLZ	Union of Soviet Socialist Republics	4WA - 4WZ	Yemen
YMA - YMZ	Turkey	4XA - 4XZ	Israel
YNA - YNZ	Nicaragua	4YA - 4YZ	International Civil Aviation Org.
YOA - YRZ	Romania	4ZA - 4ZZ	Israel
YSA - YSZ	El Salvador	5AA - 5AZ	Libya
YTA - YUZ	Yugoslavia	5BA - 5BZ	Cyprus
YVA - YYZ	Venezuela	5CA - 5GZ	Morocco
YZA - YZZ	Yugoslavia	5HA - 5IZ	Tanzania
Y2A - Y9Z	Germany	5JA - 5KZ	Colombia
ZAA - ZAZ	Albania	5LA - 5MZ	Liberia
ZBA - ZJZ	Great Britain and Northern Ireland	5NA - 5OZ	Nigeria
ZKA - ZMZ	New Zealand	5PA - 5QZ	Denmark
ZNA - ZOZ	Great Britain and Northern Ireland	5RA - 5SZ	Madagascar
ZPA - ZPZ	Paraguay	5TA - 5TZ	Mauretania
ZQA - ZQZ	Great Britain and Northern Ireland	5UA - 5UZ	Niger
		5VA - 5VZ	Togo
ZRA - ZUZ	South Africa	5WA - 5WZ	Western Samoa

Call Sign Serie	Allocation
5XA - 5XZ	Uganda
5YA - 5YZ	Kenya
6AA - 6BZ	Egypt
6CA - 6CZ	Syria
6DA - 6JZ	Mexico
6KA - 6NZ	Republic of Korea
6OA - 6OZ	Somalia
6PA - 6SZ	Pakistan
6TA - 6UZ	Sudan
6VA - 6WZ	Senegal
6XA - 6XZ	Madagascar
6YA - 6YZ	Jamaica
6ZA - 6ZZ	Liberia
7AA - 7IZ	Indonesia
7JA - 7NZ	Japan
7OA - 7OZ	Yemen
7PA - 7PZ	Lesotho
7QA - 7QZ	Malawi
7RA - 7RZ	Algeria
7SA - 7SZ	Sweden
7TA - 7YZ	Algeria
7ZA - 7ZZ	Saudi Arabia

Call Sign Serie	Allocation
8AA - 8IZ	Indonesia
8JA - 8NZ	Japan
8OA - 8OZ	Botswana
8PA - 8PZ	Barbados
8QA - 8QZ	Maldives
8RA - 8RZ	Guyana
8SA - 8SZ	Sweden
8TA - 8YZ	India
8ZA - 8ZZ	Saudi Arabia
9BA - 9DZ	Iran
9EA - 9FZ	Ethiopia
9GA - 9GZ	Ghana
9HA - 9HZ	Malta
9IA - 9JZ	Zambia
9KA - 9KZ	Kuwait
9LA - 9LZ	Sierra Leone
9MA - 9MZ	Malaysia
9NA - 9NZ	Nepal
9OA -9TZ	Zaire
9UA - 9UZ	Burundi
9VA - 9VZ	Rwanda
9YA - 9ZZ	Trinidad and Tobago

Table 1: International callsigns

Alphabetical List of Country Codes

ABBR.	Country
ABW	*Aruba*
ADM	Andaman and Nicobar Island
AFG	Afghanistan
AFS	South Africa
AGL	Angola
AIA	*Anguilla*
ALB	Albania
ALG	Algeria
ALS	*Alaska*
AMS	*Amsterdam and Saint Paul*
AND	Andorra
ANG	Anguilla
ANO	Annoban
ANT	Antarctica
AOE	*Westem Sahara*
ARG	Argentina
ARM	Armenia
ARS	Saudi Arabia
ARU	Aruba
ASC	*Ascension Island*
ATA	*Antarctic*
ATG	Antigua
ATN	*Neth. Antilles*
AUS	Australia
AUT	Austria
AVE	Aves Island
AZE	Azerbaidzhan
AZR	*Azores*
B	Brazil
BAH	Bahamas
BAL	Balleny Island
BAN	Banaba
BDI	Burundi
BEL	Belgium
BEN	Benin
BER	*Bermuda*
BFA	Burkina Faso
BGD	Bangladesh
BIH	Bosnia / Herzegovina
BHR	Bahrain
BHU	Bhutan
BIO	*British Indian Ocean Territory*
BLR	Belarus
BLZ	Belize
BOL	Bolivia

ABBR.	Country
BOT	Botswana
BRB	Barbados
BRM	Myanmar (Burma)
BRU	Brunei
BTN	Bhutan
BUL	Bulgaria
BVT	Bouvet Island
CAB	Cabinda
CAE	Canton and Enderbury Island
CAF	Central African Republic
CAN	Canada
CAR	Caroline Island (Palau)
CBG	Cambodia
CEU	Ceuta
CG7	*Cuba Guantanamo*
CHL	Chile
CHN	China
CHR	*Christmas Island*
CKH	*Cook Island*
CLM	Columbia
CLN	Sri Lanka
CME	Cameroon
CNR	*Canary Island*
COD	Democratic Republic of Congo
COG	Congo
COM	Comoro
CPT	*Clipperton*
CPV	Cape Verde
CRO	*Crozet Archipelago*
CTI	Ivory Coast
CTR	Costa Rica
CUB	Cuba
CVA	Vatican City
CYM	*Cayman Island*
CYP	Cyprus
CZE	Czech Rep.
D	Germany
DAI	Daito Island
DES	Desventurados Island
DGA	*Diego Garcia*
DJI	Djibouti
DMA	Dominica
DNK	Denmark
DOM	Dominican Republic
E	Spain
EGY	Egypt
EQA	Ecuador

ABBR.	Country
ERI	Eritrea
EST	Estonia
ETH	Ethiopia
F	France
FIN	Finland
FJI	Fiji
FLK	*Falkland Island*
FRO	*Faroe Island*
FSM	Micronesia
G	United Kingdom
GAB	Gabon
GCA	*United Kingdom R1*
GCC	*United Kingdom R3*
GDL	*Guadeloupe*
GEO	Georgia
GHA	Ghana
GIB	*Gibraltar*
GMB	Gambia
GNB	Guinea Bissau
GNE	Equatorial Guinea
GPG	Galapagos Island
GRC	Greece
GRD	Grenada
GRL	*Greenland*
GTM	Guatemala
GUF	*French Guinea*
GUI	Guinea
GUM	*Guam*
GUY	Guyana
HKG	*Hong Kong*
HMD	*Heard and McDonald Island*
HND	Honduras
HNG	Hungary
HOL	The Netherlands
HRV	Croatia
HTI	Haiti
HWA	*Hawaii*
HWL	*Howland Island*
I	Italy
ICO	*Cocos Island*
IND	India
INS	Indonesia
IRL	Ireland
IRN	Iran
IRQ	Iraq
ISL	Iceland
ISR	Israel

ABBR.	Country
ISZ	Neutral Zone IRQ - ARS
IWA	Bonin and Volcano Island
J	Japan
JAR	*Jarvis Island*
JMC	Jamaica
JON	*Johnston Island*
JOR	Jordan
JUF	Juan Fernandez Island
KAL	Kaliningrad
KAZ	Kazakhstan
KEN	Kenya
KER	*Kerguelen Island*
KGZ	Kyrgyzstan
KIR	Kiribati
KNA	Saint Kitts and Nevis
KOR	Korea Republic
KRE	Korea PDR
KWT	Kuwait
LAO	Laos
LBN	Lebanon
LBR	Liberia
LBY	Libya
LCA	St.Lucia
LIE	Lichtenstein
LSO	Lesotho
LTU	Lithuania
LUX	Luxembourg
LVA	Latvia (Lettland)
MAC	*Macau*
MAU	Mauritius
MCO	Monaco
MCS	Marcus Island
MDA	Moldova
MDG	Madagascar
MDN	Macedonia
MDR	*Madeira*
MDW	*Midway Island*
MEX	Mexico
MHL	Marshall Islands
MKD	North Macedonia
MLA	Malaysia
MLD	Maldives
MLE	Sabah, Sarawak (Malaysia)
MLI	Mali
MLT	Malta
MNG	Mongolia
MOZ	Mozambique

ABBR.	Country
MRA	*Marianas (Northern)*
MRC	Morocco
MRL	Marshall Islands
MRN	*Marion Island*
MRQ	Marquesas Island
MRT	*Martinique*
MSR	*Montserrat*
MTN	Mauretania
MWI	Malawi
MYT	*Mayotte*
NAK	Nachitschewan
NAV	Navassa Island
NCG	Nicaragua
NCL	*New Caledonia*
NFK	*Norfolk Island*
NGR	Niger
NIG	Nigeria
NIU	*Niue Island*
NMB	Namibia
NOR	Norway
NPL	Nepal
NRU	Nauru
NZL	New Zealand
OCE	*French Polynesia*
OMA	Oman
PAK	Pakistan
PAQ	*Easter Island*
PEI	Prince Edward Island
PHL	Philippines
PHX	*Phoenix Islands*
PLM	*Palmyra Island*
PLW	*Palau*
PNG	Papua New Guinea
PNR	Panama
PNZ	Panama Canal Zone
POL	Poland
POR	Portugal
PRG	Paraguay
PRU	Peru
PRV	Paresce Vela Island
PTC	*Pitcairn Island*
PTR	*Puerto Rico*
QAT	Qatar
REU	*Reunion*
RKS	*Kosovo*
ROD	*Rodrigues*
ROU	Romania

ABBR.	Country
RRW	Rwanda
RUS	Russia
S	Sweden
SAP	San Andres and Providencia Isl.
SCN	St.Kitts & Nevis
SCO	Scott Island
SDN	Sudan
SEN	Senegal
SER	Serbia
SEY	Seychelles
SHN	*St. Helena*
SLM	Solomon Island
SLO	Slovakia
SLV	El Salvador
SMA	*American Samoa*
SMO	Western Samoa
SMR	San Marino
SNG	Singapore
SOK	S. Orkney Island
SOM	Somalia
SPM	*St. Pierre & Miquelon*
SPR	Sprathly Island
SRB	Serbia
SRL	Sierra Leone
SSD	South Sudan
SSI	S. Sandwich Island
STB	St. Barthelemy Island
STP	Sao Tome and Principe Island
SUI	Switzerland
SUR	Surinam
SVK	Slovakia
SVN	Slovenia
SWN	*Swan Islands*
SWZ	Swaziland
SYR	Syria
TCA	*Turks & Caicos*
TCD	Chad
TCH	Czech & Slovak Rep.
TGO	Togo
THA	Thailand
TJK	Tadzhikistan
TKL	*Tokelau Islands*
TKM	Turkmenistan
TLS	Tomir Leste
TOB	Toubouai Island
TON	Tonga
TRC	*Tristan da Cunha*

ABBR.	Country
TRD	Trinidad and Tobago
TRI	Trinidada & Martim Vaz Island
TUN	Tunisia
TUR	Turkey
TUV	Tuvalu
TWN	Taiwan
TZA	Tanzania
UAE	United Arab Emirates
UGA	Uganda
UKR	Ukraine
URG	Uruguay
USA	USA
UZB	Uzbekistan

Table 2: Country codes

ABBR.	Country
VCT	St. Vincent & Grenadines
VEN	Venezuela
VIR	*Virgin Island (USA)*
VRG	*Virgin Island (Br.)*
VTN	Viet Nam
VUT	Vanuatu
WAB	Walvis Bay
WAK	*Wake Island*
WAL	*Wallis & Futuna*
YEM	Yemen
ZAI	Zaire
ZMB	Zambia
ZWE	Zimbabwe

ITU member states letter codes (non ITU member states are written in italic)

Abbreviations

Abbreviation	Description
2G	Second Generation
3G	Third Generation
AA	Anadolu Ajansi
AB	Airbase
ACARS	Aircraft Communications Addressing and Reporting System
ACB	Auto Correlation Bit
ACF	Auto Correlation Frequency
ACS	Automatic Channel Selection
AEROFLOT	Russian Airlines
AFAD	Afet ve Acil Durum Yönetimi Başkanlığı
AFB	Air Force Base
AFC	Area Forcast Center
AFI	MWARA Africa
AFTN	Aeronautical Fixed Telecommunication Network
AGRCRM	Austrian German Red Cross Relief Organisation
AL	Alabama
ALE	Automatic Link Establishment
ALIS	Automatic Link Setup
AM	Amplitude Modulation
AMS	Aeronautical Mobile Service
AMTOR	Amateur Microprocessor Teleprinting Over Radio
ANSA	Agenzia Nazionale Stampa Associata
AP	Associated Press
APA	Austrian Press Agency
APT	Automatic Picture Transmission
AR	Arkansas
ARQ	Synchronous transmission and automatic repetition teleprinter system
ARQ 6-70	6 character blocks simplex ARQ teleprinter system
ARQ 6-90/98	6 character blocks simplex ARQ teleprinter system
ARQ-E3	Single channel ARQ ITA 3 teleprinter system
ARQ-M2	Multiplex ARQ teleprinter system with 2 channels according CCIR 342-2
ARQ-M2 242	Multiplex ARQ teleprinter system with 2 channels according CCIR 242-2
ARQ-M4	Multiplex ARQ teleprinter system with 4 channels according CCIR 342-2
ARQ-M4 242	Multiplex ARQ teleprinter system with 4 channels according CCIR 242-2
ARQE	Single channel ARQ teleprinter system
ARQN	Single channel ARQ teleprinting system without bit inversion
ARQS	SIEMENS simplex ARQ teleprinter system
ASM	Attached Synchronisation Marker
ASECNA	Agence pour la Securite de la Navigation Aerienne en Afrique et a Madagascar
ATU	Arabic Telecommunication Union
ATU-A	Arabic ATU alphabet
ATU-80	Arabic ATU alphabet
AUTOSPEC	Automatic Single Path Error Correcting teleprinter system
AVHRR	Advanced Very High Resolution Radiometer
AWS	Air Weather Service
AZ	Arizona
B	Beacon
BA	British Army
BAS	British Antarctic Survey

Abbreviation	Description
BC	Broadcast
BCH	Bose-Chaudhuri-Hocquenghem
BD	Bd
BER	Bit Error Rate
BFBS	British Forces Broadcasting Service
BIPM	Bureau International des Poids et des Mesures
BK	Break In
BPS	Bit Per Second
BRASS	Broadcast And Ship-Shore
BSKSA	Broadcasting Service Kingdom of Saudi Arabia
BTA	Bulgarska Telegrafitscheka Agentzia
BW0...5	Burst Waveform 0...5
CA	California
CADU	Channel Access Data Unit
CAP	Civil Air Patrol
CAR	MWARA Caribbean
CCIR	Committee Consultative International Radio
CCITT	Committee Consultative International Telegraph and Telephone
CCSDS	Consultative Committee for Space Data Systems
CDMA	Code Division Multiple Access
CEP	MWARA Central East Pacific
CFARS	Canadian Forces Affiliated Radio System
CG	Coast Guard
CH	Channel
CIR	Channel Impulse Response
CIRM	Centro Internatzionale Radio Medico
CIS	Commonwealth of Independent States
CNA	Central News Agency
CO	Colorado
COSPAR	Committee on Space Research
COTHEN	Customs Over-the Horizon Enforcement Network
CPFSK	Continuous Phase Frequency Shift Keying
CPM	Continuous Phase Modulation
CPSK	Coherent Phase Shift Keying
CQ	General call to all stations
CROSS	Centre Regional Operationnel de Surveillance et de Sauvetage
CRC	Cyclic Redundancy Check
CRS	Compagnie Republicaine de Securite
CS	Coast Station
CT	Connecticut
CVCDU	Coded Virtual Channel Data Unit
CW	Continuous Wave = Morse
CWP	MWARA Central West Pacific
CZCS	Coastal Zone Colour Scanner
DC	District of Columbia
DE	Delaware
DEA	Drug Enforcement Agency
DE MPSK	Differentially Encoded Coherent MPSK
DGPT	Directorate General of Posts and Telecommunicatios
DIPLO	Direction des Services d ` Information et de Presse
DMPSK	Differential MPSK (no carrier recovery)

Abbreviation	Description
DMSK	Differential MSK
DMSP	Defense Meteorological Satellite Programme
DPR	Democratic People´s Republic
DRK	Deutsches Rotes Kreuz
DSB	Double Side Band
DSC	Digital Selective Calling
DSP	Digital Signal Processing
DTG	Date Time Group
DTRE	Direction des Telecommunications des Reseaux Exterieurs
DTS	Droit de Tirage Special
DUP-ARQ	Hungarian simplex ARQ teleprinter system
DW	Deutsche Welle
DYN	Diarios Y Noticas
E	Embassy
E	East
EA	MWARA East Asia
ECC	Error Correcting Code
EOL	End of live
EOS	Earth Observing System
EPM	Electronic Protective Measures
ETR	Electronic Tracking Radar
EUR	MWARA Europe
F	Feeder
F	Forces
FARCOS	Fast Adaptive HF Radio Communication System
FAX	Facsimile
FDM	Frequency devision multiplex
FEC	Forward Error Correction
FEC 100	One way traffic FEC teleprinter system
FECS	SIEMENS simplex FEC teleprinter system
FER	Frame Error Rate
FFSK	Fast Frequency Shift Keying
FIR	Finite Impulse Response
FL	Florida
FM	Frequency Modulation
FMCW	Frequency Modulated Continuous Wave
FMOP	Frequency Modulated On Pulse
FOV	Field of View
FQPSK	Feher-patented Quadrature Phase Shift Keying
FRC	French Red Cross
FRO	Frivilliga Radioorganisationen
FSK	Frequency Shift Keying
FSS	Fixed Satellite Services
G	Gonio
GA	Georgia
GC	Guardia Civil
GMDSS	Global Maritime Distress and Safety System
GMSK	Gaussian MSK
GMT	Greenwich Mean Time
GPS	Global Positioning System
H	Hour

Abbreviation	Description
HAB	Hamburger Abendblatt
HC-ARQ	Haegelin Cryptos simplex teleprinter system
HDL	High rate Data Link
HF	High Frequency
HRI	High Resolution Imagery
HRPT	High Resolution Picture Transmission
IA	Iowa
IAT	International Atomic Time
ICAO	International Civil Aviation Organization
ICI	Inter Carrier Interference
ICRC	International Committee of Red Cross
ID	Idaho
ID	Identification
IFL	International Frequency List
IFOV	Infrared Field of View
IFRB	International Frequency Registration Board
IIR	Infinite Impulse Response
IL	Illinois
IMO	International Maritime Organization
IN	Indiana
INA	Iraqi News Agency
INO	MWARA Indian Ocean
INTERPOL	Organisation Internationale de la Police Criminelle
IOC	Index of Cooperation
IOT	In Orbit Test
IP	Internet Protocol
IR	Infrared
IRNA	Islamic Republic News Agency
ISB	Independent Side Band
ISD	Inverted Scanning Direction
ISI	Inter Symbol Interference
ISS	International Space Station
ITA	International Telegraph Alphabet
ITU	International Telecommunication Union
JANA	Jamahiriyah News Agency
KC	Kilocycles
KCNA	Korean Central News Agency
KUP	Kwacha Unita Press
KY	Kentucky
KYODO	Kyodo Tsushin
LA	Louisana
LDL	Low Latency Data Link
LDPC	Low Density Parity Check Code
LEOP	Launch and Early Operations
LORAN	Longe Range Navigation System
LPC	Linear Predictive Coding
LPM	Lines per Minute
LRIT	Low Rate Information Transmission
LSB	Lower Side Band
M	Meteo
M/S 1.0	Relation Mark to Space equals 1

Abbreviation	Description
MA	Massachussetts
MAFOR	Marine Forecast
MAHRS	Multiple Adaptive HF Radio System
MAP	Magreb Arab Press
MARS	Military Affiliated Radio System
MASK	M-ary Amplitude Shift Keying
MC	Megacycles
MCPSK	Multi Channel PSK
MD	Maryland
ME	Maine
MENA	Middle East News Agency
MFA	Ministry for Foreign Affairs
MI	Michigan
MID	MWARA Middle East
MIL	Military
MIL-STD	Military Standard
MMS	Maritime Mobile Service
MN	Minnesota
MO	Missouri
MOD	Ministry of Defence
MODIS	Moderate Resolution Imaging Spectroradiometer
MOI	Ministry of Interior
MPDS	Mobile Packet Data Service
M_PDU	Multiplexed Protocol Data Unit
MPSK	M-ary Phase-Shift Keying
M QAM	M-point QAM
MS	Mississippi
MSA	Maritime Safety Agency
MSF	Medicines sans frontiers
MSG	Message
MSI	Maritime Safety Information
MSK	Minimum Shift Keying
MSS	Mobile Satellite Services
MT	Montana
Multi-h FM	Multi-index; correlative; duobinary FM
MUOS	Mobile User Objective System
MUX	Multiplex
MWARA	Major World Air Route
MVISR	Multichannel Visible and Infrared Scan Radiomete
N	Navy
N	North
NASA	North American Space Agency
NAT	MWARA North Atlantic
NATO	North Atlantic Treaty Organization
NAVTEX	Navigational Telex system
NBDPT	Narrow Band Direct Printing Telegraphy
NBFM	Narrow Band Frequency Modulation
NC	North Carolina
NCA	MWARA North Central Asia
ND	North Dakota
NDB	Non Directional Beacon

Abbreviation	Description
NDVI	Normalised Difference Vegetation Index
NE	Nebraska
NI	Norfolk Island
NJ	New Jersey
NOAA	National Oceanic and Atmospheric Administration
NP	MWARA North Pacific
NPOESS	National Polar-orbiting Operational Environmental Satellite System
NPP	NPOESS Preparatory Project
NR	Number
NV	Nevada
NX	News
NY	New York
OFDM	Orthogonal Frequency Division Multiplexing
OH	Ohio
OK	Oklahoma
OOK	On Off Keying
OR	Oregon
OR	Off Route flight communication
OSI	Open System Interconnection
OTHR	Over The Horizon Radar
P	Phare
PA	Pennsylvania
PACTOR	Combination of Packet Radio and SITOR
PANA	Pan African News Agency
PAP	Polska Agencja Prasowa
PDS	Production Data Sets
PDUS	Primary Data User Station
PETRA	Jordan News Agency
PIX	Press Picture
PLO	Palestine Liberation Organization
PPM	Pulse Position Modulation
PPP	Point to Point Protocol
PR	Packet Radio
PTT	Post. Telegraph and Telephone Administration
R	Radio
RAF	Royal Air Force
RAN	Royal Australian Navy
RC	Red Cross
RDARA	Regional and Domestic Air Route
RF	Radio Frequency
RFDS	Royal Flying Doctor Service
RFE	Radio Free Europe
RN	Royal Navy
RNZAF	Royal New Zealand Airforce
RNZN	Royal New Zealand Navy
ROMPRES	Agentia Romana de Presa
ROU-FEC	Romanian FEC teleprinter system
RS	Reed-Solomon
RS-ARQ	Rhode & Schwarz simplex ARQ teleprinter system
RTT	Regie des Telegraphes et des Telephones
RTTY	Radio Teletype

Abbreviation	Description
RX	Receiver
RY	Test loop ryryryry...
RYI	Test loop ryiryiryi...
S	South
SAAM	Russian Arctic and Antarctic Institute
SABS	Saudi Arabian Broadcasting Service
SALINI	Salini Costruttori
SAM	MWARA South America
SANA	Syrian Arab News Agency
SAR	Search and Rescue
SAT	MWARA South Atlantic
SAUDIA	Saudi Arabian Airlines
SC	South Carolina
SD	South Dakota
SEA	MWARA South East Asia
SEAWIFS	Sea-viewing Wide Field-of-view Sensor
SENAPRED	SERVICIO NACIONAL DE PREVENCIÓN Y RESPUESTA ANTE DESASTRES
SFSK	Sinusoidal FSK
SITOR	Simplex Teleprinting over Radio
SITOR A	Simplex Teleprinting over Radio Mode ARQ
SITOR B	Simplex Teleprinting over Radio Mode FEC
SKH	Schweizerische Katastrophenhilfe
SM QAM	Superimposed M-point QAM
SOS	Distress Signal
SP	MWARA South Pacific
SPREAD	FEC teleprinter system with 10 bit BAUER code
SQPSK	Staggered QPSK
SRK	Schweizer Rotes Kreuz
SS	Ship Station
SSB	Single Side Band
STANAG	Standardization Agreement (by NATO)
SUNA	Sudan News Agency
SW-ARQ	Adaptive Swedish simplex ARQ teleprinter system
SWL	Shortwave Listener
TAAF	Terres Australes et Antarctiques Francaises
TANJUG	Telegrafska Agencija Nova Jugoslavija
TAP	Tunis Afrique Presse
TCM	Trellis Code Modulation
TDM	Time Division Multiplex
TFC	Traffic
TFM	Tamed Frequency Modulation
TIP	TIROS Information Processor
TIROS	Television Infra-Red Orbiting Satellite
tlgr	Telegram
TS	Time signal station
TSI OQPSK	Two-Symbol Interval OQPSK
TT & C	Telemetry Tracking and Commanding
TTY	Teletype
TV	Television
TWINPLEX	Four frequency ARQ teleprinter system
TX	Transmitter

Abbreviation	Description
TX	Texas
txt	Text
UFO	UHF Follow On
UI	Unidentified
UNHCR	United Nation
UNID	Unidentified
UNIFIL	United Nations Interim Forces in Lebanon
UNO	United Nation Organization
USAF	United States Airforce
USARP	United State Antarctic Research Program
USCG	United State Coast Guard
USMILGP	United State Military Group
USN	United State Navy
USNG	United State National Guard
UT	Universal Time
UT	Utah
UTC	Universal Time Coordinated
VA	Virginia
VAFI	Africa VOLMET Area
VCAR	Caribbean VOLMET Area
VCDU	Virtual Channel Data Unit
VEUR	European VOLMET Area
VFT	Voice Frequency Telegraphy
VIIRS	Visible/Infrared Imager/Radiometer Suite
VMID	Middle East VOLMET Area
VNA	Viet Nam News Agency
VNAT	North Atlantic VOLMET Area
VNCA	North Central Asia VOLMET Area
VOA	Voice of America
VOLMET	Meteorological information for aircraft in flight
VPAC	Pacific VOLMET Area
VSAM	South America VOLMET Area
VSEA	South East Asia VOLMET Area
W	West
W I	World wide allotment area I (RDARA 1. 2. 3)
W II	World wide allotment area II (RDARA 10. 11. 12A. 12B. 12C. 12D)
W III	World wide allotment area III (RDARA 6. 8. 9. 14)
W IV	World wide allotment area IV (RDARA 12E. 12F. 12G. 12H. 12J. 13)
W V	World wide allotment area V (RDARA 4. 5. 7)
WA	Washington
WARC	World Administrative Radio Conference
WCDMA	Wideband Code Divison Multiple Access
WEFAX	Weather FAX
WERA	Wellenradar
WI	Wisconsin
WMC	World Meteorological Centre
WMO	World Meteorological Organization
WPM	Words per Minute
WV	West Virginia
WW	World Wide
WX	Weather

Abbreviation	Description
WX	Weather
WY	Wyoming
XINHUA	New China News Agency
YONHAP	Hapdong News Agency
Z	Universal Time Coordinated